GIS Fundamentals:

A First Text on Geographic Information Systems

Paul Bolstad
College of Natural Resources
University of Minnesota - St. Paul

Eider Press
White Bear Lake,
Minnesota

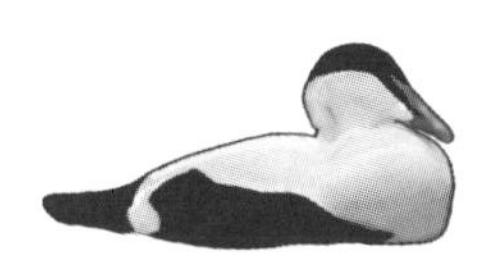

Front cover:
First globe, GOES satellite image, courtesy NASA-GSFC and NOAA
Second globe, terrrestrial and sea-floor topography, from UNEP data
Third globe, Sea-WiFS image, courtesy SeaWiFS project, NASA-GSFC
Fourth globe, political boundaries

Errata and other helpful information may be found at
http://bolstad.gis.umn.edu/GISbook.html

First printing May 2002
Second printing October 2002
Third printing, June 2003

Eider Press, 2303 4th St. White Bear Lake, MN 55110

book available from AtlasBooks, $38
www.atlasbooks.com
(800) 247-6553

ISBN 0-9717647-0-0
LCCN 2002090787

Acknowledgements

Many people have contributed in ways large and small to this book. It would not have been written had Jack Yardley not awakened my interest in geography almost 35 years ago. Bill Libby pointed the way to a life of learning and teaching, and Harold Burkhart was as good a mentor as one might find early on in a career. Tom Lillesand inspired beyond measure with his work, attitude, and actions.

Many friends and colleagues have contributed materially to this book. Tom Lillesand offered early encouragement. Lynn Usery helped clarify my thinking on many concepts in geography, and Paul Wolf on surveying and mapping. Lynn Usery, Esther Brown, Sheryl Bolstad, and Ryan Kirk spent uncounted hours reviewing early manuscripts, leading to substantial improvements in the form and content of this book. Randy Matchett, Dave Mladenoff, and Tom Patterson graciously shared their work, as did a number of businesses and public organizations.

Finally, this project would not have been possible save for the encouragement and forbearance of Holly, Sam, and Sheryl, and the support of Margaret.

While many helped in the development of this book, the errors in content are all mine. In spite of everyone's best efforts, I'm sure I've left many opportunities for improvement. If you have comments to share or suggested improvements or corrections, please send them to Eider Press, 2303 4th Street, White Bear Lake, MN 55110, or pbolstad@umn.edu.

Paul Bolstad

Table of Contents

1 An Introduction to GIS

Introduction

Geography has always been important to humans. Stone-age hunters anticipated the location of their quarry, early explorers lived or died by their knowledge of geography, and current societies live, work, and cooperate based on their understanding of who belongs where. Applied geography, in the form of maps and spatial information, has served discovery, planning, commerce, and defense for at least the past 3000 years (Figure 1-1), and maps are among the most beautiful documents of our civilization.

Most often our geographic knowledge is applied to routine tasks, as when we puzzle over a route to a child's soccer game or wonder where we might find gasoline. Spatial information has a much greater impact on our lives, often to an extent we don't realize, to help us produce the food we eat, the energy we burn, the clothes we wear, and the diversions we enjoy.

Because spatial information is so important, we have developed tools called geographic information systems (GIS) to help us develop our geographic knowledge. A GIS (we will use the abbreviation to refer to both system and systems) helps us gather and use spatial data. Some GIS components are purely technological; they include space-age data collectors, advanced communications networks, and sophisticated computing. Other GIS methods are very simple, for example, when a pencil and paper are used to field verify a map.

As with many aspects of life in the last five decades, how we gather and use spatial data has been profoundly altered by the development of modern electronics. GIS software and hardware are a primary result of these technological developments, and the capture and treatment of spatial data has quickened over the past three decades, and continues to evolve.

Key to all definitions of a GIS are the "what" and "where". GIS and spatial analyses are concerned with the absolute and relative location of features, as well as the properties and attributes of those features. The locations of important spatial objects such as rivers and streams may be recorded, and also their size, flow rate, water quality, or the kind of fish found in them. Indeed, these attributes often depend on the spatial arrangement of "important" features. A GIS aids in the analysis and display of these spatial relationships.

What is a GIS?

A GIS is a tool for making and using spatial information. Although there are many formal definitions of GIS, for practical purposes we define GIS as:

a computer-based system to aid in the collection, maintenance, storage, analysis, output, and distribution of spatial data and information.

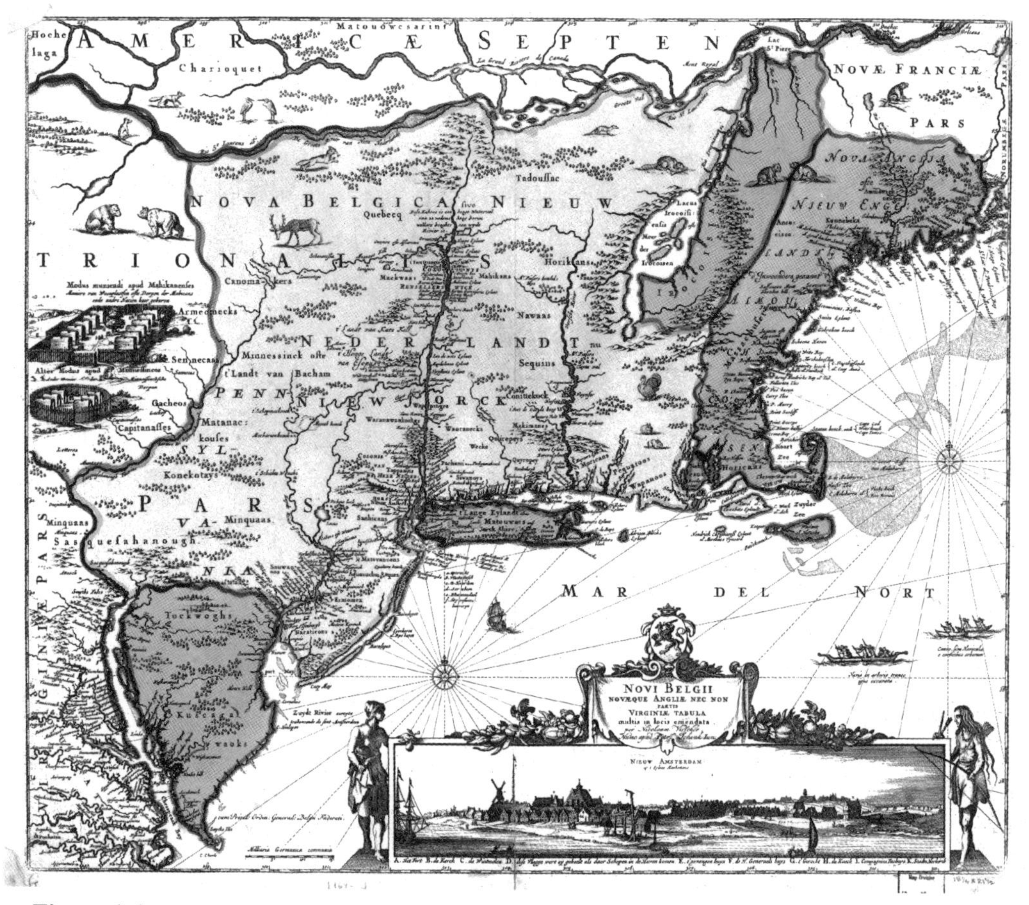

Figure 1-1: A map of New England by Nicolaes Visscher, published about 1685. Present-day Cape Cod is visible on the right, with the Connecticut and Hudson Rivers in the center of this map. Early maps were key to the European exploration of new worlds.

When used wisely GIS can help us live healthier, wealthier, and safer lives.

GIS and spatial analyses are concerned with the quantitative location of important features, as well as properties and attributes of those features. Important objects occupy space. Mount Everest is in Asia, Pierre is in South Dakota, and the cruise ship Titanic is at the bottom of the Atlantic Ocean. A GIS quantifies these locations by recording their *coordinates*, numbers that describe the position of these features. The GIS may also be used to record the height of Mount Everest, the population of Pierre, or the depth of the Titanic, as well and any other defining characteristics of the important spatial features.

Each GIS user decides what features are important, and what is important about them. For example, forests are important to many people. They protect our water supplies, provide wood, harbor wildlife, and provide spaces to recreate (Figure 1-2). We are concerned about the level of harvest, the land use around them, pollution from nearby industries, or when and where forests burn. Informed management of our forests requires at a minimum knowledge of all these related factors, perhaps above all the spatial arrangement of these factors. Buffer strips near rivers may protect water

Figure 1-2: GIS allow us to analyze the relative spatial location of important geographic features. Protection from wildfire, the preservation of scenic views, and the location of forest harvesting units may be effectively managed with the aid of spatial analysis tools in a GIS. (courtesy Space Imaging, Inc.)

supplies, clearings may prevent the spread of fire, and air pollution from downwind industries may not harm our forests, while polluters upwind might. A GIS aids immensely in the analysis of these spatial relationships and interactions among them. A GIS is also particularly useful at displaying spatial data and reporting the results of spatial analysis. In many instances GIS is the only way to solve spatially-related problems.

GIS: A Ubiquitous Tool

GIS use has become widespread during the past two decades. GIS have been used in fields from archeology to zoology, and new applications of GIS are continuously emerging. GIS use has become mandatory in many settings, and they are essential tools in business, government, education, and non-profit organizations. GIS have been used to fight crime, protect endangered species, reduce pollution, cope with natural disasters, analyze the AIDS epidemic, and to improve public health; in short, GIS have been instrumental in addressing some of our most pressing societal problems. On a more mundane level, GIS tools in aggregate have saved billions of dollars annually in the delivery of governmental and commercial goods and services. GIS regularly help in the day-to-day management of many natural and man-made resources, including sewer, water, power, and transportation networks. GIS are at the heart of one of the most important processes in U.S. democracy, the constitutionally mandated reshaping of U.S. Congressional Districts, and hence the distribution of tax dollars and other government resources.

Why Do We Need GIS?

GIS are needed in part because human population and technology have reached levels such that many resources, including air and land, are placing substantial limits on human action (Figure 1-3). Human populations have doubled in the last 50 years, reaching 6 billion, and we will likely add another 5 billion humans in the next 50 years. The first 100,000 years of human existence caused scant impacts on the World's resources, while in the past 300 years humans have permanently altered most of the Earth's surface. The atmosphere and oceans exhibit a decreasing ability to benignly absorb carbon dioxide and nitrogen, two primary waste products of humanity. Silt chokes many rivers (Figure 1-4) and there is a surfeit of localized examples where ozone, poly-chlorinated-biphenyls, or other noxious pollutants substantially harm public health. By the end of the 20th century most suitable lands had been inhabited and only a small percentage of the terrestrial surface had not been farmed, grazed, cut, built over, flooded, or otherwise altered by humans (Figure 1-5).

GIS help us identify and address environmental problems by providing crucial information on where problems occur and who are affected by them. GIS help us identify the source, location, and extent of adverse environmental impacts, and may help us devise practical plans for monitoring, managing, and mitigating environmental damage.

Human impacts on the environment have spurred a strong societal push for the

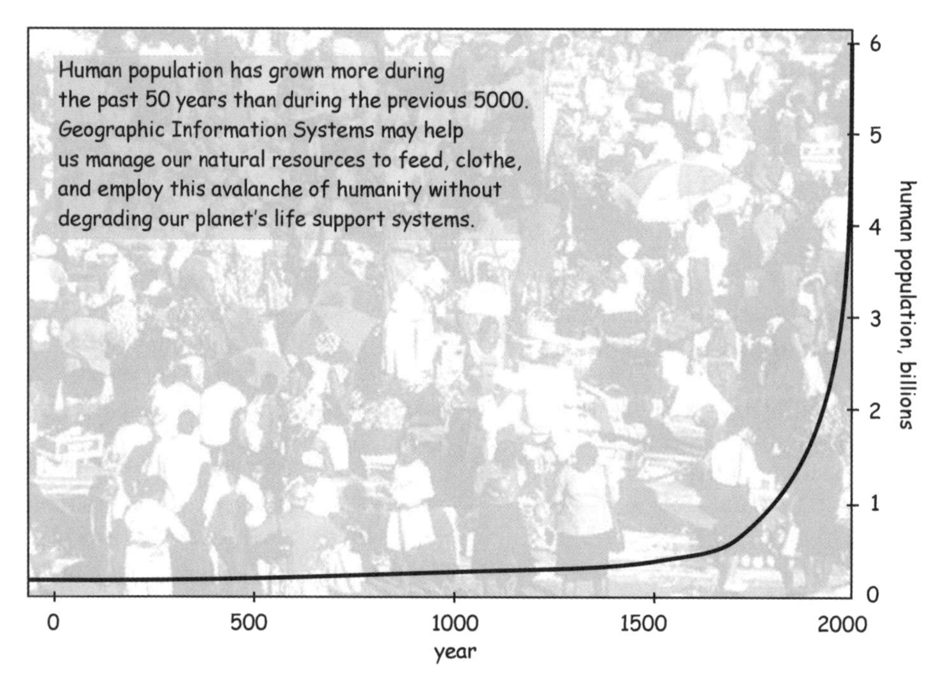

Figure 1-3: Human population growth during the past 2000 years has heightened the need for efficient resource use.

Figure 1-4: River siltation, as shown here by a satellite image of the Yangtzee River in China, is among the human impacts responsible for the societal push to adopt GIS. The river is shown as a light streak meandering through the lower middle section of the image. The silt plume is visible along the nearshore area both to the left and right of the river mouth. GIS may be used to help document, analyze, and plan for reductions in erosion and other negative environmental impacts. (courtesy NASA)

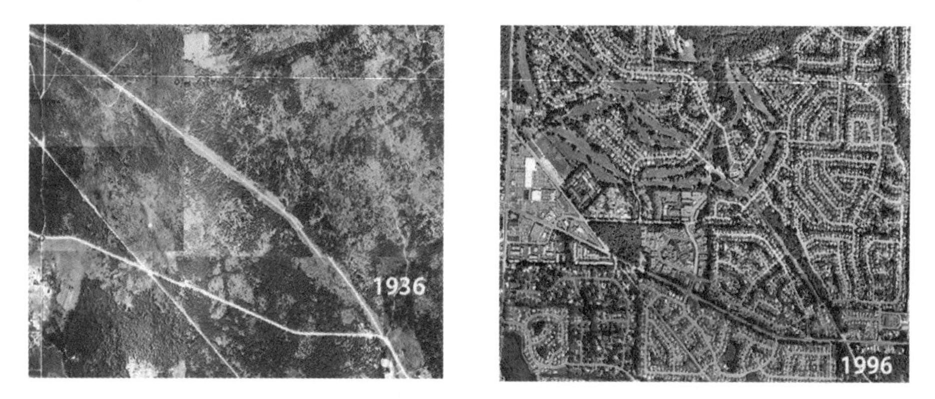

Figure 1-5: The environmental impacts wrought by humans have accelerated in many parts of the World during the past century. These photographs of the same portion of King County, Washington, give an example of how urban expansion has altered our landscapes. GIS are used to effectively plan and manage the development of our cities, and to protect our natural resources. (courtesy Washington Department of Natural Resources)

adoption of GIS. Conflicts in resource use, concerns about pollution, and precautions to protect public health have led to legislative mandates that explicitly or implicitly require the consideration of geography. The U.S. Endangered Species Act of 1973 requires adequate protection for rare and threatened organisms. Effective protection entails mapping the available habitat and the analysis of species range and migration patterns. The location of viable populations relative to current and future human land uses must be analyzed, and action taken to ensure species survival. GIS have proven to be useful tools in all of these tasks. Legislation has spurred the use of GIS in many other endeavors, including the dispensation of emergency services, protection from flooding, and the planning and development of infrastructure.

Many businesses need GIS because they provide increased efficiency in the delivery of goods and services. Retail businesses locate stores based on a number of spatially-related factors. Where are the potential customers? What is the spatial distribution of competing businesses? Where are potential properties for my store locations? What is traffic flow near the stores, and how easy is it to park near and access the stores? Spatial analyses are used every day to answer these and similar questions in business. GIS are also used in hundreds of other business applications, such as to route delivery vehicles, guide advertising, design buildings, plan construction, and sell real estate.

Public organizations have also adopted GIS because they aid in their governmental functions. Emergency service vehicles are regularly dispatched and routed using GIS. Emergency response GIS have been developed and widely installed specifically to respond to emergency service requests. Callers to E911 or other emergency response dispatchers are automatically identified by telephone number. The number is then matched to a building address and the nearest appropriate fire, police, or ambulance station identified. A map or route description is immediately generated, based on information on location and the street network, and sent to the appropriate station with a dispatch alarm.

The societal push has been complemented by a technological pull in the development and application of GIS. For more than four centuries mariners were vexed by their inability to locate their position, particularly their longitude. Thousands of lives and untold wealth were lost because ship captains could not answer the simple question, “where am I?” The methods eluded the best minds on Earth until the 19th century. Since then there has been a continual improvement in positioning technology to the point where today, anyone can locate their outdoor position to within a few meters in a few minutes. A remarkable positioning technology, known as the global positioning system (GPS), was originally developed primarily for military applications. GPS is now incorporated in cars, planes, boats, and trucks. It is an indispensable navigation and spatial data collection tool in government, business,

Figure 1-6: Portable computing is one example of the technological push driving GIS adoption. These hand-held devices substantially improve the speed and accuracy of spatial data collection. (courtesy Compaq Computer Corp.)

and recreation. Commerce, planning, and safety are improved due to the development and application of GPS and other GIS-related technologies.

The technological pull has developed on several fronts. Spatial analysis has been helped more than most fields by faster computers and larger hard disks. Most real-world spatial problems were beyond the scope of all but the largest government and business organizations until the 1990s. The requisite computing and storage capabilities were beyond any reasonable budget. GIS computing expenses are becoming an afterthought, as computing resources often cost less than a few months salary for a qualified GIS professional. Costs decrease and performance increases at dizzying rates, with predicted plateaus pushed back each year. Computer capabilities are increasing to the point that their limits on most spatial analysis are fast disappearing. Powerful field computers are becoming lighter, faster, more capable, and less expensive, so spatial data display and analysis capabilities may always be at hand (Figure 1-6). GIS on rugged, field-portable computers has been particularly useful in field data entry and editing.

In addition to the computing improvements and the development of GPS, current "cameras" deliver amazingly detailed aerial and satellite images. Advances in image collection and interpretation were spurred by World War II and then the Cold War, because access to the ground was impossible but accurate maps were required. Turned toward peacetime endeavors, imaging technologies now help us map food and fodder, houses and highways, and most other natural and human-built objects. Images may be rapidly converted to accurate spatial information over broad areas. Visible light, laser, thermal, and radar scanners are currently being developed to further increase the speed and accuracy with which we map our world. Thus, advances in these three key technologies, imaging, GPS, and computing, have substantially aided the development of GIS.

GIS in Action

Spatial data organization, analyses, and delivery are applied in a large and expanding number of ways to improve life. We will describe two examples that demonstrate how GIS are being used.

Oneida County is located in northern Wisconsin, a forested area characterized by exceptional scenic beauty. The County is in a region with among the highest concentrations of freshwater lakes in the World, a region that is also undergoing a rapid expansion in the permanent and seasonal human population. Retirees, urban exiles, and vacationers are increasingly drawn to the scenic and recreational amenities available in Oneida County. Permanent county population grew by nearly 20% from 1990 to 2000, and the seasonal influx nearly doubles the total county population each summer.

Permanent and seasonal population growth have led to a boom in construction with associated threats to the lakes that draw most people to the County. More than 1600 building permits were issued in 1993, up from about 1100 in 1989, and many of these were for near-shore houses, hotels, or businesses. Seepage from septic systems, runoff from fertilized lawns, or erosion and sediment associated with construction all threaten lake water quality. Increases in lake nutrients or sediment may lead to turbid waters, reducing the beauty and recreational value of the lakes and adjoining property.

Oneida County, the Sea Grant Institute of the University of Wisconsin, and the Land Information and Computer Graphics Facility of the University of Wisconsin have developed a Shoreland Management GIS Project to aid in the protection of valuable nearshore and lake resources, and to provide an example of GIS tools for water resource management (Figure 1-7). Oneida County has revised zoning and other ordinances to protect shoreline and lake quality and to ensure compliance without undue burden on landowners. The County has an

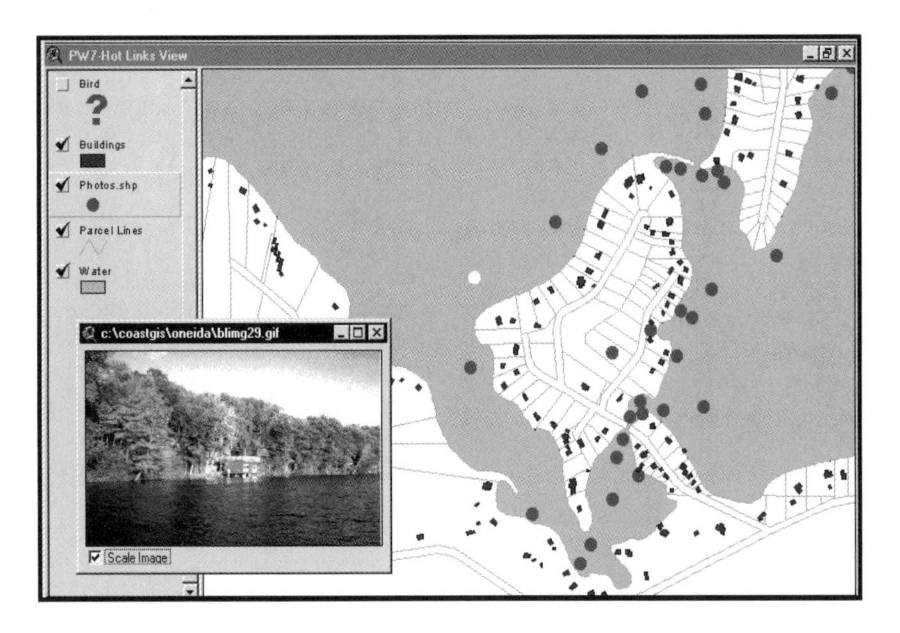

Figure 1-7: Parcel information entered in a GIS may substantially improve the maintenance and evaluation of county government. Here, images of the shoreline taken from lake vantage points are combined with digital maps of the shoreline, buildings, and parcel boundaries. The image in the lower left was obtained from the location shown as a light dot near the center of the figure. (courtesy Wisconsin Sea Grant and LICGC)

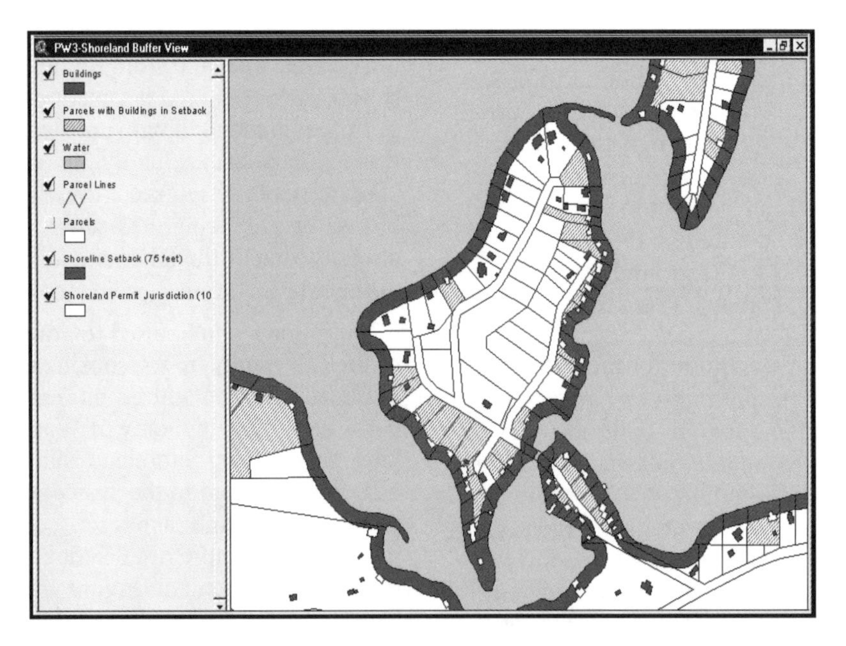

Figure 1-8: An example of the combination of spatial data in a GIS. Parcels data are combined with shoreline zoning setbacks, and non-compliant parcels (cross-hatched) are identified. (courtesy Wisconsin Sea Grant Institute and LICGC)

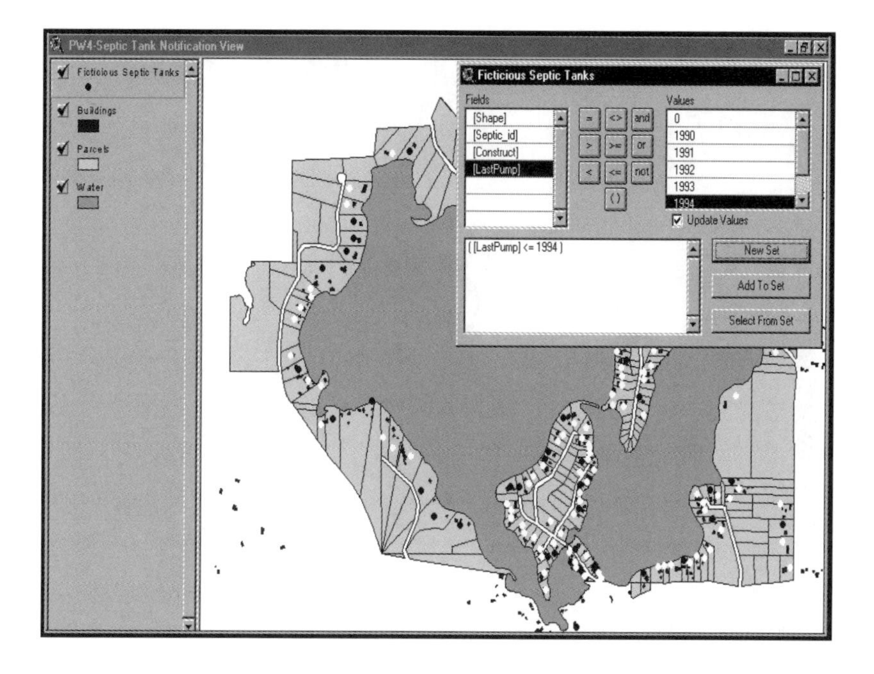

Figure 1-9: GIS may be used to streamline government function. Here, septic systems not compliant with pollution prevention ordinances are identified by white circles. (courtesy Wisconsin Sea Grant Institute and LICGC)

active land records modernization program, and may use GIS technology to assist in administration and enforcement of the zoning and shoreland protection ordinances. Specific activities include the creation of digital parcel maps, the development of parcel identification numbers (PINs) to link property attributes to parcel maps, the creation of digital aerial photographs on a regular time frame, and the incorporation of aerial or boat-based images to help detect property changes and zoning violations.

One early operation for the shoreland management GIS was the development of digital property records and associated parcel information. Parcel attributes such as the tax assessed value or owner name and address may need to be identified for many reasons, including the delivery of tax bills or for notification of nearby zoning variances or public meetings. Digital land records in a GIS may be used to streamline these and other activities.

GIS may also be used to aid in the administration of shoreline zoning ordinances. Setback requirements specify near-shore zones with special restrictions. Applications for construction or building modification may be reviewed with maps that overlay building locations with the shoreline setbacks (Figure 1-8). A GIS speeds the assessment of zoning compliance, and may be used to direct landowners to the relevant zoning ordinances.

GIS may also be used to notify landowners of routine tasks, such as septic system maintenance. Northern lakes are particularly susceptible to nitrogen pollution from near-shore septic systems (Figure 1-9). This often leads to required frequent pumping of the septic system, and verification of compliance. A GIS may be used to automatically generate notification of non-compliance. For example, landowners may be required to have their septic systems pumped every three years, and to provide proof. If not, the GIS system may automat-

Figure 1-10: A male blackfooted ferret, an endangered species. GIS are one of the tools used in attempts to save these creatures. (courtesy Randy Matchett, USFWS)

ically identify systems not in compliance and generate a letter for appropriate parcel owners.

Our second example illustrates how GIS helps us save endangered species. The blackfooted ferret is a small carnivore endemic to the western plains of North America (Figure 1-10), and is one of the most endangered mammals on the continent. The ferret lives in close association with prairie dogs, communally-living rodents once found over much of North America. Ferrets feed on prairie dogs and live in their burrows, and prairie dog colonies provide refuge from coyotes and other larger carnivores that prey on the ferret. The blackfooted ferret has become endangered because of declines in the range and number of prairie dog colonies, coupled with ferret sensitivity to canine distemper and other diseases.

The U.S. Fish and Wildlife Service (USFWS) has been charged with preventing the extinction of the blackfooted ferret. This entails establishing the number and location of surviving animals, identifying the habitat requirements for a sustainable population, and analyzing what factors are responsible for the decline in ferret numbers, so that a recovery plan may be devised.

Because blackfooted ferrets are nocturnal animals that spend much of their time underground, and because ferrets have always been rare, relatively little was initially known about their life history, habitat requirements, and the causes of mortality. For example, young ferrets often disperse from their natal prairie dog colonies in search of their own territories. Dispersal is good when it leads to an expansion of the species. However, there are limits on how far a ferret may be expected to successfully disperse. If the nearest suitable colony is too far away, the dispersing young ferret may likely die of starvation or be eaten by a coyote, eagle, large owl, or other predator.

Figure 1-11: Specialized equipment is used to collect spatial data. Here a burrow location is recorded using a GPS receiver as an interested black footed ferret looks on. (courtesy Randy Matchett, USFWS)

The dispersing ferret may reach a prairie dog colony that is too small to support it. Ferret recovery is hampered because we don't know when prairie dog colonies are too far apart, or if a colony is too small to support a breeding pair of ferrets. Because of this lack of spatial knowledge, wildlife managers may have difficulty selecting among a number of activities to enhance ferret survival. These activities include the establishment of new prairie dog colonies, fencing colonies to prevent the entry of larger predators, removing predators, captive breeding, and the capture and transport of young or dispersing animals.

GIS have been used to provide data necessary to save the blackfooted ferret (Figure 1-11). Individual ferrets are tracked, by nighttime spotlighting surveys, often in combination with radiotracking devices. Ferret locations and movements are combined with detailed data on prairie dog colony boundaries, burrow locations, surrounding vegetation and other spatial data (Figure 1-12). Individuals can be identified and vital characteristics monitored, including home range size, typical distance travelled, number of offspring, and survival. These data are combined and ana-

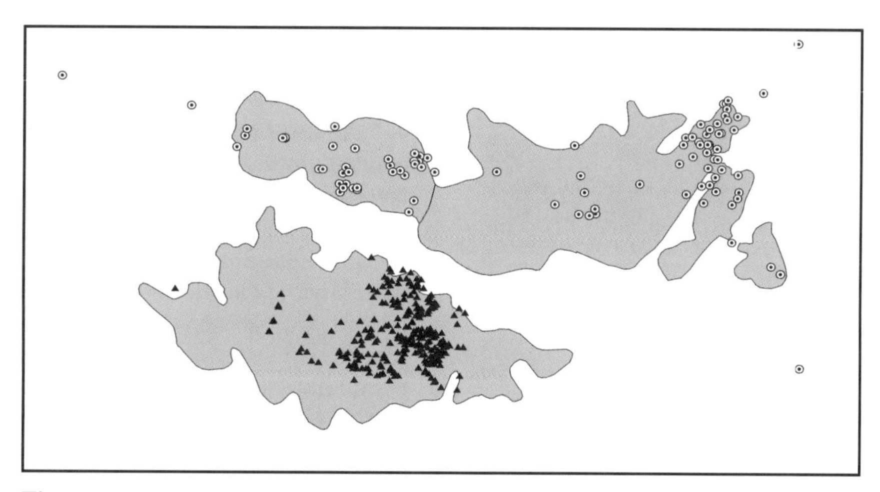

Figure 1-12: Spatial data, such as the boundaries of prairie dog colonies (gray polygons) and individual blackfooted ferret positions (triangle and circle symbols) may be combined to help understand how best to save the blackfooted ferret. (courtesy Randy Matchett, USFWS)

lyzed in a GIS to improve the likelihood of species recovery.

Geographic Information Science

While we have defined GIS as geographic information systems, there is another GIS: geographic information science. The abbreviation GIS is commonly used for the geographic information systems, while GIScience is used to abbreviate the science. The distinction is important, not the least because the future development of GIS depends on progress in GIScience.

GIScience is much broader than GIS, because GIScience forms a theoretical foundation on which GIS are based. GIS research is typically concerned with technical aspects of GIS implementation or application. GIScience includes these, but also seeks to redefine concepts in geography and geographic information in the context of the digital age. GIScience is concerned with how we conceptualize geography and how we collect, represent, store, visualize, analyze, use, and present these geographic concepts. The work draws from many fields, including traditional geography, geodesy, remote sensing, surveying, computer science, cartography, mathematics, statistics, cognitive science, linguistics, and others. GIScience investigates not only technical questions of interest to applied geographers, business-people, planners, public safety officers, and others, but GIScience is also directed at more basic questions. How do we perceive space? How might we best represent spatial concepts, given the new array of possibilities provided by our advancing technologies? How does human psychology help or hinder effective spatial reasoning?

Science has been described as a handmaiden of technology in the applied world. A more apt analogy is perhaps a parent of technology. GIS, narrowly defined, is more technology than science. Since GIS is the tool with which we solve problems, we are mistaken if we consider it as the starting and ending point in geographic reasoning. An understanding of GIScience is crucial to the further development of GIS, and in many cases, crucial to the effective application of GIS. This book focuses primarily on GIS, but provides relevant information related to GIScience as appropriate for an introductory course.

GIS Components

A GIS is comprised of hardware, software, data, humans, and a set of organizational protocols. These components must be well integrated for effective use of GIS, and the development and integration of these components is an iterative, ongoing process. The selection and purchase of hardware and software is often the easiest and quickest step in the development of a GIS. Data collection and organization, personnel development, and the establishment of protocols for GIS use are often more difficult and time-consuming endeavors.

Hardware for GIS

A fast computer, large data storage capacities, and a high-quality, large display form the hardware foundation of most GIS (Figure 1-13). A fast computer is required because spatial analyses are often applied over large areas and/or at high spatial resolutions. Calculations often have to be repeated over tens of millions of times, corresponding to each space we are analyzing in our geographical analysis. Even simple operations may take substantial time if sufficient computing capabilities are not present, and complex operations can be unbearably long-running. While advances in computing technology during the 1990s have substantially reduced the time required for most spatial analyses, computation times are still unacceptably long for a few applications.

While most computers and other hardware used in GIS are general purpose and adaptable for a wide range of tasks, there are also specialized hardware components that are specifically designed for use with spatial data. Many non-GIS endeavors require the entry of large data volumes, including inventory control in large markets, parcel delivery, and bank transactions. However, GIS is unique in the volume of coordinate data that must be entered. Specialized equipment has been developed to aid in these data entry tasks, and these devices will be described in detail in Chapter 4.

GIS Software

GIS software provides the tools to manage, analyze, and effectively display and disseminate spatial data and spatial information (Figure 1-14). GIS by necessity involves the collection and manipulation of the coordinates we use to specify location. We also must collect qualitative or quantitative information on the non-spatial attributes of our geographic features of interest. We need tools to view and edit these data, manipulate them to generate and extract the information we require, and produce the materials to communicate the information we have developed. GIS software provides the specific tools for some or all of these tasks.

There are many public domain and commercially available GIS software packages, and many of the commercial packages originated at academic or government-funded research laboratories. The Environmental Systems Research Institute (ESRI) line of products, including Arc/Info, is a good example. Much of the foundation for Arc/Info was developed during the 1960s and 1970s at Harvard University in the Laboratory of Computer Graphics and Spatial Analysis. Alumni from Harvard carried these concepts with them to Redlands, California when forming ESRI, and included them in their commercial products.

Our description below, while including most of the major or widely used software packages, is not meant to be all-inclusive. There are many additional software tools and packages available, particularly for specialized tasks or subject areas. Sources

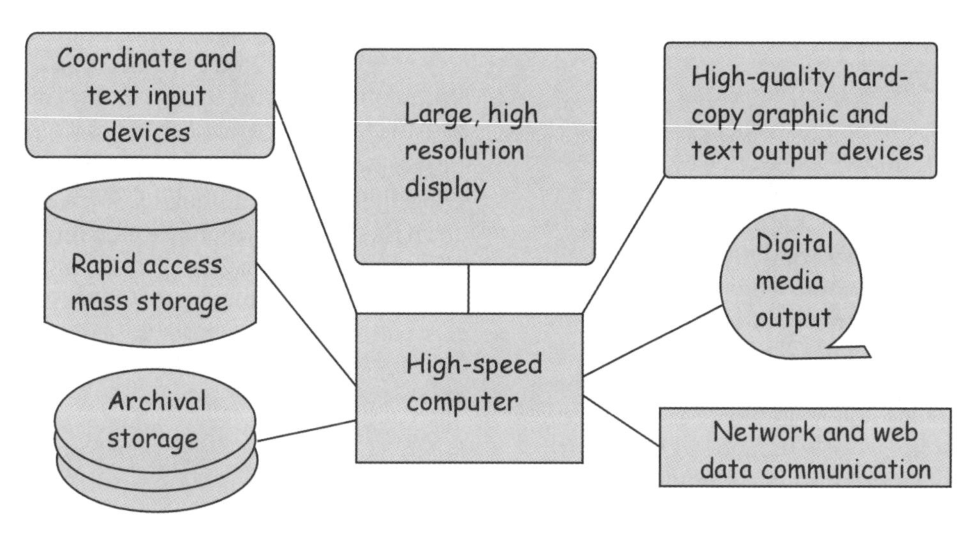

Figure 1-13: GIS are typically used with a number of general purpose and specialized hardware components.

Data entry
- manual coordinate capture
- attribute capture
- digital coordinate capture
- data import

Editing
- manual point, line and area feature editing
- manual attribute editing
- automated error detection and editing

Data management
- copy, subset, merge data
- versioning
- data registration and projection
- summarization, data reduction
- documentation

Analysis
- spatial query
- attribute query
- interpolation
- connectivity
- proximity and adjacency
- buffering
- terrain analyses
- boundary dissolve
- spatial data overlay
- moving window analyses
- map algebra

Output
- map design and layout
- hardcopy map printing
- digital graphic production
- export format generation
- metadata output
- digital map serving

Figure 1-14: Functions commonly provided by GIS software.

are provided in Appendix B that may be helpful in identifying the range of software available, and for obtaining detailed descriptions of specific GIS software characteristics and capabilities.

ArcGIS

ArcGIS and its predecessors, ArcView and Arc/Info, comprise one of the two most popular GIS software suites at the time of this writing. The Arc suite of software has a larger user base and higher annual unit sales than any other competing products. ArcGIS is a product of ESRI, a company that is based in Redlands, California but has a world-wide presence. ESRI has been developing and marketing GIS software since the early 1980s, and ArcGIS is an integrated set of software products. In addition to software, ESRI also provides substantial training, support, and fee-consultancy services at regional and international offices.

ArcGIS is popular because it provides an expandable set of capabilities. ArcView, the entry-level component of ArcGIS, may be purchased initially. GIS functions are provided for basic data entry, editing, and attribute and coordinate manipulation. Both discrete and continuous spatial data may be represented using the most common methods. Basic spatial data analyses are supported, and rapid, easy, basic map layout and printing capabilities are provided.

ArcEditor is a product which provides the next most commonly used set of spatial data manipulation functions. More complex editing tasks are possible, as are other data management functions, and more control over data base design.

Arc/Info is the comprehensive GIS toolbox from ESRI. Arc/Info is designed to provide a large set of geoprocessing procedures, from data entry through most forms of hardcopy or digital data output. As such, Arc/Info is a large, complex, sophisticated product. It supports multiple data formats, many data types and structures, and literally thousands of possible operations that may be applied to spatial data. It is not surprising that substantial training is required to master the full capabilities of Arc/Info.

ArcGIS provides substantial flexibility in how we conceptualize and model geographic features. Geographers and other GIS-related scientists have conceived of many ways to think about, structure, and store information about spatial objects. ArcGIS provides for the broadest available selection of these representations. For example, elevation data may be stored in at least four major formats, each with attendant advantages and disadvantages. There is equal flexibility in the methods for spatial data processing. This broad array of choices, while responsible for the substantial investment in time required for mastery of Arc/Info, provides concomitantly substantial analytical power.

GeoMedia

GeoMedia and the related MGE digital cartographic products are also one of the two most popular GIS suites currently in use. GIS and related products have been developed and supported by Intergraph, Inc. of Huntsville, Alabama, for over 30 years. GeoMedia offers a complete set of data entry, analysis, and output tools. A comprehensive set of editing tools may be purchased, including those for automated data entry and error detection, data development, data fusion, complex analyses, and sophisticated data display and map composition. Scripting languages may be obtained, as well as programming tools that allow specific features to be embedded in custom programs, and programing libraries to allow the modification of GeoMedia algorithms for special-purpose software.

GeoMedia is particularly adept at integrating data from divergent sources, formats, and platforms. Intergraph appears to have dedicated substantial effort toward the OpenGIS initiative, a set of standards to facilitate cross-platform and cross-software data sharing. Data in any of the common commercial databases may be integrated with spatial data from many formats. Image, coordinate, and text data may be combined.

GeoMedia also provides a comprehensive set of tools for GIS analyses. Complex spatial analyses may be performed, including queries, e.g., to find features in the database that match a set of conditions, and spatial analyses such as proximity or overlap between features. Worldwide web and mobile phone-based applications and application development are well supported.

MapInfo

MapInfo is a comprehensive set of GIS products developed and sold by the MapInfo Corporation, of Troy, New York. MapInfo products are used in a broad array of endeavors, although use seems to be concentrated in many business and municipal applications. This may be due to the ease with which MapInfo components are incorporated into other applications. Data analysis and display components are supported through a range of higher language functions, allowing them to be easily embedded in other programs. In addition, MapInfo provides a flexible, stand-alone GIS product that may be used to solve many spatial analysis problems.

Specific products have been designed for the integration of mapping into various classes of applications. For example, MapInfo products have been developed for embedding maps and spatial data into wireless handheld devices such as telephones, data loggers, or other portable devices. Products have been developed to support internet mapping applications, and serve spatial data in worldwide web based environments. Extensions to specific database products such as Oracle are provided.

Idrisi

Idrisi is a GIS system developed by the Graduate School of Geography of Clark University, in Massachusetts. Idrisi differs from the previously discussed GIS software packages in that it provides both image processing and GIS functions. Image data are useful as a source of information in GIS. There are many specialized software packages designed specifically to focus on image data collection, manipulation, and output. Idrisi offers much of this functionality while also providing a large suite of spatial data analysis and display functions.

Idrisi has been developed and maintained at an educational and research institution, and was initially used primarily as a teaching and research tool. Idrisi has adopted a number of very simple data structures, a characteristic that makes the software easy to modify in a teaching environment. Some of these structures, while slow and more space-demanding, are easy to understand and manipulate for the beginning programmer. File formats are well documented and data easy to access and modify. The developers of Idrisi have expressly encouraged researchers, students, and users to create new functions for Idrisi. The Idrisi project has then incorporated user-developed enhancements into the software package. Idrisi is an ideal package when teaching students not only to use GIS, but to develop their own spatial analysis functions.

Idrisi is relatively low cost, perhaps because of its affiliation with an academic institution, and is therefore widely used in education and in many less developed parts of the world. Low costs are an important factor in many developing countries, so Idrisi has been widely adopted there.

ERDAS

ERDAS (Earth Resources Data Analysis System) began as primarily an image processing system. The original purpose of the software was to enter and analyze satellite image data. ERDAS led a wave of commercial products for analyzing spatial data collected over large areas. Product development was spurred by the successful launch of the U.S. Landsat satellite in the 1970s. For the first time, digital images of the entire Earth surface were available to the public.

The ERDAS image processing software evolved to include other types of imagery, and to include a comprehensive set of tools for cell-based data analysis. Image data are supplied in a cell-based format. Cell-based analysis is a major focus of sections in three chapters of this book, so there will be much more discussion in later pages. For now, it is important to note that the "checkerboard" format used for image data may also be used to store and manipulate other spatial data. It is relatively easy and quite useful to develop cell-based spatial analysis tools to complement the image processing tools.

ERDAS and most other image processing packages provide data output formats that are compatible with most common GIS packages. Many image processing software systems are purchased explicitly to provide data for a GIS. The support of ESRI data formats is particularly thorough in ERDAS. ERDAS GIS components may then be used to analyze these spatial data.

AUTOCAD MAP

AUTOCAD is the world's largest-selling computer drafting and design package. Produced by Autodesk, Inc., of San Rafael, California, AUTOCAD began as an engineering drawing and printing tool. A broad range of engineering disciplines are supported, including surveying and civil engineering. Surveyors have traditionally developed and maintained the coordinates for property boundaries, and these are among the most important and often-used spatial data. AUTOCAD MAP adds substantial analytical capability to the already complete set of data input, coordinate manipulation, and data output tools provided by AUTOCAD.

The latest version, AUTOCAD MAP2000i, provides a substantial set of spatial data analysis capability. Data may be entered, verified, and output. Data may also be *queried*, searched for features with particular conditions or characteristics. More sophisticated spatial analysis may be performed, including path finding or data combination. AUTOCAD MAP2000i incorporates many of the specialized analysis capabilities of other, older GIS packages, and is a good example of the convergence of GIS software from a number of disciplines.

MicroImages

MicroImages produces TNTmips, an integrated remote sensing, GIS, and CAD software package. MicroImages also produces and supports a range of other related products, including software to edit and view spatial data, software to create digital atlases, and software to publish and serve data on the internet.

TNTmips is notable both for the breadth of tools and for the range of hardware platforms supported in a uniform manner. MicroImages recompiles a basic set of code for each platform so that the look, feel, and functionality is nearly identical irrespective of the hardware platform used. Image processing, spatial data analysis, and image, map, and data output are supported uniformly across this range.

TNTmips provides an impressive array of spatial data development and analysis tools. All common image processing tools are available, including image ingest of a broad number of formats, image registration and mosaics, reprojection, error removal, subsetting, combination, and image classification. Vector analyses are supported, including support for point, line, and area features, multi-layer combination, viewshed, proximity, and network analyses. Extensive online documentation is available, and the software is supported by an international network of dealers.

The mini-review above is in no way an exhaustive compilation of the available or useful geoprocessing software. There are many other software packages, tools, and utilities available, many of which provide unique, novel, or particularly clever combinations of geoprocessing tools. GRASS, PCI, and ENVI are just a few of the available software packages with spatial data development or analysis capabilities. In addition, there are thousands of add-ons, special purpose tools, or specific modules that complement these products. Websites for each of these products will provide more detailed descriptions, and these and other websites listed in Appendix B at the end of this book will provide more information on these and other GIS software products.

GIS in Organizations

Although new users often focus on GIS hardware and software components, we must recognize that GIS exist in an institutional context. Effective use of GIS requires an organization to support various GIS activities. Most GIS also require trained personnel to use them, and a set of protocols guiding how the GIS will be used. The institutional context determines what spatial data are important, how these data will be collected and used, and ensures that the results of GIS analyses are properly interpreted and applied. GIS share a common characteristic of many powerful technologies. If not properly used, the technology may lead to a significant waste of resources, and may do more harm than good. The proper institutional resources are required for GIS to provide all its potential benefits.

GIS are often employed as decision support tools (Figure 1-15). Data are collected, entered, and organized into a spatial database, and analyses performed to help make specific decisions. The results of spatial analyses in a GIS often uncover the need for more data, and there are often several iterations through the collection, organization, analysis, output, and assessment steps before a final decision is reached. It is important to recognize the organizational structure within which the GIS will operate, and how GIS will be integrated into the decision-making processes of the organization.

One important question that must be answered early is "what problem(s) are we to solve with the GIS?" GIS add significant analytical power through the ability to measure distances and areas, identify vicinity, analyze networks, and through the overlay and combination of different information. Unfortunately, spatial data development is often expensive, and effective GIS use requires specialized knowledge or training, so there is often considerable expense in constructing and operating a GIS. Before spending this time and money there must be a clear identification of the new questions that may be answered, or the process, product, or service that will be improved, made more efficient, or less expensive through the use of GIS. Once the ends are identified, an organization may determine the level of investment in GIS that is warranted.

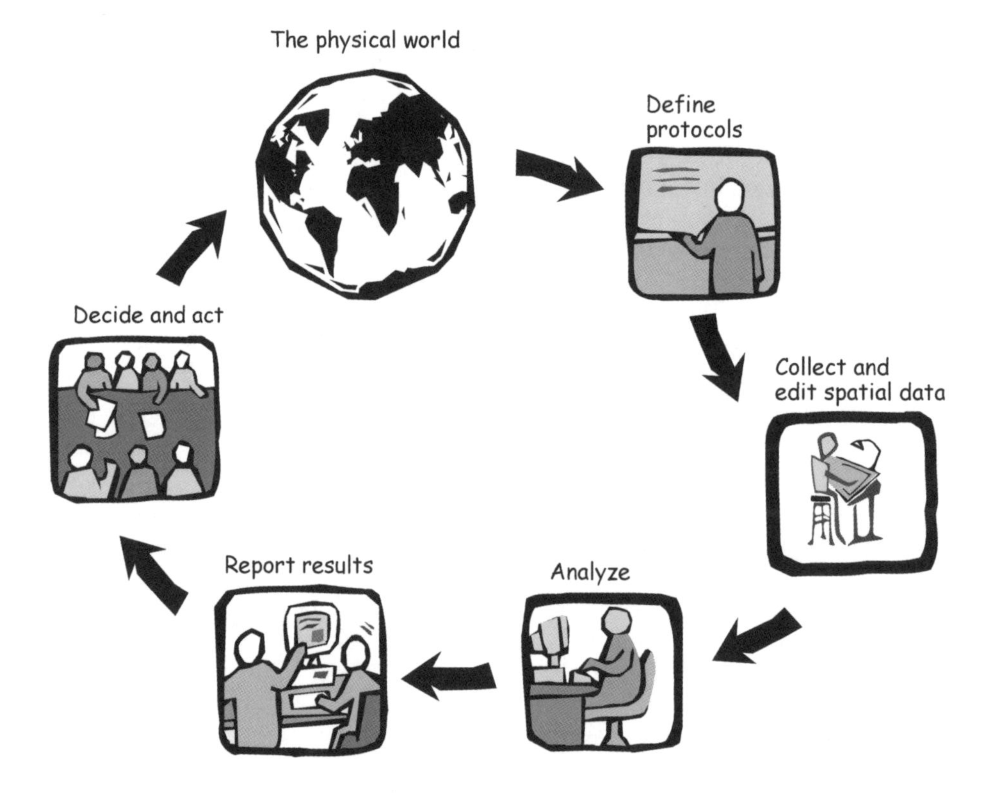

Figure 1-15: GIS exist in an institutional context. Their effective use depends on a set of protocols and an integration into the data collection, analysis, decision, and action loop of an organization.

Summary

GIS are computer-based systems that aid in the development and use of spatial data. There are many reasons we use GIS, but most are based on a societal push, our need to more effectively and efficiently use our resources, and a technological pull, our interest in applying new tools to previously insoluble problems. GIS as a technology is based on geographic information science, and is supported by the disciplines of geography, surveying, engineering, space science, computer science, cartography, statistics, and a number of others.

GIS are comprised of both hardware and software components. Because of the large volumes of spatial data and the need to input coordinate values, GIS hardware often have large storage capacities, fast computing speed, and ability to capture coordinates. Software for GIS are unique in their ability to manipulate coordinates and associated attribute data. A number of software tools and packages are available to help us develop GIS.

While GIS are defined as tools for use with spatial data, we must stress the importance of the institutional context in which GIS fit. Because GIS are most often used as decision-support tools, the effective use of GIS requires more than the purchase of hardware and software. Trained personnel and protocols for use are required if GIS are to be properly applied. GIS may then be incorporated in the question-collect-analyze-decide loop when solving problems.

The Structure of This Book

This book is designed to serve a semester-long, 15-week course in GIS at the university level. We seek to provide the relevant information to create a strong basic foundation on which to build an understanding of GIS. Because of the breadth and number of topics covered, students may be helped by knowledge of how this book is organized. Chapter 1 (this chapter), sets the stage, providing some motivation and a background for GIS. Chapter 2 describes basic data representations. It treats the main ways we use computers to represent perceptions of geography, common data structures, and how these structures are organized. Chapter 3 provides a basic description of coordinates and coordinate systems, how coordinates are defined and measured on the surface of the Earth, and conventions for converting these measurements to coordinates we use in a GIS.

Chapters 4 through 7 treat spatial data collection and entry. Data collection is often a substantial task and comprises one of the main activities of most GIS organizations. General data collection methods and equipment are described in Chapter 4. Chapter 5 describes the global positioning system (GPS), a relatively new technology for coordinate data collection. Chapter 6 describes aerial and space-based images as a source of spatial data. Most historical and contemporary maps depend in some way on image data, and this chapter provides a background on how these data are collected and used to create spatial data. Chapter 7 provides a brief description of common digital data sources available in the United States, their formats, and uses.

Chapters 8 through 13 treat the analysis of spatial data. Chapter 8 focuses on attribute data, attribute tables, database design, and analyses using attribute data. Attributes are half our spatial data, and a clear understanding of how we structure and use them is key to effective spatial reasoning. Chapters 9, 10, 11, and 12 describe basic spatial analyses, including adjacency, inclusion, overlay, and data combination for the main data models used in GIS. They also describe more complex spatio-temporal models. Chapter 13 describes various methods for interpolation. We typically find it impractical or inefficient to collect "wall-to-wall" spatial and attribute data. Interpolation allows us to extend our sampling and provide information for unsampled locations. Chapter 14 describes how we assess and document spatial data quality, while Chapter 15 provides some musings on current conditions and future trends.

We give preference to the International System of Units (SI) throughout this book. The SI system is adopted by most of the World, and is used to specify distances and locations in the most common global coordinate systems and by most spatial data collection devices. However, some English units are culturally embedded, e.g., the survey foot, or 640 acres to a Public Land Survey Section, and so these are not converted. Because a large portion of the target audience for this book is in the United States, English units of measure often supplement SI units.

Suggested Reading

Amdahl, G., Disaster Response: GIS for Public Safety, ESRI Press, Redlands, 2001.

Burrough, P.A. and Frank, A.U., Concepts and paradigms in spatial information: Are current geographical information systems truly generic?, *International Journal of Geographical Information Systems*, 1995, 9:101-116.

Burrough, P. A. and McDonnell, R. A., Principles of Geographical Information Systems, Oxford University Press, New York, 1998.

Campbell, H. J. and Masser, I., GIS in local government: some findings from Great Britain, *International Journal of Geographical Information Systems*, 1992, 6:529-546.

Commission on Geoscience, Rediscovering Geography: New Relevance for Science and Society, National Academy Press, Washington, 1997.

Goodchild, M. F., Geographical information science, *International Journal of Geographical Information Systems*, 1992, 6:31-45.

Grimshaw, D., Bringing Geographical Information Systems Into Business, 2nd Edition. Wiley, New York, 2000.

Haining, R., Spatial Data Analysis in the Social and Environmental Sciences, Cambridge University Press, Cambridge, 1990.

Huxhold, W. E., An Introduction to Urban Geographic Information Systems, Oxford University Press, Oxford, 1991.

Johnston, C., Geographic Information Systems in Ecology, Blackwell Scientific, Boston, 1998.

MaGuire, D.J., Goodchild, M.F., and Rhind, D.W., eds., Geographic Information Systems, Longman Scientific, New York, 1991.

Martin, D., Geographical Information Systems: Socio-economic Applications, 2nd Edition. Routledge, London, 1996.

McHarg, I., Design with Nature, Wiley, New York, 1995.

Peuquet, D. J. and Marble, D. F., eds., Introductory Readings in Geographic Information Systems, Taylor and Francis, Washington D.C., 1990.

Pickles, J., ed., Ground Truth: The Social Implictions of Geographic Information Systems, Guilford, New York, 1995.

Smith, D. A. and Tomlinson, R. F., Assessing costs and benefits of geographical information systems: methodological and implementation issues, *International Journal of Geographical Information Systems*, 1992, 6:247-246.

Tillman Lyle, J., Design for Human Ecosystems: Landscape, Land Use, and Natural Resources, Island Press, Washington, 1999.

Tomlinson, R., Current and potential uses of geographical information systems. The North American experience, *International Journal of Geographical Information Systems*, 1987, 1:203-218.

Unwin, D., Introductory Spatial Analysis, Methuen, London, 1981.

Study Questions

Why are we more interested in spatial data today than 100 years ago?

You have probably collected, analyzed, or communicated spatial data in one way or another during the past month. How many instances can you think of?

How are GIS software different from most other software?

How many ways are GIS hardware different from other computer hardware?

What are the limitations of using a GIS? Under what conditions might the technology hinder problem solving, rather than help?

Define a GIS in your own words. Are paper maps and paper data sheets a GIS?

2 Data Models

Introduction

Data in a GIS represent a simplified view of the real world. Physical *entities* or phenomena are approximated by data in a GIS. These data include information on the spatial location and extent of the physical entities, and information on their non-spatial properties.

Each entity is represented by a *spatial feature* or *cartogaphic object* in the GIS, and so there is an entity-object correspondence. Because every computer system has limits, only a subset of the essential characteristics are represented for each entity. As illustrated in Figure 2-1, we may represent lakes in a region by a set of polygons. These polygons are associated with a set of essential characteristics that define each lake. All other information for the area may be ignored, e.g., information on the roads, buildings, slope, or soil characteristics. Only lake boundaries and essential lake characteristics have been saved in this example.

Essential characteristics are defined by the person, group, or organization that develops the spatial data or uses the GIS. The set of characteristics used to represent an entity

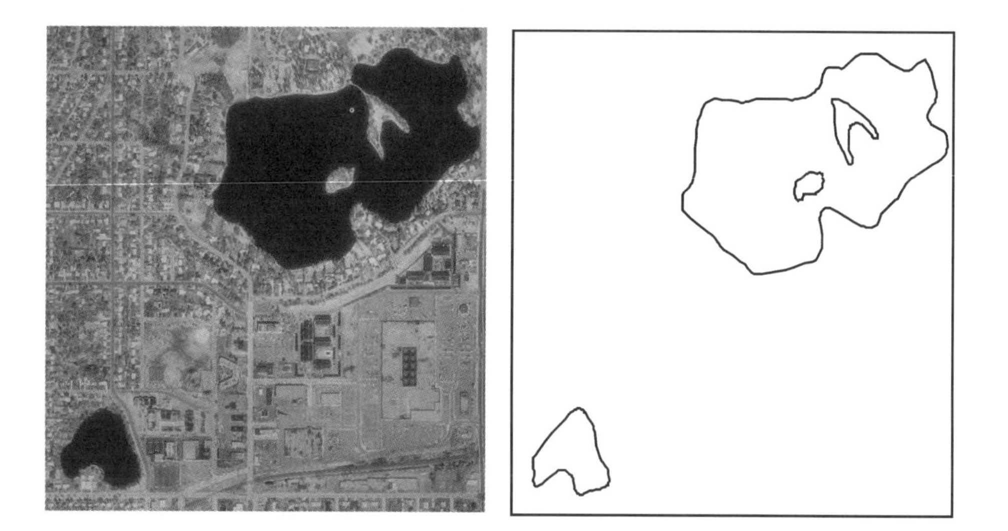

Figure 2-1: A physical entity is represented by a spatial object in a GIS. Here, the physical boundaries of lakes are represented by lines.

is subjectively chosen. What is essential to describe a forest for one person, for example a logger, would be different than what is essential to another person, such as a typical member of the Sierra Club. Objects are abstractions in a spatial database, because we can only record and maintain a subset of characteristics of any entity, and no one abstraction is universally better than any other.

A *data model* may be defined as the objects in a spatial database plus the relationships among them. The term model is fraught with ambiguity, because it is used in many disciplines to describe many things. Here the purpose of a spatial data model is to provide a formal means of representing and manipulating spatially-referenced information. In our lake example our data model consists of two parts. The first part is a set of polygons (closed areas) recording the shoreline of the lake, and the second part is a set of numbers or letters associated with each polygon. A data model may be considered the most recognizable level in our computer abstraction of the real world. Data structures and binary machine code are successively less recognizable, but more computer-compatible forms of the spatial data (Figure 2-2).

Coordinates are used to define the spatial location and extent of geographic objects. A coordinate most often consists of a pair of numbers that specify location in relation to an origin. The coordinates quan-

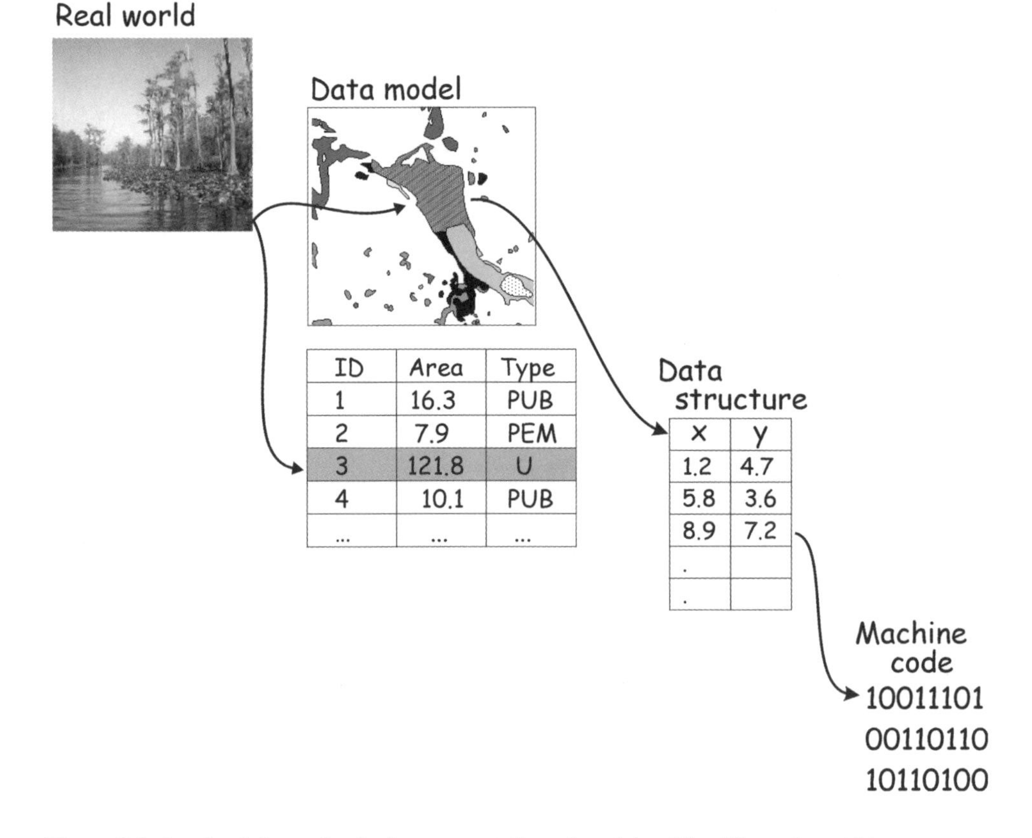

Figure 2-2: Levels of abstraction in the representation of spatial entities. The real world is represented in successively more machine-compatible but humanly obscure forms.

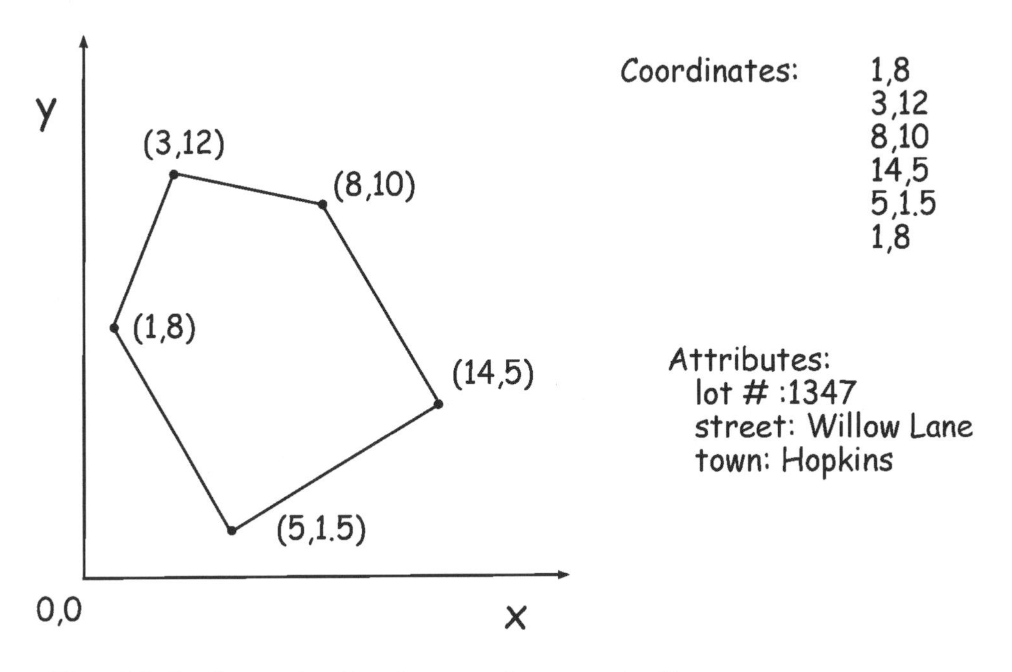

Figure 2-3: Coordinate and attribute data are used to represent entities.

tify the distance from the origin when measured along a standard direction. Single or groups of coordinates are organized to represent the shapes and boundaries that define the objects. Coordinate information is an important part of the data model, and models differ in how they represent these coordinates. Coordinates are usually expressed in one of many standard coordinate systems. The coordinate systems are usually based upon standardized map projections (discussed in Chapter 3) that unambiguously define the coordinate values for every point in an area.

Typically there are two distinct types of data used to define cartographic objects (Figure 2-3). First, coordinate or geometric data define the location and shape of the objects. Second, attribute data are collected and referenced to each object. These attribute data record the non-spatial components of an object, such as a name, color, pH, or cash value.

Attribute data are linked with coordinate data to help define each cartographic object in the GIS. The attribute data are linked to the corresponding cartographic objects in the spatial part of the GIS database. Keys, labels, or other indices are used so that the spatial and attribute data may be viewed, related, and manipulated together.

Most conceptualizations view the world as a set of layers (Figure 2-4). Each layer organizes the spatial and attribute data for a given set of cartographic objects in the region of interest. These are often referred to as *thematic layers*. As an example consider a GIS database that includes a soils data layer, a population data layer, an elevation data layer, and a roads data layer. The roads layer "contains" only roads data, including the location and properties of roads in the analysis area. There are no data regarding the location and properties of any other geographic entities in the roads layer. Information on soils, population, and elevation are

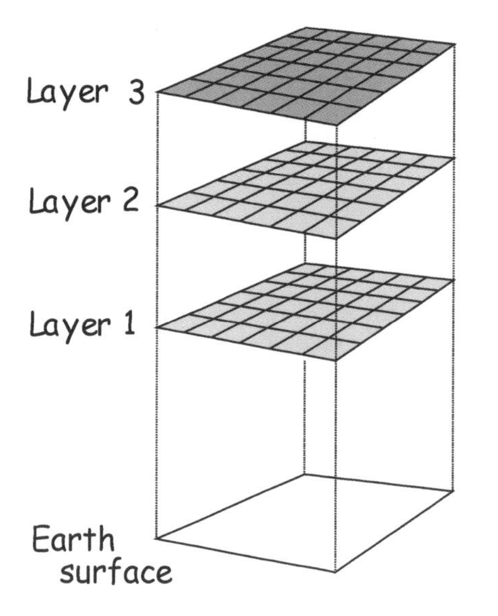

Figure 2-4: Spatial data are most often viewed as a set of thematically distinct layers.

contained in their respective data layers. Through analyses we may combine data to create a new data layer, e.g., we may identify areas that are "high" elevation and join this information with the soils data. This join creates a new data layer with a new, composite soils/elevation variable mapped.

Coordinate Data

Coordinates define location in two or three-dimensional space. Coordinate pairs, e.g., x and y, or coordinate triples, x, y, and z, are used to define the shape and location of each spatial object or phenomenon.

Spatial data in a GIS most often use a *Cartesian* coordinate system, so named after Rene Descartes, a French mathematician. Cartesian systems define two or three *orthogonal* (right-angle) axes. Two-dimensional Cartesian systems define x and y axes in a plane (Figure 2-5, left) Three-dimensional Cartesian systems define a z axis, orthogonal to both the x and y axes. An origin is defined with zero values at the inter-

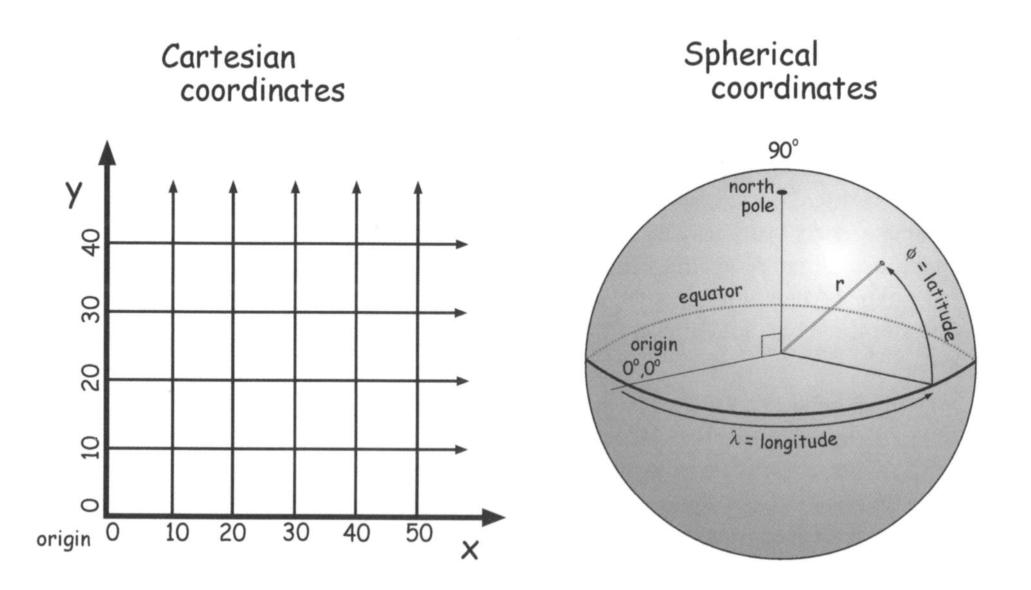

Figure 2-5: Cartesian (left) and spherical (right) coordinate systems.

section of the orthogonal axes. Cartesian coordinates are usually specified as decimal numbers, by convention increasing from bottom to top and from left to right.

Coordinate data may also be specified in a *spherical* coordinate system. Hipparchus, a Greek mathematician of the 2nd century B.C., was among the first to specify locations on the Earth using angular measurements on a sphere. The most common spherical system uses two angles of rotation and a radius distance, r, to specify locations on a modeled earth surface (Figure 2-5, right). These angles of rotation occur around a polar axis to define a longitude (λ) and with reference to an equatorial plane to define a latitude (ϕ). Latitudes increase from zero at the Equator to 90 degrees at the poles. Northern latitudes are preceded by an N and southern latitudes by an S, e.g., N90^o, S10^o. Longitudes increase east and west of an origin. Longitude values are preceded by an E and W, respectively, e.g., W110^o. Northern and eastern directions are designated as positive and southern and western designated as negative when signed coordinates are required. Spherical coordinates are most often recorded in a degrees-minutes-seconds (DMS) notation, e.g. N43^o 35′ 20″, signifying 43 degrees, 35 minutes, and 20 seconds of latitude. Minutes and seconds range from zero to sixty. Alternatively, spherical coordinates may be expressed as decimal degrees (DD). DMS may be converted to DD by:

$$DD = DEG + MIN/60 + SEC/3600 \qquad (2.1)$$

Attribute Data and Types

Attribute data are used to record the non-spatial characteristics of an entity. Attributes are also called items or variables. Attributes may be envisioned as a list of characteristics that help describe and define the features we wish to represent in a GIS. Color, depth, weight, owner, component vegetation type, or landuse are examples of variables that may be used as attributes. Attributes have values, e.g., color may be blue, black or brown, weight from 0.0 to 500.0, or landuse may be urban, agriculture, or undeveloped. Attributes are often presented in tables, with attributes arranged in rows and columns (Figure 2-2). Each row corresponds to an individual spatial object, and each column corresponds to an attribute. Tables are often organized and managed using a specialized computer program called a database management system (DBMS, described fully in Chapter 8).

Attributes of different types may be grouped together to describe the non-spatial properties of each object in the database. These attribute data may take many forms, but all attributes can be categorized as nominal, ordinal, or interval/ratio attributes.

Nominal attributes are variables that provide descriptive information about an object. Color, a vegetation type, a city name, the owner of a parcel, or soil series are all examples of nominal attributes. There is no implied order, size, or quantitative information contained in the nominal attributes.

Nominal attributes may also be images, film clips, audio recordings, or other descriptive information. Just as the color or type attributes provide nominal information for an entity, an image may also provide descriptive information. GIS for real estate management and sales often have images of the buildings or surroundings as part of the database. Digital images provide information not easily conveyed in any other manner. These image or sound attributes are sometimes referred to as BLOBs for binary large objects, but they are best considered as a special case of a nominal attribute.

Ordinal attributes imply a rank order or scale in their values. An ordinal attribute may be descriptive such as small, medium, or large, or they may be numeric, such as an erosion class which takes values from 1 through 10. The order reflects only rank, and does not specify the form of the scale. An object with an ordinal attribute that has a value of four has a higher rank for that attribute than an object with a value of two. However we cannot infer that the attribute value is twice as large, because we cannot assume the scale is linear.

Interval/ratio attributes are used for numeric items where both order and absolute difference in magnitudes are reflected in the numbers. These data are often recorded as real numbers, most often on a linear scale. Area, length, weight, value, height, or depth are a few examples of attributes which are represented by interval/ratio variables.

Common Spatial Data Models

Spatial data models begin with a conceptualization, a view of real world phenomena or entities. Consider a road map suitable for use at a statewide or provincial level. This map is based on a conceptualization that defines roads as lines. These lines connect cities and towns that are shown as discrete points or polygons on the map. Road properties may include only the road type, e.g., a limited access interstate, state highway, county road, or some other type of road. The roads have a width represented by the drawing symbol on the map, however this width, when scaled, may not represent the true road width. This conceptualization identifies each road as a linear feature that fits into a small number of categories. All state highways are represented by the same type of line, even though the state highways may vary. Some may be paved with concrete, others with bitumen. Some may have wide shoulders, others not, or dividing barriers of concrete, versus a broad vegetated median. We realize these differences can exist within this conceptualization.

There are two main conceptualizations used for digital spatial data. The first conceptualization defines discrete objects using a *vector* data model. Vector data models use discrete elements such as points, lines, and polygons to represent the geometry of real-world entities (Figure 2-6).

A farm field, a road, a wetland, cities, and census tracts are examples of discrete entities that may be represented by discrete objects. Points are used to define the locations of "small" objects such as wells, buildings, or ponds. Lines may be used to represent linear objects, e.g., rivers or roads, or to identify the boundary between what is a part of the object and what is not a part of the object. We may map landcover for a region of interest, and we categorize discrete areas as a uniform landcover type. A forest may share an edge with a pasture, and this boundary is represented by lines. The boundaries between two polygons may not be discrete on the ground, for example, a forest edge may grade into a mix of trees and grass, then to pasture; however in the vector conceptualization, a line between two landcover types will be drawn to indicate a discrete, abrupt transition between the two types. Lines and points have coordinate locations, but points have no dimension, and lines have no dimension perpendicular to their direction. Area features may be defined by a closed, connected set of lines.

The second common conceptualization identifies and represents grid cells for a given region of interest. This conceptualization employs a *raster* data model (Figure 2-6). Raster cells are arrayed in a row and column pattern to provide "wall-to-wall" coverage of a study region. Cell values are used to represent the type or quality of mapped variables. The raster model is used most commonly with variables that may change continuously across a region. Elevation, mean temperature, slope, average rainfall, cumulative ozone exposure, or soil moisture are examples of phenomena that are often represented as continuous fields. Raster representations are commonly used to represent discrete features, for example, class maps such as vegetation or political units.

Data models are at times interchangeable in that many phenomena may be represented with either the vector or raster conceptual approach. For example, elevation may be represented as a surface (continuous field) or as series of lines representing contours of equal elevation (discrete objects). Data may be converted from one conceptual view to another, e.g., the location of contour lines (lines of equal elevation) may be determined by evaluating the raster surface, or a raster data layer may be derived from a set of contour lines. These conversions entail some costs both computationally and perhaps in data accuracy.

The selection of a raster or vector conceptualization often depends on the type of operations to be performed. For example, slope is more easily determined when elevation is represented as a continuous field in a raster data set. However, discrete contours are often the preferred format for printed maps, so the discrete conceptualization of a vector data model may be preferred for this application. The best data model for a given organization or application depends on the most common operations, the experiences and views of the GIS users, the form of available data, and the influence of the data model on data quality.

In addition to the two main data models, there are other data models that may be described as variants, hybrids, or special forms by some GIS users, and as different families of data models by others. A triangulated irregular network (TIN) is an example of such a data model. This model is most often used to represent surfaces, such as elevations, through a combination of point, line, and area features. Many consider this a special, admittedly well-developed, type of vector data model. Variants or other representations related to raster data models also exist. We choose two broad categories for clarity in an introductory text, and introduce variants as appropriate later in this and other chapters.

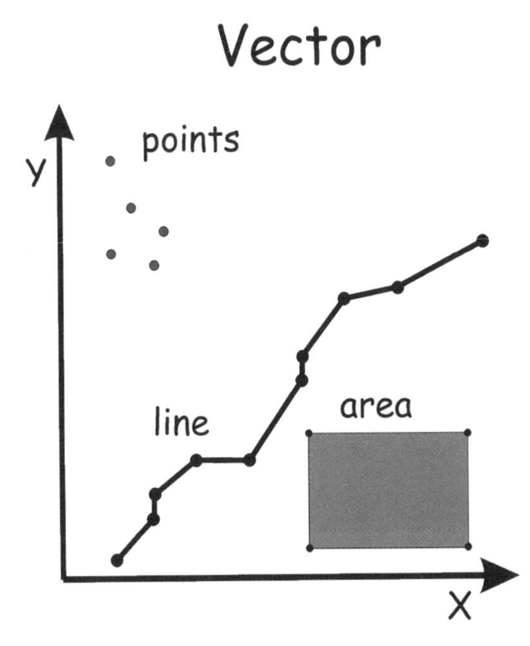

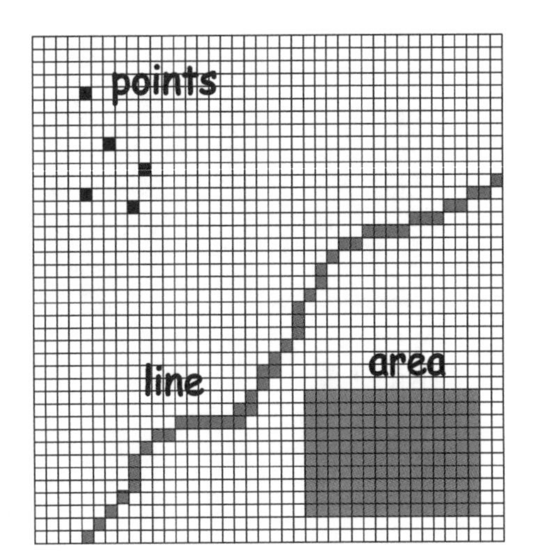

Figure 2-6: Vector and raster data models.

Vector data models will be described in the next section, including commonly found variants. Sections describing raster data models, TIN data models, and data structure then follow.

Vector Data Models

A vector data model uses sets of coordinates and associated attribute data to define discrete objects. Groups of coordinates define the location and boundaries of discrete objects, and these coordinate data plus associated attributes are used to create vector objects representing the real-world entities (Figure 2-7).

There are three basic types of vector objects: points, lines, and polygons (Figure 2-8). A point uses a single coordinate pair to represent the location of an entity that is considered to have no dimension. Gas wells, light poles, accident location, and survey points are examples of entities often represented as point objects in a spatial database. Some of these have real physical dimension, but for the purposes of the GIS users they may be represented as points. In effect, this means the size or dimension of the entity is not important spatial information, only the central location. Attribute data are attached to each point, and these attribute data record the important non-spatial characteristics of the point entities. When using a point to represent a light pole, important attribute information might be the height of the pole, the type of light and power source, and the last date the pole was serviced.

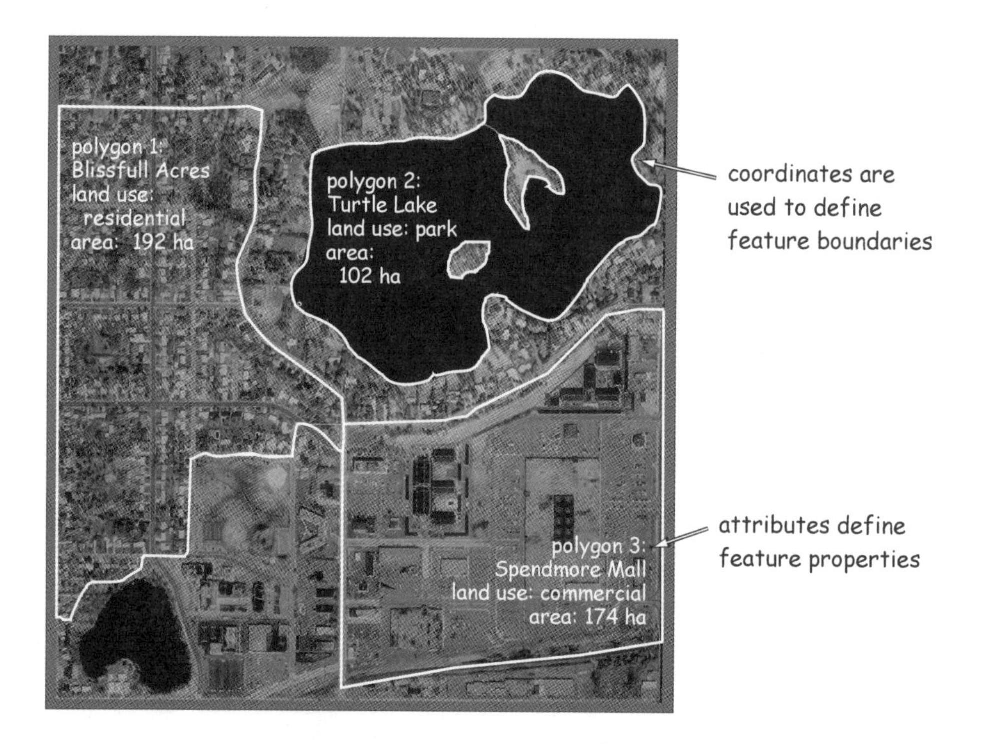

Figure 2-7: Coordinates define spatial location and shape. Attributes record the important non-spatial characteristics of features in a vector data model.

Linear features, often referred to as *arcs*, are represented as lines when using vector data models. Lines are most often represented as an ordered set of coordinate pairs. Each line is made up of line segments that run between adjacent coordinates in the ordered set (Figure 2-8). A long, straight line may be represented by two coordinate pairs, one at the start and one at the end of the line. Curved linear entities are most often represented as a collection of short, straight, line segments, although curved lines are at times represented by a mathematical equation describing a geometric shape. Lines typically have a starting point, an ending point, and intermediate points to represent the shape of the linear entity. Starting points and ending points for a line are sometimes referred to as *nodes*, while intermediate points in a line are referred to as *vertices* (Figure 2-8). Attributes may be attached to the whole line, line segments, or to nodes and vertices along the lines

Area entities are most often represented by closed polygons. These polygons are formed by a set of connected lines, either one line with an ending point that connects back to the starting point, or as a set of lines connected start-to-end (Figure 2-8). Polygons have an interior region and may entirely enclose other polygons in this region. Polygons may be adjacent to other polygons and thus share "bordering" or "edge" lines with other polygons. Attribute data may be attached to the polygons, e.g., area, perimeter, landcover type, or county name may be linked to each polygon.

The Spaghetti Vector Model

The *spaghetti* model is an early vector data model that was originally developed to organize and manipulate line data. Lines are captured individually with explicit starting and ending nodes, and intervening vertices used to define the shape of the line. The spaghetti model records each line separately. The model does not explicitly enforce or record connections of line segments when they cross, nor when two line ends meet (Figure 2-9a). A shared polygon boundary may be represented twice, with a line for each polygon on either side of the boundary. Data in this form are similar in some

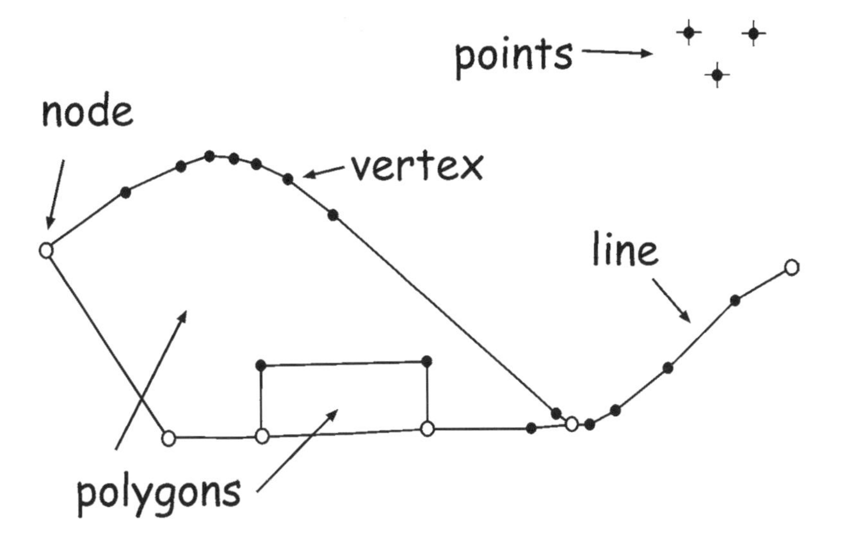

Figure 2-8: Points, nodes and vertices define points, line, and polygon features in a vector data model.

respects to a plate of cooked spaghetti, with no ends connected and no intersections when lines cross.

The spaghetti model is a relatively unstructured way of representing vector data. Because connections among lines are not enforced there may be breaks or overlaps in what should be a connected set of lines. The set of lines that defines a polygon may not form a closed area, so it is not possible to specify the region inside vs. the region outside of the polygon. Coordinates for points, lines, and polygons are often stored sequentially, such that data for nearby areas may be stored quite far apart. This significantly slows data access.

The spaghetti model severely limits spatial data analysis and is little used except when entering spatial data. Because lines often do not connect when they should, many common spatial analyses are inefficient and the results incorrect. For example, analyses such as determining an optimum set of bus routes are precluded if all street connections are not represented in a roads data layer. Area calculation, layer overlay, and many other analyses require "clean" spatial data in which all polygons close and lines meet correctly.

Topological Vector Models

Topological vector models specifically address many of the shortcomings of spaghetti data models. Early GIS developers realized that they could greatly improve the speed, accuracy, and utility of many spatial data operations by enforcing strict connectivity, by recording connectivity and adjacency, and by maintaining information on the relationships between and among points, lines, and polygons in spatial data. These early developers found it useful to record information on the *topological* characteristics of data sets.

Topology is the study of geometric properties that do not change when the forms are bent, stretched or undergo similar transformations. Polygon adjacency is an example of a topologically invariant property, because the list of neighbors to any given polygon does not change during geometric stretching or bending (Figure 2-9, b and c). *Topological vector models* explicitly record topological relationships such as adjacency and connec-

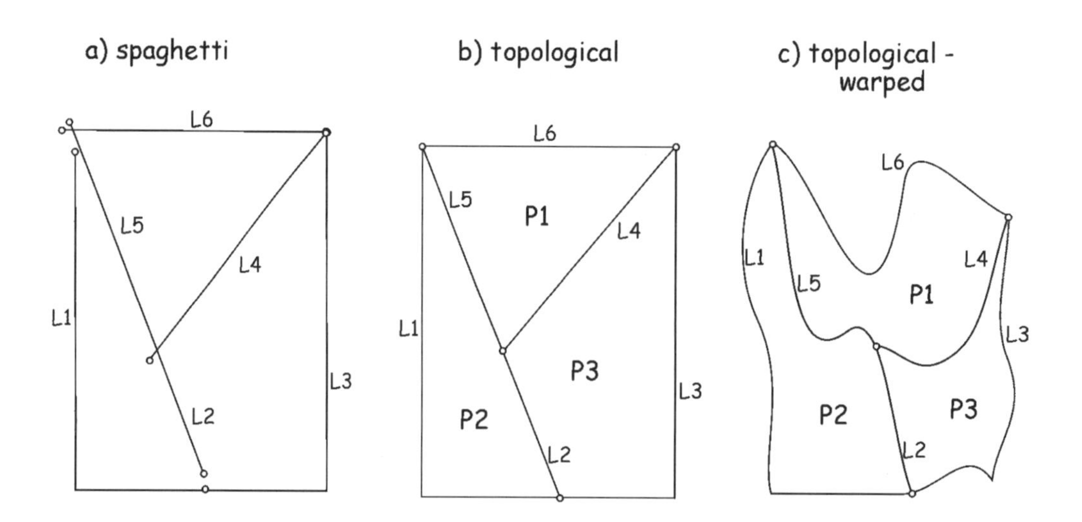

Figure 2-9: Spaghetti (a), topological (b), and topological-warped (c) vector data. Figures b and c are topologically identical because they have the same connectivity and adjacency.

tivity in the data files. These relationships may be recorded separately from the coordinate data and hence do not change when data are stretched or bent, e.g., when converting between coordinate systems.

Topological vector models may also enforce particular types of topological relationships. *Planar topology* requires that all features occur on a two-dimensional surface. There can be no overlaps among lines or polygons in the same layer (Figure 2-10). When planar topology is enforced, lines may not cross over or under other lines. At each line crossing there must be an intersection. The top left of Figure 2-10 shows a non-planar graph. Four line segments coincide. At some locations the lines intersect and a node is present, but at some locations a line passes over or under another line segment. These lines are non-planar because if forced to be in the same plane, all line crossings would intersect at a node. The top right of Figure 2-10 shows planar topology enforced for these same four line segments. Nodes, shown as white-filled circles, are found at each line crossing.

Non-planarity may also occur for polygons, as shown at the bottom of Figure 2-10. Two polygons overlap slightly at an edge. This may be due to an error, e.g., the two polygons share a boundary but have been recorded with an overlap, or there may be two areas that overlap in some way. On the left the polygons are non-planar, that is, they occur one above the other. If topological planarity is enforced, these two polygons must be resolved into three separate, non-overlapping polygons. Nodes are placed at the intersections of the polygon boundaries (lower right, Figure 2-10).

There are additional topological constructs besides planarity that may be recorded or enforced in topological data structures. For example, polygons may be exhaustive, in that there are no gaps, holes or "islands" in a set of polygons. Line direction may be recorded, so that a "from" and "to" node are identified in each line. Directionality aids the representation of river or street networks, where there may be a natural flow direction.

There is no single, uniform set of topological relationships that are included in all topological data models. Different researchers or software vendors have incorporated different topological information in their data structures. Planar topology is often included, as are representations of *adjacency* (which polygons are next to which) and *connectivity* (which lines connect to which). However, much of this information can be generated "on-the-fly", during processing. Topological relationships may be constructed only as needed, each time a data layer is accessed. Some GIS software packages create and maintain detailed topological relationships in their data. This results in more complex and perhaps larger data structures, but access is often faster, and topology provides more consistent, "cleaner" data. Other systems maintain little topological information in the data structures, but compute and act upon topology as needed during specific processing.

Topological vector models often use codes and tables to record topology. As described above, nodes are the starting and ending points of lines. Each node and line is given a unique identifier. Sequences of nodes and lines are recorded as a list of identifiers, and point, line, and polygon topology recorded in a set of tables. The vector features and tables in Figure 2-11 illustrate one form of this topological coding.

Point topology is often quite simple. Points are typically independent of each other, so they may be recorded as individual identifiers, perhaps with coordinates included, and in no particular order (Figure 2-11, top).

Line topology typically includes substantial structure, and identifies at a minimum the beginning and ending points of each line (Figure 2-11, middle). Variables record the topology and may be organized in a table. These variables may include a line identifier, the starting node, and the ending node for each line. In addition, lines may be

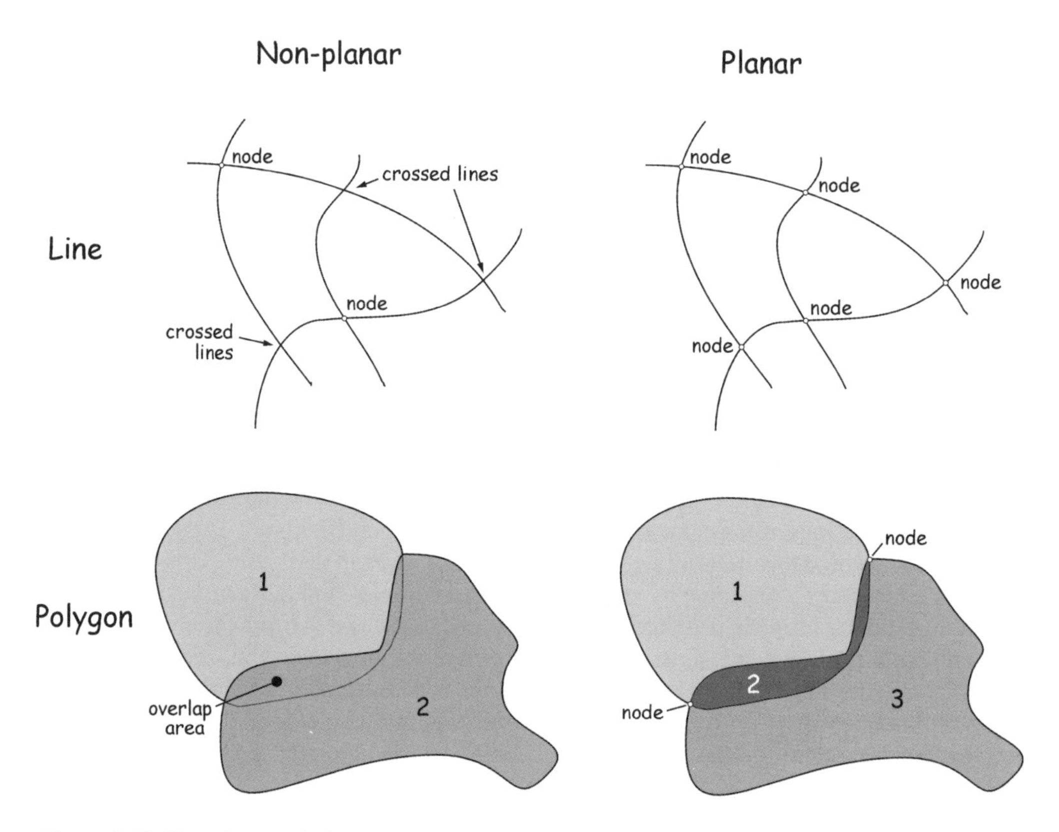

Figure 2-10: Non-planar and planar topology in lines and polygons.

assigned a direction, and the polygons to the left and right of the lines recorded. In most cases left and right are defined in relation to the direction of travel from the starting node to the ending node.

Polygon topology may also be defined by tables (Figure 2-11, bottom). The tables may record the polygon identifiers and the list of connected lines that define the polygon. Edge lines are often recorded in sequential order. The lines for a polygon form a closed loop, resulting in the starting node of the first line in the list that also serves as the ending node for the last line in the list. Note that there may be a "background" polygon defined by the outside area. This background polygon is not a closed polygon as all the rest, however it may be defined for consistency and to provide entries in the line topology table.

Finally, note that there may be coordinate tables (not shown in Figure 2-11) that record the identifiers and locations of each node, and coordinates for each vertex within a line or polygon. Node locations are recorded with coordinate pairs for each node, while line locations are represented by an identifier and a list of vertex coordinates for each line.

Figure 2-11 illustrates the inter-related structure inherent in the tables that record topology. Point or node records may be related to lines, which in turn may be related to polygons. All these may then be linked in complex ways to coordinate tables that record location.

Topological vector models greatly enhance many vector data operations. Adjacency analyses are reduced to a table look up, an operation that is relatively simple to program and quick to execute in most software systems. For example, an analyst may want to identify all polygons adjacent to a city. Assume the city is represented as a single polygon. The operation reduces to 1) scanning the polygon topology table to find the polygon labeled city and reading the list of lines that bound the polygon, and 2) scanning this list of lines for the city polygon, accumulating a list of all left and right polygons. Polygons adjacent to the city may be identified from this list. List searches on topological tables are typically much faster than searches involving coordinate data.

Topological vector models also enhance many other spatial data operations. Network and other connectivity analyses are concerned with the flow of resources through defined pathways. Topological vector models explicitly record the connections of a set of pathways and so facilitate network analyses. Overlay operations are also enhanced when using topological vector models. The mechanics of overlay operations are discussed in greater detail in Chapter 9, however we will state here that they involve identifying line adjacency, intersection, and resultant polygon formation. The interior and exterior regions of existing and new polygons must be determined, and these regions depend on polygon topology. Hence,

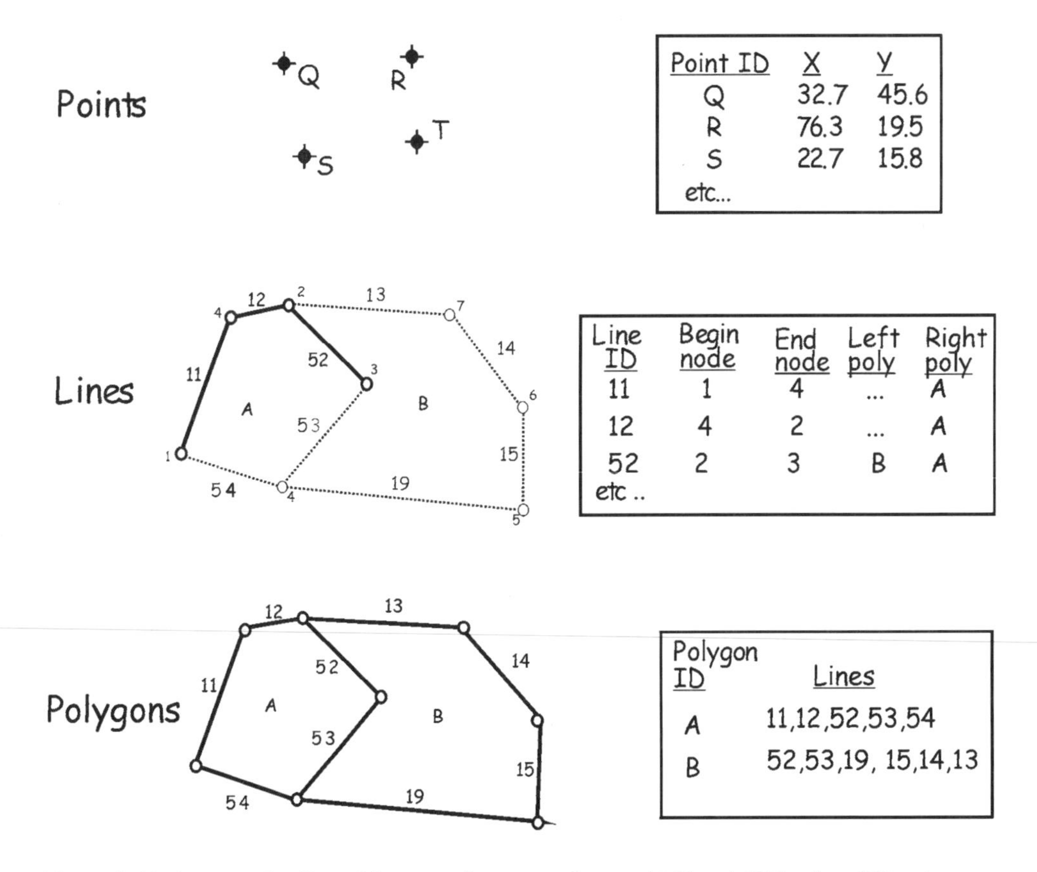

Figure 2-11: An example of possible vector feature topology and tables. Additional or different tables and data may be recorded to store topological information.

topological data are useful in many spatial analyses.

There are limitations and disadvantages to topological vector models. First, there are computational costs in defining the topological structure of a vector data layer. Software must determine the connectivity and adjacency information, assign codes, and build the topological tables. Computational costs are typically quite modest with current computer technologies.

Second, the data must be very "clean", in that all lines must begin and end with a node, all lines must connect correctly, and all polygons must be closed. Unconnected lines or unclosed polygons will cause errors during analyses. Significant human effort may be required to ensure clean vector data because each line and polygon must be checked. Software may help by flagging or fixing "dangling" nodes that do not connect to other nodes, and by automatically identifying all polygons. Each dangling node and polygon may then be checked, and edited as needed to correct errors.

These limitations are far outweighed by the gains in efficiency and analytical capabilities provided by topological vector models. Many current vector GIS packages use topological vector models in some form.

Raster Data Models

Models and Cells

Raster data models define the world as a regular set of cells in a grid pattern (Figure 2-12). Typically these cells are square and evenly spaced in the x and y directions. The phenomena or entities of interest are represented by attribute values associated with each cell location. Raster data sets have a cell dimension, defining the size of the cell. The cell dimension specifies the length and width of the cell in surface units, e.g. the *cell dimension* may be specified as a square 30 meters on each side. The cells are usually oriented parallel to the x and y directions. Thus, if we know the cell dimension and the coordinates of any one cell e.g., the lower left corner, we may calculate the coordinate of any other cell location.

Raster data models are the natural means to represent "continuous" spatial features or phenomena. Elevation, precipitation, slope, and pollutant concentration are examples of continuous spatial variables. These variables characteristically show significant changes in value over broad areas. The gradients can be quite steep (e.g., at cliffs), gentle (long, sloping ridges), or quite variable (rolling hills). Because raster data may be a dense sampling of points in two dimensions, they easily represent all variations in the changing surface. Raster data models depict these gradients by changes in the values associated with each cell.

Square raster cells have a characteristic cell dimension or cell size (Figure 2-12). This cell dimension is the edge length of each cell, and cell dimension is typically constant for a raster data layer. The cell dimension is important because it affects many properties of a raster data set, including coordinate data volume.

The volume of data required to cover a given area increases as the cell dimension gets smaller. The number of cells increases by the square of the reduction in cell dimension. Cutting the cell dimension in half causes a factor of four increase in the number of cells (Figure 2-13a and b). Reducing the cell dimension by four causes a sixteen-fold increase in the number of cells (Figure 2-13a and c). There is a trade-off between cell size and data volumes. Smaller cells may be preferred because they provide greater spatial detail, but this detail comes at the cost of larger data sets.

The cell dimension also affects the spatial precision of the data set, and hence positional accuracy. The cell coordinate is

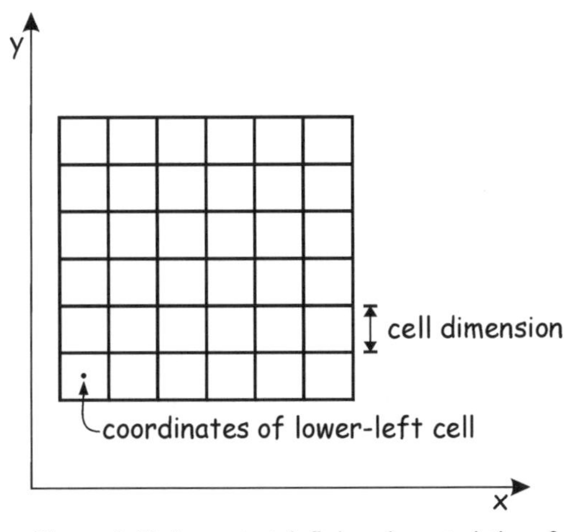

Figure 2-12: Important defining characteristics of a raster data model.

usually defined at a point in the center of the cell. The coordinate applies to the entire area covered by the cell. Positional accuracy is typically expected to be no better than approximately one-half the cell size. No matter the true location of a feature, coordinates are truncated or rounded up to the nearest cell center coordinate. Thus, the cell size should be no more than twice the desired accuracy and precision for the data layer represented in the raster, and should be smaller.

Each raster cell represents a given area on the ground and is assigned a value that may be considered to apply for the cell. In some instances the variable may be uniform over the raster cell, and hence the value is correct over the cell. However, under most conditions there is within-cell variation, and the raster cell value represents the average, central, or most common value found in the cell. Consider a raster data set representing annual weekly income with a cell dimension that is 300 meters (980 feet) on a side. Further suppose that there is a raster cell with a value of 710. The entire 300 by 300 meters area is considered to have this value of 710 dollars per week. There may be many households within the raster cell, none of which may earn exactly 710 dollars per week. However the 710 dollars may be the average, the highest point, or some other representative value for the area covered by the cell. While raster cells often represent the average or the value measured at the center of the cell, they may also represent the median, maximum, or other statistic for the cell area.

An alternative interpretation of the raster cell applies the value to the central point of the cell. Consider a raster grid containing

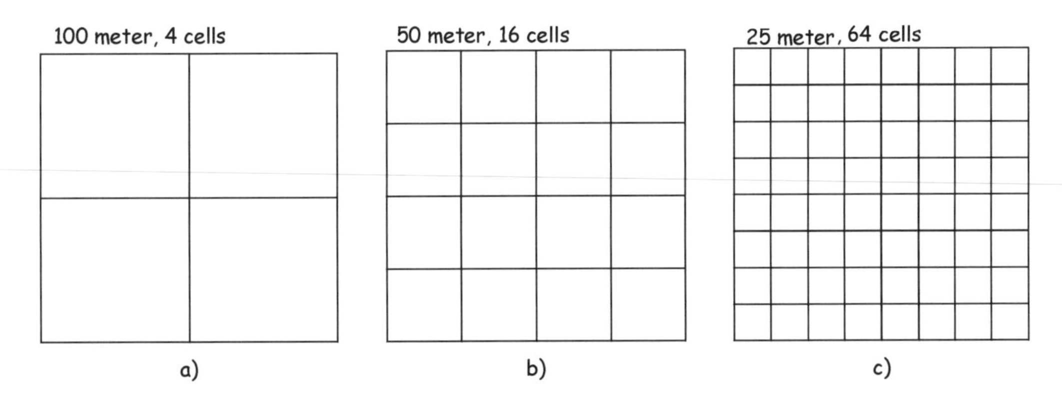

Figure 2-13: The number of cells in a raster data set depends on the cell size. For a given area, a linear decrease in cell size cause an exponential increase in cell number, e.g., halving the cell size causes a four-fold increase in cell number.

elevation values. Cells may be specified that are 200 meters square, and an elevation value assigned to each square. A cell with a value of 8000 meters (26,200 feet) may be assumed to have that value at the center of the cell, and not for the entire cell.

A raster data model may also be used to represent discrete data, e.g., to represent landcover in an area. Raster cells typically hold numeric or single-letter alphabetic characters, so some coding scheme must be defined to identify each discrete value. Each code may be found at many raster cells (Figure 2-14).

Raster cell values may be assigned and interpreted in at least seven different ways (Table 2-1). We have describe three, a raster cell as a point physical value (elevation), as a statistical value (elevation), and as discrete data (landcover). Landcover also may be interpreted as a class code. The value for any

a	a	a	a	r	f	f	a	a	a	a	a
a	a	a	a	r	f	f	a	a	a	a	a
a	a	a	f	r	f	f	a	a	a	a	a
a	a	a	r	r	f	f	a	a	a	a	a
a	a	a	r	f	f	f	a	a	a	a	a
a	f	f	r	f	f	f	a	a	a	a	a
a	f	f	r	f	u	f	a	a	a	a	a
h	h	h	h	h	h	h	h	h	h	h	h
f	f	r	u	u	u	u	a	a	a	a	a
f	f	r	f	u	u	a	a	a	a	a	a
f	f	f	r	f	f	a	a	a	a	a	a
f	f	f	f	r	f	a	a	a	a	a	a

a = agriculture
u = developed
f = forest
r = river
h = highways

Figure 2-14: Discrete or categorical data may be represented by codes in a raster data layer.

Table 2-1: Types of data represented by raster cell values. (from L. Usery, pers. comm.)

Form	Description	Example
point ID	alpha-numeric ID of closest point	nearest hospital
line ID	alpha-numeric ID of closest line	nearest road
contiguous region ID	alpha-numeric ID for dominant region	state
class code	alpha-numeric code for general class	vegetation type
table ID	numeric position in a table	table row number
physical analog	numeric value representing surface value	elevation
statistical value	numeric value from a statistical function	population density

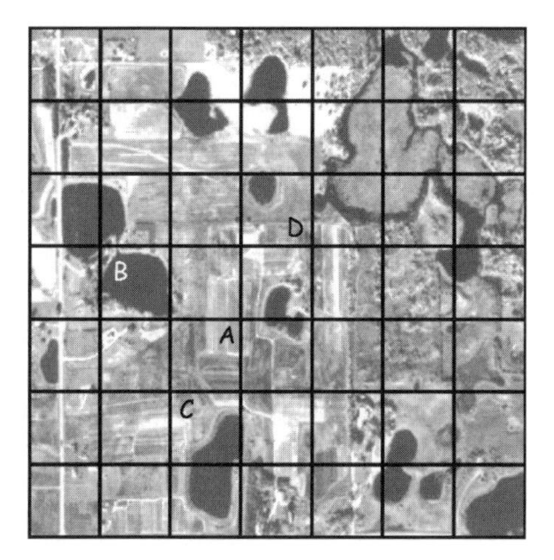

Figure 2-15: Raster cell assignment with mixed landscapes. Upland areas are lighter greys, water the darkest greys.

cell may have a given landcover value, and the cells may be discontinuous, e.g., we may have several farm fields scattered about our area with an identical landcover code. Discrete codes may also be used to identify a specific, usually continuous entity, e.g., a county, state, or country. Raster values may also be used to represent points and lines, as the IDs of lines or points that occur closest to the cell center.

Raster cell assignment may be complicated when representing what we typically think of as discrete boundaries, for example, when the raster value is interpreted as a class code or as a contiguous region ID. Consider the area in Figure 2-15. We wish to represent this area with a raster data layer, with cells assigned to one of two class codes, one each for land or water. Water bodies appear as darker areas in the image, and the raster grid is shown overlain. Cells may contain substantial areas of both land and water, and the proportion of each type may span from zero to 100 percent. Some cells are purely one class and the assignment is obvious, e.g., the cell labelled A in the Figure 2-15 contains only land. Others are mostly one type, as for cells B (water) or D (land). Some are nearly equal in their proportion of land and water, as is cell C. How do we assign classes?

One common method might be called "winner-take-all". The cell is assigned the value of the largest-area type. Cells A, C and D would be assigned the land type, cell B water. Another option places preference. If any of an "important" type is found then the cell is assigned that value, regardless of the proportion. If we specify a preference for water, then cells B, C, and D would be assigned the water type, and cell A the land type.

Regardless of the assignment method used, Figure 2-15 illustrates two considerations when discrete objects are represented using a raster data model. First, some inclusions are inevitable because cells must be assigned to a discrete class. Some mixed cells occur in nearly all raster layers. The GIS user must acknowledge these inclusions, and consider their impact on the intended spatial analyses.

Second, differences in the assignment rules may substantially alter the data layer, as shown in our simple example. More potential cell types in complex landscapes may increase the assignment sensitivity. Smaller cell sizes reduce the significance of classes in the assignment rule, but at the cost of increased data volumes.

A similar problem may occur when more than one line or point occurs within a raster cell. If two points occur, then which point ID is assigned? If two lines occur, then which line ID should be assigned?

Raster Geometry and Resampling

Raster data layers are often defined to align with cell edges parallel to the coordinate system direction. This greatly simplifies the determination of cell location. When cell edges and coordinate system axes are aligned, the calculation of a cell location is a simple process of counting and multiplication. The coordinate location of one cell is recorded, typically the lower-left or upper-left cell in the data set. With a known lower-left cell coordinate, all other cell coordinates may be determined by the formulas:

$$N_{cell} = N_{lower\text{-}left} + row * cell\ size \quad (2.2)$$

$$E_{cell} = E_{lower\text{-}left} + column * cell\ size \quad (2.3)$$

where N is the coordinate in the north direction (y), E is the coordinate in the east direction (x), and the row and column are counted from the lower left cell. Formulas are considerably more complicated when the cell edges are not parallel with the coordinate system axes.

Because cell edges and coordinate system axes are typically aligned, data often must be *resampled* when converting between coordinate systems or changing the cell size (Figure 2-16). Resampling involves re-assigning the cell values when changing raster coordinates or geometry. Cells must be resampled because the new and old raster cells represent different areas. Cell centers in the old coordinate system do not coincide with cell centers in the new coordinate system and so the average value represented by each cell must be re-computed. Common resampling approaches include the *nearest neighbor* (taking the output layer value from the nearest input layer cell center), *bilinear interpolation* (distance-based averaging of the four nearest cells), and *cubic convolution* (a weighted average of the sixteen nearest cells).

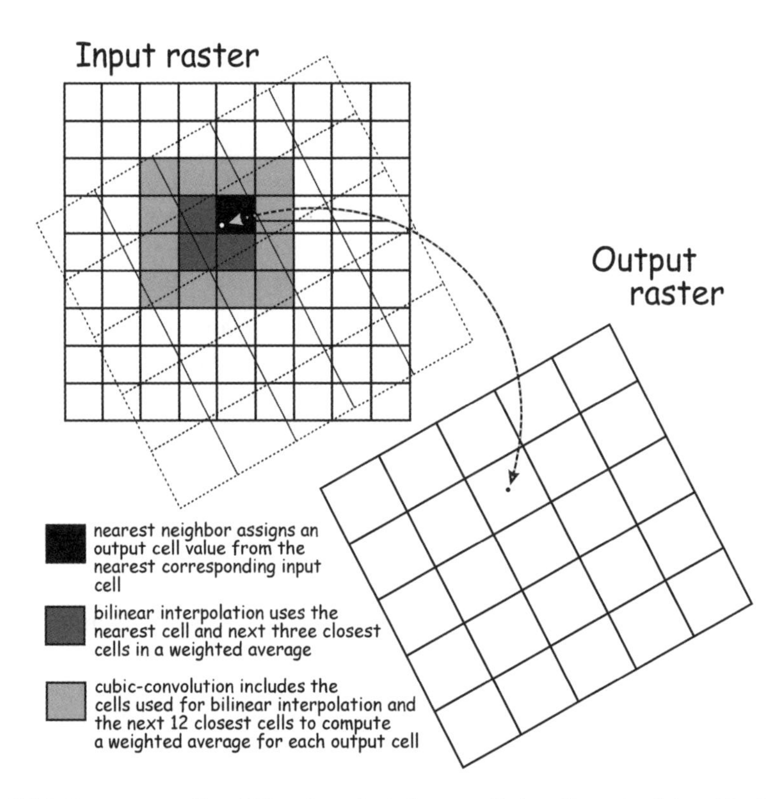

Figure 2-16: Raster resampling. When the orientation or cell size of a raster data set is changed, output cell values are calculated based on the closest (nearest neighbor), four nearest (bilinear interpolation) or sixteen closest (cubic-convolution) input cell values.

A Comparison of Raster and Vector Data Models

The question often arises, "which are better, raster or vector data models?" The answer is neither and both. Neither of the two classes of data models are better in all conditions or for all data. Both have advantages and disadvantages relative to each other and to additional, more complex data models. In some instances it is preferable to maintain data in a raster model, and in others in a vector model. Most data may be represented in both, and may be converted among data models. As an example, elevation may be represented as a set of contour lines in a vector data model or as a set of elevations in a raster grid. The choice often depends on a number of factors, including the predominant type of data (discrete or continuous), the expected types of analyses, available storage, the main sources of input data, and the expertise of the human operators.

Raster data models exhibit several advantages relative to vector data models. First, raster data models are particularly suitable for representing themes or phenomena that change frequently in space. Each raster cell may contain a value different than its neighbors. Thus trends as well as more rapid variability may be represented.

Raster data structures are generally simpler, particularly when a fixed cell size is used. Most raster models store cells as sets of rows, with cells organized from left to right, and rows stored from top to bottom. This organization is quite easy to code in an array structure in most computer languages.

Raster data models also facilitate easy overlays, at least relative to vector models. Each raster cell in a layer occupies a given position corresponding to a given location on the Earth surface. Data in different layers align cell-to-cell over this position. Thus, overlay involves locating the desired grid cell in each data layer and comparing the values found for the given cell location. This cell look-up is quite rapid in most raster data structures, and hence layer overlay is quite simple and rapid when using a raster data model.

Finally, raster data structures are the most practical method for storing, displaying, and manipulating digital image data, such as aerial photographs and satellite imagery. Digital image data are an important source of information when building, viewing, and analyzing spatial databases. Image display and analysis are based on raster operations to sharpen details on the image, specify the brightness, contrast, and colors for display, and to aid in the extraction of information.

Vector data models provide some advantages relative to raster data models. First, vector models generally lead to more compact data storage, particularly for discrete objects. Large homogenous regions are recorded by the coordinate boundaries in a vector data model. These regions are recorded as a set of cells in a raster data model. The perimeter grows more slowly than the area for most feature shapes, so the amount of data required to represent an area increases much more rapidly with a raster data model. Vector data are much more compact than raster data for most themes and levels of spatial detail.

Vector data are a more natural means for representing networks and other connected linear features. Vector data by their nature store information on intersections (nodes) and the linkages between them (lines). Traffic volume, speed, timing, and other factors may be associated with lines and intersections to model many kinds of networks.

Vector data models are easily presented in a preferred map format. Humans are familiar with continuous line and rounded curve representations in hand- or machine-drawn maps, and vector-based maps show these curves. Raster data often show a "stair-step" edge for curved boundaries, particularly when the cell resolution is large relative to the resolution at which the raster is displayed. Cell edges are often visible for lines, and the width and stair-step pattern changes as lines curve. Vector data may be plotted

Table 2-2: A comparison of raster and vector data models.

Characteristic	Raster	Vector
data structure	usually simple	usually complex
storage requirements	large for most data sets without compression	small for most data sets
coordinate conversion	may be slow due to data volumes, and may require resampling	simple
analysis	easy for continuous data, simple for many layer combinations	preferred for network analyses, many other spatial operations more complex
positional precision	floor set by cell size	limited only by quality of positional measurements
accessibility	easy to modify or program, due to simple data structure	often complex
display and output	good for images, but discrete features may show "stairstep" edges	map-like, with continuous curves, poor for images

with more visually appealing continuous lines and rounded edges.

Vector data models facilitate the calculation and storage of topological information. Topological information aids in performing adjacency, connectivity, and other analyses in an efficient manner. Topological information also allows some forms of automated error and ambiguity detection, leading to improved data quality.

Conversion between Raster and Vector Models

Spatial data may be converted between raster and vector data models. Vector-to-raster conversion involves assigning a cell value for each position occupied by vector features. Vector point features are typically assumed to have no dimension. Points in a raster data set must be represented by a value in a raster cell, so points have at least the dimension of the raster cell after conversion from vector-to-raster models. Points are usually assigned to the cell containing the point coordinate. The cell in which the point resides is given a number or other code identifying the point feature occurring at the cell location. If the cell size is too large, two or more vector points may fall in the same cell, and either an ambiguous cell identifier assigned, or a more complex numbering and assignment scheme implemented. Typically a cell size is chosen such that the diagonal cell dimension is smaller than the distance between the two closest point features.

Vector line features in a data layer may also be converted to a raster data model. Raster cells may be coded using different criteria. One simple method assigns a value to a cell if a vector line intersects with any part of the cell (Figure 2-17, left). This ensures the maintenance of connected lines in the raster form of the data. This assignment rule often leads to wider than appropriate lines because several adjacent cells may be assigned as part of the line, particularly when the line meanders near cell edges. Other assignment rules may be applied, for example, assigning a cell as occupied by a line only when the cell center is near a vector line segment (Figure 2-17, right). "Near" may be defined as some sub-cell distance, e.g., 1/3 the cell width. Lines passing through the corner of a cell will not be recorded as in the cell. This may lead to thinner linear features in the raster data set, but often at the cost of line discontinuities.

The output from vector-to-raster conversion depends on the input algorithm used. You may get a different output data layer when a different conversion algorithm is used, even though you use the same input. This brings up an important point to remember when applying any spatial operations. The output often depends in subtle ways on the spatial operation. What appear to be quite small differences in the algorithm or key defining parameters may lead to quite different results. Small changes in the assignment distance or rule in a vector-to-raster conversion operation may result in large differences in output data sets, even with the same input. There is often no clear *a priori* best method. Empirical tests or previous experiences are often useful guides to determine the best method with a given data set or conversion problem. The ease of spatial manipulation in a GIS provides a powerful and often easy to use set of tools. The GIS user should bear in mind that these tools may be more efficient at producing errors as

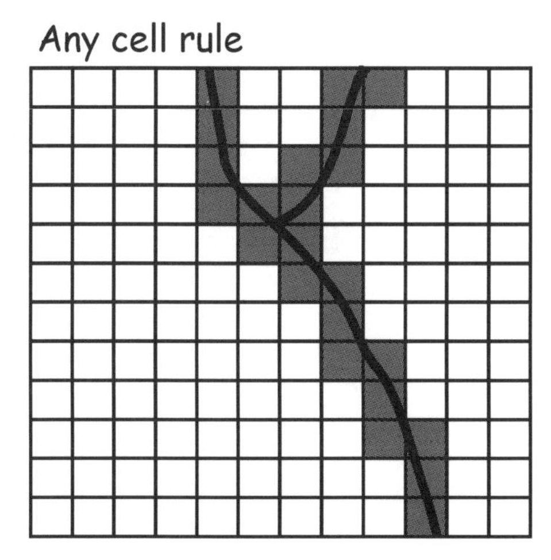

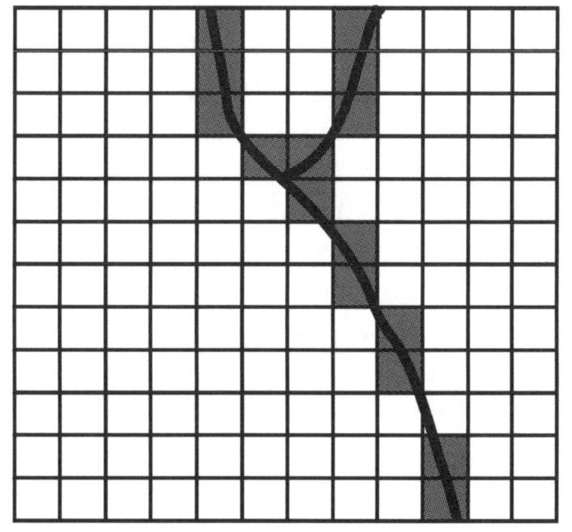

Figure 2-17: vector-to-raster conversion. Two assignment rules result in different raster coding near lines, but in this case not near points.

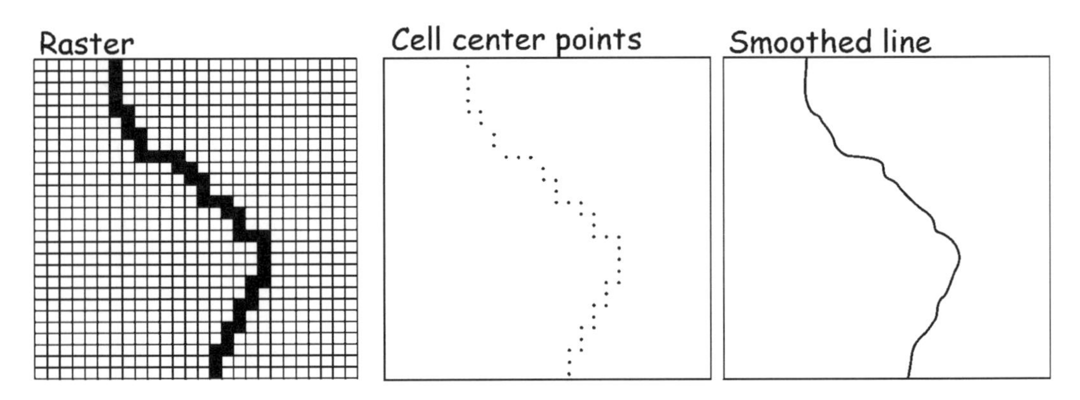

Figure 2-18: Raster data may be converted to vector formats, and may involve line smoothing or other operations to remove the "stair-step" effect.

well as more efficient at providing correct results. Until sufficient experience is obtained with a suite of algorithms, in this case vector-to-raster conversion, small, controlled tests should be performed to verify the accuracy of a given method or set of constraining parameters.

Area features are converted from vector-to-raster with methods similar to those used for vector line features. Boundaries among different polygons are identified as in vector-to-raster conversion for lines. Interior regions are then identified, and each cell in the interior region is assigned a given value. Note that the border cells containing the boundary lines must be assigned. As with vector-to-raster conversion of linear features, there are several methods to determine if a given border cell should be assigned as part of the area feature. One common method assigns the cell to the area if more than one-half the cell is within the vector polygon. Another common method assigns a raster cell to an area feature if any part of the raster cell is within the area contained within the vector polygon. Assignment results will vary with the method used.

Up to this point we have covered vector-to-raster data conversion. Data may also be converted in the opposite direction, in that raster data may be converted to vector data. Point, line, or area features represented by raster cells are converted to corresponding vector data coordinates and structures. Point features are represented as single raster cells. Each vector point feature is usually assigned the coordinate of the corresponding cell center.

Linear features represented in a raster environment may be converted to vector lines. Conversion to vector lines typically involves identifying the continuous connected set of grid cells that form the line. Cell centers are typically taken as the locations of vertices along the line (Figure 2-18). Lines may then be "smoothed" using a mathematical algorithm to remove the "stair-step" effect.

Triangulated Irregular Networks

A *triangulated irregular network* (TIN) is a data model commonly used to represent terrain heights. Typically the x, y, and z locations for measured points are entered into the TIN data model. These points are distributed in space, and the points may be connected in such a manner that the smallest triangle formed from any three points may be constructed. The TIN forms a connected network of triangles (Figure 2-19). Triangles are created such that the lines from one triangle do not cross the lines of another. Line crossings are avoided by identifying the *convergent circle* for a set of three points (Figure 2-20). The convergent circle is defined as the circle passing through all three points. A triangle is drawn only if the corresponding convergent circle contains no other sampling points. Each triangle defines a terrain surface, or facet, assumed to be of uniform slope and aspect over the triangle.

The TIN model typically uses some form of indexing to connect neighboring points. Each edge of a triangle connects to two points, which in turn each connect to other edges. These connections continue recursively until the entire network is spanned. Thus, the TIN is a rather more complicated data model than the simple raster grid when the objective is terrain representation.

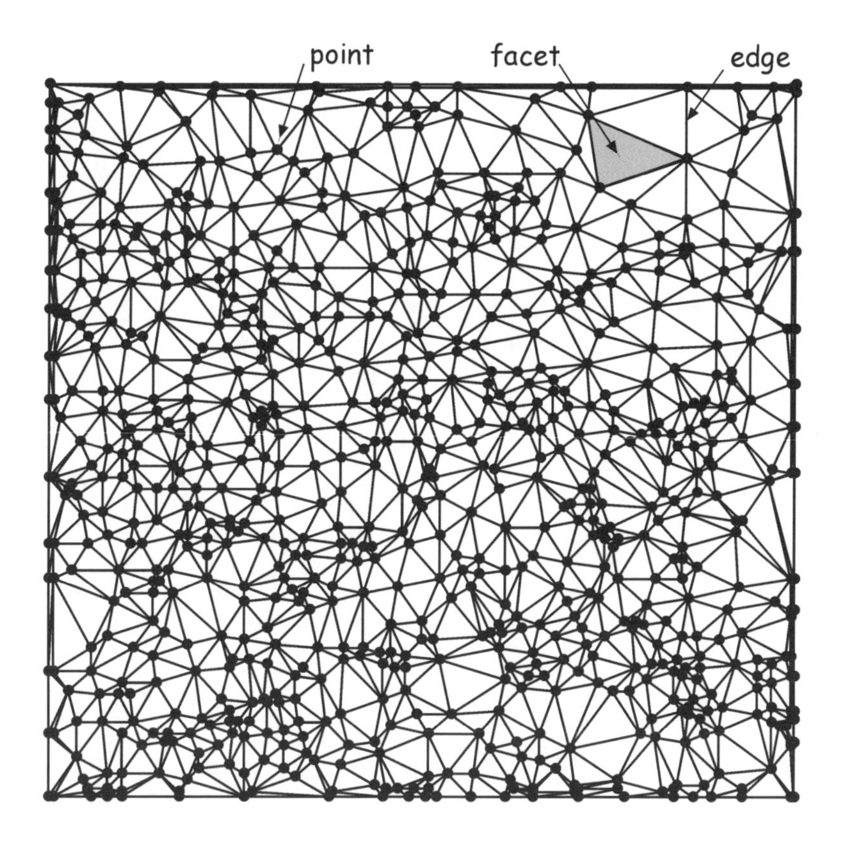

Figure 2-19: A TIN data model defines a set of adjacent triangles over a sample space. Sample points, facets, and edges are components of TIN data models.

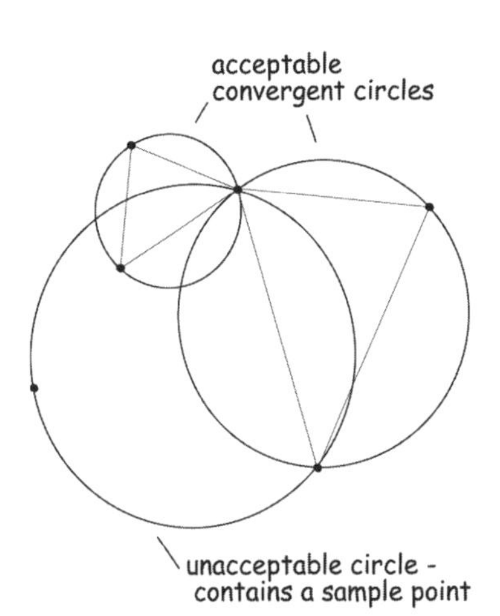

Figure 2-20: Convergent circles intersect the vertices of a triangle and contain no other possible vertices.

While the TIN model may be more complex than simple raster models, it may also be much more appropriate and efficient when storing terrain data in areas with variable relief. Relatively few points are required to represent large, flat, or smoothly continuous areas. Many more points are desirable when representing variable, discontinuous terrain. Surveyors often collect more samples per unit area where the terrain is highly variable. A TIN easily accommodates these differences in sampling density, with the result of more, smaller triangles in the densely sampled area. Rather than imposing a uniform cell size and having multiple measurements for some cells, one measurement for others, and no measurements for most cells, the TIN preserves each measurement point at each location.

Multiple Models

Digital data may often be represented using any one of several data models. The analyst must choose which representation to use. Digital elevation data are perhaps the best example of the use of multiple data models to represent the same theme (Figure 2-21). Digital representations of terrain height have a long history and widespread use in GIS. Elevation data and derived surfaces such as slope and aspect are important in hydrology, transportation, ecology, urban and regional planning, utility routing, and a number of other activities that are analyzed or modeled using GIS. Because of this widespread importance and use, digital elevation data are commonly represented in a number of data models.

Raster grids, triangulated irregular networks (TINs), and vector contours are the most common data structures used to organize and store digital elevation data. Raster and TIN data are often called digital elevation models (DEMs) or digital terrain models (DTMs) and are commonly used in terrain analysis. Contour lines are most often used as a form of input, or as a familiar form of output. Historically, hypsography (terrain heights) were depicted on maps as contour lines (Figure 2-21). Contours represent lines of equal elevation, typically spaced at fixed elevation intervals across the mapped areas. Because many important analyses and derived surfaces are more difficult using contour lines, most digital elevation data are represented in raster or TIN models.

Raster DEMs are a grid of regularly spaced elevation samples (Figure 2-21). These samples, or postings, typically have equal frequency in the grid x and y directions. Derived surfaces such as slope or aspect are easily and quickly computed from raster DEMs, and storage, processing, compression, and display are well understood and efficiently implemented. However, as described earlier, sampling density cannot be increased in areas where terrain changes are abrupt, so either flat areas will be oversampled or variable areas undersampled. A linear increase in raster resolution causes a geometric increase in the number of raster cells, so there may be significant storage and processing costs to oversampling. TINs solve these problems, at the expense of a more complicated data structure.

Other Geographic Data Models

A number of other data models have been proposed and implemented, although they are all currently uncommon. Some of these data models are appropriate for specialized applications, while others have been tried and largely discarded. Some have been partially adopted, or are slowly being incorporated into available software tools.

The *object-oriented* data model incorporates much of the philosophy of object oriented programming into a spatial data model. A main goal is to encapsulate the information and operations (often called methods) into discrete objects. These objects could be geographic features, e.g., a city might be defined as an object. Spatial and attribute data associated with a given city would be incorporated in a single city object. This may include not only information on

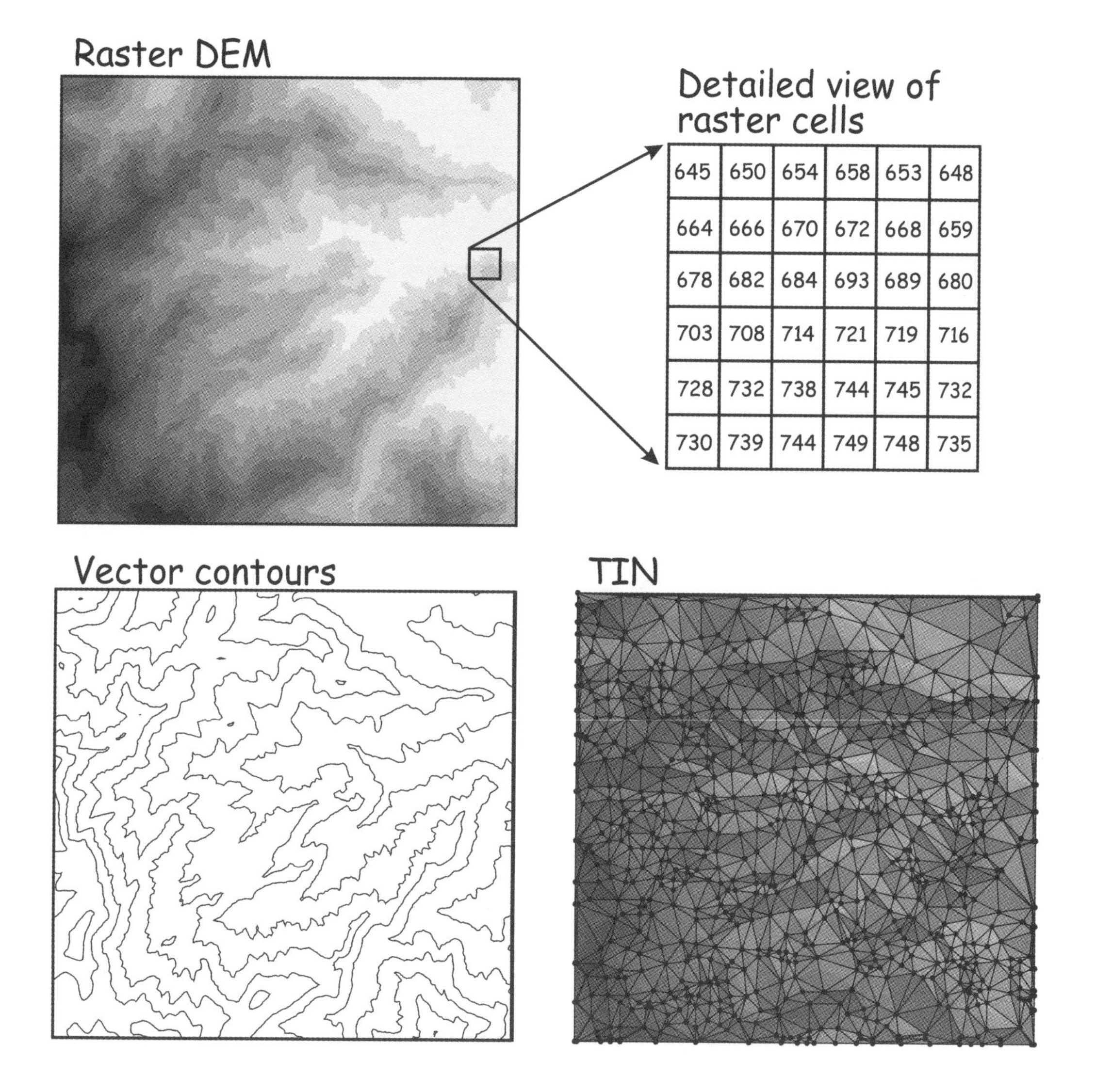

Figure 2-21: Data may often be represented in several data models. Digital elevation data are commonly represented in raster (DEM), vector (contours), and TIN data models.

the city boundary, but also streets, building locations, waterways, or other features that might be in separate data structures in a layered topological vector model. The topology could be included, but would likely be incorporated within the single object. Topological relationships to exterior objects may also be represented, e.g., relationships to adjacent cities or counties.

The object-oriented data model has both advantages and disadvantages when compared to traditional topological vector and raster data models. Some geographic entities may be naturally and easily identified as discrete units for particular problems, and so may be naturally amenable to an object-oriented approach. A power or water distribution system may be defined in this manner, where entities such as pumping stations or holding reservoirs may be discretely defined. However, it is more difficult to represent continuously varying features, such as elevation, with an object-oriented approach. In addition, for many problems the definition and indexing of objects may be quite complex. It has proven difficult to develop generic tools that may quickly and efficiently implement object-oriented models.

Data and File Structures

Binary and ASCII Numbers

No matter what spatial data model is used, the concepts must be translated into a set of numbers stored on a computer. All information stored on a computer in a digital format may be represented as a series of 0's and 1's. These data are often referred to as stored in a binary format, because each digit may contain one of two values, 0 or 1. Binary numbers are in a base of 2, so each successive column of a number represents a power of two.

We use a similar column convention in our familiar ten-based (decimal) numbering system. As an example, consider the number 47 that we represent using two columns. The seven in the first column indicates there are seven units of one. The four in the tens column indicates there are four units of ten. Each higher column represents a higher power of ten. The first column represents one (10^0=1), the next column represents tens (10^1=10), the next column hundreds (10^2=100) and upward for successive powers of ten. We add up the values represented in the columns to decipher the number.

Binary numbers are also formed by representing values in columns. In a binary system each column represents a successively higher power of two (Figure 2-22). The first (right-most) column represents 1 (2^0 = 1), the second column (from right) represents twos (2^1 = 2), the third (from right) represents fours (2^2 = 4), then eight (2^3 = 8), sixteen (2^4 = 16), and upward for successive powers of two. Thus, the binary number 1001 represent the decimal number 9: a one from the rightmost column, and eight from the fourth column (Figure 2-22).

Each digit or column in a binary number is called a *bit*, and 8 columns, or bits, are called a *byte*. A byte is a common unit for defining data types and numbers, e.g., a data file may be referred to as containing 4-byte integer numbers. This means each number is represented by 4 bytes of binary data (or 8 x 4 = 32 bits).

Several bytes are required when representing larger numbers. For example, one byte may be used to represent 256 different values. When a byte is used for non-negative integer numbers, then only values from 0 to 255 may be recorded. This will work when all values are below 255, but consider an elevation data layer with values greater than 255. If the data are not rescaled, then more than one byte of storage are required for each value. Two bytes will store a number greater than 65,500. Terrestrial elevations measured in feet or meters are all below this value, so two bytes of data are often used to store elevation data. Real numbers such as 12.19 or 865.3 typically require more bytes, and are effectively split, e.g., two bytes for the whole part of the real number, and four bytes for the fractional portion.

Binary numbers are often used to represent codes. Spatial and attribute data may then be represented as text or as standard codes. This is particularly common when raster or vector data are converted for export or import among different GIS software systems. For example, Arc/Info, a widely used GIS, produces several export formats that are in text or binary formats. Idrisi, another popular GIS, supports binary and alphanumeric raster formats.

One of the most common number coding schemes uses ASCII designators. ASCII stands for the American Standard Code for Information Interchange. ASCII is a standardized, widespread data format that uses seven bits, or the numbers 0 through 126, to represent text and other characters. An extended ASCII, or ANSI scheme, uses these same codes, plus an extra binary bit to represent numbers between 127 and 255. These codes are then used in many programs, including GIS, particularly for data export or exchange.

ASCII codes allow us to easily and uniformly represent alphanumeric characters such as letters, punctuation, other characters, and numbers. ASCII converts binary numbers to alphanumeric characters through an index. Each alphanumeric character corresponds to a specific number between 0 and 255, which allows any sequence of characters to be represented by a number. One byte is required to represent each character in extended ASCII coding, so ASCII data sets are typically much larger than binary data sets. Geographic data in a GIS may use a combination of binary and ASCII data stored in files. Binary data are typically used for

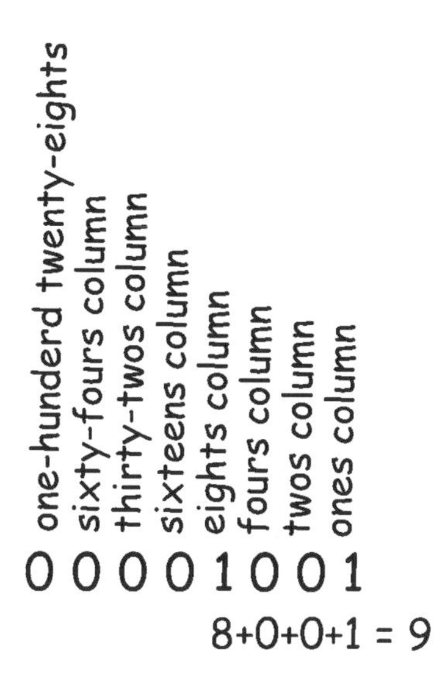

Equivalent numbers

binary	decimal
00000001	1
00000010	2
00000011	3
00000100	4
00000101	5
00000110	6
00000111	7
00001000	8
00001001	9
00001010	10
....	....

Figure 2-22: Binary representation of decimal numbers.

coordinate information, and ASCII or other codes may be used for attribute data.

Pointers

Files may be linked by file *pointers* or other structures. A pointer is an address or index that connects one file location to another. Pointers are a common way to organize information within and across multiple files. Figure 2-23 depicts an example of the use of pointers to organize spatial data. In Figure 2-23, a polygon is composed of a set of lines. Pointers are used to link the set of lines that form each polygon. There is a pointer from each line to the successive string of lines that form the polygon.

Pointers help by organizing data in such a way as to improve access speed. Unorganized data would require time-consuming searches each time a polygon boundary was to be identified. Pointers also allow efficient use of storage space. In our example, each line segment is stored only once. Several polygons may point to the line segment as it is typically much more space-efficient to add pointers than to duplicate the line segment.

Data Compression

Data compression is common in GIS. Compressions are based on algorithms that reduce the size of a computer file while maintaining the information contained in the file. Compression algorithms may be "lossless", in that all information is maintained during compression, or "lossy", in that some information is lost. A lossless compression algorithm will produce an exact copy when it is applied and then the appropriate decompression algorithm applied. A lossy algorithm will alter the data when it is applied and the appropriate decompression algorithm applied. Lossy algorithms are most often used with image data, and uncommonly applied to thematic spatial data.

Data compression is most often applied to discrete raster data, for example, when representing polygon or area information in a raster GIS. There are redundant data elements in raster representations of large homogenous areas. Each raster cell within a homogenous area will have the same code as most or all of the adjacent cells. Data compression algorithms remove much of this redundancy.

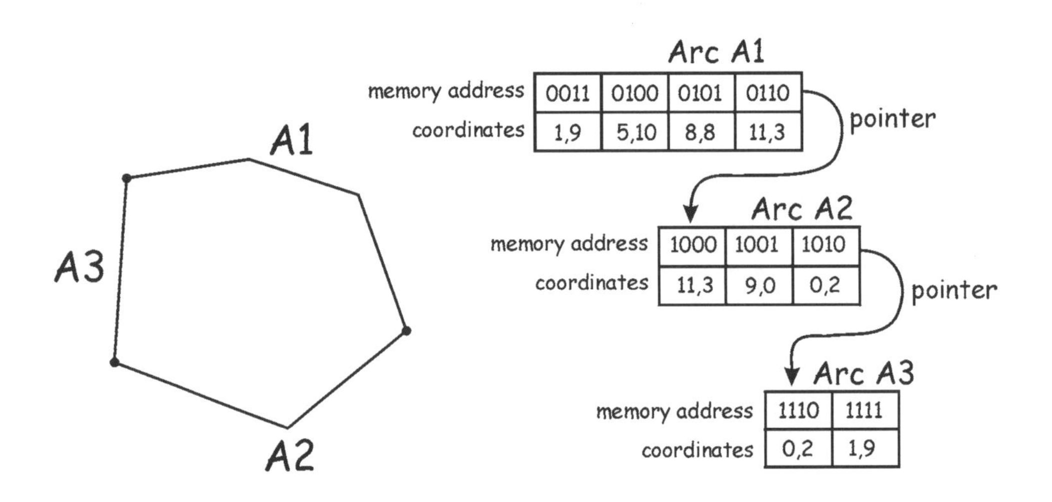

Figure 2-23: Pointers are used to organize vector data. Pointers reduce redundant storage and increase speed of access.

Raster								Run-length codes
9	9	6	6	6	6	6	7	2:9, 5:6, 1:7
6	6	6	6	6	6	6	6	8:6
9	9	6	6	6	6	7	7	2:9, 4:6, 2:7
9	8	9	6	6	7	7	5	1:9, 1:8, 1:9, 2:6, 2:7, 1:5

Figure 2-24: Run-length coding is a common and relatively simple method for compressing raster data. The left number in the run-length pair is the number of cells in the run, and the right is the cell value. Thus, the 2:9 listed at the start of the first line indicates a run of length two for the cell value 9.

Run-length coding is a common data compression method. This compression technique is based on recording sequential runs of raster cell values. Each run is recorded as the value and the run length. Seven sequential cells of type A might be listed as A7 instead of AAAAAAA. Thus, seven cells would be represented by two characters. Consider the data recorded in Figure 2-24, where each line of raster cells is represented by a set of run-length codes. In general run-length coding reduces data volume, as shown for the top three rows in Figure 2-24. Note that in some instances run-length coding increases the data volume, most often when there are no long runs. This occurs in the last line of Figure 2-24, where frequent changes in adjacent cell values result in many short runs. However, for most thematic data sets containing area information, run length coding substantially reduces the size of raster data sets.

There is also some data access cost in run-length coding. Standard raster data access involves simply counting the number of cells across a row to locate a given cell. Access to a cell in run-length coding must be computed by summing along the run-length codes. This is typically a minor additional cost, but in some applications the trade-off between speed and data volume may be objectionable.

Quad tree representations are another raster compression method. Quad trees are similar to run-length codings in that they are most often used to compress raster data sets when representing area features. Quad trees may be viewed as a raster data structure with a variable spatial resolution. Raster cell sizes are combined and adjusted within the data layer to fit into each specific area feature (Figure 2-25). Large raster cells that fit entirely into one uniform area are assigned. Successively smaller cells are then fit, halving the cell dimension at each iteration, until the smallest cell size is reached.

The dynamically varying cell size in a quad tree representation requires more sophisticated indexing than simple raster data sets. Pointers are used to link data elements in a tree-like structure, hence the name quad trees. There are many ways to structure the data pointers, from large to small, or by dividing quandrants, and these methods are beyond the scope of an introductory text. Further information on the structure of quad trees may be found in the references at the end of this chapter.

A quad tree representation may save considerable space when a raster data set includes large homogeneous areas. Each large area may be represented mostly by a few large cells representing the main body, and sets of recursively smaller cells along

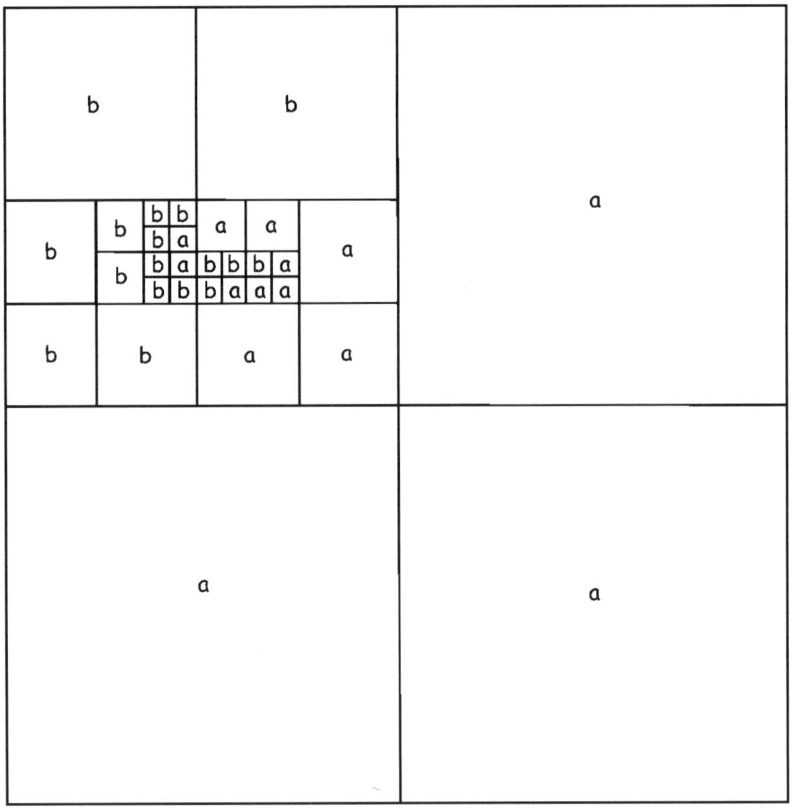

Figure 2-25: Quad tree compression.

the margins to represent the spatial detail at the edges of homogenous areas. As with most data compression algorithms, space savings are not guaranteed. There may be conditions where the additional indexing overhead requires more space than is saved. As with run-length coding, this most often occurs in spatially complex areas.

There are many other data compression methods that are commonly applied. JPEG and wavelet compression algorithms are often applied to reduce the size of spatial data, particularly image or other data. Generic bit and byte-level compression methods may be applied to any files for compression or communications. There is usually some cost in time to the compression and decompression.

Summary

In this chapter we have described our main ways of conceptualizing spatial entities, and of representing these entities as spatial features in a computer. We commonly employ two conceptualizations, also called spatial data models: a raster data model and a vector data model. Both models use a combination of coordinates, defined in a Cartesian or spherical system, and attributes, to represent our spatial features. Features are usually segregated by thematic type in layers.

Vector data models describe the world as a set of point, line, and area features. Attributes may be associated with each feature. A vector data model splits that world into discrete features, and often supports topological relationships. Vector models are most often used to represent features that are considered discrete, and are compatible with vector maps, a common output form.

Raster data models are based on grid cells, and represent the world as a "checkerboard", with uniform values within each cell. A raster data model is a natural choice for representing features that vary continuously across space, such as temperature or precipitation. Data may be converted between raster and vector data models.

We use data structures and computer codes to represent our conceptualizations in more abstract, but computer-compatible forms. These structures may be optimized to reduce storage space and increase access speed, or to enhance processing based on the nature of our spatial data.

Suggested Reading

Batty, M and Xie, Y., Model structures, exploratory spatial data analysis, and aggregation, *International Journal of Geographical Information Systems*, 1994, 8:291-307.

Bhalla, N., Object-oriented data models: a perspective and comparative review, *Journal of Information Science*, 1991, 17:145-160.

Bregt, A. K., Denneboom, J, Gesink, H. J., and van Randen, Y., Determination of rasterizing error: a case study with the soil map of The Netherlands, *International Journal of Geographical Information Systems*, 1991, 5:361-367.

Carrara, A., Bitelli, G., and Carla, R., Comparison of techniques for generating digital terrain models from contour lines, *International Journal of Geographical Information Systems*, 1997, 11:451-473.

Congalton, R.G., Exploring and evaluating the consequences of vector-to-raster and raster-to-vector conversion, *Photogrammetric Engineering and Remote Sensing*, 63:425-434.

Holroyd, F. and Bell, S. B. M., Raster GIS: Models of raster encoding, *Computers and Geosciences*, 1992, 18:419-426.

Joao, E. M., Causes and Consequences of Map Generalization, Taylor and Francis, London, 1998.

Kumler, M.P., An intensive comparison of triangulated irregular networks (TINs) and digital elevation models, *Cartographica*, 1994, 31:1-99.

Langram, G., Time in Geographical Information Systems, Taylor and Francis, London, 1992.

Laurini, R. and Thompson, D., Fundamentals of Spatial Information Systems, Academic Press, London, 1992.

Lee, J., Comparison of existing methods for building triangular irregular network models of terrain from grid digital elevation models, *International Journal of Geographical Information Systems*, 5:267-285.

Maquire, D. J., Goodchild, M. F., and Rhind, D. eds., Geographical Information Systems: Principles and Applications, Longman Scientific, Harlow, 1991.

Nagy, G. and Wagle, S. G., Approximation of polygonal maps by cellular maps, *Communications of the Association of Computational Machinery*, 1979, 22:518-525.

Peuquet, D. J., A conceptual framework and comparison of spatial data models, *Cartographica*, 1984, 21:66-113.

Peuquet, D. J., An examination of techniques for reformatting digital cartographic data. Part II: the raster to vector process, *Cartographica*, 1981, 18:375-394.

Piwowar, J. M., LeDrew, E. F., and Dudycha, D. J., Integration of spatial data in vector and raster formats in geographical information systems, *International Journal of Geographical Information Systems*, 1990, 4:429-444.

Peuker, T. K. and Chrisman, N., Cartographic Data Structures, *The American Cartographer*, 1975, 2:55-69.

Rossiter, D. G., A theoretical framework for land evaluation, *Geoderma*, 1996, 72:165-190.

Shaffer, C.A., Samet, H., and Nelson R. C., QUILT: a geographic information system based on quadtrees, *International Journal of Geographical Information Systems*, 1990, 4:103-132.

Sklar, F. and Costanza, R. Quantitative methods in landscape ecology: the analysis and interpretation of landscape heterogeneity. in: Turner, M. and Gardner, R., editors. The development of dynamic spatial models for landscape ecology: A review and prognosis. New York: Springer-Verlag; 90:239-288.

Tomlinson, R. F., The impact of the transition from analogue to digital cartographic representation, *The American Cartographer*, 1988, 15:249-262.

Wedhe, M., Grid cell size in relation to errors in maps and inventories produced by computerized map processes, *Photogrammetric Engineering and Remote Sensing*, 48:1289-1298.

Worboys, M. F., GIS: A Computing Perspective, Taylor and Francis, London, 1995.

Zeiler, M., Modeling Our World: The ESRI Guide to Geodatabase Design, ESRI Press, Redlands, 1999.

Study Questions

How is an entity different from a cartographic object?

Describe the successive levels of abstraction when representing real-world spatial phenomena on a computer. Why are there multiple levels, instead of just one level in our representation?

Define a data model and describe the two most commonly used data models.

What is topology, and why is it important? What is planar topology, and when might you want non-planar vs. planar topology?

What are the respective advantages and disadvantages of vector data models vs. raster data models?

Under what conditions are mixed cells a problem in raster data models? In what ways may the problem of mixed cells be addressed?

What is raster resampling, and why do we need to resample raster data?

What is a triangulated irregular network?

What are binary and ASCII numbers? Can you convert the following decimal numbers to a binary form: 8, 12, 244?

Why do we need to compress data? Which are most commonly compressed, raster data or vector data? Why?

What is a pointer when used in the context of spatial data, and how are they helpful in organizing spatial data?

3 Geodesy, Datums, Map Projections and Coordinate Systems

Introduction

Geographic information systems are different from other information systems because they contain spatial data. These spatial data include the coordinates defining the location, shape, and extent of geographic objects. To effectively use GIS, we must develop a clear understanding of how coordinate systems are established and coordinates measured. This chapter introduces *geodesy*, the science of measuring the shape of the Earth, and *map projections*, the transformation of coordinate locations from the curved Earth surface onto flat maps.

Defining coordinates for the Earth's surface is complicated by two main factors. First, most people best understand geography in a Cartesian coordinate system on a flat surface. Humans naturally perceive the Earth surface as flat, because at human scales the Earth's curvature is barely perceptible. Humans have been using flat maps for more than 40 centuries, and although globes are quite useful for perception and visualization at extremely small scales, they are not practical for most purposes.

A flat map must distort geometry in some way because the Earth is curved. When we plot latitude and longitude coordinates on a Cartesian system, "straight" lines will appear bent, and areas will be incorrect. This distortion may be difficult to detect on detailed maps that cover only a small portion of the Earth, but is quite apparent on large-area maps. Quantitative measurements on these maps are affected by the distortion, so we must somehow reconcile the mapping of a curved surface onto a flat surface.

The second main problem when defining a coordinate system comes from the irregular nature of the Earth's shape. We learn in our earliest geography lessons that the Earth is shaped as a sphere. This is a valid approximation for many uses, however, it is only an approximation. Several forces in the present and geologic past have resulted in an irregularly shaped Earth. These deformations affect how we best map the surface of the Earth, and how we define Cartesian coordinate systems for use in mapping and GIS.

Early Measurements

Humans have long speculated on the shape and size of the Earth. Babylonians believed the Earth was a flat disk floating in an endless ocean, a notion also adopted by Homer, one of the more widely known Greek writers. The Greeks were early champions of geometry, and they had many competing views of the shape of the Earth. One early Greek, Anaximenes, believed the Earth was a rectangular box, while Pythagoras and later Aristotle reasoned that the Earth must be a sphere. This deduction was based on many lines of evidence, and also based on a belief of divine direction by the gods. Pythagoras believed the sphere was the most perfect geometric figure, and that the gods would use this perfect shape for their greatest creation. Aristotle also observed that ships disappeared over the horizon, the moon appeared to be a sphere, the stars moved in circular patterns, and noted reports from wandering fisherman on shifts in the constellations. These observations were all consistent with a spherical Earth.

After scientific support for a spherical Earth became entrenched, greek scientists turned toward estimating the size of the sphere. Eratosthenes, a greek scholar in Egypt, performed one of the earliest well-founded measurements of the Earth's circumference. He noticed that on the summer solstice the Sun at noon shone to the bottom of a deep well in Syene. He believed that the well was located on the Tropic of Cancer, so that the Sun would be exactly overhead during the summer solstice. He also observed that at exactly the same date and time a vertical post cast a shadow when located at Alexandria, about 805 kilometers north. The shadow/post combination defined an angle which was about 7°12', or about 1/50th of a circle (Figure 3-1).

Eratosthenes deduced that the Earth must be 805 multiplied by 50 or about 40,250 kilometers in circumference. His calculations were all in stadia, the unit of measure of the time, and have been converted here to the metric equivalent, given our best idea of the length of a stadia. Eratosthenes' estimate is close to our modern measurement of the Earth circumference of 38,762 kilometers. The difference is less than 4%.

The accuracy of Eratosthenes' estimate is quite remarkable, given the equipment for measuring distance and angles at that time, and because a number of his assumptions were incorrect. The well at Syene was

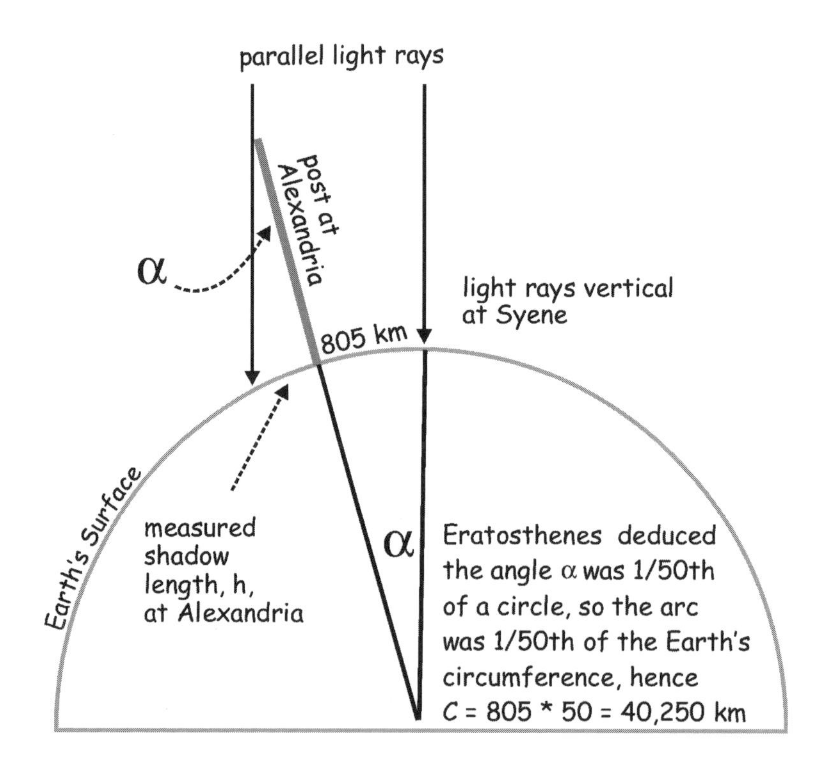

Figure 3-1: Measurements made by Eratosthenes to determine the circumference of the Earth.

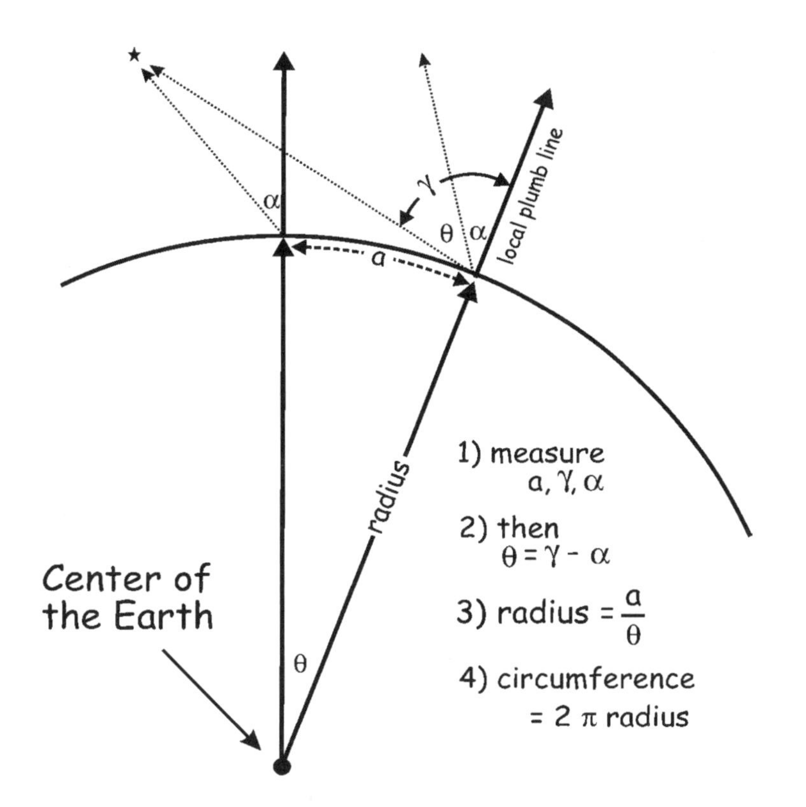

Figure 3-2: The Earth's radius and circumference may be determined by simultaneous measurement of zenith angles at two points. Two points are separated by an arc distance a measured on the Earth surface. These points also span an angle θ defined at the Earth center. The local zenith angles α and γ are related to θ, and the Earth radius is related to a and θ. Once the radius is calculated, the Earth circumference may be determined.

located about 60 kilometers off the Tropic of Cancer, so the Sun was not directly overhead. The true distance between the well location and Alexandria was about 729 kilometers, not 805, and the well was 3°3' east of the meridian of Alexandria, and not due north. However these errors either compensated for or were offset by measurement errors to end up with an amazingly accurate estimate.

Posidonius made an independent estimate of the size of the Earth sphere by measuring angles to a star near the horizon (Figure 3-2). Stars visible in the night sky define a uniform reference. The angle between a local vertical line and a star location is called a *zenith angle.* The zenith angle may be measured simultaneously at two locations on Earth, and the difference between the zenith angles used to calculate the circumference of the Earth. In Figure 3-2, the two zenith angles, α and γ, are measured to a star. The surface distance between these two locations is also measured, and the measurements combined with geometric relationships to calculate the Earth circumference (Figure 3-2). Posidonius calculated the difference in the zenith angles to a known star as about 1/48th of a circle when measured at Rhodes and Alexandria. By estimating these two towns to be about 805 miles apart, he estimated the circumference of the Earth to be 38,600 kilometers. Again there were compensating errors, but he still ended up with an accurate estimate. Another Greek scientist determined the circumference to be 28,960 kilometers, and unfortunately this shorter measurement was adopted by Ptomely for his world maps. This estimate was widely accepted until the 1500s, when Gerardus Mercator revised the figure upward.

During the 17th and 18th centuries two developments led to intense activity directed at measuring the size and shape of the Earth. Sir Isaac Newton and others reasoned the Earth must be flattened somewhat due to rotational forces. They argued that centrifugal forces cause the equatorial regions of the Earth to bulge as it spins on its axis. They proposed the Earth would be better modeled by an *ellipsoid*, a sphere that was slightly flattened at the poles. Measurements by their French contemporaries taken north and south of Paris suggested the Earth was flattened in an equatorial direction and not in a polar direction. The controversy persisted until expeditions by the French Royal Academy of Sciences between 1730 and 1745 measured the shape of the Earth near the equator in South America and in the high northern latitudes of Europe. Complex, repeated, highly accurate measurements established the curvature of the Earth was greater at the equator than the poles, and that an ellipsoid flattened at the poles was the best geometric model of the Earth's surface.

Note that the words spheroid and ellipsoid are often used interchangeably. For example, the Clarke 1880 ellipsoid is often referred to as the Clarke 1880 spheroid, because Clarke provided parameters for an ellipsoidal model of the Earth's shape. GIS software often prompts the user for a spheroid when defining a coordinate projection, and then lists a set of ellipsoids for choices.

An ellipsoid is sometimes referred to as a special class of spheroid known as an "oblate" spheroid. Thus, it is less precise but still correct to refer to an ellipsoid more generally as a spheroid. It would perhaps cause less confusion if the terms were used more consistently, but the usage is widespread. Ellipsoids are almost always used to model the Earth's shape, and they are often referred to as spheroids.

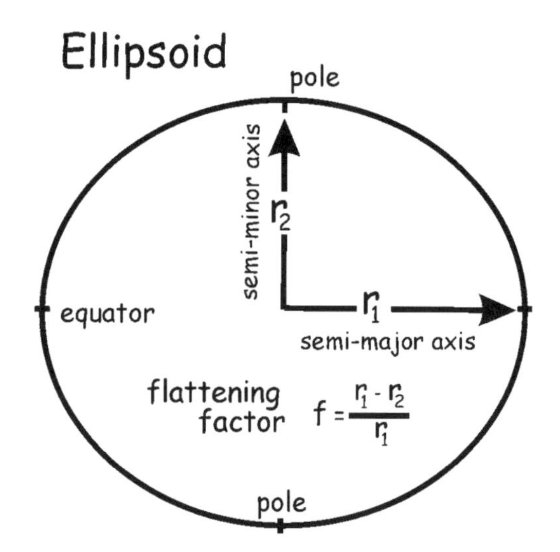

Figure 3-3: An ellipsoidal model of the Earth's shape.

Specifying the Ellipsoid

Once the general shape of the Earth was determined, geodesists focused on precisely measuring the size of the ellipsoid. The ellipsoid has two characteristic dimensions (Figure 3-3). These are the *semi-major axis*, the radius r_1 in the equatorial direction, and the *semi-minor axis*, the radius r_2 in the polar direction. The equatorial radius is always greater than the polar radius for the Earth ellipsoid. This difference in polar and equatorial radii can also be described by the flattening factor, as shown in Figure 3-3.

Earth radii have been determined since the 18th century using a number of methods, and for many parts of the globe. The most common methods until recently have involved astronomical observations similar to the those performed by Posidonius. These astronomical observations, also called celestial observations, are combined with long-distance surveys over large areas. Star and Sun locations have been observed and cataloged for centuries, and combined with accurate clocks, the positions of these celestial bodies may be measured to precisely establish the latitudes and longitudes of points on the surface of the Earth. Measurements during the 18th, 19th and early 20th centuries used optical instruments for celestial observations (Figure 3-4).

Measurement efforts through the 19th and 20th centuries led to the establishment of a set of official ellipsoids (Table 3-1). Why not use the same ellipsoid everywhere on Earth, instead of the different ellipsoids listed in Table 3-1? Different ellipsoids were adopted in various parts of the world for various reasons. Different ellipsoids were chosen primarily because there were different sets of measurements used in each region or continent.

Historically, geodetic surveys were isolated by large water bodies. For example, surveys in Australia did not span the Pacific

Table 3-1: Official ellipsoids. Radii may be specified more precisely than the 0.1 meter shown here. (from Snyder, 1987 and other sources)

Name	Year	Equatorial Radius, r_1 meters	Polar Radius, r_2 meters	Flat-tening Factor, f	Users
Airy	1830	6,377,563.4	6,356,256.9	1/ 299.32	Great Britain
Bessel	1841	6,377,397.2	6,356,079.0	1/ 299.15	Central Europe, Chile, Indonesia
Clarke	1866	6,378,206.4	6,356,583.8	1/ 294.98	Most of Africa; France
Clarke	1880	6,378,249.1	6,356,514.9	1/ 293.46	North America; Philippines
Inter-national	1924	6,378,388.0	6,356,911.9	1/ 297.00	Much of the World
Austra-lian	1965	6,378,160.0	6,356,774.7	1/ 298.25	Australia
WGS72	1972	6,378,135.0	6,356,750.5	1/ 298.26	NASA, US Defense Dept.
GRS80	1980	6,378,137.0	6,356,752.3	1/ 298.26	Worldwide
WGS84	1984 - cur-rent	6,378,137.0	6,356,752.3	1/ 298.26	Worldwide

Figure 3-4: An instrument used in the early 1900s for measuring the position of celestial bodies.

Ocean to reach Asia. Geodetic surveys relied primarily on optical instruments prior to the early 20th century. These instruments were essentially precise telescopes, and thus sighting distances were limited by the Earth's curvature. Individual survey legs greater than 50 kilometers (30 miles) were rare, so during this period there were no good ways to connect surveys between continents.

Because continental surveys were isolated, ellipsoidal parameters were fit for each continent or comparably large survey area. These ellipsoids represented continental measurements and conditions. Measurements based on Australian surveys yielded a different "best" ellipsoid than those in Europe. Europe's best ellipsoidal estimate was different from Asia's, and from South America, North America, or other regions. One ellipsoid could not be fit to all the World's survey data because during the 18th and 19th centuries there was no clear way to combine a global set of measurements.

The differences in the best fitting ellipsoid were primarily due to real differences in the shape of the Earth across the globe. The Earth's true shape diverges from the shape of an ellipsoid, often above or below any globally-defined ellipsoid. The causes and form of this divergence are described in more detail in the next section. Local undulations occur over several thousands of miles, giving different estimates of the shape of the Earth when only a portion of the Earth is surveyed.

Differences in the ellipsoidal estimates among continents were also due to differences in survey methods and data analyses employed. Computational resources, the sheer number of survey points, and the scarcity of survey points for many areas were barriers to the development of global ellipsoids. Methods for computing positions, removing errors, or adjusting point locations were not the same worldwide, and may have led to differences in ellipsoidal estimates. Different methods applied to the same data may cause differences, and it took time for the best methods to be developed, widely recognized, and adopted.

More recently, data derived from satellites, lasers, and broadcast timing signals have been used for extremely precise measurements of relative positions across continents and oceans. A global set of measurements became possible, and it has become easy to combine and compare measurements on all continents simultaneously. This has allowed the calculation of globally applicable ellipsoids. Internationally recognized ellipsoids such as WGS72, GRS80, or WGS84 have become widely used as more precise measurements have been developed over a larger portion of the globe. Continuing measurements may lead to further ellipsoid refinements.

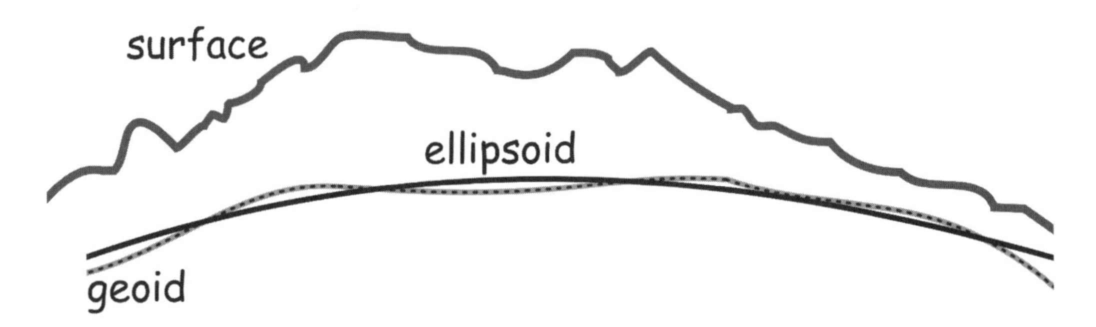

Figure 3-5: Three "surfaces" used to characterize position on the Earth.

The Geoid

As noted above, the true shape of the Earth varies slightly from the mathematically smooth surface defined by an ellipsoid. Differences in the density of the Earth cause variation in the strength of the gravitational pull, in turn causing regions to dip or bulge above or below a reference ellipsoid. This undulating shape is the *Geoid*.

The amount of geoidal deformation depends on the variation in the strength of the gravitational field, and on geologic history. Over most of the Earth the Geoid varies by less than 100 meters from the ellipsoid, and these differences between the Geoid and the ellipsoid are small relative to the differences between an ellipsoidal and a spherical model. Geoidal variation in the Earth's shape is the main cause for different ellipsoids being employed in different parts of the world.

Geodesists have defined the Geoid as the three-dimensional surface along which the pull of gravity is a specified constant. The geoidal surface may be thought of as the level of an imaginary sea, were it to cover the entire Earth and not be affected by wind, waves, the moon, or forces other than gravity. The surface of the Geoid is in this way related to mean sea level, or other references against which heights are measured. Geodesists may measure surface heights relative to the Geoid, and at any point on Earth there are three important surfaces, the ellipsoid, the Geoid, and the Earth surface (Figure 3-5). Horizontal positions on the surface of the Earth are typically defined with respect to the ellipsoid or spheroid, while heights are typically defined relative to the Geoid.

It is important to note that the Geoid is a measured and interpolated surface, and not a mathematically defined surface. The Geoid's surface is most often measured with an instrument called a gravimeter. This instrument is towed, flown, or driven over the surface of the Earth, and it measures the strength of the Earth's gravitational pull. The height to the reference gravitational surface defining the Geoid is estimated by measuring the change in gravity with elevation. Gravimetric surveys have been conducted for the entire globe, although measurement locations are denser over the continents and in the developed countries. Geoidal models have been defined which allow the approximate position of the geoidal surface to be predicted within this network of measurements.

Geographic Coordinates, Latitude, and Longitude

Once a size and shape of the reference ellipsoid has been determined, the poles and Equator are also defined. The poles are defined by the axis of revolution of the ellipsoid, and the Equator is defined as the circle mid-way between the two poles, at a right angle to the polar axis, and spanning the widest dimension of the ellipsoid. We now must go about defining a coordinate grid, a reference system by which we may specify the position of features on the ellipsoidal surface.

A geographic coordinate system has been defined for the Earth with reference to the ellipsoid, consisting of latitude, which varies from north to south, and longitude, which varies from east to west (Figure 3-6). Lines of constant longitude are called meridians, and lines of constant latitude are called parallels. Parallels run parallel to each other in an east-west direction around the Earth. The meridians are geographic north/south lines, and converge at the poles.

By convention, the Equator is taken as zero degrees latitude, and latitudes increase to the north and south (Figure 3-6). Latitudes are thus designated by their magnitude and direction, for example 35°N or 72°S. When signed values are required, northern latitudes are designated positive and southern latitudes designated negative. An international meeting in 1884 placed the longitudinal origin, designated as the *Prime Meridian*, through the Royal Greenwich Observatory in England. Directional longitudes with a west and east designator may be specified as an angle of rotation away from the Prime Meridian. When required, west is considered negative and east positive.

There is often confusion between magnetic north and geographic north. Magnetic north is the point towards which a compass points. The geographic North Pole is located at one of the poles of the Earth's axis of rotation. Magnetic north and the geographic North Pole do not coincide (Figure 3-7). If you were standing on the geographic North Pole with a compass, it would point approximately in the direction of northern Canada, towards magnetic north some 600 kilometers away.

Because magnetic north and the geographic North Pole do not coincide, magnetic north and the geographic North Pole

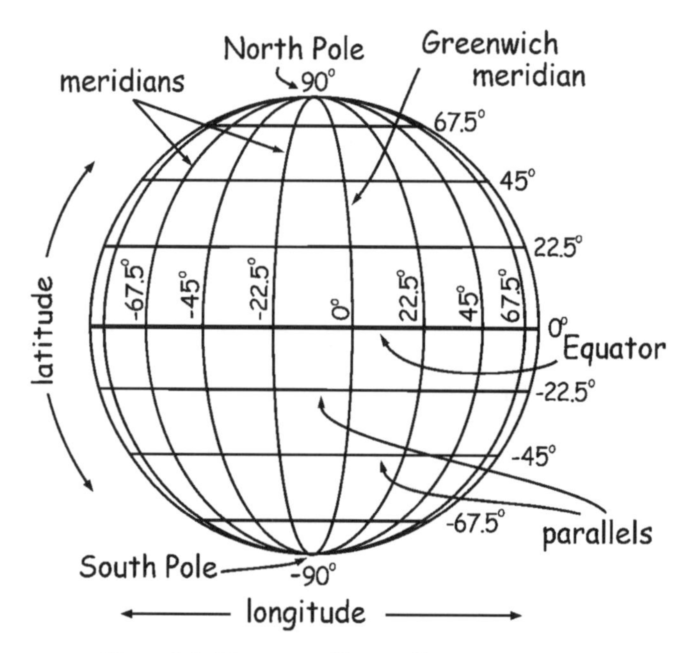

Figure 3-6: The geographic coordinate system.

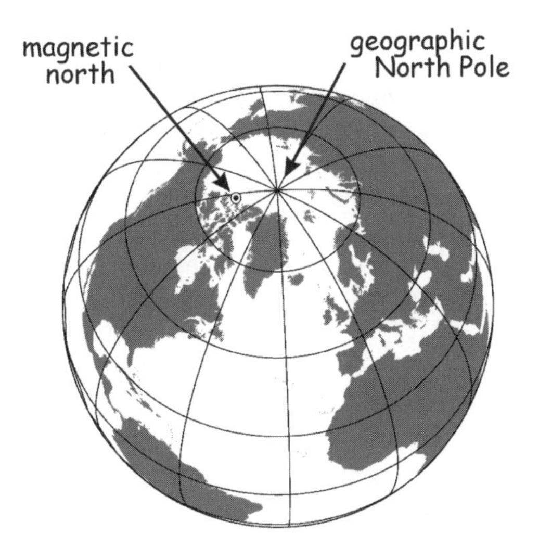

Figure 3-7: Magnetic north and the geographic North Pole.

are in slightly different directions most places on Earth. This angular difference is called the magnetic *declination* and varies across the globe. The specification of map projections and coordinate systems is always in reference to the geographic North Pole, not magnetic north.

Geographic coordinates do not form a Cartesian system (Figure 3-8). As previously described in Chapter 2, a Cartesian system defines lines of equal value to form a right-angle grid. Geographic coordinates are defined on a curved surface, and the longitudinal lines converge at the poles. Therefore there are no Cartesian parallels in the north-south direction, and the Earth surface length of a degree of longitude varies from approximately 111.3 kilometers at the equator to 0 kilometers at the poles. This distortion is seen on the left in Figure 3-8. Circles with a fixed 5 degree radius appear distorted near the poles when drawn on a globe. The circles become "flattened" in the east-west direction. In contrast, circles appear as circles when the geographic coordinates are plotted in a Cartesian system, as at the right of Figure 3-8, but the underling geography is distorted; note the erroneous size and shape of Antarctica. While the ground distance spanned by a degree of longitude changes markedly across the globe, ground distance for a degree of latitude varies only slightly, from 110.6 kilometers at the equator to 111.7 kilometers at the poles.

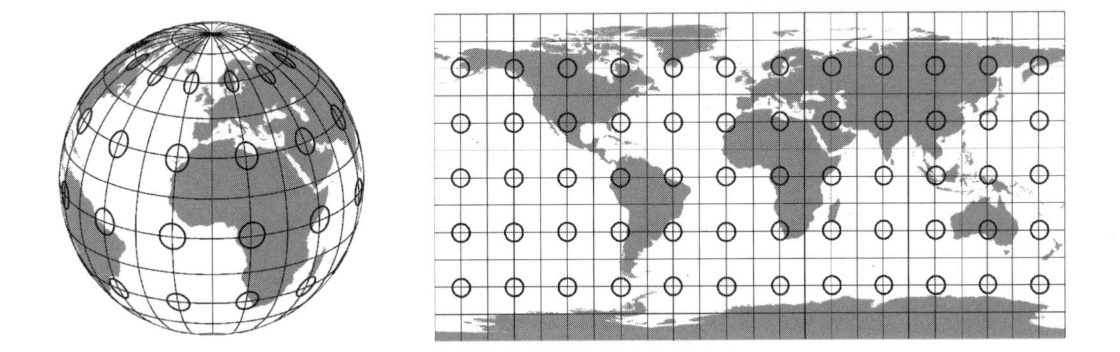

Figure 3-8: Geographic coordinates on a spherical (left) and Cartesian (right) representation. Notice the circles with a 5 degree radius appear distorted on the spherical representation, illustrating the change in surface distance represented by a degree of longitude from the Equator to near the poles.

Datums

We have defined a geographic coordinate system that provides for specifying locations on the Earth. This gives us the exact longitude of only one location, the Greenwich Observatory. By definition the longitude of the Greenwich meridian is 0 degrees. We must estimate the longitudes and latitudes of all other locations through measurements. We establish a set of points for which the horizontal and vertical positions have been accurately determined. All other coordinate locations we use are measured with reference to this set of precisely surveyed points, including the coordinates we enter in our GIS to represent spatial features.

Many countries have a government body charged with making precise geodetic surveys. For example, nearly all horizontal locations in the United States are traceable or related to a national network of highly accurate survey points established and/or maintained by the National Geodetic Survey (NGS). This unit of the U.S. federal government establishes geodetic latitudes and longitudes of known, monumented points. These points, taken together, form the basis for a *geodetic datum*, upon which most subsequent surveys and positional measurements are based.

A geodetic datum consists of two major components. The first component is the previously described specification of an ellipsoid with a spherical coordinate system and an origin. The second part of a datum consists of a set of points and lines that have been painstakingly surveyed using the best methods and equipment. Different datums are specified through time because new points are added and survey methods improve. When a sufficiently large number of new, precise survey points have been measured we periodically update our datum by re-estimating the coordinates of our datum points after including these newer, better measurements.

Historically, the relative positions of a set of datum points was determined using celestial measurements in combination with high-accuracy ground measurements. Most early measurements involved precise field surveys with optical instruments (Figure 3-9). These methods have been replaced in recent years by sophisticated electronic and satellite-based surveying systems. Measurements from an extensive set of surveys were combined, errors and inconsistencies identified and removed, and the geographic location for each datum point estimated.

Precisely surveyed points used in developing a datum are often monumented, and these monumented points are known as *benchmarks*. Benchmarks usually consist of a brass disk embedded in rock or concrete (Figure 3-10), although they also may consist of squares or circles chiseled in rocks, iron posts, or other long-term marks.

Early geodetic surveys combined horizontal measurements with repeated, excruciatingly precise astronomical observations to

Figure 3-9: An early geodetic survey, used in fixing datum locations. (courtesy NCGS)

Figure 3-10: A brass disk used to monument a survey benchmark.

determine latitude and longitude of a subset of points. Only a few datum points were determined using astronomical observations. Astronomical observations were typically used at the starting point, a few intermediate points, and near the end of geodetic surveys. This is because star positions required repeated measurements over several nights. Clouds, haze, or a full moon often lengthened the measurement times. In addition, celestial measurements required correction for atmospheric refraction, a process which bends light and changes the apparent position of stars. Refraction depends on how high the star is in the sky at the time of measurement, as well as temperature, atmospheric humidity, and other factors.

Horizontal measurements were as precise and much faster than astronomical measurements when surveys originated at known locations. Thus, most campaigns involved the establishment of a few starting, intermediate, and ending points using astronomical measurements. Horizontal surface measurements were then used to connect these astronomically surveyed points and thereby create an expanded, well-distributed set of known datum points. Figure 3-11 shows an example survey, where open circles signify points established by astronomical measurements and filled circles denote points established by surface measurements.

Triangulation surveys (Figure 3-11) were used to establish datum points via horizontal surface measurements. Triangulation surveys utilize a network of interlocking triangles to determine positions at survey stations. Because there are multiple measurements to each survey station, the location at each station may be computed by various paths. The survey accuracy can be field checked, because large divergence in a calculated station location via different paths may quickly be identified. There are always some differences in the measured locations when traversing different paths. An acceptable error limit was set, usually as a proportion, e.g., differences in the measured location of more than 1 part in 100,000 would be considered unacceptable. When unacceptable errors were found, survey lines were re-measured.

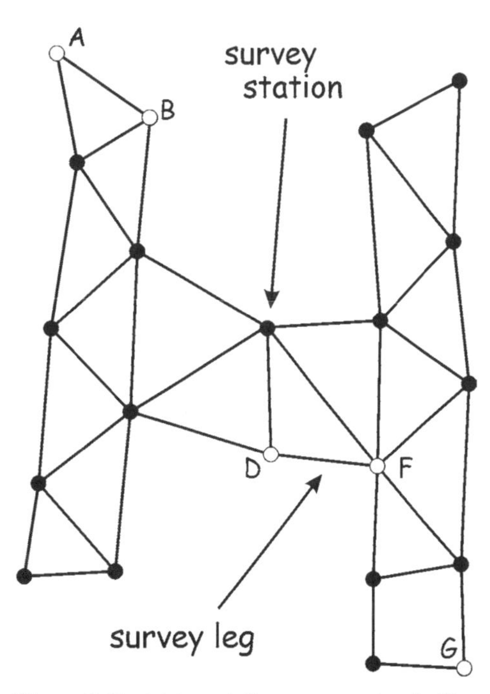

Figure 3-11: A triangulation survey network. Stations may be measured using astronomical (open circles) or surface surveys (filled circles).

Triangulation surveys were also adopted because they required few surface distance measurements. This was an advantage between the late 18th and early 20th century, the period during which the widespread networks of datum points were established. During that time it was much easier to accurately measure horizontal and vertical angles than to measure horizontal distances. Direct measurements of surface distances were less certain because they relied on metal tapes, chains, or rods. Tapes and chains sagged by varying amounts. Metal changed length with temperature and stretched with use. Ingenious bi-metallic compensation bars were developed that combined two metals along their length. These metals changed length at different rates with temperature. A scale engraved in the bars indicated the amount of expansion and facilitated a compensation to offset the temperature-caused error. Nonetheless, laying the bars end-to-end over great distances was a painstaking, slow process, and so surface measurements were to be avoided.

Triangulation surveying requires the surface measurement of an initial baseline, and then relies on geometric relationships and angle measurements to calculate all subsequent distances. Figure 3-12 shows a sequence of measurements for a typical triangulation survey conducted in the 19th or early 20th centuries. The positions of stations S1 and S2 (Figure 3-12, top) were determined by celestial observations, and the length A measured using compensation bars or equally precise methods. A surveying instrument was placed at stations S1, S2, and S3 and used to measure angles a, b, and

If we measure the initial baseline length A, and measure the angles a, b, and c, we are then able to calculate the lengths B and C:

by the law of sines, $\frac{A}{B} = \frac{\text{sine (a)}}{\text{sine (b)}}$

then $B = A\frac{\text{sine (b)}}{\text{sine (a)}}$ and $C = A\frac{\text{sine (c)}}{\text{sine (a)}}$

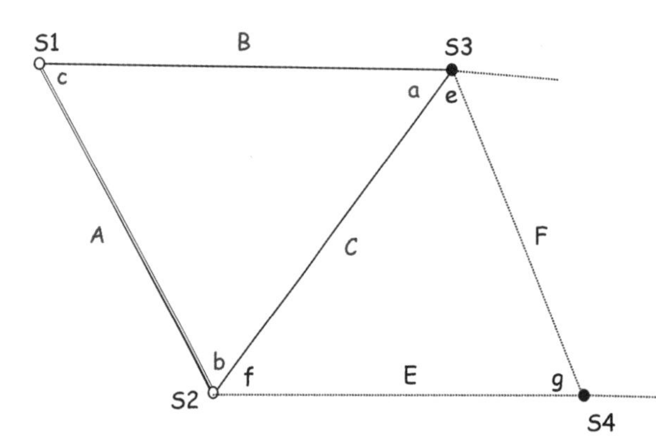

With the length C known, angles e, f, and g may then be measured. The law of sines may be used with the now known distance C to calculate lengths E and F. Successive datum points may be established to extend the network using primarily angle measurements.

Figure 3-12: Early geodetic surveys used triangulation networks to reduce the number of surface distance measurements.

c. The law of sines allows the calculation of lengths B and C, and combined with the angle measurements allows precise estimates of the location of station S3. Angles e, f, and g (Figure 3-12, bottom) were then measured, and the precise location of station S4 estimated. The geodetic survey could then be extended indefinitely by repeated application of angle measurements and the law of sines, with only occasional surface distance measurements as a check on positions.

Survey stations were typically placed on high vantage points to lengthen the distance between triangulation stations. Longer distances meant fewer stations were required to cover a given area. Since the triangulation networks spanned continents, there was a strong impetus to keep costs down. The survey stations were quite far apart, up to 10's of kilometers, and measurements were often made from tall objects such as mountaintops, church steeples, or specially constructed Bilby towers to give long sight lines and hence long survey legs (Figure 3-13). Besides use in developing datums, these monumented points may be used as starting points for precise local surveys.

The positions of all points in a reference datum are estimated in a network-wide *datum adjustment*. "Highest accuracy" measurements are collected for each angle or distance in the triangulation surveys of datum points. However, some errors are inevitable, and small errors in each measurement are allowed. For example, distance of a 30 kilometer (20 mile) survey leg may differ by 30 centimeters (1 foot) when measured using two independent methods. While quite small, these small discrepancies in the distances or angles for redundantly measured points still exist, even though the best available methods have been used. The survey network must be adjusted, that is, the error reduced and distributed across the network in some "optimum" fashion. A datum adjustment resolves and distributes these errors across the survey network.

Periodic datum adjustments result in multiple regional or global reference datums. For example, there are several reference datums for North America. Several datums exist because new surveys are added to regions that had been poorly measured before. In addition, survey instruments are continuously improving, and computational and survey adjustment methods are also getting better. When enough new survey points have been collected, a new datum is estimated. This means new coordinate locations

Figure 3-13: A Bilby tower near Bozeman, Montana, USA, used as a platform for a triangulation survey. Towers or other high vantage points increased the distance between survey stations. (courtesy NCGS)

are estimated for all the datum points. The datum points don't move, but our best estimates of the datum point coordinates will change. Differences between the datums reflect differences in the control points, survey methods, and mathematical models and assumptions used in the datum adjustment.

The calculation of a new datum requires that all surveys must be simultaneously adjusted to reflect our current "best" estimate of the true positions of each datum point. Generally a statistical least-squares adjustment is performed, but this is not a trivial exercise, considering the adjustment may include survey data for tens of thousands of old and newly surveyed points from across the continent, or even the globe. Because of their complexity, these continent-wide or global datum calculations have historically been quite infrequent. Computational barriers to datum adjustments have diminished in the past few decades, and so datum adjustments and new versions of datums are becoming more common.

Three main horizontal datums have been used widely in North America. The first of these is the *North American Datum of 1927* (NAD27). NAD27 is a general least-squares adjustment that included all horizontal geodetic surveys completed at that time. The geodesists used the Clarke Ellipsoid of 1866 and held fixed the latitude and longitude of a survey station in Kansas. NAD27 yielded adjusted latitudes and longitudes for approximately 26,000 survey stations in the United States and Canada.

The *North American Datum of 1983* (NAD83) is the immediate successor datum to NAD27. It was undertaken by the National Coast and Geodetic Survey to include the large number of geodetic survey points established between the mid-1920s and the early 1980s. Approximately 250,000 stations and 2,000,000 distance measurements were included in the adjustment. The GRS80 ellipsoid was used as reference. NAD83 uses an Earth-centered reference, rather than the fixed station selected for NAD27.

The *World Geodetic System of 1984* (WGS84) is also commonly used, and was developed by the U.S. Department of Defense (DOD). It was developed in 1987 based on Doppler satellite measurements of the Earth, and is the base system used in most DOD maps and positional data. WGS84 has been updated based on more recent satellite measurements and is specified using a version designator, e.g., the update based on data collected up to January 1994 is designated as WGS84 (G370). The WGS84 ellipsoid is very similar to the GRS80 ellipsoid.

Since different datums are based on different sets of measurements and ellipsoids, the coordinates for benchmark datum points typically differ between datums. That is to say, the latitude and longitude location of a given benchmark in the NAD27 datum will likely be different from the latitude and longitude of that same benchmark in the NAD83 or WGS84 datums. This is described as a *datum shift*. The monumented points do not move. Physically, the points remain in the same location; our estimates of the point locations change. As survey measurements improve through time, and there are more of them, we obtain better estimates of the true locations of the monumented datum points.

Coordinates across different datums may differ slightly, because of the differences in the ellipsoid and in the measurements and methods used to estimate the locations of datum points. Differences may be small, e.g., the shift in coordinate locations from WGS84 to NAD83 is often less than a meter, but shifts may be quite large. Care should be taken to not mix spatial data across datums unless the magnitude for the datum shift has been established for the area of interest, and this magnitude is considered small relative to overall data use and accuracy specifications.

Datum shifts between NAD27 and NAD83 are often quite large, up to 100's of meters. Figure 3-14 indicates the relative size of shifts across the country between NAD27 and NAD83, based on estimates

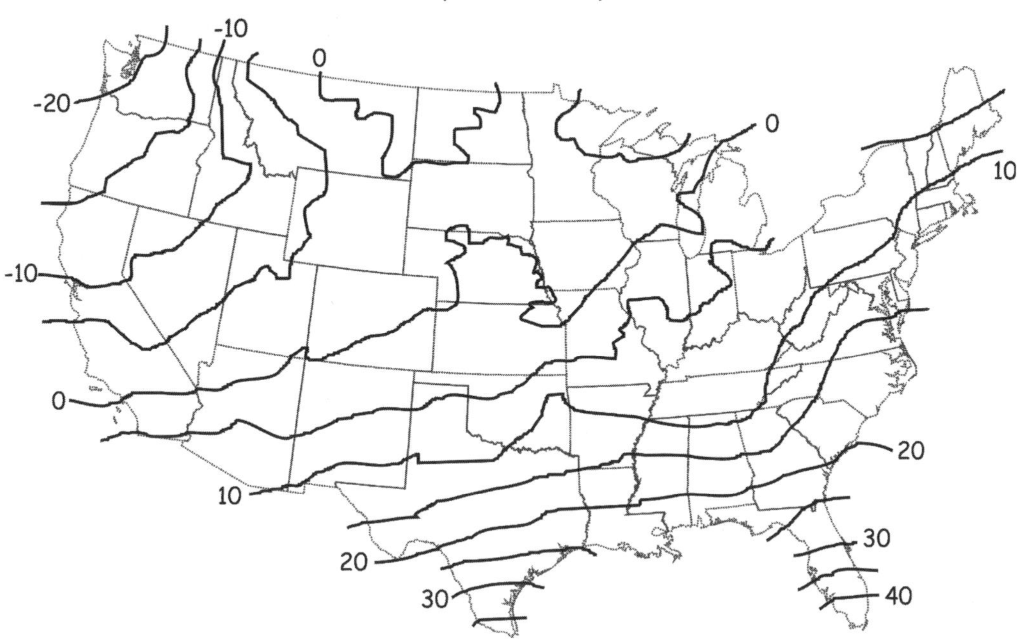

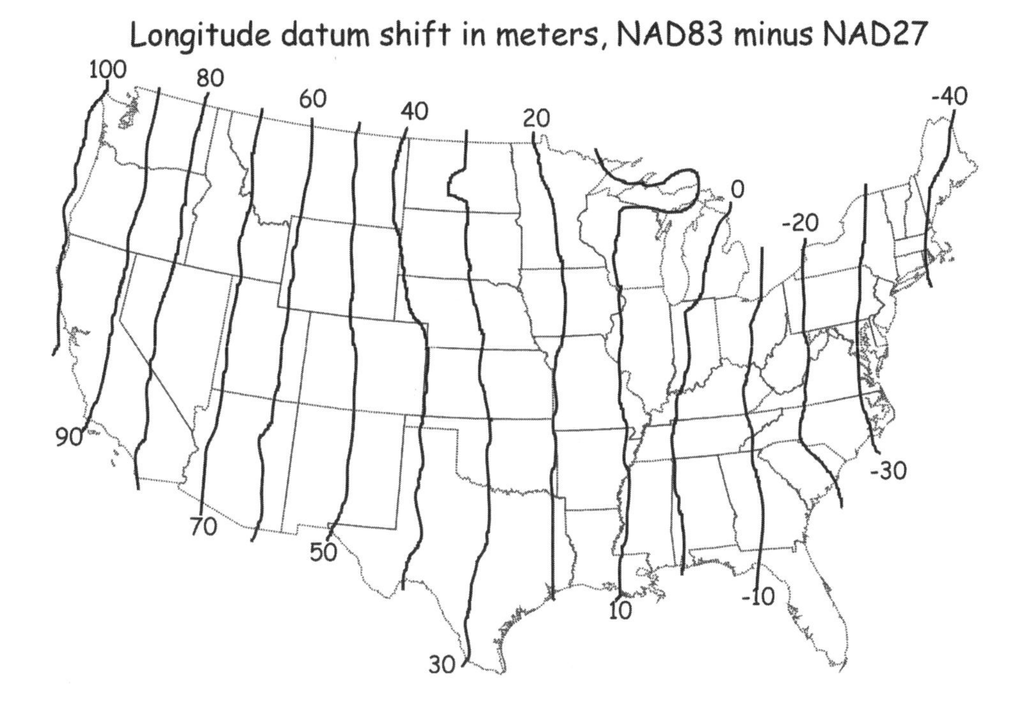

Figure 3-14: Datum shift in meters between the North American Datum of 1927 (NAD27) and the North American Datum of 1983 (NAD83).

provided by the National Geodetic Survey. Notice that datum shifts between NAD27 and NAD83 are approximately -20 to 40 meters (70 to 140 feet) northward and between -40 and 100 meters (140 to 330 feet) east-west. Because of this datum shift it is important to note the datum underpinning a GIS development effort. If source data or maps are referenced to different datums, then they must be converted before they will overlay correctly.

Different vertical datums also exist. The two most common in the U.S. are the *National Geodetic Vertical Datum* of 1929 (NGVD29) and the *North American Vertical Datum of 1988* (NAVD88). The 1929 Vertical Datum was derived from a best fit of the average elevation of 26 stations in North America. The 1988 datum is based on over 600,000 kilometers (360,000 miles) of control leveling performed since 1929, and also reflects geologic crustal movements or subsidence that may have changed benchmark elevation. NAVD88 adjusted the estimate of mean sea level slightly. Differences are observed when comparing NGVD29 and NAVD88 elevations for the same benchmarks.

Control Density

In most cases the datum control points are too sparse to be sufficient for all needs in GIS data development. For example, precise point locations may be required when setting up a GPS receiving station, to georegister a scanned photograph or other imagery, or as the basis for a detailed subdivision or highway survey. It is unlikely there will be more than one or two datum points within the work area for each of these activities. In many instances a denser network of known points is desirable. The monumented, precisely determined point locations that are part of a datum may be used as a starting point for additional surveying. These smaller area surveys increase the density of precisely known points. The quality of the point locations depends on the quality of the intervening survey, and a set of standards has been established for reporting survey quality.

The Federal Geodetic Control Committee of the United States (FGCC) has published a detailed set of survey accuracy specifications. These specifications set a minimum acceptable accuracy for surveys and establish procedures and protocols to ensure the advertised accuracy has been obtained. The FGCC specifications establish a hierarchy of accuracy. First order survey measurements are accurate to within 1 part in 100,000. This means the error of the survey is no larger than one unit of measure for each 100,000 units of distance surveyed. The maximum horizontal measurement error of a 5,000 meter baseline (about 3 miles) would be no larger than 5 centimeters (about 2 inches). Accuracies are specified by Class and Orders, down to a Class III, 2nd order point with an error of no more than 1 part in 5,000. Federal, state, provincial, and county surveyors typically have a set of identified points that have been precisely surveyed to augment the local control network. The accuracy of these points is usually available, and point description and location may be obtained from the appropriate surveying authority. A list of precisely surveyed points may be obtained from state, county, city, or other surveyors, along with a physical description of the points and their location relative to nearby features. These control points may then be used as starting locations in the development of additional spatial data.

Map Projections and Coordinate Systems

Datums tell us the latitudes and longitudes of a set of points on an ellipsoid. We need to transfer the locations of features measured with reference to these datum points from the curved ellipsoid to a flat map. A *map projection* is a systematic rendering of locations from the curved Earth surface onto a flat map surface. Points are "projected" from the Earth surface and onto the map surface.

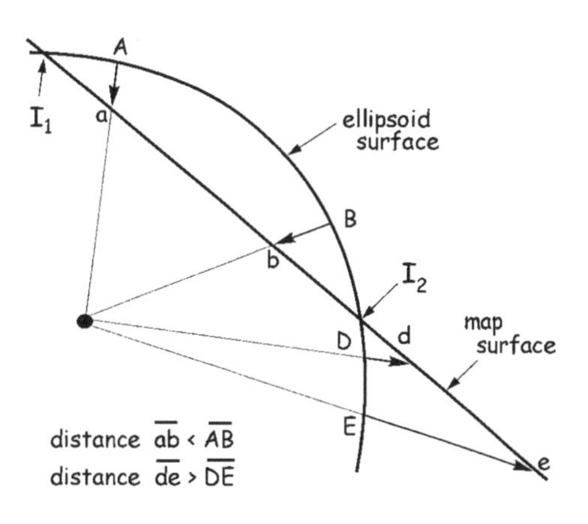

Figure 3-16: Distortion during map projection.

Many map projections may be envisioned as a process by which a light source shines rays from a projection center onto a map surface (Figure 3-15). Light rays radiate out from a source to intersect both the ellipsoid surface and the map surface. The rays specify where each point on the ellipsoid surface is to be placed on the map surface. Other projection methods exist, for example in some projections the source is not a single point, however the basic process involves the systematic transfer of points from the curved ellipsoidal surface to a flat map surface.

Distortions are unavoidable when making flat maps. This is because locations are projected from a complexly curved Earth surface to a flat or simply curved map surface. Portions of the rendered Earth surface must be compressed or stretched to fit onto the map. This is illustrated in a side view of a projection from an ellipsoid onto a plane (Figure 3-16). The map surface intersects the Earth at two locations, I_1 and I_2. Points toward the edge of the map surface, such as D and E, are stretched apart. The scaled map distance between d and e is greater than the distance from D to E measured on the surface of the Earth. More simply put, the distance along the line (map) is greater than the corresponding distance along the curve (Earth surface). Conversely, points such as A and B that lie in between I_1 and I_2 would appear compressed together. The scaled map distance from a to b would be less than the surface-measured distance from A to B. Distortions at I_1 and I_2 are zero.

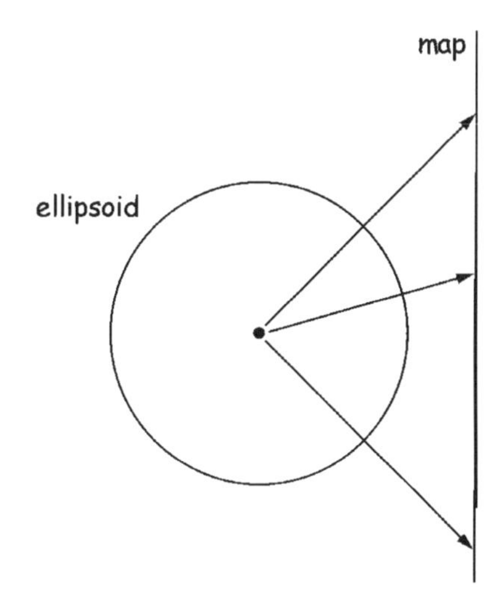

Figure 3-15: A conceptual view of a map projection.

Figure 3-16 demonstrates a few important facts. Distortion may take different forms in different portions of the map. In one portion of the map features may be compressed and exhibit reduced areas or distances relative to Earth surface measurements, while in another portion of the map areas or distances may be expanded. Second, there are often a few points or lines where distortions are zero, where length, direction, or some other geometric property is preserved. Finally, distortion is usually less near the points or lines of intersection, where the map surface intersects the imaginary globe. Distortion usually increases with increasing distance from the intersection points or lines.

Different map projections may distort the globe in different ways. The projection source, represented by the point at the middle of the circle in Figure 3-16, may change locations. The surface onto which we are projecting may change in shape, and we may place the projection surface at different locations at or near the globe. If we change any of these three factors, we will change how or where our map is distorted.

Many map projections are based on a *developable surface*, a geometric shape onto which the Earth surface locations are projected. Cones, cylinders, and planes are the most common types of developable surfaces. A plane is already flat, and cones and cylinders may be mathematically "cut" and "unrolled" to develop a flat surface (Figure 3-17). Projections may be characterized by the developable surface, e.g., as *conic* (cone), *cylindrical* (cylinder), and *azimuthal* (plane). The orientation of the developable surface may also change among projections, e.g., the axis of a cylinder may coincide with the poles (equatorial) or the axis may pass through the equator (transverse).

Many map projections are not based on developable surfaces. Projections with names such as pseudocylindrical, Mollweide, sinusoidal, and Goode homolosine are examples. These projections often specify a direct mathematical projection from an ellipsoid onto a flat surface. They use mathematical forms not related to cones, cylinders, planes, or other three-dimensional figures, and may change the projection surface for different parts of the globe. For example, projections such as the Goode homolosine projection may be formed by fusing other projections, and "interrupting" segments of the Earth to give a global view. These projections use complex rules and breaks to reduce distortion for many continents, and are not based on developable surfaces.

We typically have to specify several characteristics when we specify a map projection. For example, for an azimuthal projection we must specify the location of the

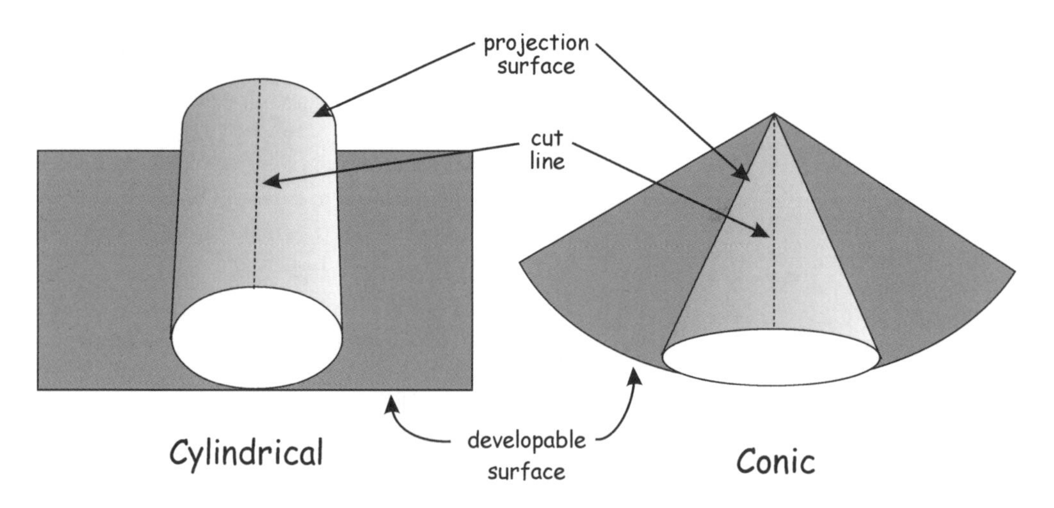

Figure 3-17: Projection surfaces are derived from curved "developable" surfaces that may be mathematically "unrolled" to a flat surface.

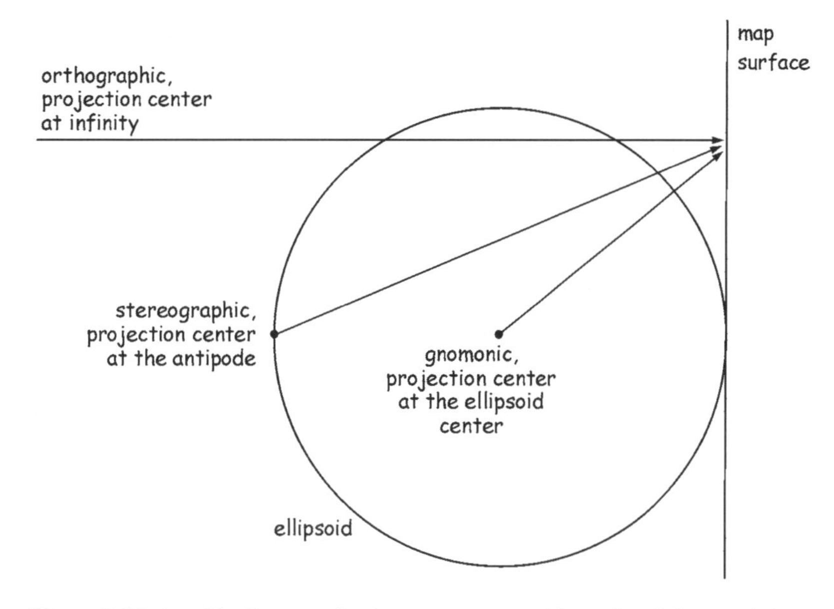

Figure 3-18: As with all map projections, we must specify a suite of characteristics when defining an azimuthal map projection. One characteristic is the location of the projection center, most often placed at the center of the ellipsoid, or the antipode, or at infinity.

projection center (Figure 3-18) and the location and orientation of the plane onto which the globe is projected. Azimuthal projections are often tangent to (just touch) the ellipsoid at one point, and we must specify the location of this point. A projection center ("light" source location) must also be specified, most often placed at one of three locations. The projection center may be at the center of the ellipsoid (a *gnomonic* projection), at the antipodal surface of the ellipsoid (diametrically opposite the tangent point, a *stereographic* projection), or at infinity (an *orthographic* projection). Scale factors, the location of the origin of the coordinate system, and other projection parameters may be required. Defining characteristics must be specified for all projection, e.g., we must specify the size and orientation of a cone in a conic projection, or the size, intersection properties, and orientation of a cylinder in a cylindrical projection.

Common Map Projections in GIS

There are hundreds of map projections used throughout the world, however most spatial data in GIS are specified using projections from a relatively small number of projection types.

The Lambert conformal conic and transverse Mercator are among most common projection types used for spatial data in North America and much of the World (Figure 3-19). Standard sets of projections have been established from these two basic types. The Lambert conformal conic projection may be conceptualized as a cone intersecting the surface of the Earth, with points on the Earth's surface projected onto the cone. The cone in the Lambert conformal conic intersects the ellipsoid along two arcs, typically parallels of latitude, as shown in Figure 3-19 (top left). These lines of intersection are known as *standard parallels*.

Distortion in a Lambert conformal conic projection is typically smallest near the standard parallels, where the developable surface intersects the Earth. Distortion increases in a complex fashion as distance to these intersection lines increases. This characteristic is illustrated at the top of Figure 3-19. Circles of a constant 5 degree radius are drawn on the projected surface. These circles are distorted, depending on the location on the projection surface. Circles nearer the standard parallels are less distorted. Those farther away tend to be more distorted. Distortions can be quite severe, as illustrated by the apparent expansion of southern South America.

Note that sets of circles in an east-west row show similar distortion properties in the Lambert conformal conic projection (Figure 3-19, top right). Those circles that fall between the standard parallels exhibit a uniformly lower distortion than those in other portions of the projected map. One property of the Lambert conformal conic projection is a low-distortion band running in an east-west direction between the standard parallels. Thus, the Lambert conformal conic projection may be used for areas which are larger in an east-west than a north-south direction, as there is little added distortion when extending the mapped area in the east-west direction.

The transverse Mercator is another common map projection. This map projection may be conceptualized as enveloping the Earth in a horizontal cylinder, and projecting the Earth surface onto the cylinder (Figure 3-19, bottom). The transverse Mercator commonly intersects the Earth ellipsoid along a single north-south tangent, or along two *secant* lines where the cylinder intersects the ellipsoid. A line parallel to and midway between the secants is often known as the central meridian. The central meridian extends north and south through transverse Mercator projections.

As with the Lambert conformal conic, the transverse Mercator projection has a band of low distortion, but this band runs in a north-south direction. Distortion is least near the line(s) of intersection. At the bottom of Figure 3-19 the intersection occurs at zero degrees longitude, traversing western Africa, eastern Spain, and England. Distortion is lowest near this line and increases markedly with distance east or west away from this line, e.g., the shape of South America is severely distorted in the bottom right of Figure 3-19. For this reason transverse Mercator projections are often used for areas that extend in a north-south direction, as there is little added distortion extending in that direction.

Different projection parameters may be used to specify an appropriate coordinate system for a region of interest. Specific standard parallels or central meridians are used to minimize distortion over a mapping area. An origin location, measurement units, x and y (or northing and easting) offsets, a scale factor, and other parameters may also be required to define a specific projection. Once a projection is defined then the coordinates of every point on the surface of the Earth may be determined, usually by a closed-form or approximate mathematical formula.

The State Plane Coordinate System

The State Plane Coordinate System is a standard set of projections for the United States. The State Plane coordinate system specifies positions in a Cartesian coordinate system over several county to whole-state areas. There are typically one or more State Plane coordinate systems defined for each state in the United States. More than one State Plane zone may be required to limit distortion errors due to the map projection.

State Plane systems greatly facilitate surveying, mapping, and spatial data development in a GIS, particularly when county or larger areas are involved. Over relatively small areas the surface of the Earth can be assumed to be flat without introducing much distortion. However, Earth curvature causes area, distance, and shape distortion to increase when we assume a flat surface over larger areas. The State Plane system provides a common coordinate reference for

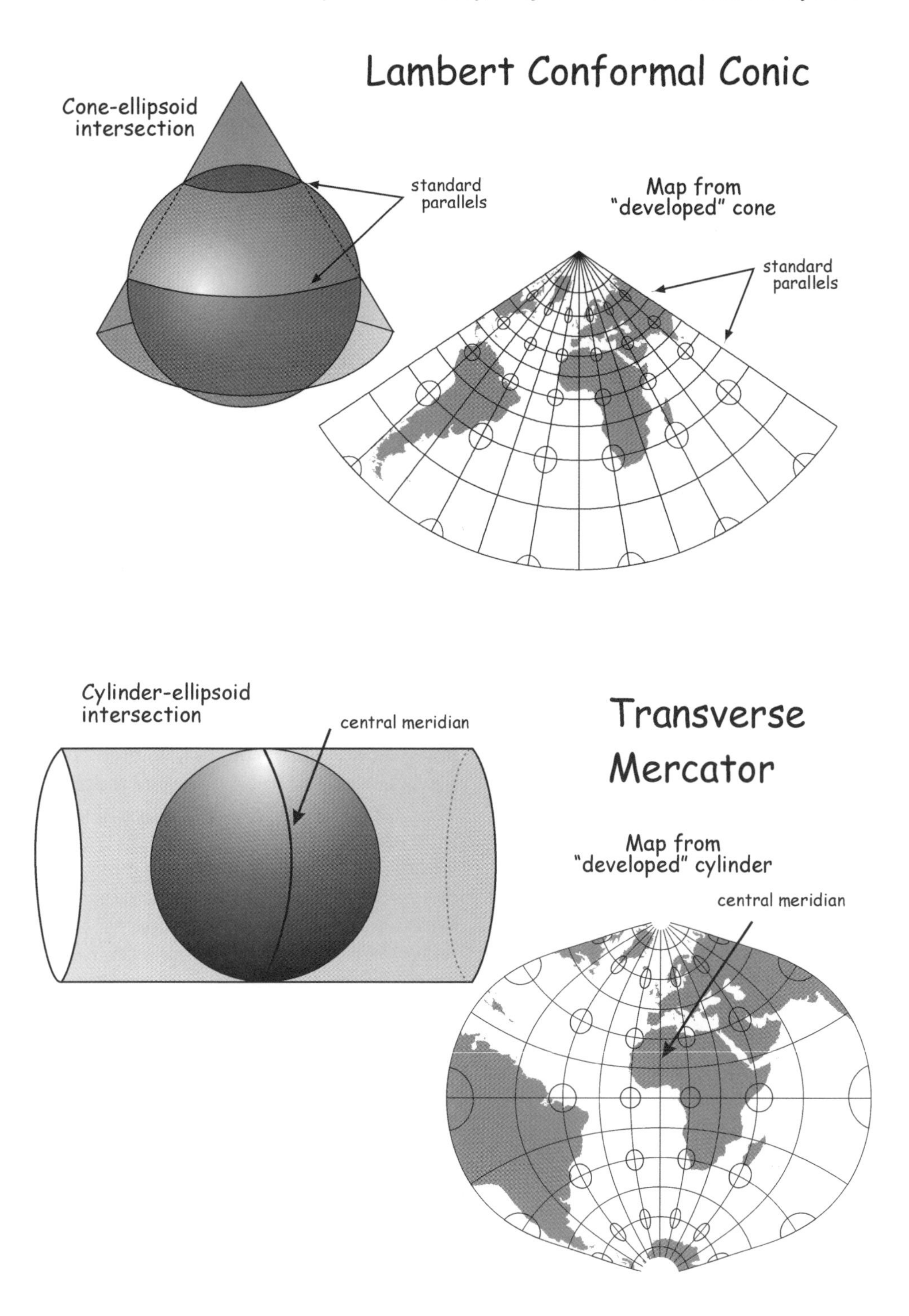

Figure 3-19: Lambert conformal conic (LCC, top) and transverse Mercator (TM, bottom) projections. The LCC is derived from a cone intersecting the ellipsoid along two standard parallels (top left). The "developed" map surface is mathematically unrolled from the cone (top right). In a similar manner the TM typically intersects the ellipsoid near a central meridian (bottom left), and the flat map comes from "unrolling" the developable surface (bottom right). Distortion is illustrated in the developed surfaces by the deformation of the 5-degree diameter geographic circles and by the bent shape of the latitude/longitude lines.

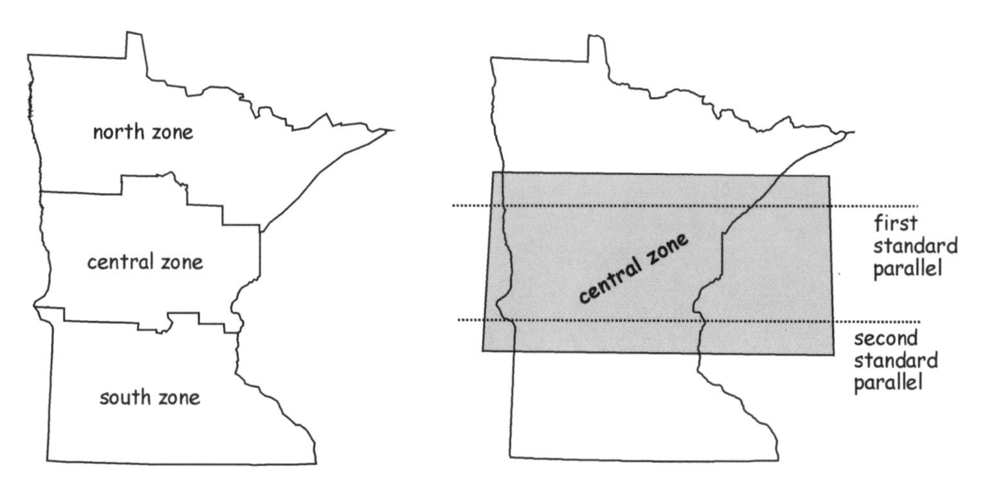

Figure 3-20: The State Plane zones of Minnesota, and details of the standard parallel placement for the Minnesota central State Plane zone.

horizontal coordinates over large areas while limiting error to specified maximum values. Zones are specified in each state such that projection distortions are kept below one part in 10,000. State Plane coordinate systems are used in many types of work, including property surveys, property subdivisions, large-scale construction projects, and photogrammetric mapping, and are often adopted for GIS.

One State Plane projection zone may suffice for small states. Larger states commonly require several zones, each with a different projection, for each of several geographic zones of the state. For example Delaware has one State Plane coordinate zone, while California has 7, and Alaska has 10 State Plane coordinate zones, each corresponding to a different projection within the state. Zones are added to a state to ensure acceptable projection distortion within all zones (Figure 3-20, left). Within each zone, the distance on the curving Earth surface differ by less than one part in 10,000 from the distance on the flat projection surface. Zones are defined by county, parish, or other municipal boundaries. For example, the Minnesota south/central zone boundary runs approximately east-west through the state along defined county boundaries (Figure 3-20, left).

The State Plane coordinate system is based on two basic types of map projections: the Lambert conformal conic and the transverse Mercator projections. Different projections are used for each state or sub-area within a state, and are chosen to minimize distortion within a given state and zone. Because distortion in a transverse Mercator increases with distance from the central meridian, this projection type is most often used with states that have a long north-south axis (e.g., Illinois or New Hampshire). Conversely, a Lambert conformal conic projection is most often used when the long axis of a state is in the east-west direction (e.g. North Carolina and Virginia). When computing the State Plane coordinates, points are projected from their geodetic latitudes and longitudes to x and y coordinates in the State Plane systems.

The Lambert conformal conic projection is specified in part by two standard parallels that run in an east-west direction. A different set of standard parallels are defined for each State Plane zone. These parallels are placed at one-sixth of the zone width from the north and south limits of the zone (Figure 3-20, right). The zone projection is defined by specifying the standard parallels and a central meridian that has a longitude near the center of the zone. This central meridian points in the direction of geographic north, however all other meridians converge to this central meridian, so they do not point to geographic north. The Lambert conformal conic is used to specify projections for State Plane zones for 31 states.

The transverse Mercator projection is used to define State Plane coordinates for states or zones where the long axis is north-south, such as Illinois. Distortion increases in an east-west direction when using a transverse Mercator. By centering the projection on "tall", narrow zones, we may avoid excessive distortion while requiring few zones. The transverse Mercator is used for 22 State Plane systems (the sum of states is greater than 50 because both the transverse Mercator and Lambert conformal conic are used in some states, e.g., Florida).

As noted earlier, the transverse Mercator specifies a central meridian. This central meridian defines grid north in the projection. A line along the central meridian points to geographic north, and specifies the Cartesian grid direction for the map projection. All parallels of latitude and all meridians except the central meridian are curved for a transverse Mercator projection, and hence these lines do not parallel the grid x or y directions.

Finally, note that more than one version of the State Plane coordinate system has been defined. Changes were introduced with the adoption of the North American Datum of 1983. Prior to 1983 the State Plane projections were based on the NAD27 datum. Changes were minor in some cases, and major in others, depending on the state and

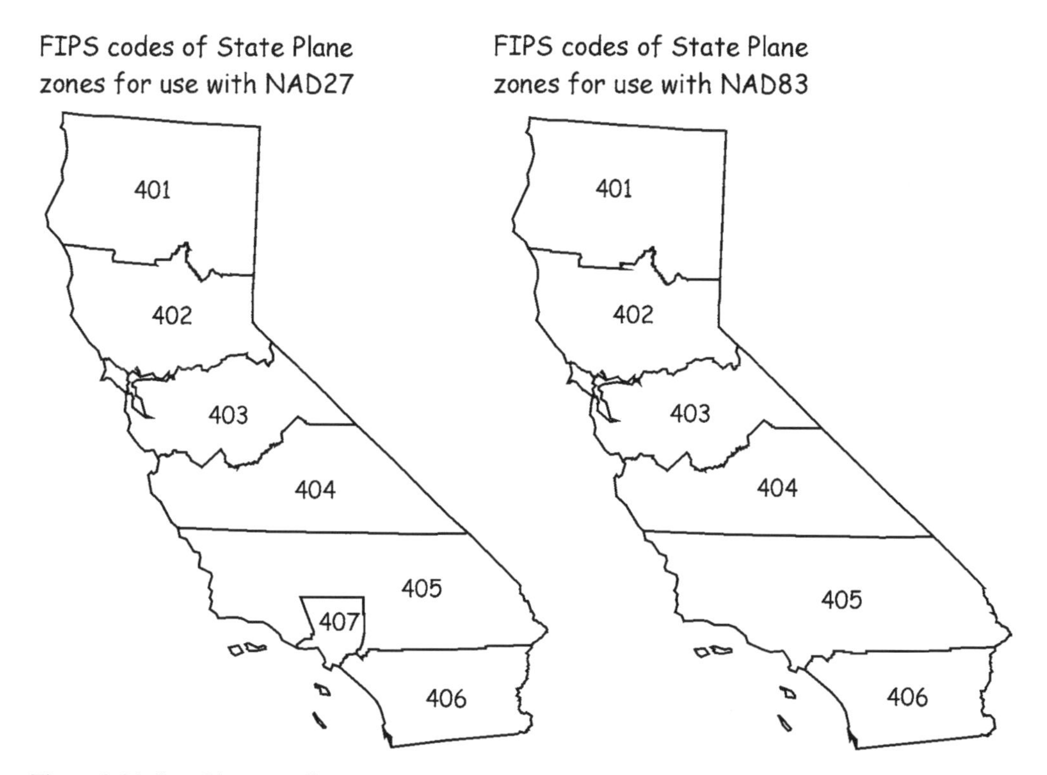

Figure 3-21: State Plane coordinate system zones and FIPS codes for California based on the NAD27 and NAD83 datums. Note the inclusion of zone 407 from NAD27 into zone 405 in NAD83.

State Plane zone. Some states, such as South Carolina, Nebraska, and California dropped zones between NAD27 and NAD83 versions of the State Plane coordinate system (Figure 3-21). Others maintained the same number of State Plane zones, but changed the projection by the placement of the meridians, or by switching to a metric coordinate system rather than one using feet, or by shifting the origin. State Plane zones are sometimes identified by the Federal Information Processing System (FIPS) codes, and most codes are similar across NAD27 and NAD83 versions of the State Plane coordinate system. Care must be taken when using older data to identify the version of the State Plane coordinate system used because the FIPS and State Plane zone designators may be the same, but the projection parameters may have changed between NAD27 and NAD83.

You might ask, how do I convert between geographic and State Plane systems, or between different versions of the State Plane system? Exact or approximate mathematical formulas have been developed to convert to and from geographic (latitude and longitude) to State Plane coordinates. Given a coordinate pair in the State Plane system, you can calculate the corresponding geographic coordinates. Conversely, given a geographic coordinate, you can calculate the appropriate State Plane coordinate. Since these formulas are known for both NAD27 and NAD83 versions of the State Plane system, we can convert among coordinate systems with ease. Any maps or measurements in the State Plane system can be converted to geographic coordinates and back. This facilitates conversion among State Plane versions, but also conversion to other, unrelated coordinate systems.

Universal Transverse Mercator Projection

Another standard coordinate system has been based on the transverse Mercator projection, distinct from the State Plane system. This system is known as the Universal Transverse Mercator (UTM) coordinate system. The State Plane system is defined only for the United States. The UTM is a global coordinate system. It is widely used in the U.S.A. and other parts of North America, and is also used in many other countries worldwide.

The UTM system divides the Earth into zones that are six degrees in longitude wide, and extend from 80 degrees south latitude to 84 degrees north latitude. UTM zones are numbered from 1 to 60 in an easterly direction, starting at longitude 180 degrees West (Figure 3-22). Zones are further split north and south of the equator. Therefore, the zone containing most of England is identified as UTM Zone 30 North, while the zones containing most of New Zealand are designated UTM Zones 59 South and 60 South. Directional designations are often abbreviated, e.g., 30N in place of 30 North.

The UTM coordinate system is common for data and study areas spanning large regions, e.g., several State Plane zones. Many data from U.S. federal government sources are in a UTM coordinate system because many agencies manage land spanning large areas, and the UTM is a well-known, standard system. Because the UTM zones are 6 degrees wide there are many large-area projects that would not fit in one State Plane zone but do fit in one UTM zone (Figure 3-23). Utah, Indiana, Alabama, and other states fit entirely into one UTM zone. Many other states are predominantly in one UTM zone, for example, most of Colorado is included in UTM zone 13 (Figure 3-23).

As indicated before, all regions for an analysis area must be in the same coordinate system if they are to be analyzed together. If not, the data will not co-occur as they should. The large width of the UTM zones accommodates many large-area analyses, and many states, national forests, or multi-county agencies have adopted the dominant UTM coordinate system as a standard.

We must note that the UTM coordinate system is not always compatible with regional analyses. Because coordinate values are discontinuous across UTM zone bound-

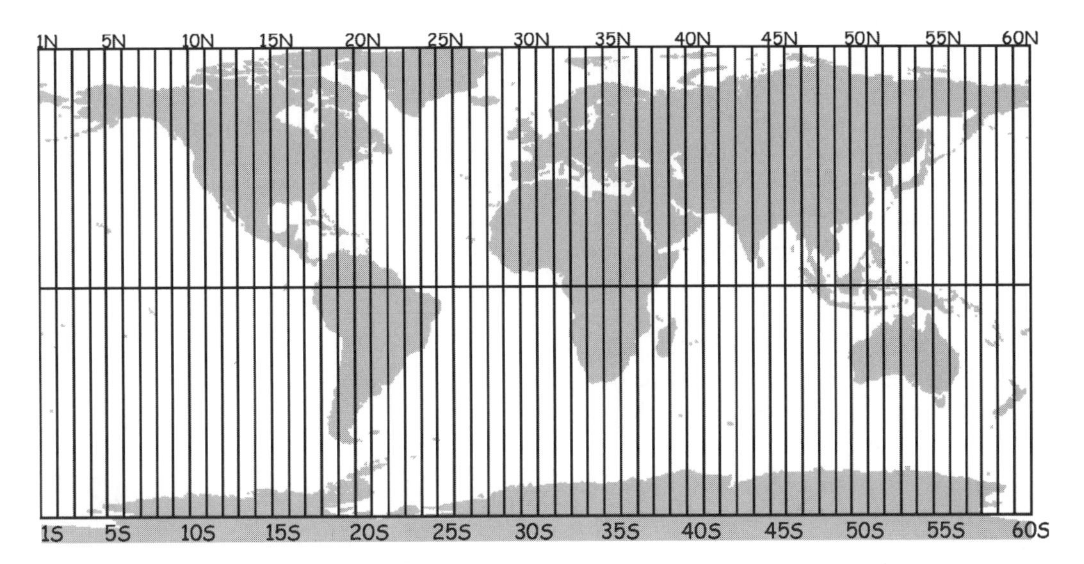

Figure 3-22: UTM zone boundaries and zone designators. Zones are six degrees wide and numbered from 1 to 60 from the International Date Line, 180°W. Zones are also identified by their position north and south of the equator, e.g., Zone 7 North, Zone 16 South.

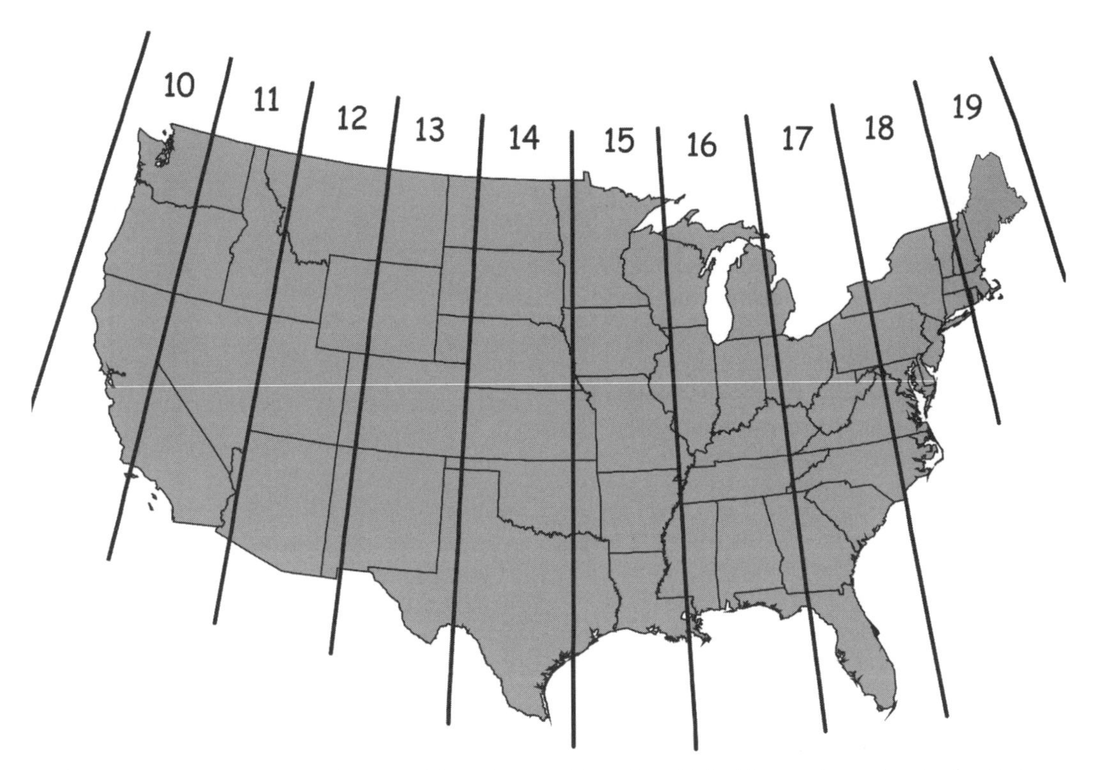

Figure 3-23: UTM zones for the lower 48 contiguous states of the United States of America. Each UTM zone is 6 degrees wide. All zones in the Northern Hemisphere are north zones, e.g., Zone 10 North, 11 North,...19 North.

aries, analyses are difficult across these boundaries. UTM zone 15 is a different coordinate system than UTM zone 16. The state of Wisconsin approximately straddles these two zones, and the state of Georgia straddles zones 16 and 17. If a uniform state-wide coordinate system is required, the choice of zone is not clear, and either one zone, or another, or some compromise zone must be chosen. For example, statewide analyses in Georgia and in Wisconsin are often conducted using UTM-like projections with central meridians running through the middle of each State, three degrees offset from the standard UTM system.

The UTM system specifies a unique transverse Mercator projection for each UTM zone. Each projection is centered on the zone central meridian and extends east and west to cover the six degree wide region (Figure 3-24). Distances in the UTM system are specified in meters north and east of a zone origin. The y values increases in a northerly direction (northings) and the x values increases in an easterly direction (eastings).

The origins of the UTM coordinate system are defined differently depending on whether the zone is north or south of the equator. In either case the UTM coordinate system is defined so that all coordinates are positive. Zone easting coordinates are all greater than zero because the central meridian for each zone is assigned an easting value of 500,000 meters. This effectively places the origin (E = 0) at a point 500,000 meters west of the central meridian. All zones are less than 1,000,000 meters wide, ensuring that all eastings will be positive.

The Equator is used as the northing origin for all north zones. Thus, the equator is assigned a northing value of zero for north zones. This avoids negative coordinates, because all of the UTM north zones are defined to be north of the equator. However, this will not work for UTM south zones. A zero value at the equator would yield negative northings, because the projection is set up with values increasing to the north. UTM south zone projections are defined with a *false northing* equal to 10,000,000 meters. This false northing may be considered the northing (N) coordinate value at the equator. The false northing is added to all computed values during the projection. This ensures that only non-negative coordinate values are observed for all northings in UTM south zones.

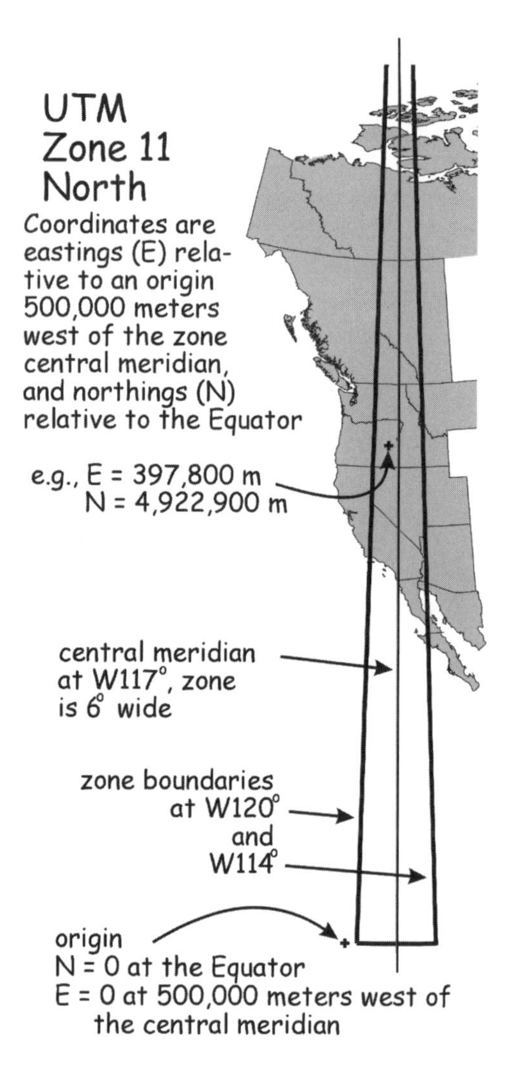

Figure 3-24: UTM zone 11N.

Continental and Global Projections

There are map projections that are commonly used when depicting maps of continents, hemispheres, or regions. Just as with smaller areas, map projections for continental or larger areas may be selected based on the distortion properties of the resultant map. Sizeable projection distortion in area, distance, and angle are observed in most large-area projections. Angles, distances, and areas are typically not measured or computed from these projections, as the differences between the map-derived and surface-measured values are too great for most uses. Large-area maps are most often used to display or communicate data for continental or global areas.

There are a number of projections that have been or are widely used for the World. These including variants of the Mercator, Goode, Mollweide, and Miller projections, among others. There is a trade-off that must be made in global projections, between a continuous map surface and distortion. If a single, uncut surface is mapped, then there is severe distortion in some portion of the map. Figure 3-25 shows a Miller cylindrical projection, often used in maps of the World. This projection is similar to a Mercator projection, and is based on a cylinder that intersects the Earth at the Equator. Distortion increases towards the poles, although not as much as with the Mercator.

Distortion in world maps may be reduced by using a cut or interrupted surface. Different projection parameters or surfaces may be specified for different parts of the globe. Projections may be mathematically constrained to be continuous across the area mapped.

Figure 3-26 illustrates an interrupted projection in the form of a Goode homolosine. This projection is based on a sinusoidal projection and a Mollweide projection. These two projection types are merged at the parallels of identical scale. The parallel of identical scale is set near the mid-northern latitude of 44° 40' N.

Continental projections may also be established. Generally, the projections are chosen to minimize area or shape distortion for the region to be mapped. Lambert conformal conic or other conic projections are often chosen for areas with a long east-west dimension, for example when mapping the

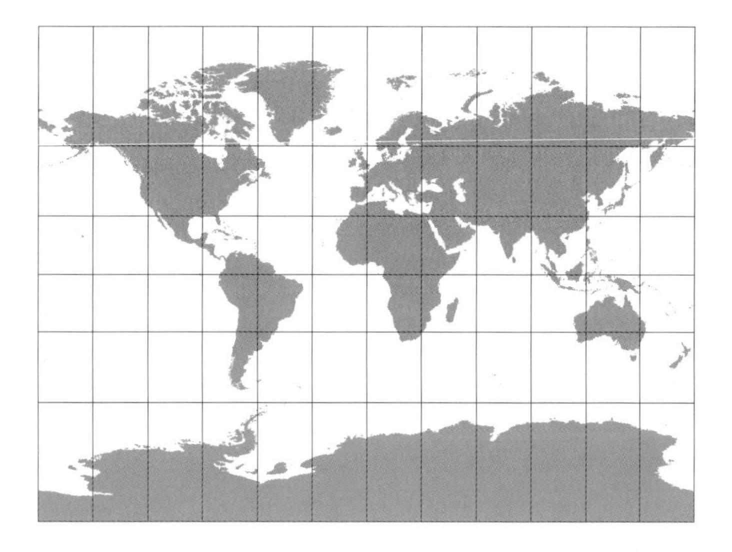

Figure 3-25: A Miller cylindrical projection, commonly used for maps of the World. This is an example of an uninterrupted map surface.

contiguous 48 United States of America, or North America (Figure 3-27). Standard parallels are placed near the top and bottom of the continental area to reduce distortion across the region mapped. Transverse cylindrical projections are often used for large north-south continents.

None of these worldwide or continental projections are commonly used in a GIS for data storage or analysis. Uninterrupted coordinate systems show too much distortion to be of use in measuring most spatial quantities, and interrupted projections do not define a Cartesian coordinate system that defines positions for all points on the Earth surface. Worldwide data are typically stored in geographic coordinates (latitudes and longitudes). These data may then be projected to a specific coordinate system for display or document preparation.

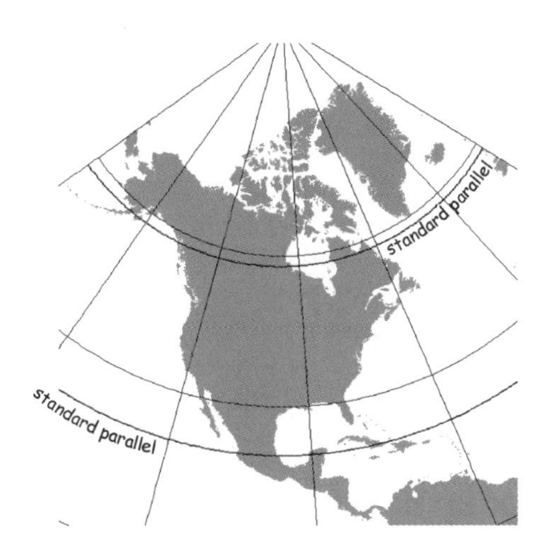

Figure 3-27: A Lambert conformal conic projection of North America. Standard parallels are placed to reduce distortion within the projected continent.

Summary

In order to enter coordinates in a GIS, we need to uniquely define the location of all points on Earth. We must develop a reference frame for our coordinate system, and locate positions on this system. Since the Earth is a curved surface and we work with

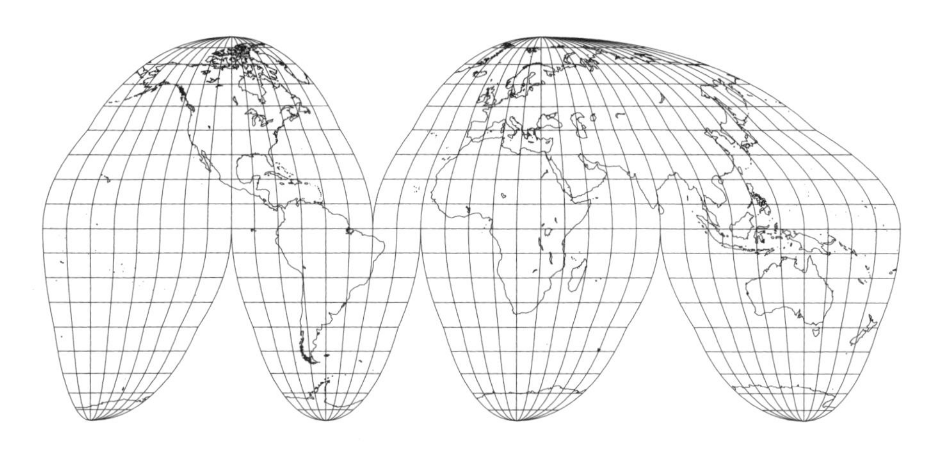

Figure 3-26: A Goode homolosine projection. This is an example of an interrupted projection, often used to reduce some forms of distortion when displaying the entire Earth surface. (from Snyder and Voxland, 1989)

flat maps, we must somehow reconcile these two views of the World. We define positions on the globe via geodesy and surveying. We convert these locations to flat surfaces via map projections.

We begin by modeling the Earth's shape with an ellipsoid. An ellipsoid differs from the Geoid, a gravitationally-defined Earth surface, and these differences caused some early confusion in the adoption of standard global ellipsoids. There is a long history of ellipsoidal measurement, and we have arrived at our best estimates of global and regional ellipsoids after collecting large, painstakingly-developed sets of precise surface and astronomical measurements. These measurements are combined into datums, and these datums are used to specify the coordinate locations of points on the surface of the Earth.

Map projections are a systematic rendering of points from the curved Earth surface onto a flat map surface. While there are many purely mathematical or purely empirical map projections, the most common map projections used in GIS are based on developable surfaces. Cones, cylinders, and planes are the most common developable surfaces. A map projection is constructed by passing rays from a projection center through both the Earth surface and the developable surface. Points on the Earth are projected along the rays and onto the developable surface. This surface is then mathematically unrolled to form a flat map.

Standard sets of projections are commonly used for spatial data in a GIS. In the United States, the UTM and State Plane coordinate systems define a standard set of map projections that are widely used. Other map projections are commonly used for continental or global maps, and for smaller maps in other regions of the World.

Suggested Reading

Brandenburger, A. J. and Gosh, S. K., The world's topographic and cadastral mapping operations, *Photogrammetric Engineering and Remote Sensing*, 1985, 51:437-444.

Colvocoresses, A.P., The gridded map, *Photogrammetric Engineering and Remote Sensing*, 1997, 63:371-376.

Doyle, F.J., Map conversion and the UTM Grid, *Photogrammetric Engineering and Remote Sensing*, 1997 63:367-370.

Iliffe, J. C., Datums and Map Projections for Remote Sensing, GIS, and Surveying, CRC Press, Boca Raton, 2000.

Keay, J., The Great Arc, Harper Collins, London, 2000.

Maling, D. H., Coordinate Systems and Map Projections, George Phillip, London, 1992.

Schwartz, C.R., North American Datum of 1983, NOAA Professional Paper NOS 2, National Geodetic Survey, Rockville, 1989.

Smith, J., Introduction to Geodesy: The History and Concepts of Modern Geodesy, Wiley, New York, 1997.

Sobel, D., Longitude, Penguin Books, New York, 1995.

Snyder, J., Flattening the Earth: Two Thousand Years of Map Projections, University of Chicago Press, Chicago, 1993.

Snyder, J. P., Map Projections, A Working Manual, USGS Professional Paper No. 1396, United States Government Printing Office, Washington D.C., 1987.

Snyder, J.P., and Voxland, P.M., An Album of Map Projections, USGS Professional Paper No. 1453, United States Government Printing Office, Washington D.C., 1989.

Tobler, W.R., A classification of map projections, *Annals of the Association of American Geographers*, 1962 52:167-175.

Welch, R., and Homsey, Datum shifts for UTM coordinates, *Photogrammetric Engineering and Remote Sensing*, 1997, 63:371-376.

Wolf, P. R., and Brinker, R.C., Elementary Surveying, 8th Ed., Harper and Row, New York, 1989.

Study Questions

Can you describe how Eratosthenes estimated the circumference of the Earth? What value did he obtain?

How did the method of Eratosthenes and Posidonius differ? Were their estimates similar, and how accurate were their estimates relative to current measurements?

What is an ellipsoid? How does an ellipse differ from a sphere? What is the equation for the flattening factor?

Why do different ellipsoids have different radii? Can you provide three reasons?

Can you define the Geoid? How does it differ from the ellipsoid, or the surface of the Earth? How do we measure the position of the Geoid?

Can you define a parallel or meridian in a geographic coordinate system? Where do the "horizontal" and "vertical" zero lines occur?

How does magnetic north differ from the geographic North Pole?

Can you define a datum? Can you describe how datums are developed?

Why are there multiple datums, even for the same place on Earth? Can you define what we mean when we say there is a datum shift?

What is a triangulation survey, a Bilby tower, and a benchmark?

What is a map projection? Can you define a gnomonic, stereographic, or orthographic projection?

What is a developable surface? What are the most common shapes for a developable surface?

Can you describe the State Plane coordinate system? What type of projections are used in a State Plane coordinate system?

Can you define and describe the Universal Transverse Mercator coordinate system? What type of developable surface is used with a UTM projection? What are UTM zones, where is the origin of a zone, and how are negative coordinates avoided?

4 Data Sources and Data Entry

Building a GIS Database

Introduction

Spatial data entry and editing are early and frequent activities for many GIS users. This is due to the inherently large number of coordinates that are required to represent features in a GIS. Data entry efforts are often large because feature geometry must be specified, and each coordinate value must be entered into the GIS database. Sometimes this entry is painstakingly slow, as when points are manually entered. But even with automated techniques and the most recent digital data entry methods, spatial data entry and editing takes significant time for most organizations.

There are many spatial data sources, and they are conveniently categorized as *hardcopy* and *digital* forms. Hardcopy forms are any drawn, written, or printed documents, including hand-drawn maps, manually measured survey data, legal records, and coordinate lists with associated tabular data.

These hardcopy data are an important source of geographic information for a number of reasons. First, maps are a valuable record of historical knowledge. Electronic computers were invented in the 1950s and nearly all geographic information produced prior to the 1960s was recorded in hardcopy form. Mapmaking has at least a 4000-year history, and during the 20th century there was a large increase in the number and coverage of *cartometric* quality maps. Cartometric maps are those that faithfully represent the relative position of objects and may be suitable for entry into a spatial database.

Historical and current photographs are also a valuable source of geographic data. Although photographs do not typically provide an orthographic (flat, undistorted) view, they are a rich source of geographic information, and standard techniques may be used to remove major systematic distortions. Surveyor's notes and coordinate lists may also provide positional information in a hardcopy format that may be entered into a GIS.

Digital forms of spatial data are those provided in a computer-compatible format. These include text files, lists of coordinates, digital images, and coordinate and attribute data in structured file formats. Previously digitized data are a very common source. These data may be from hardcopy maps that have already been converted to digital formats. Files and export formats may be used to transfer them to a local GIS system. The Global Positioning System (GPS), described in Chapter 5, is a direct measurement system which may be used to record coordinates in the field and report them directly into digital formats. Most modern surveying instruments also may be used to take direct measurements, reduce these measurements to coordinates using onboard computers, and output digital coordinate or attribute data in specific GIS formats. Finally, a number of digital image sources are available, e.g., satellite or airborne images which are collected in a digital raster format, or hardcopy aerial photographs that have been scanned to produce digital images.

Our objective in this chapter is to introduce the forms, methods, and equipment typically used in spatial data entry. We will also cover basic editing methods and data documentation.

Hardcopy Maps

Hardcopy maps and tables were the most common storage medium for spatial data until the widespread adoption of GIS in the 1980s. Prior to this time, nearly all spatial data were collected with the aim of recording the numerical coordinates on paper and/or plotting them on hardcopy maps. Maps were and still are a relatively stable, permanent, familiar, and useful way to summarize spatial data, and because hardcopy maps are a source of so much digital data, most GIS users should be familiar with basic map properties.

Most maps contain several components (Figure 4-1). A data area or pane occupies the largest part of the map, and contains most of the depicted spatial data. A neatline is often included to provide a frame around all map elements, and insets may contain additional map elements. Scalebars, legends, titles, and other graphic elements such as a north arrow are often included.

Many types of maps are produced, and the types are often referred to by the way features are depicted on the map. *Feature maps* are the simplest, in that they map points, lines, or areas and provide nominal information (Figure 4-2, upper left). A road may be plotted with a symbol defining the type of road or a point may be plotted indicating the location of a city center, but the width of the road or number of city dwellers are not provided in the shading or other symbology on the map. Feature maps are per-

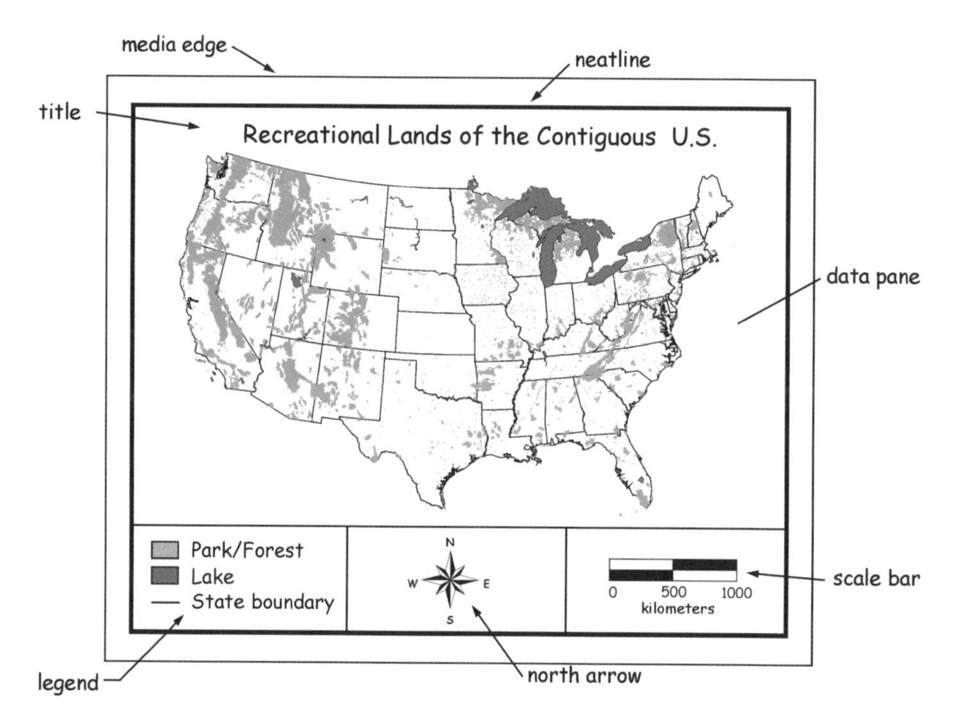

Figure 4-1: An example of a map and its components.

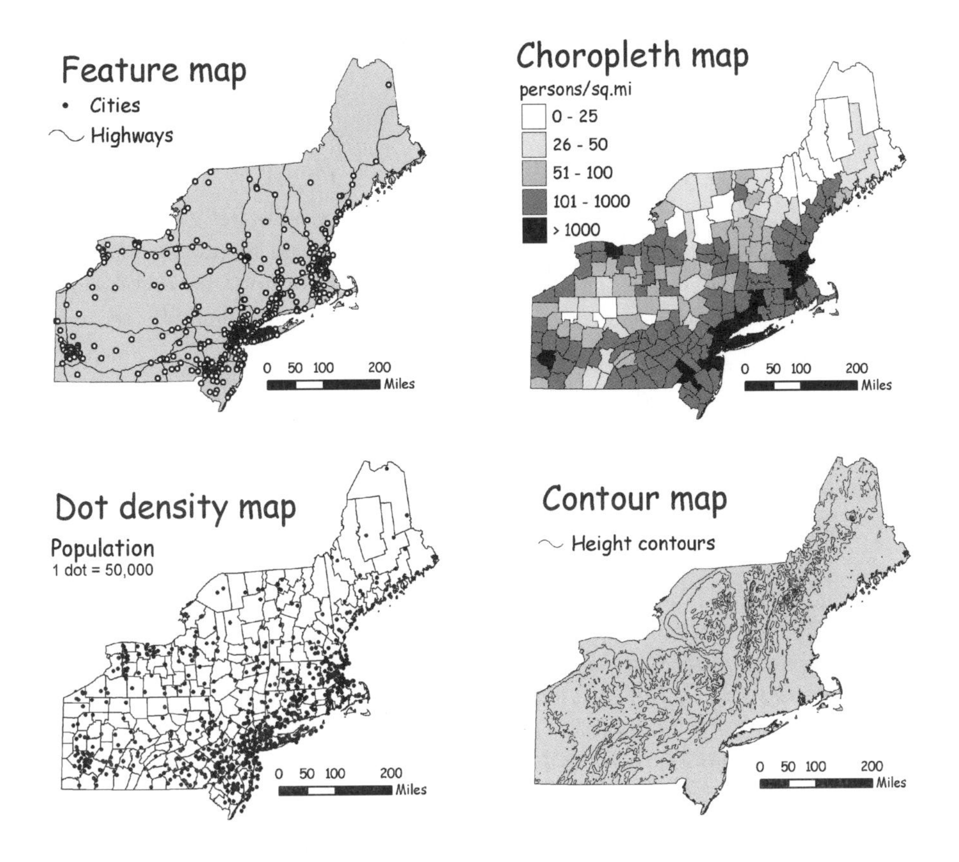

Figure 4-2: Common hardcopy map types depicting New England, in the northeastern United States.

haps the most common map form, and examples include standard map series such as the 7.5 minute topographic maps produced by the U.S. Geological Survey.

Choropleth maps depict quantitative information for areas. A mapped variable such as population density is represented in the map (Figure 4-2, top right). Polygons define area boundaries, such as counties, states, census tracts, or other standard administrative units. Each polygon is given a color or pattern corresponding to values for a mapped variable, e.g., in Figure 4-2, top right, the darkest polygons have a population density greater than 1000 persons per square mile.

Dot-density maps are another map form commonly used to show quantitative data (Figure 4-2, bottom left). Dots or other point symbols are plotted to represent values. Dots are placed in the polygon for which a quantity is to be depicted such that the number of dots equals the total value for the polygon. Note that the dots are typically placed anywhere in the polygon, and are typically spread randomly within the polygon area. Each dot on the map in the lower left of Figure 4-2 represents 50,000 people, however each point is not a city or other concentration of inhabitants. Note the position of points in the dot-density map relative to the city locations in the feature map directly above in Figure 4-2.

Isopleth maps, also known as *contour maps*, display lines of equal value (Figure 4-2, bottom right). Isopleth maps are used to represent continuous surfaces. Rainfall, elevation, and temperature are features that are commonly represented using isopleth maps. A line on the isopleth map represents a specified value, e.g., a 10°C isopleth defines the position on the landscape that is at that temperature. Lines typically do not cross, in that it cannot be two different temperatures at the same location. However, isopleth maps are commonly used to depict elevation, and cliffs or overhanging terrain do have multiple elevations at the same location. In this case the lower elevations typically pass "under" the higher elevations, and the isopleth is labeled with the tallest height. Note that the isopleths are typically interpolated surfaces and are not measured on the ground.

Not all maps are appropriate as a source of information for GIS. The type of map, how it was produced, and the intended purpose must be considered when interpreting the information on maps. Consider the dot-density map described above. Population is depicted by points, but the points are plotted with random offsets or using some method that doesn't reflect the exact location of the population within each polygon. In truth the population may be distributed across the polygon in isolated houses and small villages. Dot density maps use a point symbol to represent a value that is aggregated from the entire polygon. If we digitized the point locations of each dot when entering data into a GIS we would record unwarranted positional information in our data. The map should be interpreted correctly, in that the number of dots within a polygon should be counted, this number multiplied by the population per dot, and the population value associated with the entire polygon.

Map Scale

Most hardcopy maps have a fixed *map scale*. The scale may be defined as the ratio of a distance on the map to the equivalent distance on the Earth. For example, a 1:24,000-scale map indicates 1 inch on the map equals 24,000 inches on the Earth. Scales may be reported as a unitless ratio, as in the previous example, as a distance ratio (four inches to one mile), or as a graphical scalebar plotted on the map.

The notion of large vs. small scale is often confused because scale is often reported as a ratio. A larger ratio signifies a large-scale map, so a 1:24,000-scale map is considered large-scale relative to a 1:100,000-scale map. Many people mistakenly refer to a 1:100,000-scale map as larger scale than a 1:24,000-scale map because it covers a larger area. For example, a 1:100,000-scale map that is 20 inches on a side covers more ground than a 1:24,000-scale map that is 20 inches on a side. However, it is the size of the ratio or fraction, and not the area covered that determines the map scale. It is helpful to remember that features appear larger on a large-scale map (Figure 4-3). It is also helpful to remember that large scale maps show more detail. The larger the ratio (and smaller the denominator), the larger the map scale.

Table 4-1: The surface error caused by a one millimeter (0.039 inch) map error will change as map scale changes. Note the larger error at smaller map scales.

Map Scale	Error (m)	Error (ft)
1:24,000	24	79
1:50,000	50	164
1:62,500	63	205
1:100,000	100	328
1:250,000	250	820
1:1,000,000	1,000	3,281

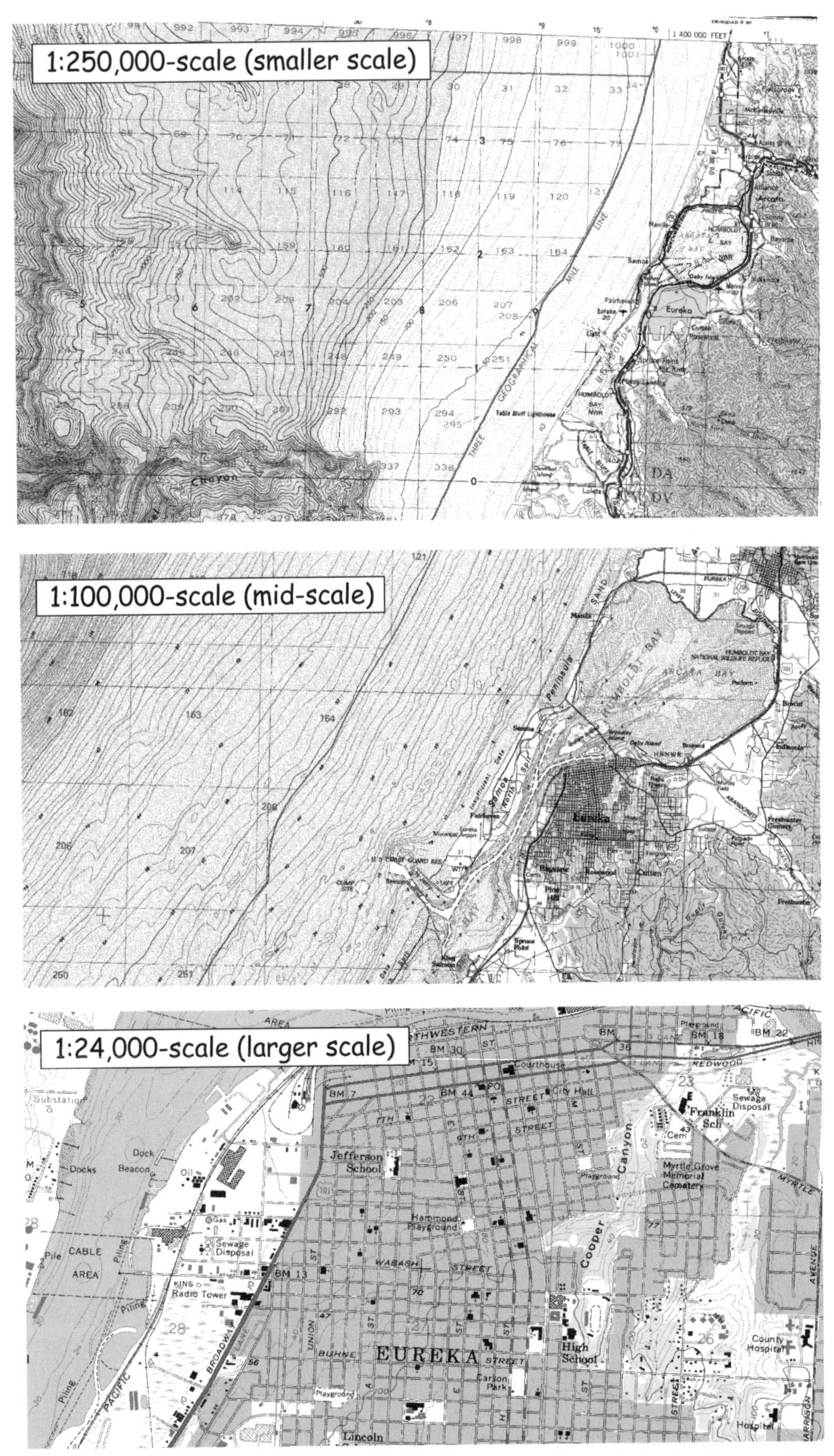

Figure 4-3: Coverage, relative distance, and detail change from smaller-scale (top) to larger scale (bottom) maps. Note the changes in the depiction of Eureka, the city at the right center of the top two panels.

The scale is effectively uniform over the entire area of most mid- to large-scale maps. Because maps have a fixed scale, and because there are upper limits on the accuracy with which data can be plotted on a map, large scale maps generally have less geometric error than small-scale maps if the same methods were used to produce them. Small errors in measurement, plotting, printing, and paper deformation are magnified by the scale factor. Thus, these errors which occur during map production are magnified more on a small-scale map than a large-scale map.

Table 4-1 illustrates the effects of map scale on data quality. Errors of one millimeter (0.039 inches) on a 1:24,000-scale map correspond to 24 meters (79 feet) on the surface of the Earth. This same one millimeter error on a 1:1,000,000-scale map correspond to 1000 meters (3281 feet) on the Earth surface. Thus, small errors or intentional offsets in map plotting or printing may cause significant positional errors when scaled to distances on the Earth, and these errors are greater for smaller-scale maps.

Map Generalization

Maps are abstractions of reality, as are spatial data in a GIS database. Not all the geometric or attribute detail of the physical world are recorded on a map, and only the most important characteristics are included. The set of features that are most important is subjectively defined and will differ among users. The mapmaker determines the set of features to place on the map, and selects the methods to collect and represent the shape and location of these features on the map. These choices unavoidably set limits on the size and shape of features that may be represented. Consider a lake mapping project that uses image data with a 250 meter cell resolution as the primary data source (Figure 4-4). The abstraction of the shoreline will not represent bays and peninsulas less than approximately 250 meters across, by conscious choice of the mapmakers.

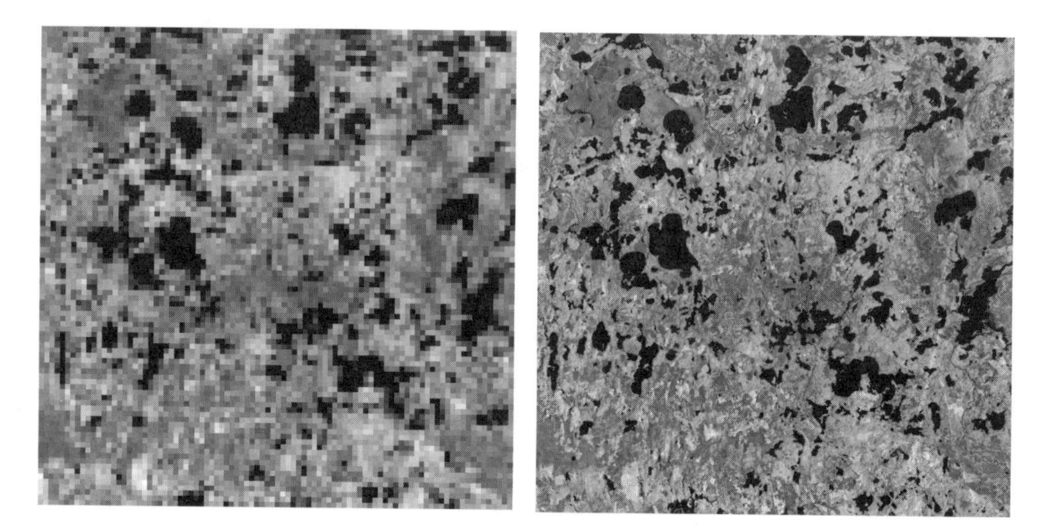

Figure 4-4: A mapmaker chooses the materials and methods used to produce a map, and so imposes a limit on spatial detail. Here, the choice of an input image with a 250 meter resolution (left) renders it impossible to represent all the details of the real lake boundaries (right). In this example, features smaller than approximately 250 meters on a side may not be faithfully represented on the map.

Feature generalization is one common form of abstraction. Generalization is a modification of features when representing them on a map. The geographic aspects of features are generalized because there are limits on the time, methods, or materials available when collecting geographic data. These limits also apply when compiling or printing a map. These generalizations, depicted in Figure 4-5, may be classed as:

Simplified: boundary or shape details are lost or "rounded off",

Omitted: Small features in a group may be excluded from the map,

Fused: multiple features may be grouped to form a larger feature,

Displaced: features may be offset to prevent overlap or to provide a standard distance between mapping symbols, or

Exaggerated: standard symbol sizes are often chosen, e.g., standard road symbol widths, which are much larger when scaled than the true road width.

Generalization is present at some level in every map, and should be recognized and evaluated for each map that is used as a source for data in a GIS (Figure 4-6). Large-scale maps typically cover smaller areas and show more detail. This usually results in less map generalization on large-scale maps, as shown by fewer omissions, less simplification, and fewer fused or aggregated features in larger-scale maps relative to smaller-scale maps. If generalization results in omission or degradation of data beyond acceptable levels, then the analyst or organization should switch to a larger-scale map if appropriate and available, or return to the field or original source materials to collect data of the requisite precision.

Map Media

Most maps have been printed on paper, and hence there are some errors because paper is not a dimensionally stable medium. Paper is ubiquitous, inexpensive, and easily printed. However paper is most often composed of a mat of wood fibers, and these fibers shrink and swell with changes in humidity. Because many fibers are oriented in a common direction and because fiber shrinkage or expansion is greater longitudi-

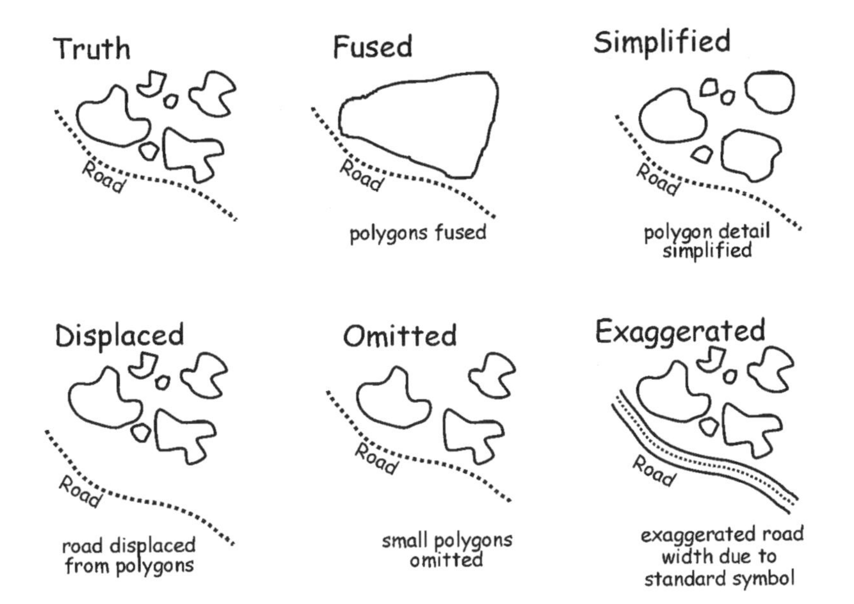

Figure 4-5: Generalizations common in maps.

Figure 4-6: Examples of map generalization. Portions are shown for three maps for an area in central Minnesota. A large scale (1:24,000), intermediate scale (1:62,500), and small scale (1:250,000) map are shown. **Note that the maps are not shown at true scale to facilitate comparison**. Each map has a different level of map generalization. Generalizations increase with smaller-scale maps, and include omissions of smaller lakes, successively greater road width exaggerations, and increasingly generalized shorelines.

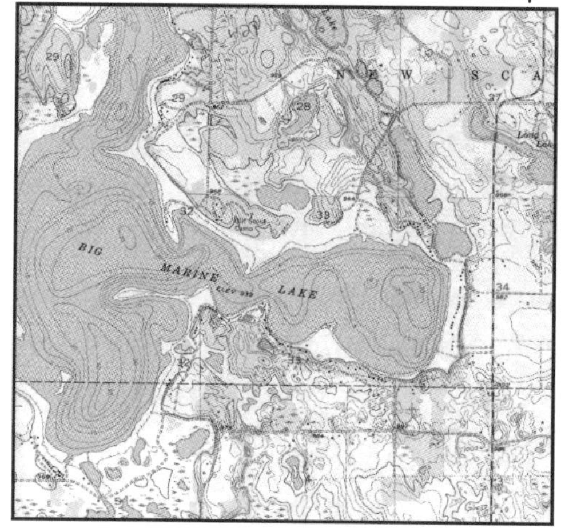

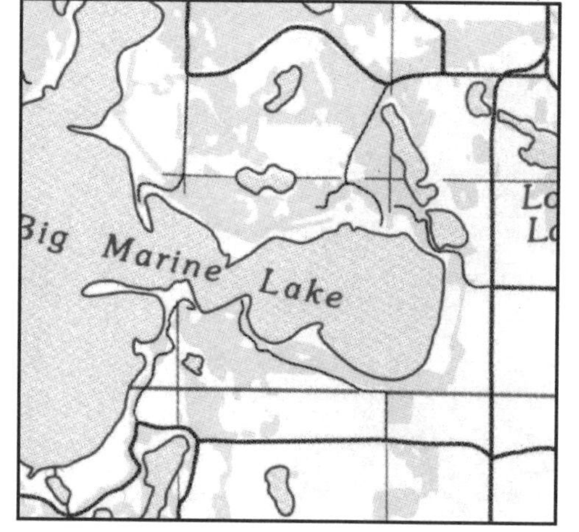

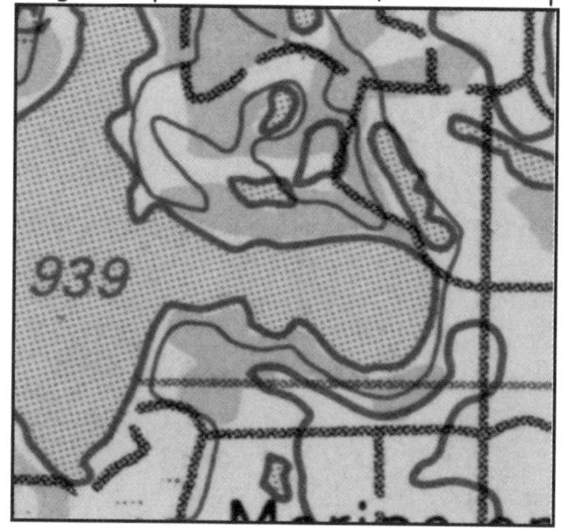

nally than laterally, the paper does not change shape uniformly. Under controlled environmental conditions this is often not a significant problem, because deformation is usually quite small. Shrinkage and swelling may be significant, particularly under extreme environmental conditions or when the maps are constrained in some manner. For example, when maps are manually converted to digital coordinates they may be taped to the rigid surface of a digitizing board or tablet. Changes in humidity may cause the maps to expand and "pucker", or rise off the digitizing surface, introducing horizontal uncertainty during digitization. Creases, folding, and wrinkling can also lead to non-uniform deformation of paper maps.

Maps are sometimes printed on dimensionally stable materials to avoid the deformations that occur when using paper. Plastic or other hydrocarbon-based media are sometimes used. These materials are highly resistant to expansion or contraction over broad ranges of temperature and humidity. They typically do not fold easily, although when crushed the materials will retain the deformation, as with paper. The materials are more likely to split or splinter than paper. However stable-base media are not commonly used because of higher costs relative to paper. These media are more expensive to produce per unit area due to lower production volumes. Printing on plastic or other dimensionally stable media is also more expensive because specialized inks and equipment are often required.

Map Boundaries and Spatial Data

One final characteristic of maps impacts their use as a source of spatial data: maps have edges, and discontinuities often occur at these edges. Large-scale, high-quality maps generally cover small areas. This is because of the trade-off between scale and area coverage, and because of limits on the practical size of a map. Cartometric maps larger than a meter in any dimension have proven to be impractical for most organizations. Maps above this size can be produced by relatively few printers, they are difficult to store without folding, and it is often difficult to find a flat surface that is large enough to display or unroll them. Thus, human ergonomics set a practical limit on the physical size of a map.

The fixed maximum map dimension when coupled with a fixed map scale defines the area coverage of the map. Larger scale maps generally cover smaller areas. A 1:100,000-scale map that is 18 inches (47 centimeters) on a side spans approximately 28 miles (47 kilometers). A 1:24,000-scale map that is 18 inches on a side represents 9 miles (15 kilometers) on the Earth surface. Because spatial data in a GIS often span several large-scale maps, these map boundaries may occur in the area of the spatial database. Problems may arise at these boundaries when maps are entered into a spatial database, or when maps of different scales are combined.

Differences in the time of data collection for adjacent map sheets may lead to inconsistencies across the border. Landscape change through time is one major source of differences across map boundaries. For example, the U.S.Geological Survey has produced 1:24,000-scale map sheets for all of the lower 48 United States of America. The original mapping took place over several decades, and there were inevitable time lags between mapping some adjacent areas. As much as two decades passed between mapping or updating adjacent map sheets. Thus, many created features, such as roads, canals, buildings, or municipal boundaries are discontinuous or inconsistent across map sheets.

Different interpreters may also cause differences across map sheet boundaries. Large-area mapping projects typically employ several interpreters, each working on different map sheets for a region. All professional, large-area mapping efforts should have protocols specifying the scale, sources, equipment, methods, classification, keys, and cross-correlation to ensure consistent mapping across map sheet boundaries. In spite of these efforts, some differences due to human interpretation occur. Feature placement, category assignment, and generalization vary among interpreters. These problems are compounded when extensive checking and guidelines are not enforced across map sheet boundaries, especially when adjacent areas are mapped at different times or by two different organizations.

Finally, differences in coordinate registration can lead to spatial mismatch across map sheets. *Registration*, discussed later in this chapter, is the process of converting digitizer or other coordinate data to an Earth surface coordinate system. These registrations contain unavoidable errors that translate into spatial uncertainty. There may be mismatches when data from two separate registrations are joined along the edge of a map sheet.

Spatial data stored in a GIS are not bound by the same constraints that limit the physical dimensions of hardcopy maps. Digital storage allows seamless digital maps of large areas. However, the inconsistencies that exist on hardcopy maps may only be transferred to the digital data. Inconsistencies at map sheet edges need to be identified and resolved when maps are converted to digital formats.

Digitizing: Manual Coordinate Capture

Digitizing is the process by which coordinates from a map, image, or other sources are converted into a digital format in a GIS. Points, lines, and areas on maps represent real-world entities or phenomena, and these must be recorded in digital forms before they can be used in a GIS. The coordinate values that define the locations and shapes of entities must be captured, that is, recorded as numbers and structured in the spatial database. There is a wealth of spatial data in existing maps and photographs, and new imagery and maps add to this source of information on a nearly continuous basis.

Manual Map Digitization

Manual digitizing is human-guided coordinate capture. An operator securely attaches a hardcopy map to a digitizing surface and traces lines or points with an electrically sensitized puck (Figure 4-7).The puck typically has cross-hairs and multiple input buttons. When a button is pressed, a signal is sent to the digitizing device to record a coordinate location. Points are captured individually. Line locations are recorded by tracing over the line, capturing coordinate locations along the line at frequent intervals so that the line shape is faithfully represented. Areas are identified by digitizing the coordinates for all bounding lines.

The most common digitizing devices are "digitizing tables" or "digitizing tablets" (Figure 4-8). The digitizing table typically has a hard, flat surface, although portable digitizing mats are available which are flexible enough to roll up and easily transport. Digitizer designs have employed several types of electrical and mechanical input devices, however most common designs are based on a wire grid embedded in or under a table. Depressing a button specifies the puck location relative to the digitizer coordinate system defined by the wire grid. The location of the puck at the time the button is pressed is determined and sent to the computer to be recorded. Often there are buttons to erase the last digitized point or perform other editing functions. Digitizing tables may be quite accurate, with a resolution of between 0.25 and 0.025 millimeters (0.01 and 0.001 inches). If a puck is held stationary and points captured repeatedly, they will differ by less than this resolution.

While manual digitizing can be slow, labor intensive, tedious, and inconsistent among human operators, manual digitizing is among the most common methods for hardcopy data entry. There are many reasons for this. Manual digitizing provides sufficiently accurate data for many, if not most, applications. Manual digitizing with precision digitizing equipment may record data to at least the accuracy of most maps, so the equipment, if properly used, does not add substantial error. Manual digitizing also requires lower initial capital outlays than

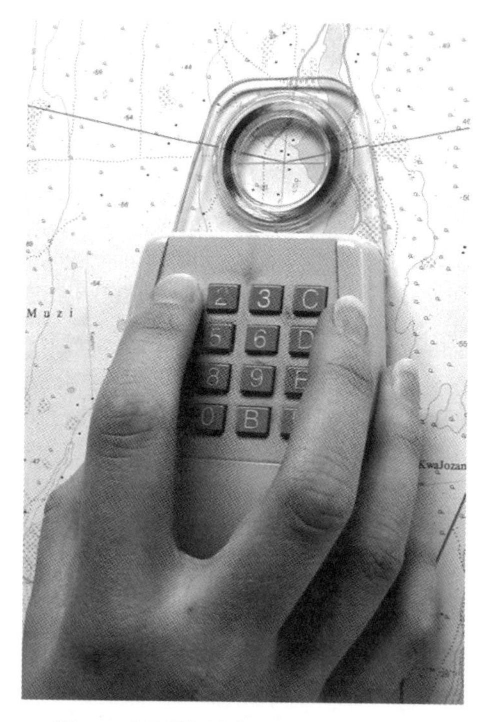

Figure 4-7: Digitizing puck.

Figure 4-8: Manual digitizing on a digitizing table.

most alternative digitizing methods, particularly for larger maps. Not all organizations can afford the high cost of precise, large-format map scanners, or digitize enough maps to justify the cost of purchasing such a scanner. Another limitation is the condition of the source material. Humans are usually better than machines at interpreting the information contained in faded, multicolor, or poor quality maps. Finally, manual digitizing is often best because short training periods are required, data quality may be frequently evaluated, and digitizing tablets are commonly available. For these reasons manual digitization is likely to remain an important data entry method for some time to come.

There are a number of characteristics of manual digitization that may negatively affect the positional quality of spatial data. As described earlier, map scale impacts the spatial accuracy of digitized data. Data collected from small-scale maps typically contain larger positional errors than data collected from large-scale maps.

Equipment characteristics also affect data accuracy. There is an upper limit on the precision of each digitizing tablet, and tablet precision reflects the digitizer resolution. Precision may be considered the minimum distance below which points cannot be effectively digitized as separate locations. The precision is often reported as a repeatability: how close points are clustered when the digitizing puck is not moved. Although these points should be placed at the same location, many are not. There will be some variation in the position reported by the electronic or mechanical position sensors, and this affects the digitizing accuracy.

Both device precision and maps scales should be considered when selecting a digitizing tablet. Map scale and repeatability both set an upper limit on the positional quality of digitized data. The most precise digitizers may be required when attempting to meet a stringent error standard while digitizing small-scale maps.

The abilities and attitude of the person digitizing (the "operator") may also affect the geometric quality of manually digitized data. Operators vary in their visual acuity, steadiness of hand, attention to detail, and ability to concentrate. Hence, some operators will more accurately capture the coordinate information contained in maps. The abilities of any single operator will also vary through time, due to fatigue or difficulty maintaining focus on a repetitive task. Frequent breaks from digitizing, comparisons among operators, and quality and consistency checks should be integrated into any manual digitization process to ensure accurate and consistent data collection.

The combined errors from both operators and equipment have been well-characterized and may be quite small. One test using a high-precision digitizing table revealed digitizing errors averaging approximately 0.067 millimeters (Figure 4-9). Errors followed a random normal distribution, and varied significantly among operators. These average errors translated to approximately 1.6 meter error when scaled from the 1:24,000 map to a ground-equivalent distance. This average error is less than the acceptable production error for the map, and is suitable for many spatial analyses.

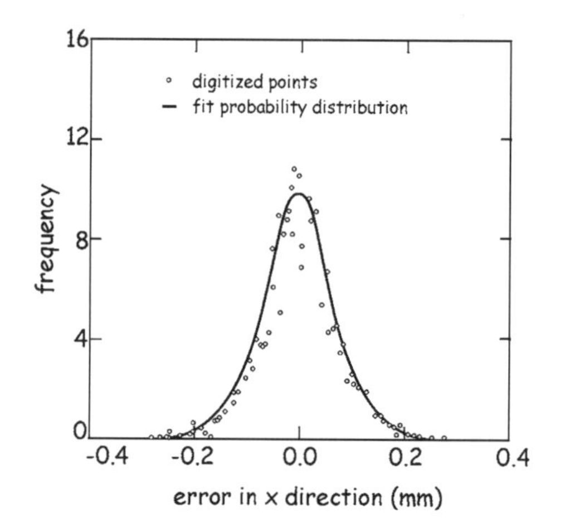

Figure 4-9: Digitizing error, defined by repeat digitizing. Points repeatedly digitized cluster around the true location, and follow a normal probability distribution. (from Bolstad et al., 1990)

The Manual Digitizing Process

Manual digitizing involves placing a map on a digitizing surface and tracing the location of feature boundaries. The map is securely fixed to the digitizing surface so that it will not move during digitizing. Maps may be taped to the surface, usually attached each corner and each edge. Typically, one corner is taped and the map smoothed by hand; a slight downward pressure is applied to the map while moving the hand from the taped edge to the opposite corner. This opposite corner may then be taped, and the map smoothed from the middle outwards to the remaining opposing corners. Opposing edges are then taped in a similar manner. This taping sequence ensures a secure map surface, important because even small shifts of the map during digitizing can result in large errors that are difficult to remove. As an example, consider the error introduced with a shift of 2.5 millimeters (0.1 inches) while digitizing a 1:100,000-scale map. This shift would result in an error equal to 250 meters (800 feet) measured on the Earth surface. The Earth surface coordinate error would be less for an equivalent map shift when digitizing from a larger scale map, however it may still be quite large even when digitizing from a 1:24,000-scale map or larger, underscoring the need to firmly fix the map to the digitizing surface.

Digitizing tablets may include a mechanism for securing the map. Some tablets are built with a transparent plastic mat attached along an edge of the digitizing surface. The plastic mat may be lifted off of the surface,

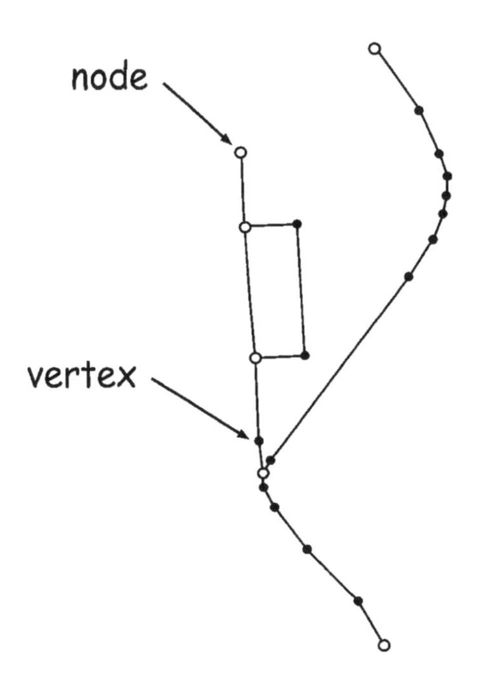

Figure 4-10: Nodes define the starting and ending points of lines. Vertices define line shape.

ing the line shape, and an ending *node* (Figure 4-10). Hence, lines may be viewed as a series of straight line segments connecting vertices and nodes.

Digitizing may be in *point mode*, where the operator must depress a button or otherwise signal to the computer to sample each point, or in *stream mode*, where points are automatically sampled at a fixed time or distance frequency, e.g., once each second. Stream mode is not appropriate when digitizing point features, because it is usually not possible to find and locate points at a uniform rate. Stream mode may be advantageous when large numbers of lines are digitized, because points may be sampled more quickly and there may be less operator fatigue. The stream sampling rate must be specified with care to avoid over- or undersampled lines. Too rapid a collection frequency results in redundant points not needed to accurately represent line or polygon shape (Figure 4-11). Too slow a collection frequency in stream mode digitizing

the map placed on the tablet, and the mat placed back down into position. The mat then holds the map securely to the surface. Other digitizing tablets are built with a dense pattern of small perforations in the table surface. A pump creates a partial vacuum just below the tablet surface. This pressure causes a suction at each perforation, pulling the map down onto the digitizing table, and ensuring the map does not move during digitizing.

Coordinate data are sampled by manually positioning the puck over each target point and collecting coordinate locations. This position/collect step is repeated for every point to be captured, and in this manner the locations and shapes of all required map features defined. Features that are viewed as points are represented by digitizing a single location. Lines are represented by digitizing an ordered set of points, and polygons by digitizing a connected set of lines. Lines have a starting point, often called a *starting node*, a set of *vertices* defin-

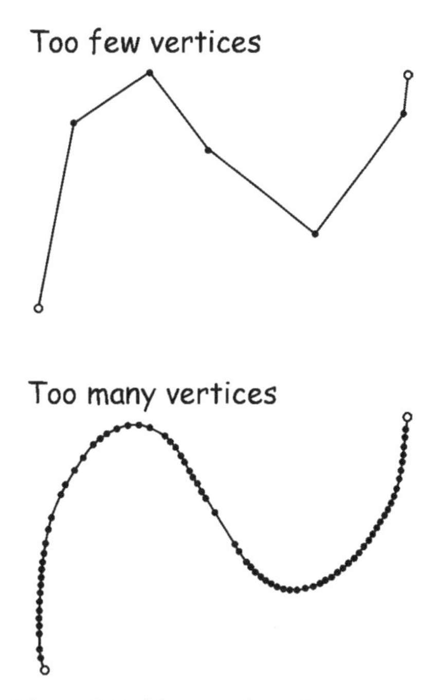

Figure 4-11: Lines may be under- or over-sampled during digitizing, resulting in too few or too many vertices.

may result in the loss of important spatial detail. In addition, when using time-triggered stream digitizing the operator must remember to continuously move the digitizing puck; if the operator rests the digitizing puck for a period longer than the sampling interval there will be multiple points that redundantly represent a portion of the line. Point mode digitizing allows the operator to specify the location of each point, vertex, and node, and hence precisely control the sampling frequency.

Digitizing Errors, Node and Line Snapping

Positional errors are inevitable when data are manually digitized. These errors may be "small" relative to the intended use of the data, for example the positional errors may be less than 2 meters when only 5 meter accuracy is required. However these relatively small errors may still cause problems when utilizing the data. These small errors may prevent the generation of correct networks or polygons. For example, a data layer representing a river system may not be correct because major tributaries may not connect. Polygon features may not be correctly defined because their boundaries may not completely close. These small errors must be removed or avoided during digitizing. Figure 4-12 shows some common digitizing errors.

Undershoots and *overshoots* are common errors that occur when digitizing. Undershoots are nodes that do not quite reach a node or line, and overshoots are lines that cross over existing nodes or lines (Figure 4-12). Undershoots cause unconnected networks and unclosed polygons in the examples cited above (Figure 4-12). Overshoots typically do not cause problems when defining polygons, but overshoots may cause difficulties when defining and analyzing line networks.

Node snapping and *line snapping* are used to reduce undershoots and overshoots while digitizing. Snapping relies on a *snap tolerance* or *snap distance.* This distance may be interpreted as a minimum distance, within which nodes or vertices are considered to occupy the same location (Figure 4-

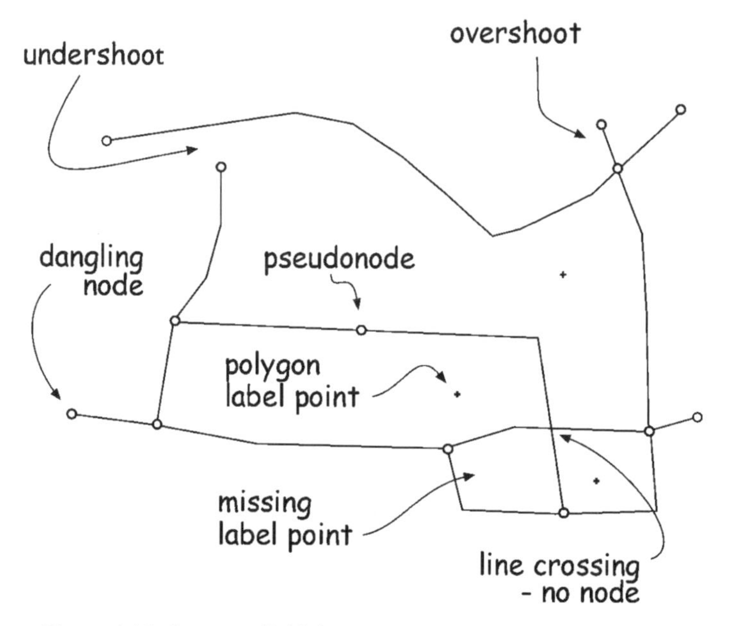

Figure 4-12: Common digitizing errors.

13). Node snapping prevents a new node from being placed within the snap distance of an already existing node; instead, the node is joined or "snapped" to the existing node (Figure 4-13). Remember that nodes are used to define the ending points of a line. By snapping two nodes together, we ensure a connection between digitized lines. *Line snapping* may also be specified. Line snapping inserts a node at a line crossing and clips the end when a small overshoot is digitized. Line snapping forces a node to connect a nearby line while digitizing, but only when the undershoot or overshoot is less than the snapping distance. Line snapping requires the calculation of an intersection point on an already existing line. The snap process places a new node at the intersection point, and connects the digitized line to the existing line at the intersection point. This splits the existing line into two new lines. When used properly, line and node snapping reduce the number of undershoots and overshoots. Closed polygons or intersecting lines are easier to digitize accurately and efficiently when node and line snapping are in force.

The snap distance must be carefully selected for snapping to be effective. If the snap distance is too short, then snapping has little effect. Consider a system where the operator may digitize with better than 5 meter accuracy only 10% of the time. This means 90% of the digitized points will be more than 5 meters from the intended location. If the snap tolerance is set to the equivalent of 0.1 meters, then very few nodes will be within the snap tolerance, and snapping has little effect. Another problem comes from setting the snap tolerance too large. If the snap tolerance in our previous example is set to 100 meters, and we want the data accurate to the nearest 5 meters, then we may lose significant spatial information that is contained in the hardcopy map. Lines less than 100 meters apart cannot be digitized as

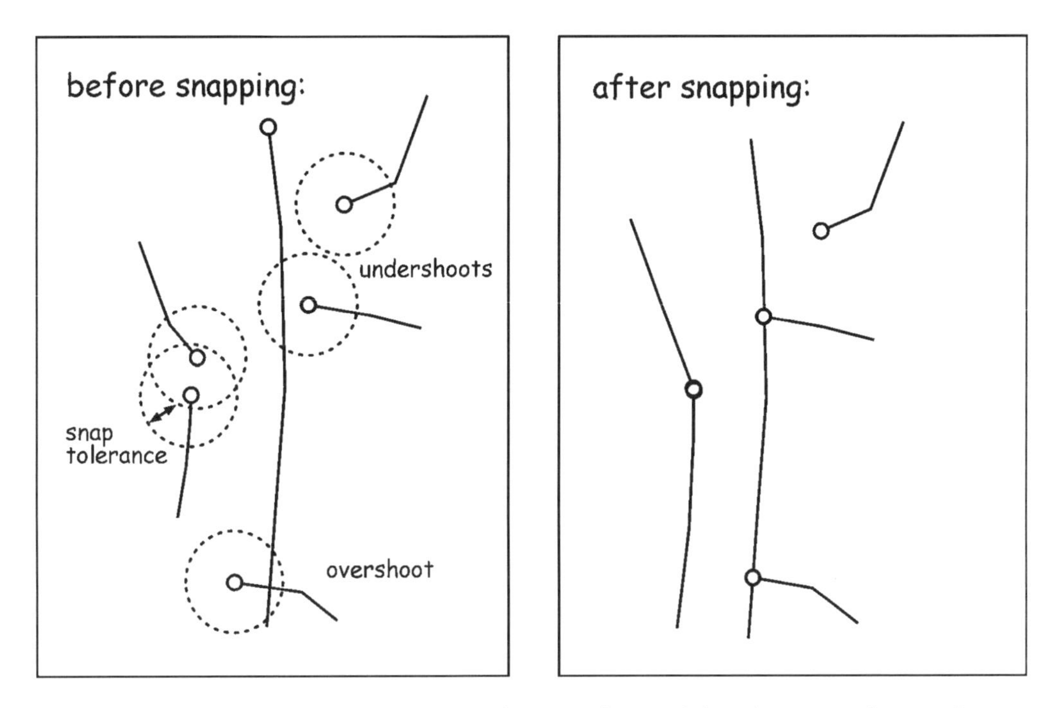

Figure 4-13: Undershoots, overshoots, and snapping. Snapping may join nodes, or may place a node onto a nearby line segment. Snapping does not occur if the nodes and/or lines are separated by more than the snap tolerance.

separate objects. Many features may not be represented in the digital data layer. The snap distance should be smaller than the desired positional accuracy, and such that significant detail contained in the digitized map is recorded. However the snap distance should not be below the capabilities of the system and operator used for digitizing. Careful selection of the snap distance may reduce digitizing errors and significantly reduce time required for later editing.

Reshaping: Line Smoothing and Thinning

Digitizing software may provide tools to smooth, densify, or thin points while entering data. One common technique uses *spline* functions to smoothly interpolate curves between digitized points and thereby both smooth and densify the set of vertices used to represent a line. A spline is a connected set of polynomial functions with constraints between functions (Figure 4-14). Polynomial functions are fit to successive sets of points along the vertices in a line, for example, a function may be fit to points 1 through 5, and a separate polynomial function fit to points 5 through 11 (Figure 4-14). A constraint may be placed that these functions connect smoothly, usually by requiring the first and second derivatives of the functions be continuous at the intersection point. This means the lines have the same slope at the intersection point, and the slope is changing at the same rate for both lines at the intersection point. Once the spline functions are calculated they may be used to add vertices, for example, several new vertices may be automatically interpolated on the line between digitized vertices 8 and 9, leading to the "smooth" curve shown in Figure 4-14.

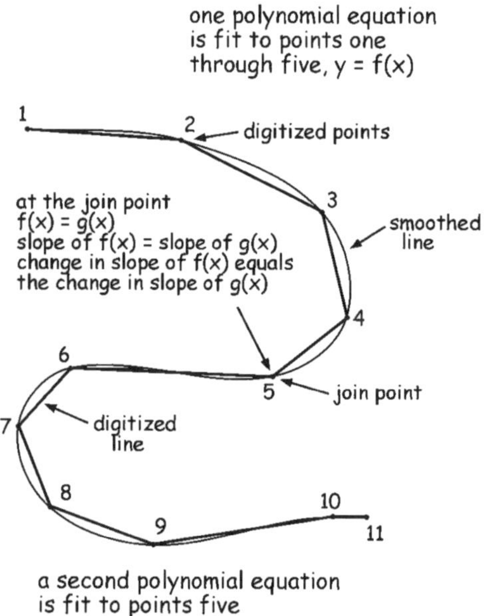

Figure 4-14: Spline interpolation to smooth digitized lines.

Data may also be digitized with too many vertices. High densities may occur when data are manually digitized in stream mode, and the operator pauses or moves slowly relative to the time interval. High vertex densities may also be found when data are derived from spline or smoothing functions that specify too high a point density. Finally, automated scanning and then raster-to-vector conversion may result in coordinate pairs spaced at absurdly high densities. Many of these coordinate data are redundant and may be removed without sacrificing data accuracy. Too many vertices may be a problem in that they slow processing, although this has become less of an issue as computing power has increased. Point thinning algorithms have been developed to reduce the number of points while maintaining the line shape.

Many point thinning methods use a perpendicular "weed" distance, measured from a spanning line, to identify redundant points (Figure 4-15). The Lang method exemplifies this approach. A spanning line connects two non-adjacent vertices in a line. A pre-determined number of vertices are initially spanned. The initial spanning number has been set to 4 in the Figure 4-15, meaning four points will be considered at each starting point. Areas closer than the weed distance are shown in gray in the figure. A straight line is drawn between a starting point and an endpoint that is the 4th point down the line (Figure 4-15a). Any intermediate points that are closer than the weed distance are marked for removal. In Figure 4-15a no points are within the weed distance, therefore none are marked. The endpoint is then moved to the next closest remaining point (Figure 4-15b), and all intermediate points tested for removal. Again, any points closer than the weed distance are marked for removal. Note that in Figure 4-15b one point is within the weed distance, and is removed. Once all points in the initial spanning distance are checked, the last remaining endpoint becomes the new starting point, and a new spanning line drawn to connect 4 points (Figure 4-15c,d). The process may be repeated for successive sets of points in a line segment until all vertices have been evaluated (Figure 4-15e to h). All close vertices are viewed as not recording a significant change in the line shape, and hence are expendable. Increasing the weed distance thins more vertices, and at some upper weed distance too many vertices are removed. A balance must be struck between the removal of redundant vertices and the loss of shape-defining points, usually through a careful set of test cases with successively larger weed distances.

There are many variants on this basic concept. Some look only at three immediately adjacent points, testing the middle point against the line spanned by its two neighboring points. Others constrain or expand the search based on the complexity of the line. Rather than always looking at four points, as in our example above, more points are scrutinized when the line is not complex (nearly straight), and fewer when the line is complex (many changes in direction).

Global methods, such as the Douglas-Peucker algorithm, begin by using all vertex points in a test (Figure 4-16). The Douglas-Peuker method splits a line recursively until lines can get no smaller. Initially, the first and last points in a line are spanned (Figure 4-16a), and if all intermediate points are

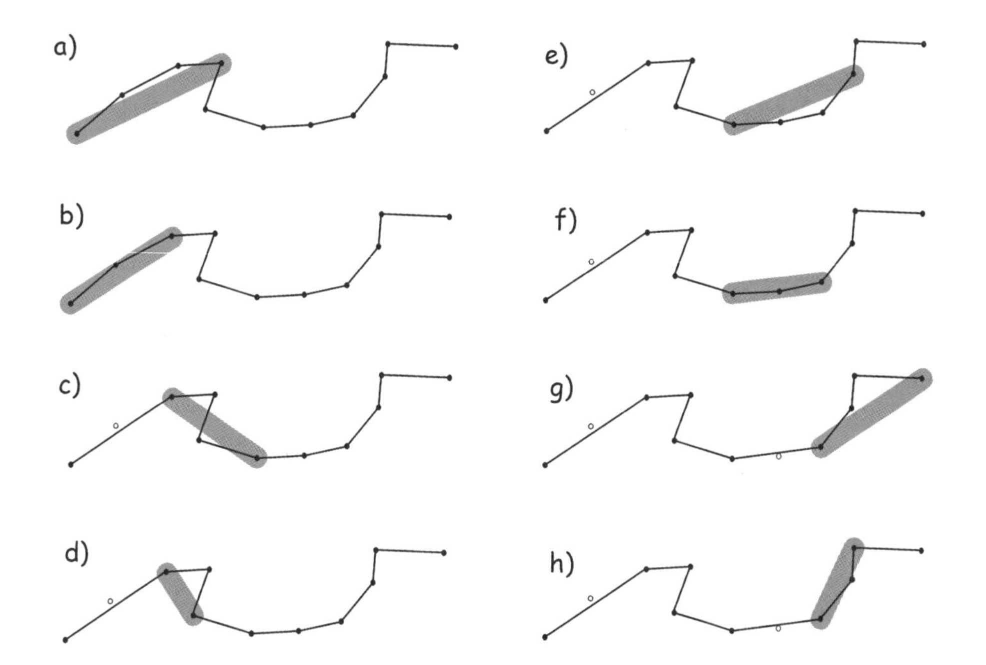

Figure 4-15:The Lang algorithm, a common line-thinning method. Vertices are removed, or thinned, when they are within a weed distance to a spanning line. Thinned points are shown as open circles (adapted from Weibel, 1997).

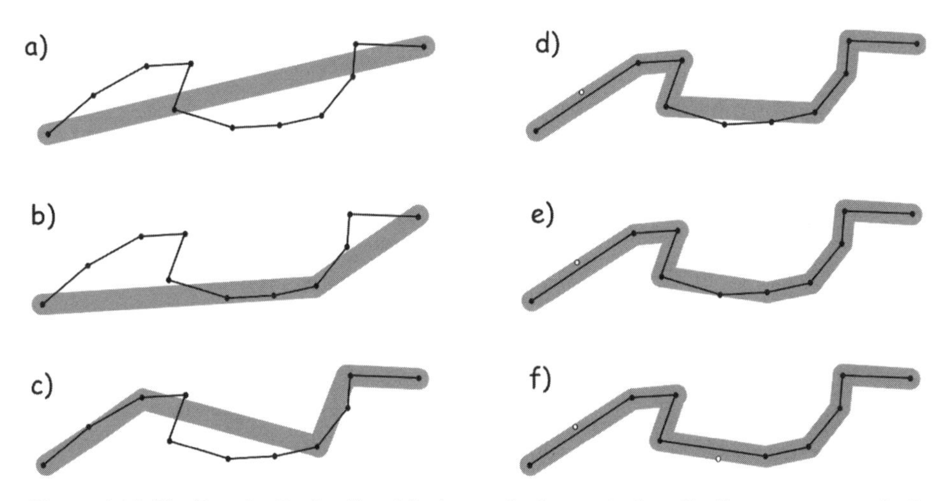

Figure 4-16: The Douglas-Peuker line-thinning method recursively splits line segments until all intervening points are within the weed distance, or there are no intermediate points (adapted from Weibel, 1997).

within the weed distance, then they are all deleted. If there is at least one point further than the weed distance, the line is split into two lines (Figure 4-16b). The split occurs at the point that is farthest away from the initial spanning line. The algorithm is then applied to the two new line segments (Figure 4-16c). Spanning lines are drawn from the first to last points for each new line segment. Again, the rule is applied: if all points for a new line segment are within the weed distance, delete them, or else split this new line into two smaller lines. This process is repeated for each succeedingly smaller line segment until each line segments can get no smaller (Figure 4-16d to f).

Scan Digitizing

Optical scanning is another method to convert hardcopy documents into digital formats. Scanners have light emitting and sensing elements. Most scanners pass a sensing element over the map. This sensing element measures both the precise location of the point being sensed and the strength of the light reflected or transmitted from that point. Reflectance values are converted to numbers.

A threshold is often applied to determine if the sensed point is part of a feature to be recorded. For example, a map may consist of dark lines on a white background. A threshold might be set such that if less than 10% of the light striking the map is returned to the sensor, the sensed point is considered part of a line. If 10% or more of the energy is reflected back to the sensor, the point is considered part of the white space between lines. The scanner then produces a raster representation of the map. Values are recorded where points or lines exist on the map and null or zero values are recorded in the intervening spaces.

Most scanners are either bed or drum type. Bed scanners provide a flat surface on which the map is placed (Figure 4-17). A mat or hinged cover is then placed on top of the map, flattening and securing the map to the bed. On some bed scanners an optical train is passed over the map, emitting light and sensing the light reflected back from the map. Sensing arrays are typically used to measure the reflectance so that one to several rows of cells may be scanned simultaneously. A motor then moves the optical train to the adjacent lines and the process is repeated. Positional accuracy depends on an optical device in one direction (along the sensing array) and on a mechanical device in the other direction (as the optical head travels down the scanning bed). Positional accuracy is generally greater in the direction of the sensing array. Significant distortion may be introduced by the motor or other drive mechanism, due to variations in the scanning speed or orientation of the optical train. These errors must be considered when selecting a flatbed scanning device.

Drum scanners differ from flatbed scanners in that they employ a rotating cylinder. A map is fixed onto the surface of a cylinder, and the cylinder set to rotate at a uniform velocity. The angular velocity of a rotating cylinder is easier to control than the straight-line motion of a bed scanner, so many of the early high-precision scanners used drums. Many drum scanners are similar to bed scanners in that they use optical detection of reflected light to sense map elements. Alternative designs are also available, using light sensitive lines or arrays on a drum. Sensors are pointed at the rotating drum and moved down the drum in a set of steps. Output is sensed and stored in a raster data set as with flatbed scanners.

Scanners work best when very clean map materials are available. Even the most expensive scanners may report a significant number of spurious lines or points when old, marked, folded, or wrinkled maps are used. These spurious features must be subsequently removed via manual editing, thus

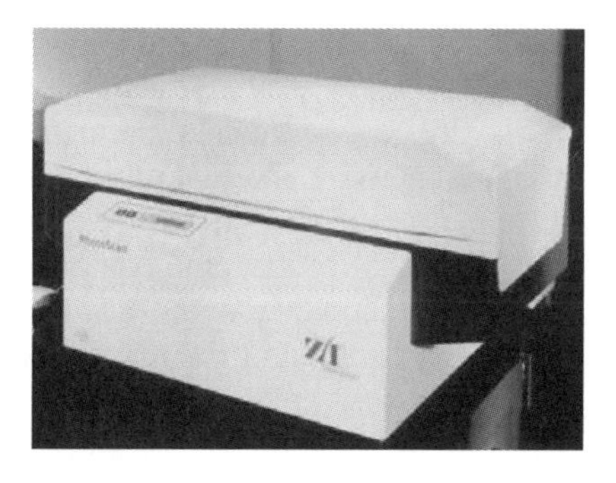

Figure 4-17: A flatbed scanner.

negating the speed advantage of scanning over manual digitizing. Scanning also works best when maps are available as map separates, with one thematic feature type on each map. Scan editing takes less time when maps do not contain writing or other annotation. Strongly contrasting colors are preferred, e.g. black lines on a white background, rather then dark grey on light grey. Finally, scanning is most advantageous when a large number of cartographic elements are found on the maps.

Scan digitization usually requires some form of *skeletonizing*, particularly if the data are to be converted to a vector data format. Scanned lines are often wider than a single pixel (Figure 4-18). One of several pixels may be selected to specify the position of a given portion of the line. The same holds true for points. A pixel near the "center" of the point or line is typically chosen, with the center of a line defined as the pixel nearest the center of the local perpendicular bisector of the line. Skeletonizing reduces the widths of lines or points to a single pixel.

On-screen Digitizing

On-screen digitizing, also known as heads-up digitizing, may be considered a combination of manual digitizing and scanning. It involves manually digitizing on a computer screen, using a scanned map or image as a backdrop. Digitizing software allows the operator to trace the points, lines, or polygons that are identified on the scanned map (Figure 4-19). On-screen digitizing may be used for recording information from scanned aerial photographs, digital photographs, satellite, or other images.

On-screen digitizing offers advantages over both map and scan digitizing. Manual map digitization is often limited by the visual acuity and pointing ability of the operator. The pointing precision of the operator and digitizing systems translates to a fixed ground distance when manually digitizing a hardcopy map. For example, consider an operator that can reliably digitize a location to the nearest 0.4 millimeters (0.01 inch) on

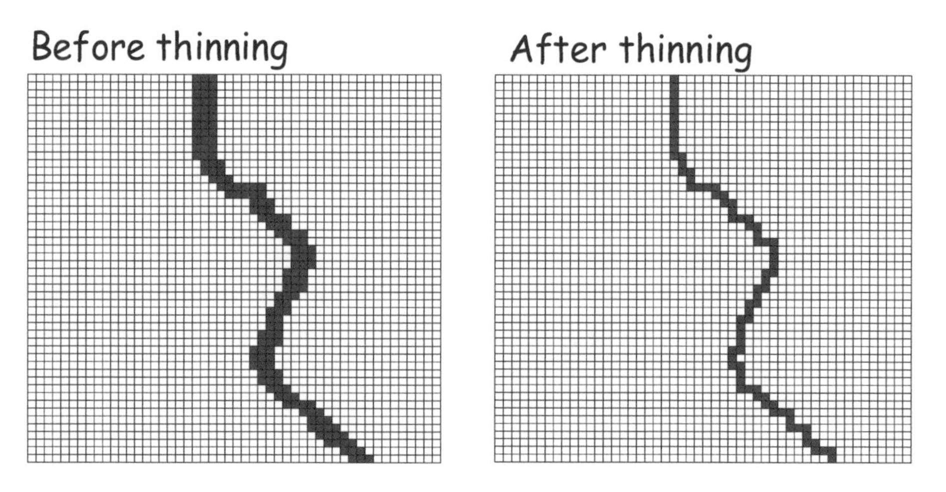

Figure 4-18: Skeletonizing, a form of line thinning that is often applied after scan-digitizing.

a 1:20,000-scale map. Also assume the best digitizer available is being used, and we know the observed digitizing error is larger than the error in the map. The 0.4 millimeter precision translates to approximately 8 meters of error on the Earth surface. The precision cannot be appreciably improved using manual digitization alone, because a majority of the imprecision is due to operator abilities. However once the map is scanned, the image may be displayed on the screen at any map scale. The operator may zoom to a 1:5,000-scale or greater on-screen, and digitizing improved. While other factors remain that limit the accuracy of the derived spatial data (for example map plotting or production errors, or scanner accuracy), on-screen digitizing may be used to limit operator-induced positional error when manually digitizing.

Figure 4-19: An example of on-screen digitizing. Images or maps are displayed on a computer screen and feature data digitized manually. Roads, shown in white, were digitized near the left and lower edges of this photograph.

On-screen digitizing also removes or reduces the need for a digitizing table. Large digitizing tables are an additional piece of equipment and require significant space. Digitizing tablets are specialized for a single use. Operator expenses are typically higher than hardware costs; if manual digitizing is an infrequent activity, the cost of digitizing hardware per unit map digitized may be significant. High quality scanning equipment is quite expensive, however maps may be sent to a third-party for scanning at relatively low cost.

Editing Geographic Data

Spatial data may be edited, or changed, for several reasons. Errors and inconsistencies are inevitably introduced during spatial data entry. Undershoots, overshoots, missing or extra lines, missing or extra points or labels are all errors that must be corrected. Spatial data may change over time. Parcels are subdivided, roads extended or moved, forests grow or are cut, and these changes may be entered in the spatial database through editing. New technologies may be developed that provide more accurate positional information, and even though existing data may be consistent and current, the more accurate data may be more useful, leading to data editing.

Error identification is the first step in editing. Errors may be identified by printing a map of the digitized data and verifying that each point, line, and polygonal feature is present and correctly located. Plots are often printed both at a similar scale and a significantly larger scale than the original source materials. The large-scale plots are often paneled with some overlap among panels. Plots at scale are helpful for identifying missing features, and large-scale plots aid in identifying undershoots, overshoots, and small omissions or additions. Annotations are made on these plots as they are checked systematically for each feature.

Software may aid in identifying potential errors. Line features typically begin and end with a node, and nodes may be classified as connecting or dangling. A connecting node joins two or more lines, while a dangling node is attached to only one line. Some dangling nodes may be intentional, e.g., a cul-de-sac in a street network, while others will be the result of under- or overshoots. Dangling nodes that are plotted with unique symbols may be quickly evaluated, and if appropriate, corrected.

Attribute consistency may also be used to identify errors. This technique identifies areas in which contradictory theme types occur in different data layers. The two layers are either graphically or cartographically overlain. Contradictory co-occurrences are identified, such as water in one layer and upland areas in a second. These contradictions are then either resolved manually, or automatically via some pre-defined precedence hierarchy.

Editing typically includes the ability to select, split, update, and add features. Selection may be based on geometric attributes, or with a cursor guided by the operator. Selections may be made individually, by geographic extent (select all features in a box, circle, or within a certain distance of the pointer) or by geometric attributes (e.g., select all nodes that connect to only one line). Once a feature is selected, various operations may be available, including erasing all or part of the feature, changing the coordinate values defining the feature, and in the case of lines, splitting or adding to the feature. A line may be split into parts, either to isolate a segment for future deletion, or to modify only a portion of the line. Coordinates are typically altered by interactively selecting and dragging points, nodes or vertices to their proper conformation. Points or line segments are added as needed.

Groups of features in an area may be adjusted through interactive *rubbersheeting*. Rubbersheeting involves fitting a local equation to adjust the coordinates of features. Polynomial equations are often used due to their flexibility and ease of application. Anchor points are selected, again on the graphics screen, and other points selected by dragging interactively on the screen to match point locations. All lines and points except the anchor points are interactively adjusted. One common application involves adjusting linework representing cultural features, such as a road network, when higher geometric-accuracy photo or satellite image data are available. The linework is overlain on an image backdrop and subsequently adjusted.

These edits should be made with due attention to the magnitude of positional change introduced during editing. On-screen editing to eliminate undershoots should only be performed when the "true" locations of features can unambiguously be identified, and the new features confidently placed in the correct location. Automatic removal of "short" undershoots may be performed without introducing additional spatial error in most instances. A short distance for an undershoot is subjectively defined, but typically it is below the error inherent in the source map, or at least a distance that is insignificant when considering the intended use of the spatial data.

Features Common to Several Layers

One common problem in digitizing derives from representation of features that occur on different maps. These features rarely have identical locations on each map, and often occur in different locations when digitized into their respective data layers (Figure 4-20). For example, water boundaries on soil survey maps rarely correspond to water boundaries found on USGS topographic quads. Features may be different on different maps for many reasons. Perhaps the maps were made for different purposes or at different times. Features may differ because the maps were from different source materials, for example, one map may have been based on ground surveys while another was based on aerial photographs. Digitizing may compound the problem due to differences among layers in digitizing methods or operators.

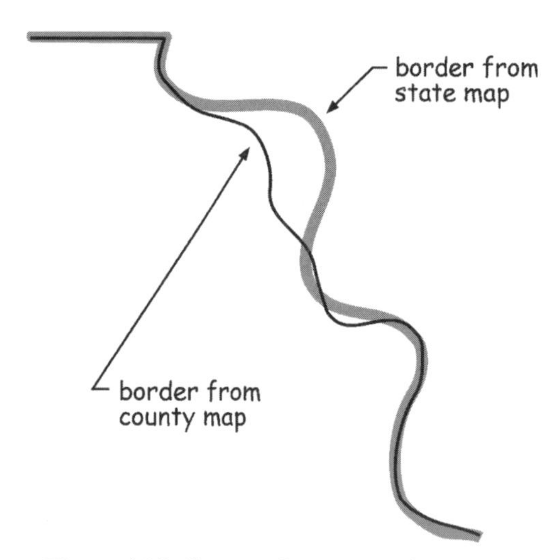

Figure 4-20: Common features may be spatially inconsistent in different spatial data layers.

There are several ways to remove this "common feature" inconsistency. One involves re-drafting the data from conflicting sources onto one base map. Inconsistencies are removed at the drafting stage. For example, vegetation and roads data may show stand boundaries at road edges that are inconsistent with the road locations. Both of these data layers may be drafted onto the same base, and the common boundaries fixed by a single line. This line is digitized once, and used to specify the location of both the road and vegetation boundary when digitizing. Re-drafting, although labor intensive and time consuming, forces a resolution of inconsistent boundary locations. Re-drafting also allows the combination of several maps into a single data layer.

A second, often preferable method involves establishing a "master" boundary which is the highest accuracy composite of the available data sets. A digital copy or overlay operation establishes the common features as a base in all the data layers, and this base may be used as each new layer is produced. For example, water boundaries might be extracted from the soil survey and USGS quad maps and these data combined in a third data layer. The third data layer would be edited to produce a composite, high quality water layer. The composite water layer would then be copied back into both the soils and USGS quad layers. This second approach, while resulting in visually consistent spatial data layers, is in many instances only a cosmetic improvement of the data. If there are large discrepancies ("large" is defined relative to the required spatial data accuracy), then the source of the discrepancies should be identified and the most accurate data used.

Coordinate Geometry

While maps are a common source of digital data, most of the maps were themselves developed at least in part from surveying measurements. Spatial data layers may be produced directly from field surveys, or from field surveys combined with measurements on aerial photographs. Surveying is particularly common when a highly valued data layer is to be developed or when very precise coordinates are required. Property lines are a good example of data that are often required to be of very high accuracy. Other commonly surveyed features include power lines, sewers, pipelines, and other utilities. These survey measurements have historically been performed with optical or optical-electronic devices such as transits, theodolites, and electronic distance meters (Figure 4-21).

A common output from survey measurements, known as *coordinate geometry* or COGO, consist of a starting point (a *station*) and a list of directions (*bearings*) and distances to subsequent stations. The COGO defines a connected set of points from the starting station to each subsequent station. Basic trigonometric functions may be used to calculate the coordinates for each survey station. These stations are located at the vertices that define lines or areas of interest. In the past these distance and bearing data were manually plotted onto paper maps. Most survey data are now transferred directly to spatial data formats.

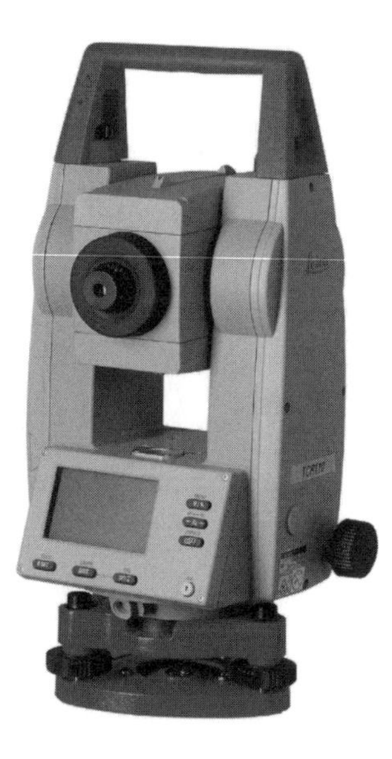

Figure 4-21: A surveying instrument used for the collection of coordinate geometry data. (courtesy Leica Corporation)

Field measurements may be directly entered and coordinate locations derived in the GIS software, or the coordinate calculations may be performed in the surveying instrument first. Many current surveying instruments contain an integrated computer and provide for digital data collection and storage. Coordinates may be tagged with attribute data in the field, at the measurement location. These data are then downloaded directly from a coordinate measuring device to a computer. Specialized surveying programs may be used for error checking and other processing. Many of these surveying packages then output data in formats designed for import into various GIS software systems.

Survey methods vary, but plane surveying used in developing coordinate geometry input are often based on measuring distances and angles for a set of points (Figure 4-22). These points are usually connected in a *traverse*, a combined set of distance and *bearing angle* measurements between *traverse stations*.

COGO calculations are illustrated on the left of Figure 4-22. Starting from a known coordinate, x_o, y_o, we measure a distance L and an angle θ. We may then calculate the distances in the x and y directions to another set of coordinates, x_1 and y_1. The coordinates of x_1 and y_1 are obtained by addition of the appropriate trigonometric functions. COGO calculations may then be repeated, using the x_1 and y_1 coordinates as the new starting location for calculating the position of the next traverse station.

The right side of Figure 4-22 shows a sequence of measurements for a traverse. Starting at x_s, y_s, the distance A and bearing angle, here 45°, are measured to station x_m, y_m. The bearing and distance are then measured to the next station, with coordinates x_n, y_n. Distances and angles are measured for all subsequent stations. Starting with the known coordinates at the starting station, x_s, y_s, coordinates for all other stations are calculated using COGO formulas.

Past survey records of bearings and distances may serve as a source of COGO input. Most of these measurements were recorded for property boundary locations, and are stored in notebooks, on deeds, and in plat books. These paper records must be converted to an electronic format prior to conversion to the coordinate locations. As described above, trigonometric functions may then be used to calculate the position of each station.

The direct entry of COGO data, when available and where practical, will usually lead to more accurate digital databases than the digitization of cartometric maps. Spatial errors, approximations, and positional uncertainty are introduced during the transition from survey measurement to hardcopy maps. Limits on plotting precision, printing alignment and distortion, deformation in the paper or other media, map generalization, and other factors compound to add uncertainty to the plotted locations on maps. Manual or scan digitizing may introduce additional uncertainty or error. However the survey data, converted to the coordinate geometry, contain none of these errors.

Another form of direct coordinate data entry exists, and is known as the Global Positioning System (GPS). Satellite-based measurements are used to determine positions and these may become part of the spatial data in a GIS. The GPS system receives an extended discussion in Chapter 5.

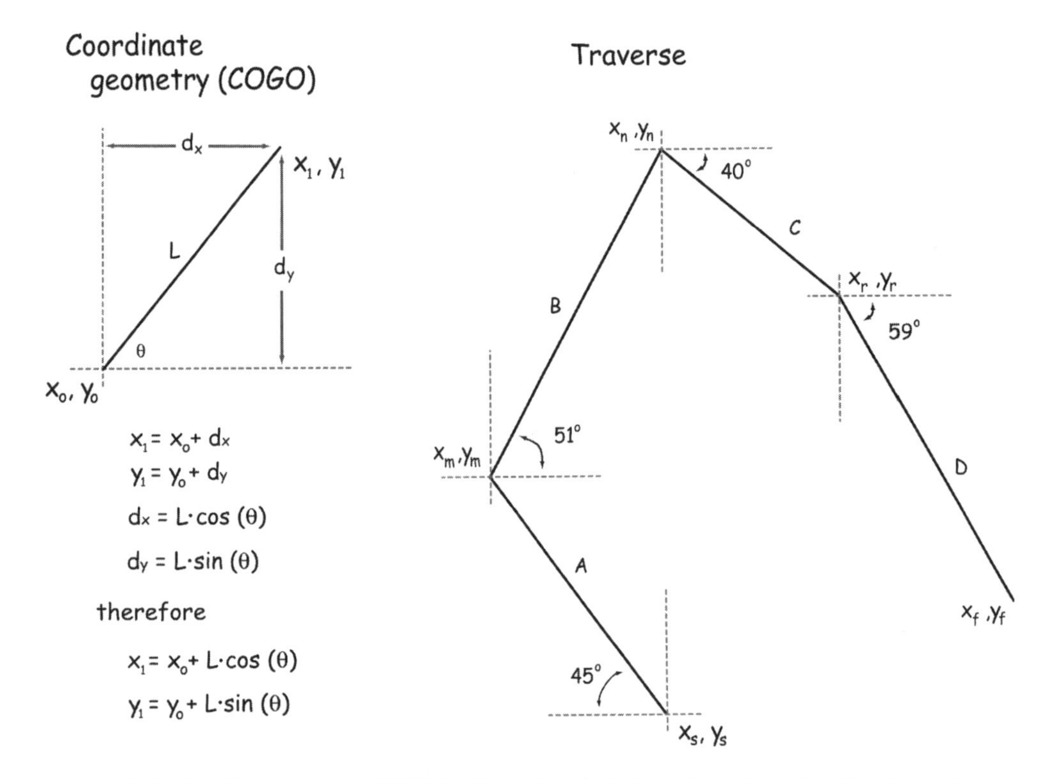

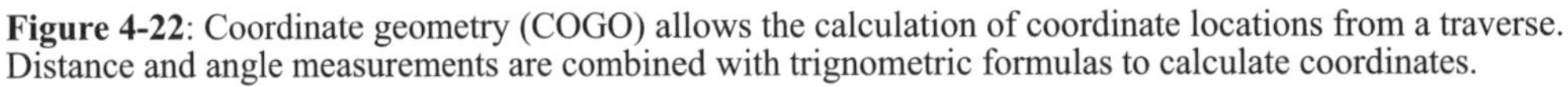
Figure 4-22: Coordinate geometry (COGO) allows the calculation of coordinate locations from a traverse. Distance and angle measurements are combined with trignometric formulas to calculate coordinates.

Coordinate Transformation

Coordinate transformation is a common operation in the development of spatial data for GIS. A coordinate transformation brings spatial data into an Earth-based map coordinate system so that each data layer aligns vertically with every other data layer. This vertical alignment ensures features fall in their proper relative position when digital data from different layers are combined. Within the limits of data accuracy, a good transformation helps avoid inconsistent spatial relationships such as farm fields on freeways, roads under water, or cities in the middle of swamps. Coordinate transformation is also referred to as *registration*, because it "registers" the layers to a map coordinate system.

Coordinate transformation is most commonly used to convert newly digitized data from the digitizer coordinate system to a standard map coordinate system (Figure 4-23). The input coordinate system is usually based on the digitizer or scanner-assigned values. A hardcopy map may be taped to a digitizing table and coordinates recorded as a puck is moved across the map surface. These coordinates are usually recorded in inch or centimeter units relative to an origin located near the lower left corner of the digitizing table. The absolute values of the coordinates depend on where the map happened to be placed on the table prior to digitizing, but the relative position of digitized points does not change as long as the map is not deformed or moved. Before these newly dig-

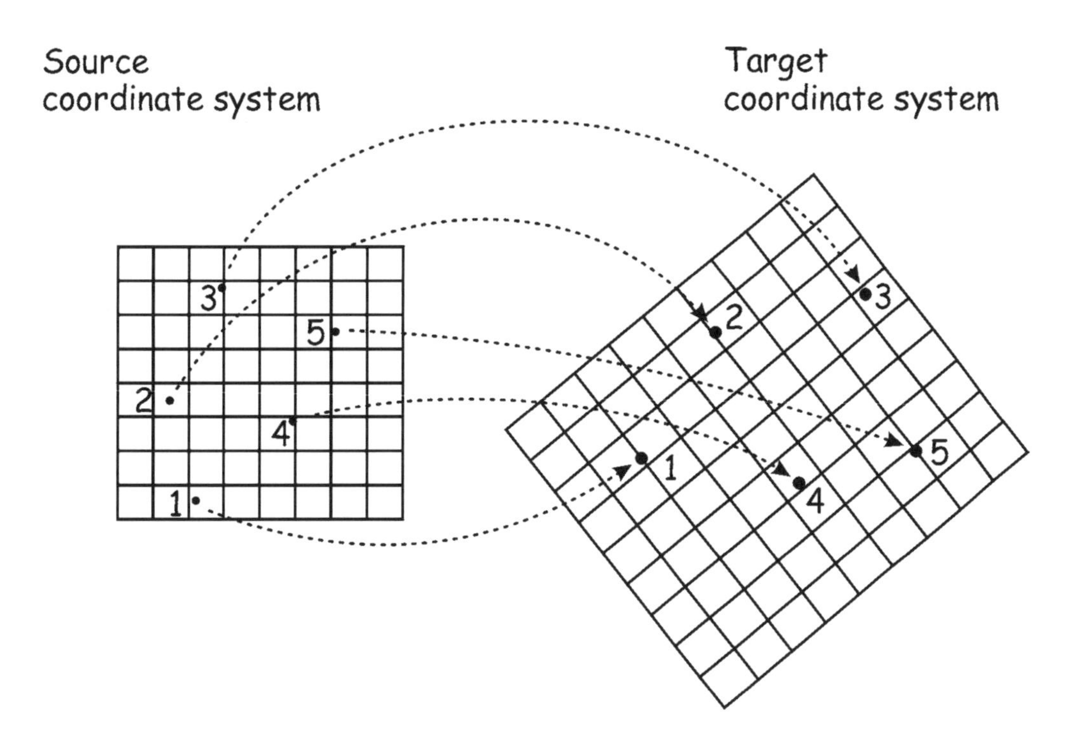

Figure 4-23: Control points in a coordinate transformation. Control points are used to guide the transformation of a source or input set of coordinates to a target or output set of coordinates. There are five control points in this example. Corresponding positions are shown in both coordinate systems.

itized data may be used with other data, these "inch-space" or "digitizer" coordinates must be transformed into an Earth-based map coordinate system.

Control Points

A set of *Control Points* are used to transform the digitized data from the digitizer coordinate system to a map-projected coordinate system. Control points are different from other digitized features. When we digitize most points, lines, or areas, we do not know the map projection coordinates for these features. We simply collect the digitizer x and y coordinates that are established with reference to some arbitrary origin on the digitizing tablet. Control points differ in that we know the map projection coordinates for these points as well as the digitizer coordinates.

These two sets of coordinates for each control point, one for the map projection and one for the digitizer system, are used to estimate a coordinate transformation. Control points are used to estimate the coefficients for transformation equations, usually through a statistical, least-squares process. The transformation equations are then used to convert coordinates from the digitizer system to the map projection system.

The transformation may be estimated in the initial digitizing steps, and applied as the coordinates are digitized from the map. This "on-the-fly" transformation allows data to be output and analyzed with reference to map-projected coordinates. A previously registered data layer or image may be displayed on screen just prior to digitizing a new map. Control points may then be entered, the new map attached to the digitizing table, and the map registered. The new data may then be displayed on top of the previously registered data. This allows a quick check on the location of the newly digitized objects against corresponding objects in the study area.

In contrast to on-the-fly transformations, data may be recorded in digitizer coordinates and the transformation applied later. All data are digitized, including the control point locations. The digitizer coordinates of the control point may then be matched to corresponding map projection coordinates, and transformation equations estimated. These transformation equations are then applied to convert all digitized data to map projection coordinates.

There are many sources of control points. A source should be chosen that provides control points with the highest feasible coordinate accuracy, and control point accuracy should be at least as good as the desired overall positional accuracy required for the spatial data. Control points should be as evenly distributed as possible throughout the data area, and the minimum number of points depends on the mathematical form of the transformation. Additional control points above the minimum number are usually collected; this increases the power of the statistical method used to estimate the transformation functions. The statistical method also provides an estimate of the positional accuracy of the transformation.

The x, y (horizontal), and sometimes z (elevation) coordinates of control points are known to a high degree of accuracy and precision. Because "high" precision and accuracy are subjectively defined, there are many methods to determine control point locations. Sub-centimeter accuracy may be required for control points used in property boundary layers, while accuracies of a few meters may be acceptable for large-area vegetation mapping. Common sources of control point measurements are traditional transit and distance surveys, global positioning system measurements, existing cartometric quality maps, or existing digital data layers on which suitable features may be identified.

Control Point Sources: Surveying

Traditional ground surveys based on optical surface measurements are a common, although decreasingly used method for determining control point locations. Modern surveys use complex instruments such as transits and theodolites to precisely measure the relative location of points. If the survey starts from a known point, then the coordinate location of any survey station may be determined via simple trigonometric functions. Federal, state, county, and local governments all maintain information on accurately surveyed locations (Figure 4-24), and these points may be used as control points or as starting points for additional surveys. Most of these known points have been established using traditional surveying techniques. Indeed, the development of this "control network" infrastructure is one of the first and most important responsibilities of government. These survey points form the basis for distance, location, and area measurements used to define property, political, and municipal boundaries, and hence this control network underlies most commerce, transportation, and land ownership and management. Coordinates, general location, and descriptions are documented for these control networks, and may be obtained from a number of government sources. In the United States these sources include county surveyors, state surveyors and departments of transportation, and the National Geodetic Survey (NGS).

Figure 4-24: Previous surveys are a common source of control points.

The ground survey network is often quite sparse and insufficient for registering many large-scale maps. Even when there is a sufficient number of ground-surveyed points in an area, many may not be suitable for use as control points in a coordinate transformation of spatial data. The control points may not be visible on the maps or images to be registered. For example, a surveyed point may fall along the edge of a road. If the control point is at a mapped road intersection, we may use the easting and northing coordinates of the road intersection as a control point during map registration. However if the surveyed point is along the edge of a road that is not near any mapped feature such as a road intersection, building, or water tower, then it may not be used as a control point. We require the point be visible on the map, data layer, or image and that we also have precise ground coordinates in our target map projection.

GPS Control Points

The global positioning system (GPS) is a relatively new technology that allows us to establish control points. GPS, discussed in detail in Chapter 5, may help us obtain the coordinates of control points that are visible on a map or image. GPS is particularly useful for determining control point coordinates because it is a quick, relatively inexpensive way to obtain the location of widely-spaced points. GPS is often preferred because it is typically faster and often less expensive than traditional surveying methods. GPS positional accuracy depends on the technology and methods employed; it typically ranges from sub-centimeter to a few meters. Most points recently added to the NGS and other government-maintained networks were measured using GPS technologies.

Control Points from Existing Maps and Digital Data

Existing maps are another common source of control points. Point locations are plotted and coordinates often printed on maps, for example the corner location coordinates are printed on USGS quadrangle maps (Figure 4-25). Road intersections and other well-defined locations are often represented on maps. If enough recognizable features can be identified, then control points may be obtained from the maps. Control points derived in this manner typically come only from cartometric quality maps, those maps produced with the intent of giving an accurate, map-projected representation of features on the Earth's surface.

Digital image data are a common source of ground control points, particularly when natural resource or municipal databases are to be developed for managing large areas. Digital images often provide a richly detailed depiction of surface features (Figure 4-26), Digital image data may be obtained that are registered to a known coordinate system. Typically, the coordinates of a cor-

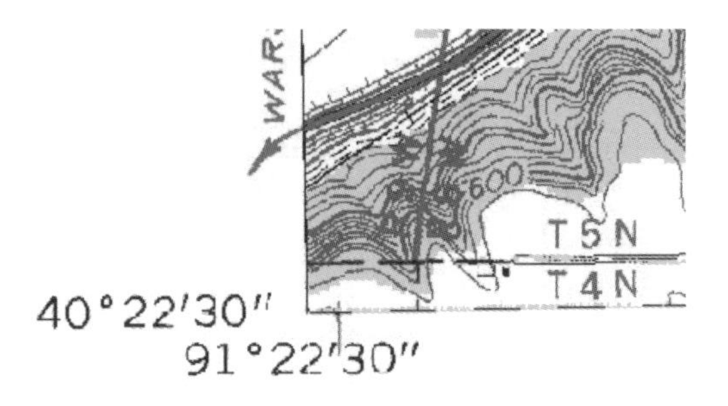

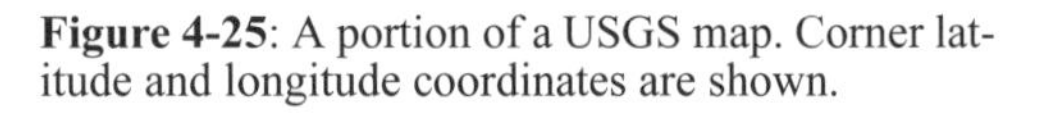

Figure 4-25: A portion of a USGS map. Corner latitude and longitude coordinates are shown.

Figure 4-26: Potential control points may be identified on a digital image. Road intersections or other permanent, well-defined features are identified. Coordinates may be determined from information provided with the digital image. This image was available at a one meter resolution, and appears blurred when viewed at this high magnification.

ner pixel are provided, and the lines and columns for the image run parallel to the easting (E) and northing (N) direction of the coordinate system. Because the pixel dimensions are known, the calculation of a pixel coordinate involves the multiplication of row and column number by the pixel size, and the application of the corner offset, either by addition or subtraction. In this manner, the image row/column may be converted to an E, N coordinate pair, and control point coordinates determined.

Finally, existing digital data may also provide control points. A short description of these digital data sources are provided here, and expanded descriptions of these and other digital data are provided in Chapter 7. The coordinates of any identifiable feature may be obtained from the digital data layer. For example, the USGS has produced Digital Raster Graphics (DRG) files that are scanned images of the 1:24,000-scale quadrangle maps. These DRGs come referenced to a standard coordinate system, so it is a simple and straightforward task to extract the coordinates of road intersections or other well-defined features that have been plotted on the USGS quadrangle maps. Digital line graph data (DLGs) are also available that contain vector data, and control points may be identified at road intersections and other identifiable features.

Control points are used in coordinate transformation, irrespective of source. Typically a number of control points are identified for a study area. The x and y coordinates for control points are obtained from the digitized map, and the map projection coordinates, E and N, are determined from survey, GPS, or other sources (Figure 4-27). These coordinate pairs are then used with a transformation to convert data layers into a desirable map coordinate system. There are several different types of coordinate transformations.

The Affine Transformation

The *affine coordinate transformation* employs linear equations to calculate map coordinates. Map projection coordinates are often referred to as eastings (E) and northings (N), and are related to the x and y digitizer coordinates by the equations:

$$E = T_E + a_1x + a_2y \qquad (4.1)$$

$$N = T_N + b_1x + b_2y \qquad (4.2)$$

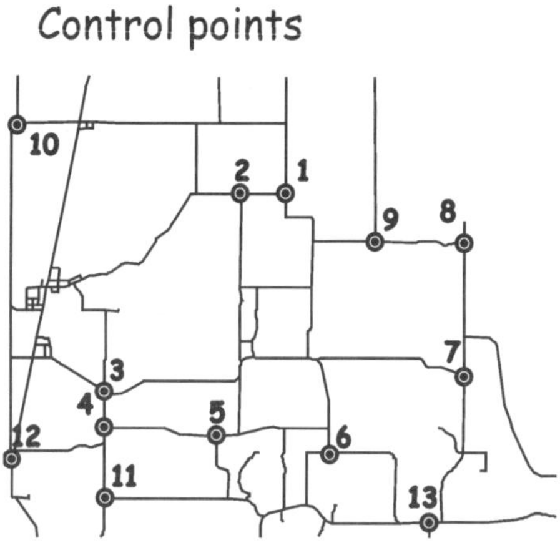

Control points

Digitizer coordinates

ID	x	y
1.0	103.0	-100.1
2.0	0.8	-69.1
3.0	-20.0	-69.0
4.0	-60.0	-47.0
5.0	-102.0	-47.2
6.0	-101.7	10.8
7.0	-86.0	75.8
8.0	-40.0	45.7
9.0	11.0	36.8
10.0	63.0	34.0
11.0	63.0	17.7
12.0	63.0	64.3
13.0	106.0	47.7

Projection coordinates (UTM)

E	N
500,083.4	5,003,683.5
504,092.3	5,002,499.5
504,907.5	5,002,499.5
506,493.3	5,001,673.5
508,101.3	5,001,651.0
508,090.1	4,999,384.0
507,475.9	4,996,849.0
505,689.2	4,998,022.0
503,679.2	4,998,368.0
501,657.9	4,998,479.5
501,669.1	4,999,116.0
501,680.3	4,997,296.0
500,005.3	4,997,943.5

Figure 4-27: Control points locations, and corresponding digitizer and map projection coordinates.

T_E and T_N are translation changes between the coordinates, and can be thought of as shifts in the origins from one coordinate system to the next. The a_i and b_i parameters incorporate the change in scales and rotation angle between one coordinate system and the next. The affine is the most commonly applied coordinate transformation because it provides for these three main effects of translation, rotation, and scaling when converting from a digitizer to map coordinates, and because it requires relatively fewer control points and often introduces less error than higher-order polynomial transformations.

The affine system of equations has six parameters to be estimated, T_E, T_N, a_1, a_2, b_1, and b_2. Each control point provides E, N, x, and y coordinates, and allows us to write two equations. For example, we may have a control point consisting of a precisely surveyed center of a road intersection. This point has digitizer coordinates of x=103.0 centimeters and y = -100.1 centimeters, and corresponding Earth-based map projection coordinates of E = 500,083.4 and N = 4,903,683.5. We may then write two equations based on this control point:

$$500{,}083.4 = T_E + a_1(103.0) + a_2(-100.1) \quad (4.3)$$

$$4{,}903{,}683.5 = T_N + b_1(103.0) + b_2(-100.1) \quad (4.4)$$

We cannot find a unique solution to this system of equations, because there are six unknowns (T_E, T_N, a_1, a_2, b_1, b_2) and only two equations. We need as many equations as unknowns to solve this system. Each control point gives us two equations, one for the easting and one for the northing, so we need a minimum of three control points to estimate the parameters of an affine transformation. If we wish to use a statistical technique to estimate the transformation parameters, we need at least one additional control point, for a total of four control points. As with all statistical estimates, more control points are better than fewer, but we will reach a point of diminishing returns after some number of points, typically somewhere between 15 and 30 control points.

Figure 4-28: Examples of control points, predicted control locations, and residuals from coordinate transformation.

The affine coordinate transformation is usually fit using a statistical method that minimizes the *root mean square error* RMSE. The RMSE is defined as:

$$RMSE = \sqrt{\frac{e_1^2 + e_2^2 + e_3^2 \ldots + e_n^2}{n}} \quad (4.5)$$

where the e_i are the residual distances between the true Earth-based E and N coordinates and the E and N coordinates in the output data layer. Figure 4-28 shows examples of this lack of fit. Individual residuals may be observed at each control point location. The location of the control points predicted from the transformation equations do not equal the true locations of the control points, and this difference is the residual or positional error at that control point.

A statistical method for estimating transformation equations is preferred because it provides us with an indication of the quality of the transformation. Measurement and digitizing uncertainty introduce unavoidable spatial errors into the control point coordinate values. These uncertainties result in (hopefully) small differences in the transformed and true locations of transformed coordinates. A comparison of the measured and predicted control point coordinates will provide some measure of the transformation quality. Transformations are often fit iteratively, and control points with large errors inspected, corrected, or deleted until an acceptable RMSE is obtained (Figure 4-29). The RMSE will be less than the true transformation error at a randomly selected point, because we are actively minimizing the N and E residual errors when we statistically fit the transformation equations. However the RMSE will be an index of transformation accuracy, and a lower RMSE generally indicates a more accurate affine transformation.

Estimating the coordinate transformation parameters is often an iterative process for a number of reasons. First, the control point x and y coordinates may not be precisely digitized. Manual digitization requires the operator place a pointer at the control point and record the location. Poor eyesight, a shaky hand, fatigue, lack of attention, misidentification of the control location, or a blunder may result in erroneous x and y coordinate values. Control point locations may not be accurately or precisely represented on the map, for example when road intersections are used as control and wide road symbols or offsets are applied. There may also be uncertainties or errors in the E and N coordinates, all of which will introduce error into the transformation. Typically, control points are entered, the affine transformation parameters estimated, and the overall RMSE and individual point E and N errors evaluated (Figure 4-28, Figure 4-29). Suspect points are identified and the transformation re-estimated and errors evaluated. This process continues until a satisfactory transformation is fit. The transformation is then applied to all features to convert them from digitizer to map coordinates.

Other Coordinate Transformations

Other coordinate transformations are sometimes used. The conformal coordinate transformation is similar to the affine, and has the form:

$$E = T_E + cx - dy \qquad (4.6)$$

$$N = T_N + dx + cy \qquad (4.7)$$

The coefficients T_E, T_N, c and d are estimated from control point data. As with the affine, the conformal transformation is also a first-order polynomial. The conformal transformation requires equal scale changes in the x and y directions. In the affine, scale changes in the x and y directions can be different. Note the symmetry in the equations 4.6 and 4.7, in that the x and y coefficients match across equations, and there is a change in sign for the d coefficient. This results in a system of equations with only four unknown parameters, and so the conformal may be estimated when only two control points are available.

Higher-order polynomial transformations are sometimes used to transform among coordinate systems. An example of a 2nd-order polynomial is:

$$E = b_1+b_2x+b_3y+b_4x^2+b_5y^2+b_6xy \qquad (4.8)$$

Note that the combined powers of the x and y variables may be up to 2. This allows for curvature in the transformation in both the x and y directions. A minimum of six control points is required to fit this 2nd-order polynomial transformation, and seven are required when using a statistical fit. The estimated parameters T_E, T_N, a_1, a_2, b_1, and b_2 will be different in equations 4.1 and 4.2 when compared to (4.8), even if the same set of control points is used for both statistical fits. We change the form of the equations by including the higher-order squared and xy

Goal : fit transformation with a RMSE < 5 meters

ID	Residual E	Residual N
1.0	-16.9	-17.4
2.0	1.1	-0.5
3.0	4.3	1.4
4.0	27.6	30.0
5.0	-4.0	-3.7
6.0	-4.7	-9.7
7.0	-9.8	-9.2
8.0	0.4	-5.3
9.0	-18.3	-1.1
10.0	-9.5	3.0
11.0	2.4	3.2
12.0	11.6	2.8
13.0	15.9	6.5

Iteration 1: fit all control points, examine residuals and RMSE

E = 504,125.4 -39.04 x + 0.0414y
N = 4,999,806.3 -0.079x -38.99y
RMSE = 16.4 meters
remove control poinst # 4, #1, and #9 and re-fit model

ID	Residual E	Residual N
2.0	-1.7	-2.4
3.0	3.0	1.0
5.0	1.1	2.6
6.0	1.5	-1.6
7.0	-3.5	-0.1
8.0	2.8	-0.5
10.0	-14.7	-0.1
11.0	-3.2	-0.4
12.0	6.9	0.7
13.0	7.8	0.8

Iteration 2: re-fit with remaining control points, examine residuals and RMSE

E = 504,126.8 -38.9x + 0.022y
N = 4,999,805.9 -0.007x -39.02y
RMSE = 6.2 meters
remove control point # 10 and re-fit model

ID	Residual E	Residual N
2.0	-3.0	-2.4
3.0	2.1	1.0
5.0	1.9	2.6
6.0	1.8	-1.6
7.0	-4.1	-0.1
8.0	1.5	-0.5
11.0	-6.6	-0.4
12.0	3.2	0.7
13.0	3.2	0.7

Iteration 3: re-fit with remaining control points, examine residuals and RMSE

E = 504,128.6 -38.94x + 0.0299y
N = 4,999,805.9 - 0.007x - 39.02y
RMSE = 3.7 meters
below 5 meter RMSE, after visual inspection of each residual on map and ensuring remaining control points are well distributed, accept transformation

Figure 4-29: Iterative fitting of an affine transformation. The data from Figure 4-27 were used. Four control points were removed in the first two iterations. The final model fit in iteration three meets the RMSE criteria. It is better to examine control points with large residuals to determine if the cause for the error may be identified. If so, the control point coordinates may be modified, and the control points retained while fitting the transformation.

cross-product terms, and all estimated parameters will vary.

The RMSE is typically lower for a 2nd and other higher-order polynomials than an affine transformation, but this does not mean the higher order polynomial provides a more accurate transformation. The higher-order polynomial will introduce more error than an affine transformation on most orthographic maps, and an affine transformation is preferred. The RMSE is a useful tool when comparing among transformations that are the same model form (e.g., when comparing one affine to another affine as in Figure 4-29), but is not useful when comparing among different model forms. Due to the nature of the statistical fitting process, the RMSE and related measures of fit will decrease as higher order polynomials are used. High-order polynomials allow more flexibility in warping the surface to fit the control points. Unfortunately, this warping may significantly deform the non-control-point coordinates, and add large non-linear errors when the transformation is applied to all data in a layer. Thus, higher-order polynomials and other "rubber-sheeting" methods should be used with caution, and the accuracy of the transformation tested with an evaluation that includes independent check points. These check points are withheld when estimating the transformation, and their transformed vs. measured coordinates compared.

Map Projection vs. Transformation

Map transformations should not be confused with map projections. A map transformation typically employs a linear equation to convert coordinates from one Cartesian coordinate system to another. A map projection, described in Chapter 3, differs from a transformation in that it is an analytical, formula-based conversion between coordinate systems, usually from a curved, latitude/longitude coordinate system to a Cartesian coordinate system. No statistical fitting process is used with a map projection.

Transformations should rarely be used in place of projection equations when converting geographic data between map projections. Consider an example when data are delivered to an organization in Universal Transverse Mercator (UTM) coordinates and are to be converted to State Plane coordinates prior to integration into a GIS database. Two paths may be chosen. The first involving projection from UTM to geographic coordinates (latitude and longitude), and then from these geographic coordinates to the appropriate State Plane coordinates. The software may hide the intermediate geographic coordinates, but most projections among coordinate systems go through an intermediate set of geographic coordinates. This is because we know the forward and inverse conversions to and from latitude/longitude coordinates are known for all analytical map projections, but do not know the projection to projection equations for most. The correct method of applying a projection rather than a transformation may involve substantial computation because the forward and inverse projections may be quite complex, but this path doesn't require the identification of control points.

An alternative approach involves using a linear or polynomial transformation to convert between different map projections. In this case a set of control points would be identified and the coordinates determined in both UTM and State Plane coordinate systems. The transformation coefficients would be estimated and these equations applied to all data in the UTM data layer. This new output data layer would be in State Plane coordinates. This transformation process should be avoided, as a transformation may introduce additional positional error.

Transforming between projections is used quite often, inadvertently, when digitizing data from paper maps. For example, USGS 1:24,000-scale maps are cast on a polyconic projection. If these maps are digitized, it would be preferable to register them to the appropriate polyconic projection, and then re-project these data to the desired end projection. This is often not done, because the error in ignoring the projection over the size of the mapped area is typically less than the positional error associated with digitizing. Experience and specific calculations have shown that the spatial errors in using a transformation instead of a projection are small at these map scales under typical digitizing conditions.

This second approach, using a transformation when a projection is called for, should not be used until it has been tested as appropriate for each new set of conditions. Each map projection distorts the surface geometry. These distortions are complex and nonlinear. Affine or polynomial transformations are unlikely to remove this non-linear distortion. Exceptions to this rule occur when the area being transformed is small, particularly when the projection distortion is small relative to the random uncertainties, transformation errors, or errors in the spatial data. However, there are no guidelines on what constitutes a sufficiently "small" area. In our example above, USGS 1:24,000 maps are often digitized directly into a UTM coordinate system with no obvious ill effects, because the errors in map production and digitizing are often much larger than those in the projection distortion for the map area. However, you should not infer this practice is appropriate under all conditions, particularly when working with smaller-scale maps.

Summary

Spatial data entry is a common activity for many GIS users. Although data may be derived from several sources, maps are a common source, and care must be taken to choose appropriate map types and to interpret the maps correctly when converting them to spatial data in a GIS.

Maps are used for spatial data entry due to several unique characteristics. We have a long history of hardcopy map production, so centuries of spatial information are stored there. Maps are inexpensive, widely available, and easy to convert to digital forms, although the process is often time consuming, and may be costly. Maps are usually converted to digital data through a manual digitization process, whereby a human analyst traces and records the location of important features. Maps may also be digitized via a scanning device.

The quality of data derived from a map depends on the type and size of the map, how the map was produced, the map scale, and the methods used for digitizing. Large-scale maps generally-provide more accurate positional data than comparable small-scale maps. Large-scale maps often have less map generalization, and small horizontal errors in plotting, printing, and digitizing are magnified less during conversion of large-scale maps.

Snapping, smoothing, vertex thinning, and other tools may be used to improve the quality and utility of digitized data. These methods are used to ensure positional data are captured efficiently and at the proper level of detail.

There are other common sources of digital spatial data, including COGO and GPS. COGO involves the input of coordinate geometry data, and may be used to build spatial data from surveyor's measurements of distance and direction. GPS, discussed in-depth in the next chapter, is a system that allows rapid, accurate measurement of coordinates in the field.

Map and other data often need be converted to a target coordinate system via a map transformation. A transformation is different from a map projection, discussed in Chapter 3, in that a transformation uses an empirical, least-squares process to convert coordinates from one Cartesian systems to another. Transformations are often used when registering digitized data to a known coordinate system. Map transformations should not be used when a map projection is called for.

Suggested Reading

Aronoff, S., Geographic Information Systems, A Management Perspective, WDL Publications, Ottawa, 1989.

Bolstad, P., Gessler, P.V., and Lillesand, T.M., Positional uncertainty in manually digitized map data, *International Journal of Geographical Information Systems*, 1990, 4:399-412.

Burrough, P. A. and Frank, A. U., eds., Geographical Objects with Indeterminate Boundaries, Taylor and Francis, London, 1996.

Chrisman, N. R., The role of quality information in the long-term functioning of a geographic information system, *Cartographica*, 1984, 21:79-87.

Chrisman, N. R., Efficient digitizing through the combination of appropriate hardware and software for error detection and editing, *International Journal of Geographical Information Systems*, 1987, 1:265-277.

DeMers, M., Fundamentals of Geographic Information Systems, 2nd Edition. Wiley, New York, 2000.

Douglas, D. H. and Peuker, T. K., Algorithms for the reduction of the number of points required to represent a digitized line or its caricature, *Canadian Cartographer*, 1973, 10:112-122.

Holroyd, F. and Bell, S. B. M., Raster GIS: Models of raster encoding, *Computers and Geosciences*, 1992, 18:419-426

Joao, E. M., Causes and Consequences of Map Generalization, Taylor and Francis, London, 1998.

Laurini, R. and Thompson, D., Fundamentals of Spatial Information Systems, Academic Press, London, 1992.

Maquire, D. J., Goodchild, M. F., and Rhind, D., eds., Geographical Information Systems: Principles and Applications, Longman Scientific, Harlow, 1991.

Muehrcke, P.C. and Muehrcke, J.P., Map Use: Reading, Analysis, and Interpretation, 3rd Edition. J.P. Publications, Madison, 1992.

Nagy, G. and Wagle, S. G., Approximation of polygonal maps by cellular maps, *Communications of the Association of Computational Machinery*, 1979, 22:518-525.

Peuquet, D. J., A conceptual framework and comparison of spatial data models, *Cartographica*, 1984, 21:66-133.

Peuquet, D. J., An examination of techniques for reformatting digital cartographic data. Part II: the raster to vector process, *Cartographica*, 1981, 18:21-33.

Peuker, T. K. and Chrisman, N., Cartographic Data Structures, *The American Cartographer*, 1975, 2:55-69.

Shaeffer, C. A., Samet, H., and Nelson R. C., QUILT: a geographic information system based on quadtrees, *International Journal of Geographical Information Systems*, 1990, 4:103-132.

Shea, K.S., and McMaster, R.B., Cartographic generalization in a digital environment: when and how to generalize, *Proceedings AutoCarto 9*, 1989, pp.56-67.

Warner, W. and Carson, W., Errors associated with a standard digitizing tablet, *ITC Journal*, 1991, 2:82-85.

Weibel, R., Generalization of spatial data: principles and selected algorithms, in van Kreveld, M., Nievergelt, J., Roos, T., and Widmayer, P. (eds.), Algorithmic Foundations of Geographic Information Systems, Springer-Verlag, Berlin, 1997.

Zeiler, M., Modeling Our World: The ESRI Guide to Geodatabase Design, ESRI Press, Redlands, 1999.

Study Questions

Why have so many digital spatial data been derived from hardcopy maps?

Which is a larger scale map, 1:20,000 or 1:1,000,000?

Can you describe three different types of generalization?

What are the most common map media? Why?

Is media deformation more problematic with large scale maps or small scale maps?

Which map typically shows more detail - a large scale map or a small scale map? Can you give three reasons why?

Why is manual map digitization so commonly used?

Can you describe the process of manual map digitization?

What features are typically digitized? What is the difference between a node and a vertex?

What is snapping in the context of digitizing? What are undershoots and overshoots, and why are they undesirable?

What is a spline, and how are they used during digitizing?

Why is line thinning sometimes necessary? Can you describe a line-thinning algorithm?

Can you contrast manual digitizing to the various forms of scan digitizing? What are the advantages and disadvantages of each?

What is the "common feature problem" when digitizing, and how might it be overcome?

What is COGO?

Can you describe the general goal and process of map registration?

What are control points, and where do they come from?

Can you define an affine transformation, including the form of the equation? Why is it called a linear transformation?

What is the root mean square error (RMSE), and how does it relate to a coordinate transformation?

Why are higher order (polynomial) projections to be avoided under most circumstances?

5 The Global Positioning System

GPS Basics

The Global Positioning System (GPS) is a satellite-based technology that gives precise positional information, day or night, in most weather and terrain conditions. GPS technologies help navigate and track large and small boats, planes, trucks, and automobiles, and small, lightweight GPS units have been developed that are easily carried by an individual. Because it is inexpensive, accurate, and easy to use, GPS has significantly changed surveying, navigation, shipping, airline, transportation, and other fields, and is also having a pervasive impact in the geographic information sciences. GPS has become the most common method for field data collection in GIS.

As of 2001 there are two functioning satellite GPS systems, and a third is planned. The U.S. Department of Defense operates NAVSTAR, the most commonly used system. A Russian system named GLONASS also exists, but is little used outside the former Soviet Union. A third system, Galileo, is planned as a non-military positioning service to be financed, designed, and built by a consortium of European governments and industries. We will describe the U.S. GPS system here, as it is representative of all three.

There are three main components, or segments, to GPS (Figure 5-1). The first is the *satellite segment.* This segment consists of a constellation of satellites orbiting the

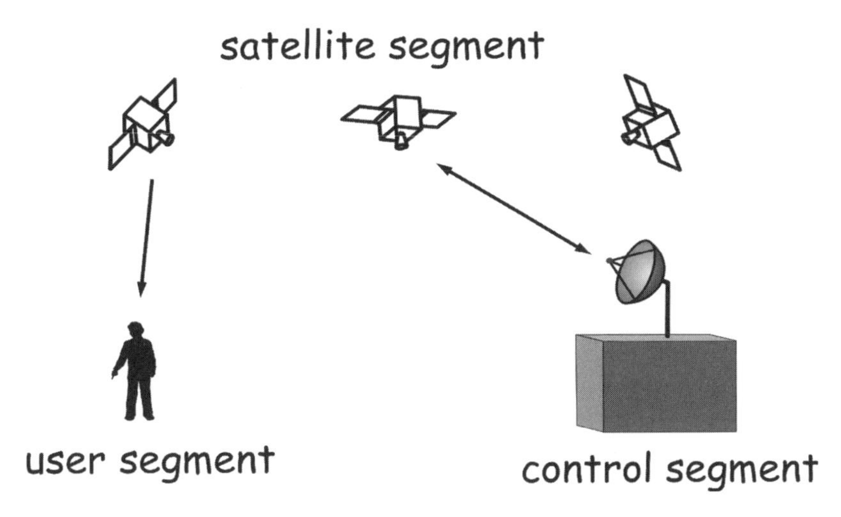

Figure 5-1: The three segments that comprise the Global Positioning System.

Earth at an altitude of approximately 20,000 kilometers. The system was designed to operate with 21 active GPS satellites and three spares. These satellites are distributed among six offset orbital planes (Figure 5-2). Every satellite orbits the Earth twice daily, and each satellite is usually above the flat horizon for 8 or more hours each day. Experimental and operational blocks of satellites were planned, and both types have been outlasting their design life, so there have typically been more than 24 satellites in orbit simultaneously. Between four to eight active satellites are typically visible from any unobstructed viewing location on Earth.

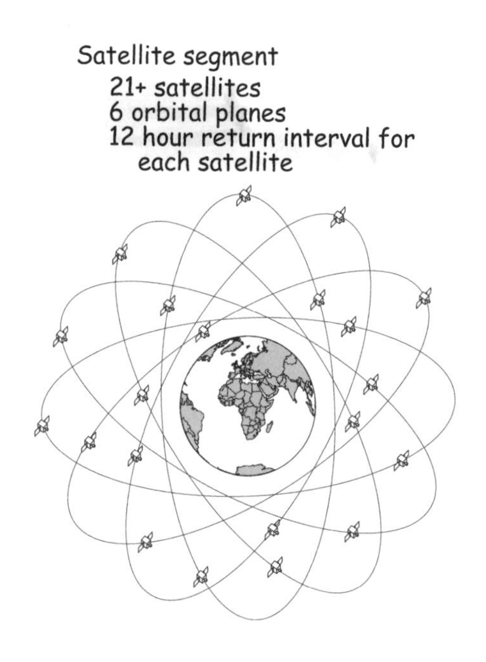

Figure 5-2: Satellite orbit characteristics for the GPS constellation.

The second component of the GPS is the *control segment*. The control segment consists of the tracking, communications, data gathering, integration, analysis, and control facilities (Figure 5-3). These are used to observe, maintain, and manage the GPS satellites and system. There are five tracking stations spread across the Earth, with a Master Control Station in Colorado Springs,

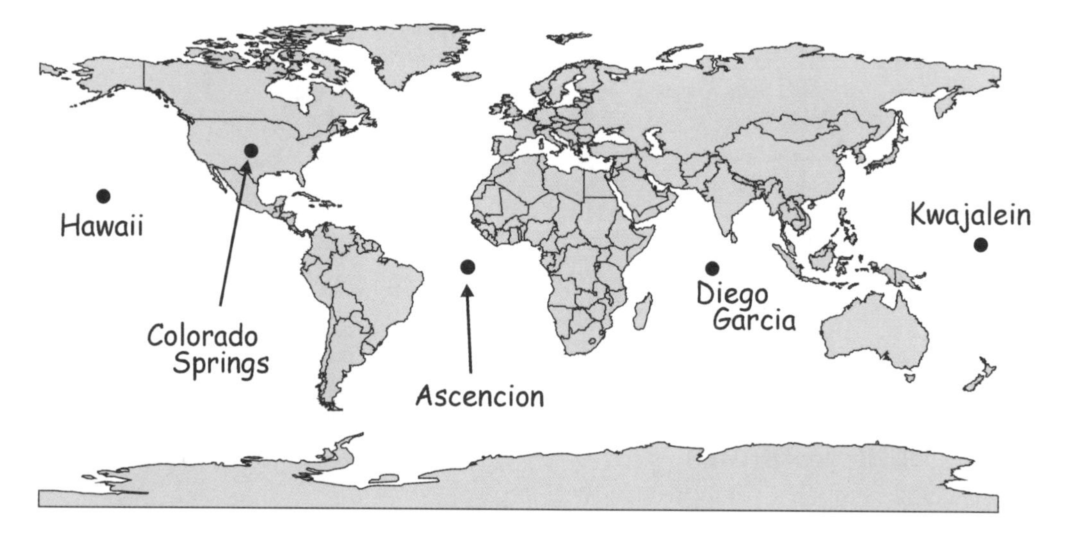

Figure 5-3: A master control station at Colorado Springs and four tracking stations are used to monitor, manage, and maintain the constellation of GPS satellites.

USA. Data are gathered from a number of sources by the Master Control Station. These data include satellite health and status information from each GPS satellite, tracking information from each tracking station, timing data from the U.S. Naval Observatory, and Earth data from the U.S. Defense Mapping Agency. The Master Control Station synthesizes this information and broadcasts navigation, timing, and other data to each satellite. The Master Control Station also signals each satellite as appropriate for course corrections, changes in operation, or other maintenance.

The third part of the GPS is the *user segment*. The user segment comprises the set of individuals with one or more GPS receivers. A GPS receiver is a device that records data transmitted by each satellite and processes these data to obtain three-dimensional coordinates (Figure 5-4). There is a wide array of receivers and methods for determining position.

GPS Broadcast Signals

GPS positioning is based on radio signals broadcast by each satellite. The satellites broadcast at a fundamental frequency of 10.23 MHz (MHz = Megahertz, or millions of cycles per second). The satellites also broadcast at other frequencies that are integer multiples or divisors of the fundamental frequency (Table 5-1). There are two *carrier signals*, L1 at 1575.42 MHz and L2 at 1227.6 MHz. These carrier signals are modulated to produce two *coded signals*, the C/A code at 1.023 MHz and the P code at 10.23 MHz. The L1 signal carries both the C/A and P codes, while the L2 carries only the P code.

The coded signals are sometimes referred to as the *pseudo-random code*. The code appears quite similar to random noise. However, short segments of the code are unique in that the signal waveform for each satellite and time is different from other time periods. The coded signal does repeat, but the repeat interval is long enough to not cause problems in positioning. Algorithms

Figure 5-4: A hand-held GPS receiver (left) and in use (right). (courtesy CMT Inc.)

Table 5-1 : GPS Signals

Name	Frequency (MHz)
L1	1575.42
L2	1227.6
P	10.23
C/A	1.023

for code generation and code interpretation are built into each GPS receiver.

Positions based on carrier signal measurements are inherently more accurate than those based on the code signal measurements. The mathematics and physics of carrier measurement are inherently better suited for making positional measurements. However this added accuracy comes at a cost. Carrier measurements, also known as carrier phase measurements, require more sophisticated and expensive receivers. Perhaps a greater constraint derives from the requirement that carrier phase instruments must receive uninterrupted signals for longer periods of time than C/A code receivers. If the satellite passes behind an obstruction, e.g., a building, mountain, or tree, the signal may be momentarily lost, and carrier phase measurements begun anew. Satellite signals are frequently lost in heavily obstructed environments, substantially reducing the efficiency of carrier phase data collection.

Each GPS satellite also broadcasts data on satellite status and location. Information includes an *almanac*, data used to determine the location of every satellite in the GPS constellation. The broadcast also includes *ephemeris data* for the broadcasting satellite. These ephemerides allow a GPS receiver to accurately calculate the position of the broadcasting satellite. Satellite health, clock corrections, and other data are also transmitted.

GPS Range Distances

GPS positioning is based primarily on *range distances* determined from the carrier and coded signals. A range distance, or *range*, is a distance between two objects, in this case the distance between a GPS satellite in space and a GPS receiver on or above the Earth surface (Figure 5-5). GPS signals travel at the speed of light. The range distance from the receiver to each satellite is calculated based on signal travel time from the satellite to the receiver:

$$\text{Range} = \text{speed of light} * \text{travel time} \quad (5.1)$$

C/A coded signals are used to calculate signal travel time by matching sections of the code (Figure 5-6). Timing information is sent with the coded signal, allowing the GPS receiver to calculate the precise transmission time for each code fragment. The GPS receiver also observes the reception time for each code fragment. The difference between transmission and reception times is the travel time, which is then used to calculate a range distance. Range measurements can be repeated quite rapidly, typically up to a rate of one per second; therefore several range

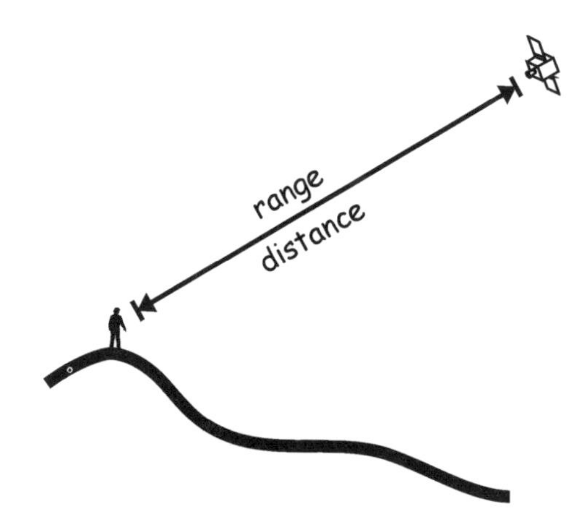

Figure 5-5: A single GPS range measurement.

measurements may be made for each satellite in a short period of time.

Carrier phase GPS is also based on a set of range measurements. In contrast to C/A signals, the carrier signal has no code, so the phase of the satellite signal is measured. Each wave transmitted at a given frequency is identical, and there is some unknown integer number of waves plus a partial wave that fit in the distance between the satellite and the receiver at any one time. The long, unbroken observations on a satellite required by carrier phase positioning are needed to resolve these integer ambiguities and calculate a satellite range.

Simultaneous range measurements from multiple satellites are used to establish receiver position. A range measurement is combined with information on satellite location to define a sphere. A range measurement from a single satellite restricts the receiver to a location somewhere on the surface of a sphere centered on the satellite (Figure 5-7a). Range measurements from two satellites identify two spheres, and the receiver is located on the circle defined by the intersection of the two spheres (Figure 5-7b). Range measurements from three satellites define three spheres, and these three spheres will intersect at two points (Figure 5-7c). A sequence of range measurements through time from three satellites will reveal that one of the points remains stationary, while the other point moves rapidly through space. The second intersection point moves through space because the size and relative geometry of the spheres changes through time as the satellites change position on their orbital paths. One of the intersection points moves large distances between position fixes, while the other at the receiver location does not. If system and receiver clocks were completely accurate it would be possible to determine the position of a stationary receiver by taking measurements from three satellites over a short time interval. Simultaneous measurements from four satellites (Figure 5-7d) are usually required to reduce receiver clock errors and to allow instantaneous position measurement with a moving receiver, e.g., on a plane, in a car, or while walking. Data may be collected from more than four satellites at a time, and this usually improves the accuracy of position measurements.

Positional Uncertainty

Errors in range measurements and satellite location introduce errors into GPS-determined positions. Range errors vary on short time intervals and so cause the intersection

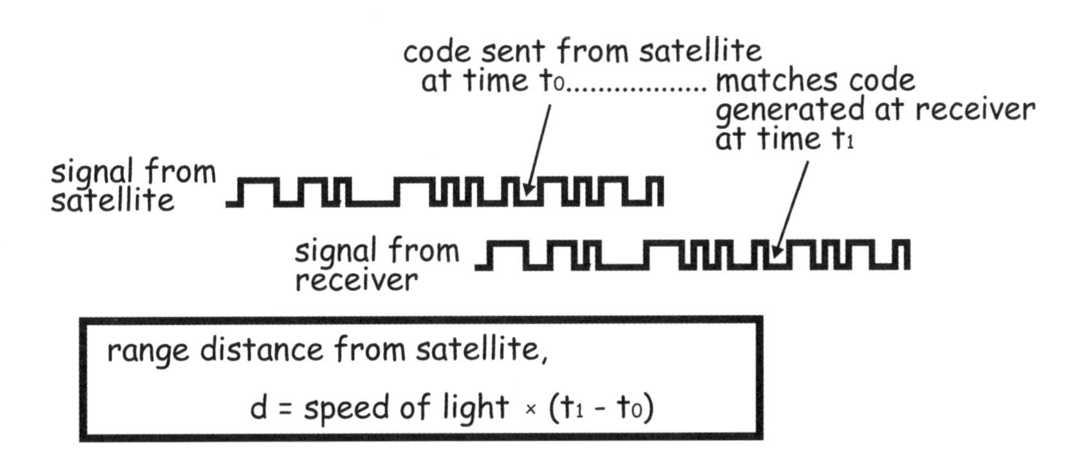

Figure 5-6: A decoded C/A satellite signal provides a range measurement.

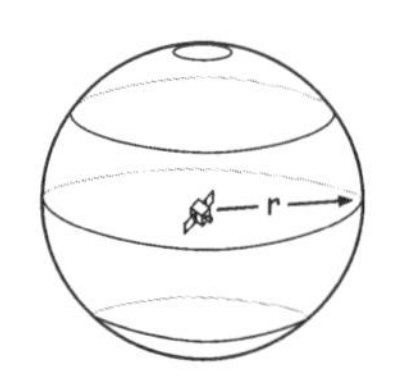

a) with a range measurement from one satellite, the receiver is positioned somewhere on the sphere defined by the satellite position and the range distance, r

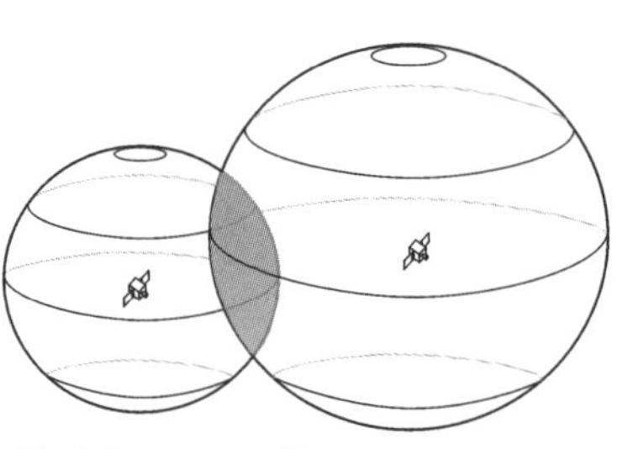

b) with two satellites, the receiver is somewhere on a circle where the two spheres intersect

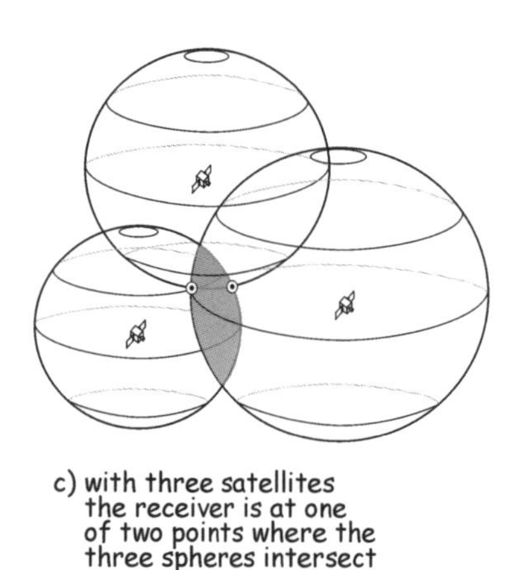

c) with three satellites the receiver is at one of two points where the three spheres intersect

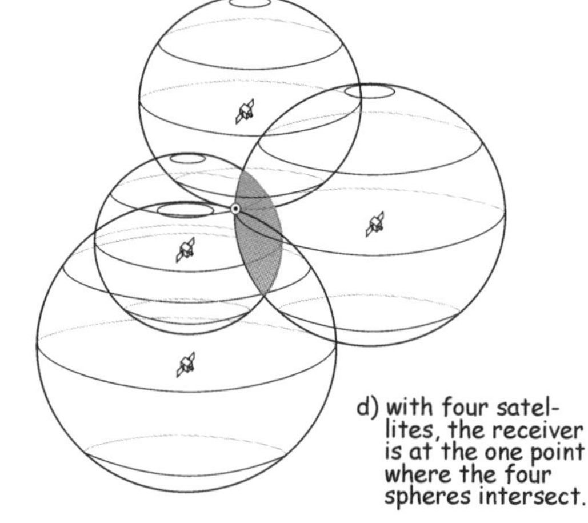

d) with four satellites, the receiver is at the one point where the four spheres intersect.

Figure 5-7: Range measurements from multiple GPS satellites. Range measurements are combined to narrow down the position of a GPS receiver. Range measurements from more than four satellites may be used to improve the accuracy of a measured position. (adapted from Hurn, 1989)

of the range spheres to change through time, even when the GPS receiver is in a fixed location. This results in a band of uncertainty encompassing the GPS receiver position (Figure 5-8). Errors in the ephemeris data lead to erroneous estimates of the satellite position, and also result in an offset or positioning error.

Several methods are used to reduce positional error. One common method involves collecting many position fixes while remaining stationary. These multiple position fixes are then averaged, yielding a mean position estimate. Multiple fixes also provide an estimate of the variation in the position fixes, so both a mean and standard deviation may be calculated. While the standard deviation provides no information on the absolute error, it does allow some estimate of the precision of the mean GPS position fix. The magnitude and properties of this variation may be useful in evaluating GPS measurements and the source of errors. However, collecting multiple position fixes is not possible when data collection must take place while moving, e.g., when determining the location of an airborne plane. Also, averaging does not remove any bias in the calculated position. Alternative methods for reducing positional error rely on reducing the several sources of range errors.

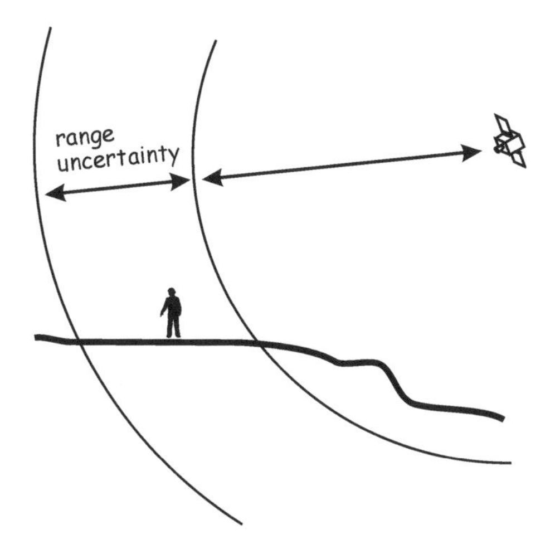

Figure 5-8: Uncertainty in range measurements leads to positional errors in GPS measurements.

Sources of Range Error

Atmospheric and *ionospheric delays* are major sources of GPS positional error. Range calculations depend on the speed of light. While we usually assume the speed of light is a constant, this is not true. The speed of light is constant only while passing through a vacuum and uniform electromagnetic field. The Earth is surrounded by a blanket of charged particles called the ionosphere. These charged particles are formed by incoming solar radiation, which strips electrons from elements in the upper atmosphere. Changes in the charged particle density through space and time result in a changing electromagnetic field surrounding the Earth. The atmosphere is below the ionosphere, and atmospheric density is significantly different from that of a vacuum. Variation in the atmospheric density is due largely to changes in temperature, atmospheric pressure, and water vapor, and atmospheric density is highly variable through space and time. The GPS signal must pass through the ionosphere and atmosphere, resulting in range errors.

We may reduce atmospheric and ionospheric errors through adjustments in the value used for the speed of light. Physical models may be developed which incorporate measurements of the ionospheric charge density. Because the charge density varies both around the globe and through time, and because there is no practical way to measure and disseminate the variation in charge density in a timely manner, the average charge density may be used. These physical models reduce the range error somewhat. Alternatively, correction could be based on the observation that the change in the speed of light depends on the frequency of the light. Specialized *dual frequency* receivers collect information on the L1 and L2 and other GPS signals simultaneously, and use sophisticated physical models to remove most of the ionospheric errors. Dual frequency receivers are limited though, because they are typically much more expensive than code-based receivers, and because they must maintain a continuous fix for a longer period of time. Finally, there are no good models for atmospheric effects, thus there is no analytical method to remove range errors due to atmospheric delays.

System operation and delays are other sources of range uncertainty. Small errors in satellite tracking cause errors in satellite positional measurements. Timing and other signals are relayed from sources to the Master Control Center and up to the satellites, but there are uncertainties and delays in signal transmission, so timing signals may be slightly offset. Atomic clocks on the satellite may be un-synchronized or in error, although this is typically one of the smaller contributions to positional errors.

Receivers also introduce errors into GPS positions. Receiver clocks may contain biases or may use algorithms that do not precisely calculate position. Signals may reflect off of objects prior to reaching the antenna. These reflected, or *multipath* signals travel a further distance than direct GPS signals, and so introduce an offset into GPS positions. Multipath signals often have lower power than direct signals, so some multipath signals may be screened by setting a threshold signal-to-noise ratio. Signals with high noise relative to the mean signal strength are ignored. Multipath signals may also be screened by properly designed antennas. Multipath signals are most commonly a problem in urban settings with an abundance of corner reflectors, such as the sides of buildings and streets.

Satellite Geometry and Dilution of Precision

The geometry of the GPS satellite constellation is another factor that affects positional error. Range errors create an area of uncertainty perpendicular to the transmission direction of the GPS signal. These areas of uncertainty may be visualized as a set of nested spheres, with the true position somewhere within the volume defined between these nested spheres (Figure 5-9). These areas of uncertainty from different satellites intersect, and the smaller the intersection area, the more accurate the position fixes are likely to be (Figure 5-10). Signals from widely spaced satellites are complementary because they result in a smaller area of uncertainty. Signals from satellites in close proximity overlap over broad areas, resulting in large areas of positional uncertainty. Widespread satellite constellations provide more accurate GPS position measurements.

Satellite geometry is summarized in a number called the *Dilution of Precision*, or DOP. There are various kinds of DOPs, including the Horizontal (HDOP), Vertical (VDOP), and Positional (PDOP) Dilution of Precision. The PDOP is commonly used and is defined as the ratio of the volume of an ideal tetrahedron to the volume of the tetrahedron formed by four ideally widespread satellites. The ideal tetrahedron is formed by one satellite overhead and three satellites spaced at 120-degree intervals around the horizon. This constellation is assigned a PDOP of one, and closer groupings of satellites have higher PDOPs. Lower PDOPs are

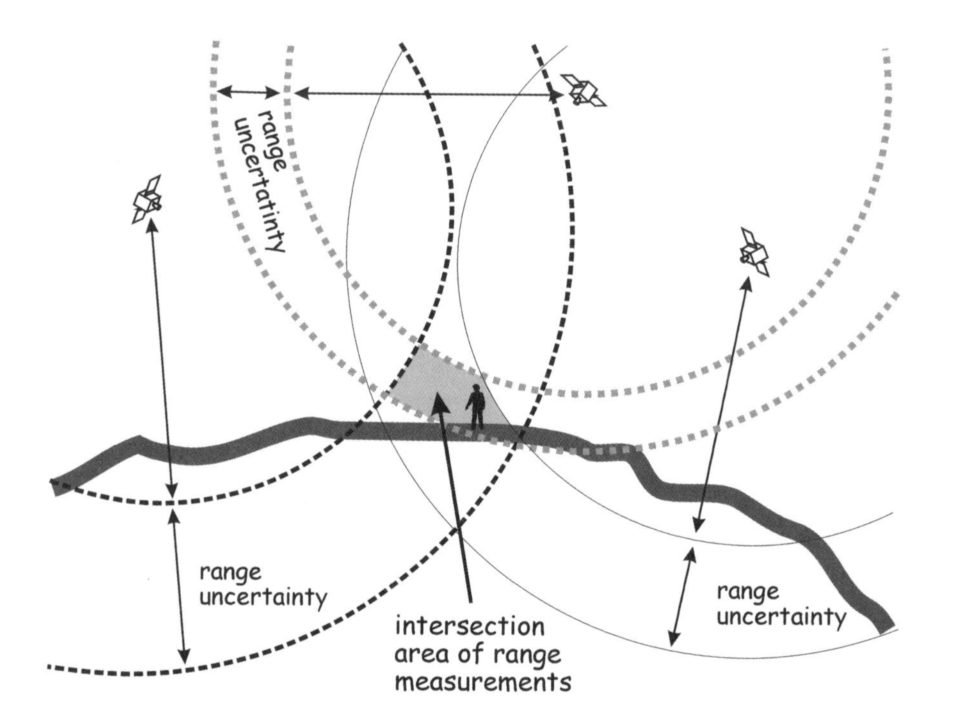

Figure 5-9: Relative GPS satellite position affects positional accuracy. Range uncertainties are associated with each range measurement. These combine to form an area of uncertainty at the intersection of the range measurements.

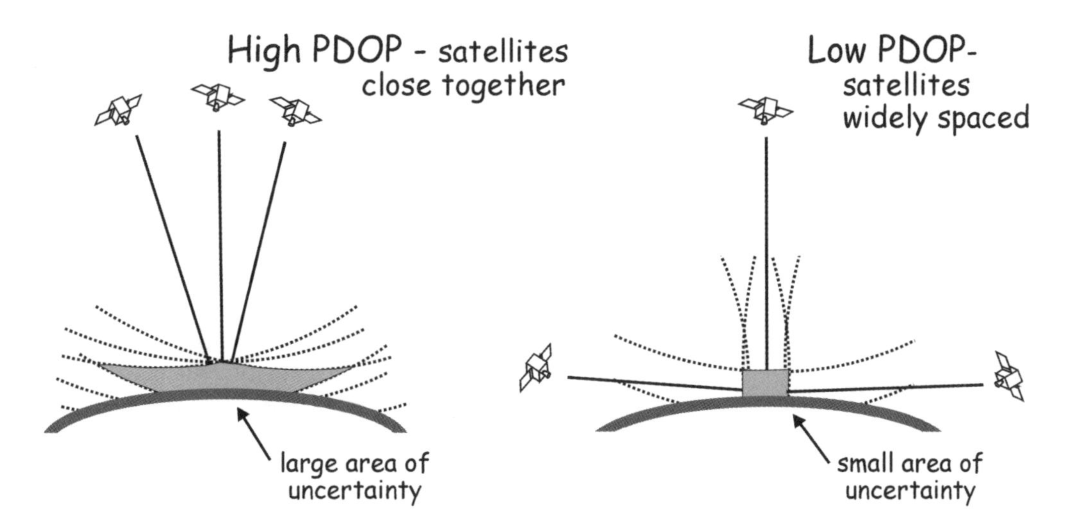

Figure 5-10: GPS satellite distribution affects positional accuracy. Closely-space satellites result in larger positional errors than widely-spaced satellites. Satellite geometry is summarized by PDOP with lower PDOPs indicating better satellite geometries.

better. Most GPS receivers review the almanac transmitted by the GPS satellites and attempt to select the constellation with the lowest PDOP. If this best constellation is not available, e.g., some satellites are not visible, successively poorer constellations are tested until the best available constellation is found. The receivers typically provide a measurement of PDOP while data are collected, and a maximum PDOP threshold may be specified. Data are not gathered when the PDOP is above the threshold value.

Range errors and DOPs combine to affect GPS position accuracies. There are many sources of range error, and these combine to form an overall range uncertainty for the measurement from each visible GPS satellite. These range measurements are combined to determine a position, but the accuracy of the position depends on both the quality of the range measurements and the satellite geometry, determined by the DOP. If more precise coordinate locations are required, then the choices are to use equipment that makes more precise range measurements, and/or to collect data when DOPs are low.

GPS accuracies depend on the type of receiver, atmospheric and ionospheric conditions, the number of range measurements, the satellite constellation, and the algorithms used for position determination. Current C/A code receivers typically provide accuracies between 3 and 30 meters for a single fix. Errors larger than 100 meters for a single fix occur occasionally. Accuracies may be improved substantially, to between 2 and 15 meters, when multiple fixes are averaged. The longer the data collection time the greater the accuracy. Improvements come largely from reducing the impact of rarer, large errors, but average accuracies are rarely below one meter when using a single C/A code receivers.

Accuracies when using carrier phase or similar receivers are much higher, on the order of a few centimeters. These accuracies come at the cost of longer data collection times, and are most often obtained when using differential correction, a process described in the following section.

Differential Correction

The previous sections have focused on position measurements using a single receiver. This operating mode is known as autonomous GPS positioning. An alternative method, known as *differential GPS positioning*, employs two or more receivers. Differential positioning is used primarily to remove most of the range errors and thus greatly improve the accuracy of GPS positional measurements.

Differential GPS positioning entails establishing a *base station* receiver over a known coordinate point (Figure 5-11). The true coordinate location of the base station is typically determined using high-accuracy survey methods, for example, repeated astronomical observations or precise ground surveys, as described in Chapter 3. The location of the base station is also estimated using the GPS satellite signals. The GPS location may be determined with a high frequency, for example, separate position fixes may be calculated each second. The difference between the true and GPS positions may be computed each time a GPS fix is taken at the base station. These differences between the GPS and known positions at a base station gives differential GPS positioning its name.

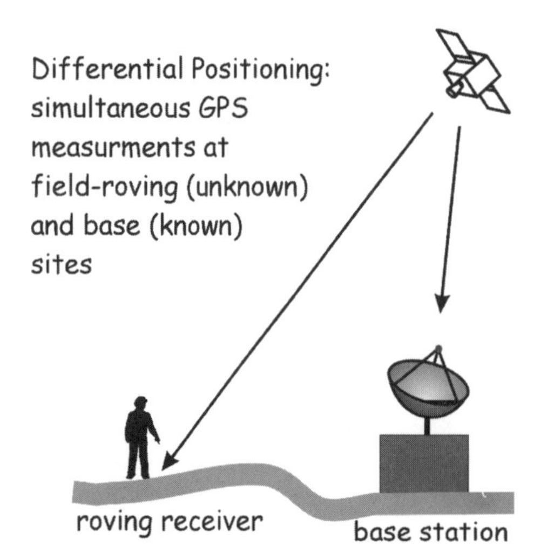

Figure 5-11: Differential GPS positioning.

A difference between a true and GPS position defines an error vector for each GPS fix taken at the base station (Figure 5-12). This error vector is characterized by a distance and a direction. The complement of this error vector has the same distance but is in the opposite direction. The complement is calculated for each position fix. Thus, for each set of satellites we may calculate an error vector between the GPS and true location for the exact time that the fix was taken. These vectors change in distance and direction over a very short time interval, e.g., an error for a fix taken now will likely be quite different for an error taken 10 seconds later. However, if I know the precise time of a fix, I can specify the error.

We use the calculated error vectors to correct GPS positions measured with *roving* GPS receivers (Figure 5-12). These roving receivers are used to measure GPS positions at remote locations in the field. Each error vector at a field location is the same as the error vector at the base station, provided the rover and base station positions are calculated based on the same set of GPS satellites and the measurements are taken at the same time. We may apply the difference vector measured at the base station to correct the data collected by the rover. This removes many errors and substantially improves each position fix taken with the roving field receivers.

There are limits on the application of differential correction. First, the base station and roving receivers must collect data from the same set of satellites. The position estimate for any given fix depends heavily on the set of satellites in the constellation used to estimate the position. Positional errors when using one set of satellites are not likely to be the same as positional errors when using a different set of satellites, even if the sets differ by just one satellite.

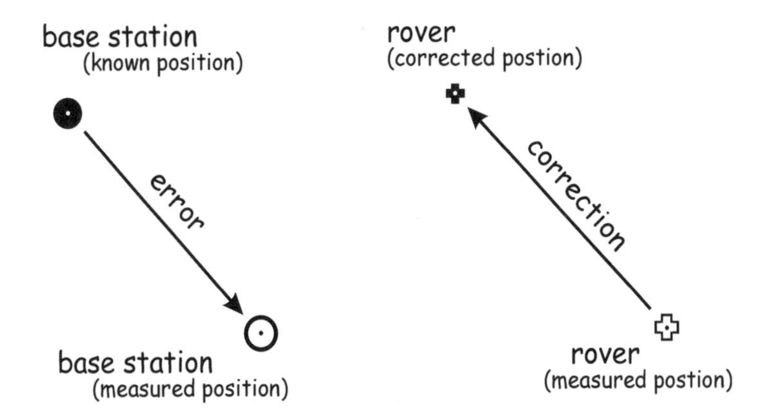

Figure 5-12: Differential correction is based on measuring a GPS position error at a base station, and applying the error as a correction to a simultaneously measured rover position.

The simultaneous viewing requirement limits the distance between the base and roving receivers. The farther apart the receivers, the more likely they will not view the same set of satellites. While differential positioning has been successful at distances of over 1500 kilometers (1000 miles), best performance is found when the roving receiver is within 300 kilometers (180 miles) of the base station.

Successful differential correction also requires near simultaneity in the base and rover measurements. Errors change rapidly through time. If the base and rover measurements are collected more than a few tens of seconds apart they do not correspond to the same set of errors, and thus the difference at the base station cannot be used to correct the rover data. Many systems allow data collection to be synchronized to a standard timing signal, thereby ensuring a good match when the error vectors are applied to correct the roving-receiver GPS data.

Base station data and roving receiver data must be combined before the differential correction may be applied. A base station correction may be calculated for each fix, but this correction must somehow be joined with the roving receiver data to apply the correction. Many receivers allow large amounts of data to be stored, either internal to the receiver, or to an attached computer. Files may be downloaded from the base station and roving units to a common computer. Software provided by most GPS system vendors is then used to combine the base and rover data and compute and apply the differential corrections to the position fixes. This is known as post-processed differential correction, as corrections are applied after, or post, data collection (Figure 5-13, top).

Post-processed differential positioning is suitable for many field digitization activities. Road locations may be digitized with a GPS receiver mounted to the top of a vehicle. The vehicle is driven over the roads to be digitized, and rover data recorded simultaneously with a base station. Base and rover data are downloaded to a process computer and differential corrections computed and applied. These roads data may then be further processed to create a data layer in a GIS.

Post-processing differential positioning has one serious limitation. Because precise positions are not known when the rover is in the field, post-processing technologies are useless when precise navigation is required. A surveyor recovering buried or hidden property corners often needs to navigate to within a meter of a position while in the field, so that monuments, stakes, or other markers may be recovered. When using

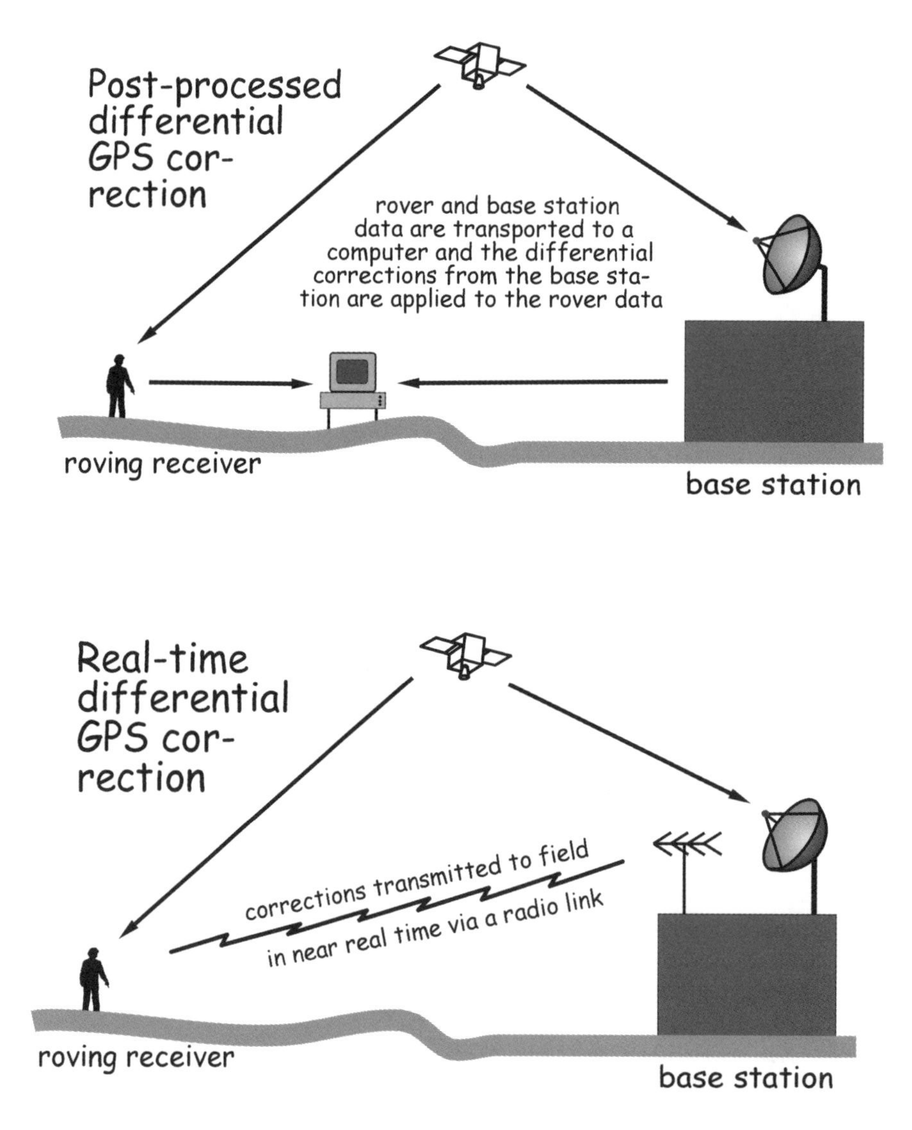

Figure 5-13: Post-processed and real-time differential GPS correction.

post-processed differential GPS the field receiver is operating as an autonomous positioning device, and accuracies of a few meters to tens of meters are expected. This is not acceptable for many navigation purposes because too much time will be spent searching for the final location when one is close to the destination. Alternative methods, described in the next section, provide more precise in-field position determination.

Real-Time Differential Positioning

An alternative GPS method, known as *real-time differential correction*, may be appropriate when precise navigation is required. Real-time differential correction requires some extra equipment and there is some cost in slightly lower accuracy when compared to post-processed differential GPS. However, the accuracy of real-time differential correction is substantially better than autonomous GPS, and accurate locations are determined while still in the field.

Real-time differential GPS positioning requires a communications link between the base station and the roving receiver (Figure 5-13, bottom). Typically the base station is connected to a radio transmitter and antenna. FM radio links are often used due to their longer range and good transmission through vegetation, into canyons or deep valleys, or into other constrained terrain. The base station collects a GPS signal and calculates position fixes. The error is calculated for each position fix. The magnitude and direction of each error is passed to the radio transmitter, along with information on the timing and satellite constellation used. This continuous stream of corrections is broadcast via the base-station radio and antenna.

Roving GPS receivers are outfitted with a receiving radio, and any receiver within the broadcast range of the base station may receive the correction signal. The roving receiver is also recording GPS data and calculating position fixes. Each position fix by the roving receiver is matched to the corresponding correction from the base-station radio broadcast. The appropriate correction is then applied to each fix and accurate field locations computed in real time.

Real-time differential correction requires a broadcasting base station; however, every user is not required to establish a base station and radio or other communication system. The U.S. Coast Guard has established a set of *GPS radio beacons* in North America that broadcast a standardized correction signal. Any roving GPS receiver within the broadcast range of these radio beacons may use the standardized signal for differential correction. These GPS *beacon receivers* typically have an additional antenna and electronics for receiving and decoding the radio beacon signal. The radio beacons were originally intended to aid in ship navigation and control in coastal and major inland waters, so beacons are concentrated near the Atlantic, Pacific, and Gulf Coasts of the United States, the Great Lakes/ St. Lawrence Seaway, and along the Mississippi River. There are plans to extend the network to the entire U.S. so that the beacons may form the basis for real-time differential correction anywhere in the country. Many GPS manufacturers sell beacon receiver packages that support real-time correction using the Coast Guard beacon signal.

WAAS and Satellite-based Corrections

There are alternatives to ground-based differential correction for improving the accuracy of GPS observations. One alternative, known as the Wide Area Augmentation System (WAAS), is administered by the U.S. Federal Aviation Administration to provide accurate, dependable aircraft navigation. WAAS is designed to provide real-time accuracies from single fixes to within seven meters or less. Accuracies should be better when multiple fixes are averaged.

WAAS is based on a network of ground reference stations scattered about North America. Signals from GPS satellites are received at each station and errors calculated, as with differential GPS. A generalized correction is calculated based on location and this correction is transmitted to a geo-stationary satellite. The correction signal may then be broadcast. The correction is collected by a WAAS-compatible roving receiver and applied to each position measurement. Preliminary tests indicate individual errors are less than seven meters 95 percent of the time, and average errors when collecting for 30 minutes are between one and three meters. This is a substantial improvement over uncorrected C/A code, where errors above 15 meters are common. WAAS is not as accurate as differential correction, but may be a substantial improvement over uncorrected GPS.

Commercial, satellite-based differential correction systems are also available. These systems, such as Omnistar and Landstar, consist of a set of GPS base stations distributed across a region. Stations may be widely spaced. Data are gathered from these base stations, including GPS positioning signals and satellite location, and correction vectors calculated. These vectors are then packaged and relayed to communications satellites, from where they are broadcast back to roving GPS receivers. Satellite-based differential correction systems require the use of a compatible GPS receiver. The primary modification is an additional antenna to receive the satellite signal, plus electronics to decode and apply error vectors. These services claim to provide sub-meter real-time accuracies.

GPS Applications

Navigation, tracking, field digitizing, and surveying are the main applications of GPS. Tracking may be seen as a GIS application. It involves noting the location of objects through time. A common example is delivery vehicle tracking in near real time. Large delivery and distribution organizations frequently require information on the location of a fleet of vehicles. Vehicles equipped with a GPS receiver and a radio broadcast link may report back to a dispatch office every few seconds. In effect, the dispatcher may have a real-time map of vehicle location. Icons on a digital map are used to represent vehicle location. A quick glance may reveal which vehicle is nearest a delivery or retrieval site, or which driver overly frequents a donut shop.

Navigation is a second common GPS application. GPS receivers have been developed specifically for navigation, with digital maps or compasses set into on-screen displays (Figure 5-14). Digital maps may be uploaded to these GPS receivers from larger databases, and streets, water features, topography, or other spatial data shown as a back-

Figure 5-14: A GPS receiver developed for field navigation. (courtesy Garmin Corp.)

ground. Directions to identified points may be displayed, either as a route on the digital map, or as a set of instructions, e.g., directions to turn at oncoming streets. These GPS receivers and digital maps may be considered as an extremely specialized GIS system. These systems are useful when collecting or verifying spatial data, e.g., to navigate to the approximate vicinity of communications towers for which a set of attribute measurements will be collected.

Field digitization is a primary application of GPS in GIS. Data may be recorded directly in the field to update point, line, or area locations. Features are visited or traversed in the field, and an appropriate number of GPS fixes collected. GPS receivers have been carried in automobiles, on boats, bicycles, or helmets, or by hand to capture the coordinate locations of points and boundaries (Figure 5-15).

GPS data are often more accurate than data collected from the highest-quality cartometric maps. For example, differentially corrected C/A code GPS data typically have accuracies below five meters, and often below two meters, while accuracies are often near 15 to 20 meters for data collected via manual digitization of 1:24,000-scale maps. Precise differential correction of carrier-phase GPS data often yield centimeter-level accuracies, far better than can be obtained from digitizing all but the largest-scale maps.

Figure 5-15: Line features may be field-digitized via GPS, as in this example of a GIS/GPS system mounted in an automobile. (courtesy USDI)

Digitization with GPS often involves the capture of both coordinate and attribute data in the field. Typically the GPS receiver is activated and acquires a set of satellites. A file is opened and position fixes are logged at some fixed rate, e.g., every 2 seconds. Attribute data may also be entered, either while the position fixes are being collected, or before or after positional data collection. In some software the position fixes may be tagged or identified. For example, a specific corner may be tagged while digitizing a line. Multiple features may be collected in one file and the identities maintained via attached attributes. Data are processed as needed to improve accuracy, and converted to a format compatible with the GIS system in use. GPS data collection and data reduction tools often provide the ability to convert multiple fixes into a single point average. These functions may be applied for all position fixes in a file, or for a subset of position fixes embedded in a GPS file.

GPS field digitization is most commonly used for point and line feature data collection. Lines and points can be unambiguously digitized using GPS. Multiple position fixes provide higher accuracies, so multiple fixes are often collected for point locations. Multiple position fixes are often collected for important vertices in line data. However GPS data collection for area features suffer from a number of unique difficulties. First, it takes considerable time to traverse an area, so relatively large parcels or many small parcels may be impractical to digitize in the field. Second, the problem of multiple representations of the same boundary occurs when digitizing polygonal features. It is impossible to walk exactly the same line and record the same coordinates when GPS-digitizing a new polygon that is adjacent to an already digitized polygon. Attempting to retrace the common boundary wastes time and provides redundant and conflicting data. The alternative is to digitize only the new lines, and snap to "field-nodes", much as when capturing data using a coordinate digitizer (see Chapter 4). Ensuring that lines connect properly is difficult when digitizing with a GPS receiver, because it is often difficult or impractical to mark a boundary or starting point already digitized. On-screen displays, particularly with real-time differential corrections, provide one means to ensure connection.

Imaginative uses of GPS are arising almost daily as this technology revolutionizes the collection of positional data. GPS equipment has been interfaced with grain harvesting equipment. Grain production is recorded during harvest, so that yield and quality is mapped every few meters in a farm field. This allows the farmer to analyze and improve production on a site-specific basis, for example, by tailoring fertilizing applications for each square meter in the field. The mix of fertilizers may change with position, again controlled by a GPS receiver and software carried aboard a tractor.

GPS tracking of animal movement is another new and innovative application. GPS is facilitating a revolution in animal movement analysis because of the frequency and density of points that may be collected (Figure 5-16). More position fixes may be collected in a month using GPS equipment than may be collected in a decade using alternative methods.

Animal movement analysis has long been based on observation of recognizable individuals. Each time a recognizable animal is seen the location is noted. The number of position fixes is often low, however, because some animals are difficult to spot, elusive, or live in areas of dense vegetation or varied terrain. Alternative methods have been developed that are based on *radiotelemetry*.

Figure 5-16: A wildebeest fit with a GPS tracking collar. The antenna is visible as the white patch on top of the collar, and the power supply and data logging housing is visible at the bottom of the collar. Animal position is tracked day and night, yielding substantially improved information on animal activity and habitat use. (Courtesy Gordon T. Carl, Lotek Wireless Inc.)

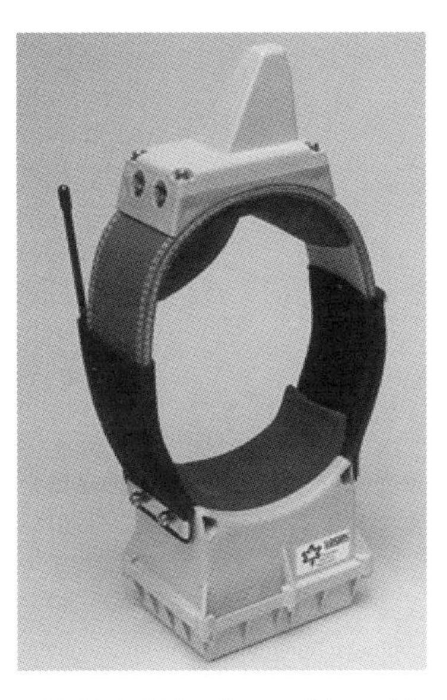

Figure 5-17: A GPS collar used in tracking animal movement. (courtesy Lotek Wireless, Inc.)

Radiotelemetry involves the use of a transmitting and receiving radio unit to determine animal location. A transmitting radio is attached to an animal, and a technician in the field uses a radio receiver to determine the position of the animal. Measurements from several directions are combined and the approximate location of the animal may be plotted.

GPS animal tracking is a substantial improvement over previous methods. GPS units are fit to animals, usually by a harness or collars (Figure 5-17). The animals are released, and positional information recorded by the GPS receiver. Logging intervals are variable, from every few minutes to every few days, and data may be periodically downloaded via a radio link. Systems may be set up with an automatic or radio-activated drop mechanism, so that data may be downloaded and the receiver re-used. While only recently developed, GPS-based animal tracking units are currently in use on all continents in the study of threatened, endangered, or important species.

GPS has other applications in GIS. GPS is particularly helpful when collecting control point data. Remember that control points are used to correct and transform map or image data to real-world coordinates. Control points may be difficult or impossible to obtain directly from the information plotted on a map, particularly when graticule or gridlines are absent. GPS offers a direct method for measuring the coordinates for potential control points represented on the map. Road intersections or other points may be identified on the map and then visited with a GPS receiver.

GPS-measured control points are also often used in analytical correction of aerial imagery (see Chapter 6). Most image data are not initially in a map coordinate system, yet images are often particularly useful for developing or updating spatial data. Aerial photographs contain detailed information. However, aerial photographs are subject to geometric distortion. These errors may be analytically corrected through suitable methods (see Chapter 6), but these methods require several control points per image, or at least per project when stereopairs are used. GPS significantly reduces the cost of control point collection, thereby making single- or multi-photo correction a viable alternative for most organizations that collect spatial data.

Summary

GPS is a satellite-based positioning system. It is composed of user, control, and satellite segments, and allows precise position location both quickly and with high accuracy.

GPS is based on range measurements. These range measurements are derived from measurements of a broadcast signal that may be either coded or uncoded. Range measurements from multiple satellites may be combined to estimate position.

GPS positional estimates contain error due to uncertainties in satellite position, atmospheric and ionospheric interference, multipath reflectance, and poor satellite geometry. Position estimates may be improved via differential correction.

Suggested Reading

Bergstrom, G., GPS in forest management, *GPS World*, 1990, 10:46-49.

Bobbe, T., Real-time differential GPS for aerial surveying and remote sensing, *GPS World*, 1992, 4:18-22.

Deckert, C. J. and Bolstad, P. V., Forest canopy, terrain, and distance effects on global positioning system point accuracy, *Forest Science*, 1996, 62:317-321.

Fix, R. A. and Burt, T. P., Global Positioning Systems: an effective way to map a small area or catchment, *Earth Surface Processes and Landforms*, 1995, 20:817-827.

Hurn, J., GPS, a guide to the next utility, Trimble Navigation Ltd., Sunnyvale, 1989.

Kennedy, M., The Global Positioning System and GIS, Ann Arbor Press, Ann Arbor, 1996.

Welch, R., Remillard, M., and Alberts, J., Integration of GPS, remote sensing, and GIS techniques for coastal resource management, *Photogrammetric Engineering and Remote Sensing*, 1992, 58:1571-1578.

Wilson, J.P., Spangrud, D.S., Nielsen, G.A., Jacobsen, J.S., and Tyler, D.A., GPS sampling intensity and pattern effects on computed terrain attributes, Soil Science Society of Marica Journal, 1998, 62:1410-1417.

Zygmont, J., Keeping tabs on cars and trucks, *High Technology*, 1986, 18-23.

Study Questions

Can you describe the general components of the global positioning system, including the three common segments and what they do?

What is the basic principle behind GPS positioning? What is a range measurement, and how does it help you locate yourself?

Can you describe the GPS signals that are broadcast, and the basic difference between carrier and coded signals?

How many satellites must you measure to obtain a 3-dimensional position fix?

What are the main sources and relative magnitudes of uncertainty in GPS positioning?

How accurate is GPS positioning? Be sure you specify a range, and describe under what conditions accuracies are at the high and low end of the range.

What is a dilution of precision (DOP)? How does it affect GPS position measurements?

Can you describe the basic principle behind differential positioning?

What is the difference between post-processed and real-time differential positioning? Can you list three pros and cons of each?

What is WAAS? Is it better or worse than ground-based differential positioning?

6 Aerial and Satellite Images

Introduction

Aerial and satellite images are a valuable and common source of data for GIS. These images are collected remotely, in that they record information from a distance; thus, photos and satellite images are often referred to as *remotely-sensed* data. Remotely-sensed data come in many forms, however in the context of GIS we usually use the term to describe images that have been recorded with aerial cameras or satellite scanners. While photographs are occasionally obtained from satellite-borne cameras, and aerial image scanners are currently in operation, aerial cameras and satellite scanners are the two primary image sources. Whatever their origin, images are a rich source of spatial information and have been used as a basis for mapping for more than seven decades.

Remotely sensed images provide a number of positive attributes when used as sources of spatial data. These positive attributes include:

Large area coverage – images capture data at a relatively low cost and in a uniform manner (Figure 6-1). For example, it would take months to collect enough ground survey data to accurately produce a topographic map for 10 square kilometers. Photographs of a region this size may be collected in a few minutes and the topographic data extracted and interpreted in a few weeks.

Extended spectral range – photos and scanners can detect light from wavelengths outside the range of human eyesight. Some kinds of aerial photographs are sensitive into the infrared region, a portion of the light spectrum that the human eye cannot sense. Satellite scanners sense even broader spectral ranges, up to thermal wavelengths and beyond. This expanded spectral range allows us to detect features or phenomena that would otherwise not be visible.

Geometric accuracy – remotely-sensed data are or may be converted to geometrically accurate spatial data. Aerial photographs are the source of many of our most accurate large-area maps. Under some conditions aerial photographs contain geometric distortion. These distortions occur because of imperfections in the camera, lens, or film systems, or due to camera tilt or terrain variation in the area being imaged. Satellite scanners may also contain errors due to the imaging equipment or satellite platform. However, distortion removal methods are well known, and techniques and equipment have been developed to provide highly accurate spatial data from images. Cameras and imaging scanners have been developed specifically for the purpose of quantitative mapping. These systems are combined with techniques for identifying and removing most of the spatial error in aerial or satellite images, so spatially accurate data may be collected from images.

Permanent record – an image is fixed in time, so the conditions at the time of the photograph may be analyzed many years hence. Comparison of conditions over multiple dates, or determination of conditions at a specific date in the past are often quite valuable, and remotely-sensed images are often the most accurate source of historical information.

Figure 6-1: Images are a valuable source of spatial data. The upper image, centered on northeastern Egypt, illustrates the broad-area coverage provided by satellite data. The lower image of pyramids in Egypt illustrates the high spatial detail that may be obtained. (courtesy NASA, top, and Space Imaging, bottom)

Basic principles

The most common forms of remote sensing are based on reflected electromagnetic energy. When energy from the Sun or other source strikes an object, a portion of the energy is reflected. Different materials reflect different amounts of incoming energy, and this differential reflectance gives objects a distinct appearance. We use these differences to distinguish among objects.

Light energy is the principal form detected in remote sensing for GIS. Light energy is characterized by its *wavelength*, the distance between peaks in the electromagnetic stream. Each "color" of light has a distinctive wavelength, e.g., we perceive light with wavelengths between 0.4 and 0.5 micrometers (μm) as blue. Light emitted by the Sun is composed of several different wavelengths, and the full range of wavelengths is called the *electromagnetic spectrum.* A graph of this spectrum may be used to represent the amount of incident light energy across a range of wavelengths (Figure 6-2). The amount of energy in each wavelength is typically plotted vs. each wavelength value, yielding a curve depicting total electromagnetic energy reaching any object. Notice in Figure 6-2 that the amount of energy emitted by the Sun increases rapidly to a maximum between 0.4 to 0.7 μm, and drops off at higher wavelengths. Some regions of the electromagnetic spectrum are named, e.g. X-rays have wavelengths of approximately 0.0001 μm, visible light is between 0.4 and 0.7 μm, and near-infrared light is between 0.7 and 1.1 μm.

Our eyes perceive light in the visible portion of the spectrum, between 0.4 and 0.7 μm. We typically identify three base colors: blue, from approximately 0.4 to 0.5 μm, green from 0.5 to 0.6 μm, and red from 0.6 to 0.7 μm. Other colors are often described as a mixture of these three colors at varying levels of brightness. For example, an equal mixture of blue, green, and red light at a high intensity is perceived as "white" light. This same mixture but at lower intensities produces various shades of gray. Other colors

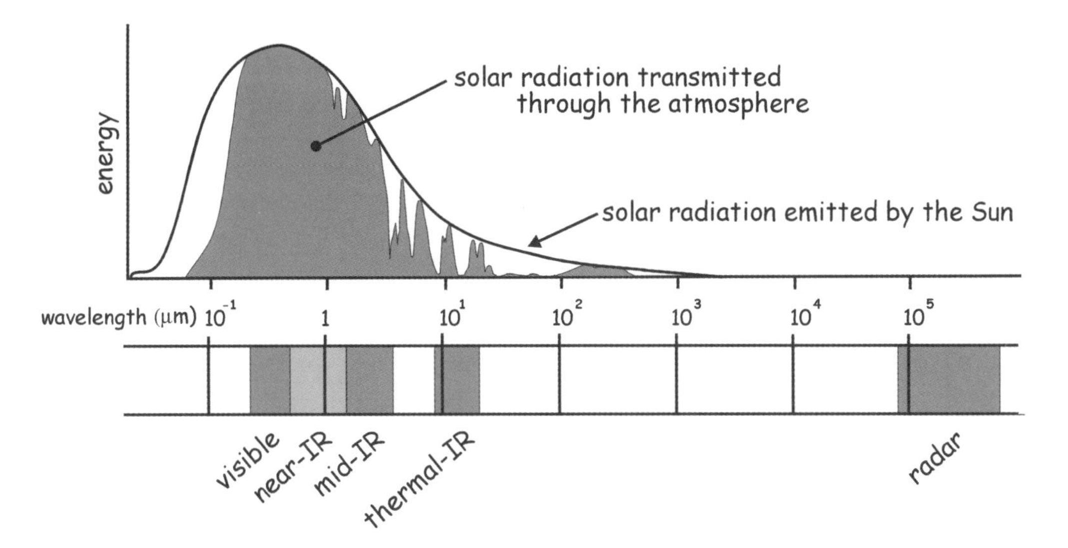

Figure 6-2: Electromagnetic energy is emitted by the Sun and transmitted through the atmosphere (upper graph). Solar radiation is partially absorbed as it passes through the atmosphere. This results in variable surface radiation in the visible and infrared (IR) wavelength regions (lower graph)

are produced with other mixes, e.g., equal parts red and green light are perceived as yellow. The specific combination of wavelengths and their relative intensities produce all the colors visible to the human eye.

Electromagnetic energy striking an object is reflected, absorbed, or transmitted. Most solid objects absorb or reflect incident electromagnetic energy and transmit none. Liquid water and atmospheric gasses are the most common natural materials that transmit light energy as well as absorb and reflect it.

Transmittance through the atmosphere is most closely tied to the amount of water vapor in the air. Water vapor absorbs energy in several portions of the spectrum, and higher atmospheric water content results in lower transmittance (Figure 6-2). Carbon dioxide, other gasses, and particulates such as dust also contribute to atmospheric absorption.

Natural objects appear to be the color they most reflect, e.g., green leaves absorb more red and blue light and reflect more green light; hence, they appear green. We use these differences in reflectance properties across a range of wavelengths to distinguish among objects. While we perceive differences in the visible wavelengths, these differences also extend into other portions of the electromagnetic spectrum (Figure 6-3). For example, individual leaves of many plant species appear to be the same shade of green, however some reflect much more energy in the infrared portion of the spectrum, and thus appear to have a different "color" when viewed at these wavelengths.

Most remote-sensing systems are *passive*, in that they use energy generated by the Sun and reflected off of the target objects. Aerial photographs and most satellite data are collected using passive systems. Because they rely on an external energy source, the images from these passive systems may be impacted by atmospheric conditions in multiple ways. Figure 6-4 illustrates the many paths by which energy reaches a remote sensing device. Note that only one of the energy paths is useful, in that only the surface-reflected energy provides information on the feature of interest. The other paths result in no or only diffuse radiation reaching the sensor, and provide no useful information about the target objects. Most passive systems are not useful during cloudy or extremely hazy periods because nearly all the energy is scattered and no directly

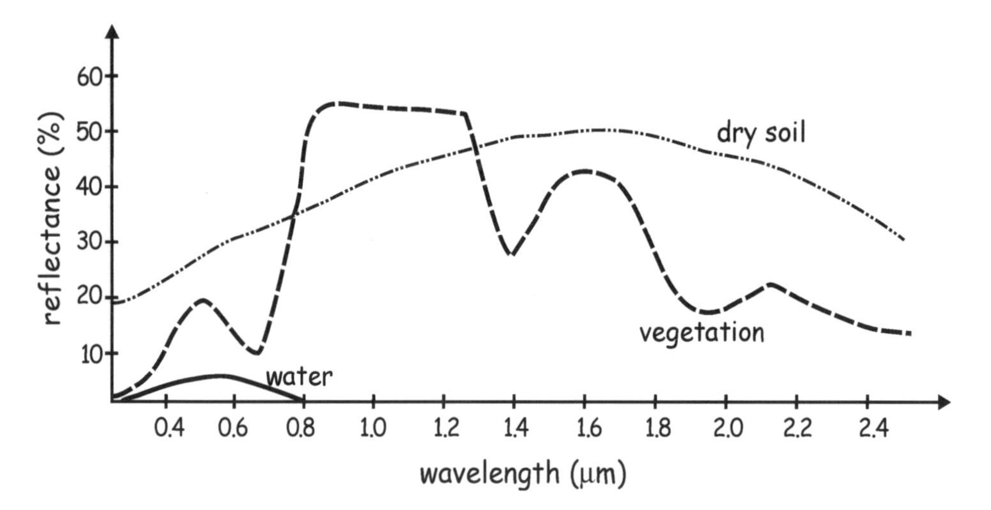

Figure 6-3: Spectral reflectance curves for some common substances. The proportion of incoming radiation that is reflected varies across wavelengths. (adapted from Lillesand and Kiefer, 1999)

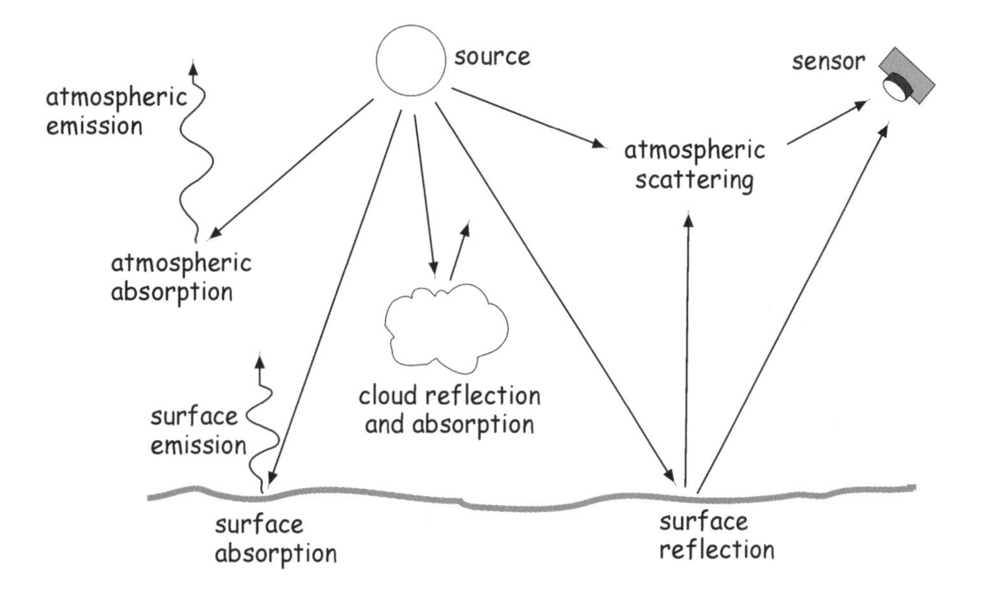

Figure 6-4: Energy pathways from source to sensor. Light and other electromagnetic energy may be absorbed, transmitted, or reflected by the atmosphere. Light reflected from the surface and transmitted to the sensor is used to create an image. The image may be degraded by atmospheric scattering due to water vapor, dust, smoke, and other constituents. Incoming or reflected energy may be scattered. In addition, absorbed energy may be re-radiated at longer wavelengths, forming the basis of thermal remote sensing.

reflected energy may reach the sensor. Most passive systems rely on the Sun's energy, so they have limited use at night, although some of these systems have been used to monitor urban and industrial lights.

Active systems are an alternative for gathering remotely sensed data under cloudy or nighttime conditions. Active systems generate an energy signal and detect the energy returned. Differences in the quantity and direction of the returned energy are used to identify the type and properties of features in an image. RADAR (radio detection and ranging) is the most common active remote sensing system, while LIDAR systems (light detection and ranging), based on lasers, are finding increasing applications. Radar focuses a beam of energy through an antenna, and then records the reflected energy. These signals are swept across the landscape, and the returns assembled to produce a radar image. Because a given radar system is typically restricted to one wavelength, radar images are usually monochromatic (in shades of gray). These images may be collected day or night, and most radar systems penetrate clouds because water vapor does not absorb the relatively long radar wavelengths.

Image scale and *extent* are important attributes of remotely sensed data. Image scale, as in map scale, is defined as the relative distance on the image to the corresponding distance on the ground. For example, one inch on a 1:15,840-scale photograph corresponds to 15,840 inches on the Earth surface. As we shall see later, photographs and other images may contain distortions typically not found in maps, and these distortions often result in a non-uniform scale across the photograph. Thus, we usually specify an average scale for a photograph.

Image extent is the area covered by an image. The extent depends on the physical size of the image, e.g., a 9 x 9 inch photograph. Image extent also depends on the image scale. The scale multiplied by the edge sizes of the photograph gives the dimensions of the image extent. For example, a 1:15,840-scale photograph on 9 x 9 inch paper has an image extent of approximately 2.25 by 2.25 miles, or 5.1 square miles. In metric equivalents this corresponds to a 22.5 by 22.5 centimeter photograph, or an area of 12.7 square kilometers on the surface of the Earth.

Image resolution is another important property. The resolution is the smallest object that can reliably be detected on the image. Typically the object color or brightness is assumed to contrast substantially with the background, e.g., a black object on a white background. The resolution on aerial photographs depends on the film grain size, contrast, and exposure properties, and is often tested via photographs of alternating patterns of black and white lines. At some line width the difference between black and white lines cannot be distinguished, and so the film resolution may be reported as this threshold number of lines per millimeter. Digital image resolution is often equated with the raster cell size.

Aerial Photographs

Images taken from airborne cameras are and have historically been a primary source of geographic data. Aerial photography quickly followed the invention of portable cameras in the mid-19th century, and became a practical reality with the development of dependable airplanes in the early 20th century. *Photogrammetry*, the science of measuring geometry from photographs, was well-developed by the early 1930s, with continuous refinements since. Aerial photographs underpin most large-area maps and surveys in most countries. For example, aerial photographs form the basis of the most detailed national map series in the United States, the USGS 1:24,000-scale quadrangle maps. Aerial photographs and photogrammetry are routinely used in urban planning and management, construction, engineering, agriculture, forestry, wildlife management, and other mapping applications.

Primary Uses of Aerial Photographs

Although there are hundreds of applications for aerial photographs, most applications in support of GIS may be placed into three main categories. First, aerial photographs are often used as a basis for surveying and topographic mapping, to measure and identify the horizontal and vertical locations of objects in a study area. Measurements on photographs offer a rapid and accurate way to obtain geographic coordinates, particularly when photo measurements are combined with ground surveys. Second, photographic interpretation may be used to categorize or assign attributes to surface features. For example, photographs are often used as the basis for landcover maps, and to assess the extent of fire or flood damage. Finally, photographs are often used as a backdrop for maps of other features, as when photographs are used as a background layer for soil survey maps produced by the U.S. National Resource Conservation Service.

Photographic Film

Most aerial photographs are based on the exposure of film. Film consists of a sandwich of light-sensitive coatings spread on a thin plastic sheet. Film may be black and white, with a single layer of light-sensitive material, or color, with several layers of light-sensitive material. Each film layer is sensitive to a different set of wavelengths (Figure 6-5). These layers, referred to as the *emulsions*, undergo chemical reactions when exposed to light. More light energy falling on the film results in a more complete chemical reaction, and hence a greater film exposure.

Most black and white films use a thin emulsion containing silver halide crystals (Figure 6-5). These crystals change when exposed to light. After exposure the film is passed through a *developer*, to make the exposed silver halide crystals visible. The film is then passed through a *stop bath* to stop development, a *fixer*, to wash away unexposed crystals, and then through one to several rinsing and drying steps.

Color films are a more complex combination of several dye layers, each sensitive to a different set of wavelengths (Figure 6-5). The basic principles of exposure and development apply. Normal color film has blue, green, and red sensitive layers. In addition, there is typically a filter layer that blocks blue light, because the green and red sensitive dyes are also sensitive to blue light. This blue-light filter is between the blue and other data layers. Because colors are a mixture of blue, green, and red light, light striking the film exposes each layer to some degree. Color is represented by the differences in densities of color in the dye layers when the film is developed.

Films may be categorized by the wavelengths of light they respond to. The most commonly used films are sensitive to light in the visible portion of the spectrum, from 0.4 to 0.7 μm (Figure 6-6). Black and white films sensitive to this range of wavelengths are often referred to as *panchromatic* films. Panchromatic films are often used for aerial photography because they are relatively inexpensive and have a wide exposure range relative to some color films. This wider range means we may obtain a useful image over a wider range of light conditions. *True color* film is also sensitive to light across the visible spectrum, but in three separate colors.

Infrared films have been developed and are widely used when differences in vegetation type or density are of interest. Black and white infrared films are sensitive through the visible spectrum and longer infrared wavelengths, up to approximately 0.95 μm (Figure 6-6). In black and white infrared film these wavelengths are recorded on a single layer of emulsion, resulting in shades of gray from black (little reflected energy) to white (much reflected energy). Healthy vegetation typically appears as a range of grays on

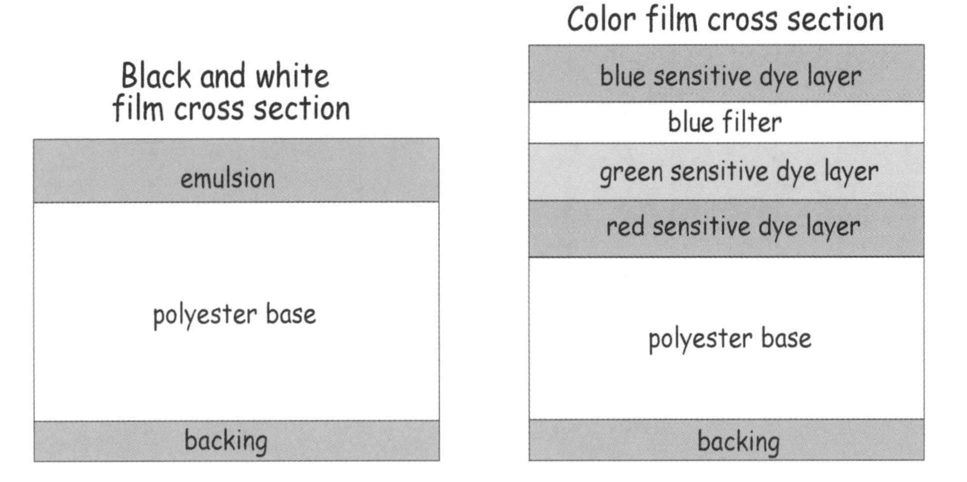

Figure 6-5: Black-and-white and color film are composed of a sandwich of layers. Emulsions on the film change in response to light, resulting in an image. (adapted from Kodak, 1982)

black and white infrared film. Small differences in the internal leaf structure and in canopy shape and texture are accentuated on infrared film, so differences among vegetation types are more easily distinguished when using this film type. Color infrared film is also sensitive through infrared wavelengths, to about 0.95 µm. Differences in infrared reflectance are particularly apparent when using color infrared film, although color infrared photographs at first appear strange to many users.

Vegetation usually appears as shades of red on infrared film. This is because although infrared film is a sandwich of dye layers, the various layers do not match the colors of the incoming wavelengths. Incoming light in the green portion of the spectrum is recorded on the blue dye layer of the film, the red portion of the spectrum is recorded on the green dye layer, and infrared portion

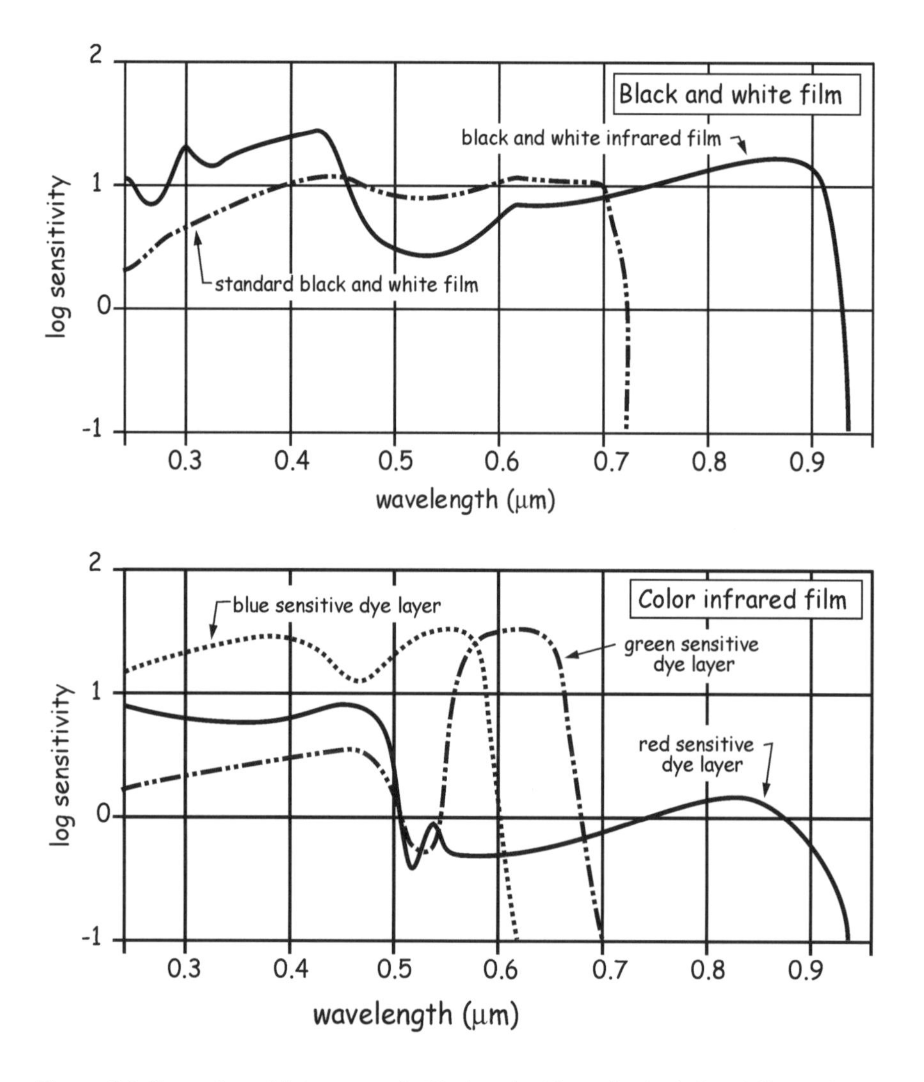

Figure 6-6: Spectral sensitivity curves for black-and-white and color infrared films. Higher sensitivity means a greater response to incoming light at the specified wavelength. (adapted from Kodak, 1982)

of the spectrum is recorded on the red dye layer. Vegetation that reflects strongly in the infrared portion of the spectrum appears bright red on a color infrared photograph. Infrared films are also quite sensitive to differences in brightness and thus have high *contrast*. High contrast films accentuate differences between low and high reflectances, so slight differences in infrared reflectance can change the shade of red substantially. Vegetation mapping projects most often specify infrared film taken during the period when leaves are present.

Camera Formats and Systems

A simple camera consists of a lens and a body (Figure 6-7). The lens is typically made of several individual glass elements, with a *diaphragm* to control the amount of light reaching the film. A *shutter* within the lens or camera controls the length of time the film is exposed to light. The *optical axis* of the lens is the central direction of the incoming image, and is precisely oriented to intersect the film in a perpendicular direction. Film is typically wound on a supply reel (unexposed film) and a take-up reel (exposed film). Film is exposed as it is fed across the camera *focal plane*. Aerial photographs are

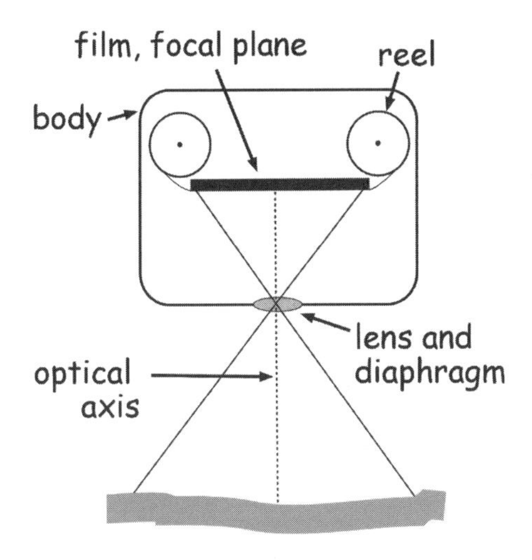

Figure 6-7: A simple camera.

Figure 6-8: A large-format camera. (courtesy Z/I Imaging Systems)

commonly advanced by a motor drive because photographs are taken in quick succession. The time, altitude, and other conditions or information regarding the photographs or mapping project may be recorded by the camera on the *data strip* in the margin of the photograph.

Photographs come in various *formats*, or sizes. These formats are usually specified by the edge dimension of the imaged area, e.g., a 240 mm (9-inch) format specifies a square photograph 240 millimeters inches on a side, and a 35mm format specifies a film with an imaging area 24 millimeters by 36 millimeters. These are the two most common film formats, the former most often used for precise geometric mapping, the latter for inexpensive qualitative surveys, although 70 millimeter systems are also used. Camera systems based on 35mm and 70 mm film are considered *small-format*, while cameras capable of using 240 mm film are considered *large-format*. All other things being equal, larger formats cover more ground area for each photograph.

Large-format cameras (Figure 6-8) are most often used to take photographs that will be used for spatial data development. These cameras are carried aboard specialized aircraft designed for photographic mapping projects. These aircraft typically have an instrument bay or hole cut in the floor, through which the camera is mounted. The camera mount and aircraft control systems are designed to maintain the camera optical axis as near vertical as possible. Aircraft navigation and control systems are specialized to support aerial photography, with precise positioning and flight control.

The largest differences between small- and large-format camera systems are geometric accuracy and cost. Large format camera systems are most often specifically designed for mapping, so the camera and components are built to minimize geometric distortion. These camera systems are precisely made, sophisticated, highly specialized, and expensive. Small-format camera systems are typically general-purpose, lightweight, relatively inexpensive, simple, and portable. However small format cameras typically introduce significant geometric distortion in the images, and this may cause significant geometric error in derived maps or digital spatial data. Accurate mapping systems have been developed that are based on small-format cameras, however most mapping projects that require geometrically accurate spatial data should use photographs from large-format mapping cameras.

Geometric Quality of Aerial Photographs

Aerial photographs are a rich source of spatial information for use in GIS, but aerial photographs inherently contain geometric distortion. Most geometrically precise maps are *cartometric,* also known as *orthographic*. An orthographic map plots the position of objects after they have been projected onto a common plane, often called a datum plane (Figure 6-9, center). Objects above or below the plane are vertically projected down or up onto the horizontal plane. Thus, the top and bottom of a vertical wall are projected onto the same location in the datum plane. An orthographic view is equivalent to looking vertically downward onto a scene from an infinite height. The tops of all vertical structures are visible, and all portions of the orthographic surface are visible (Figure 6-9, right). The only exception is when there is an overhanging cliff, where the ground below is vertically obscured.

Unfortunately, most aerial photographs provide a non-orthographic *perspective view* (Figure 6-9, left). Perspective views give a geometrically distorted image of the Earth surface. Distortion affects the relative positions of objects, and uncorrected data derived from aerial photographs may not directly overlay data in an accurate orthographic map. The amount of distortion may be reduced by selecting the appropriate camera, lens, film, flying height, and type of aircraft. Distortion may also be controlled by collecting photographs under proper weather conditions, e.g., on calm days, and by employing skilled pilots and operators. However, some aspects of the distortion may not be controlled, and no camera system is perfect, so there is geometric distortion in every aerial photograph. The real question becomes "is the distortion and geometric error below acceptable limits, given the intended use of the spatial data?" This question is not unique to aerial photographs, it applies equally well to satellite images, spatial data derived from GPS and traditional ground surveys, or data from any other source.

Distortion in aerial photographs comes primarily from six sources: terrain, camera tilt, film deformation, the camera lens, other camera errors, and atmospheric bending. The first two sources of error, terrain variation and camera tilt, are usually the largest sources of geometric distortion when using a large-format, mapping camera. The last four are relatively small when a large-format mapping camera is used, but they may still be unacceptable, particularly when the highest quality data are required. Established methods may be used to reduce the usually small errors due to lens, camera, and atmospheric distortion when using a large-format camera.

Camera and lens distortions may be quite large when non-metric, small-format cameras are used, such as 35mm or 70mm

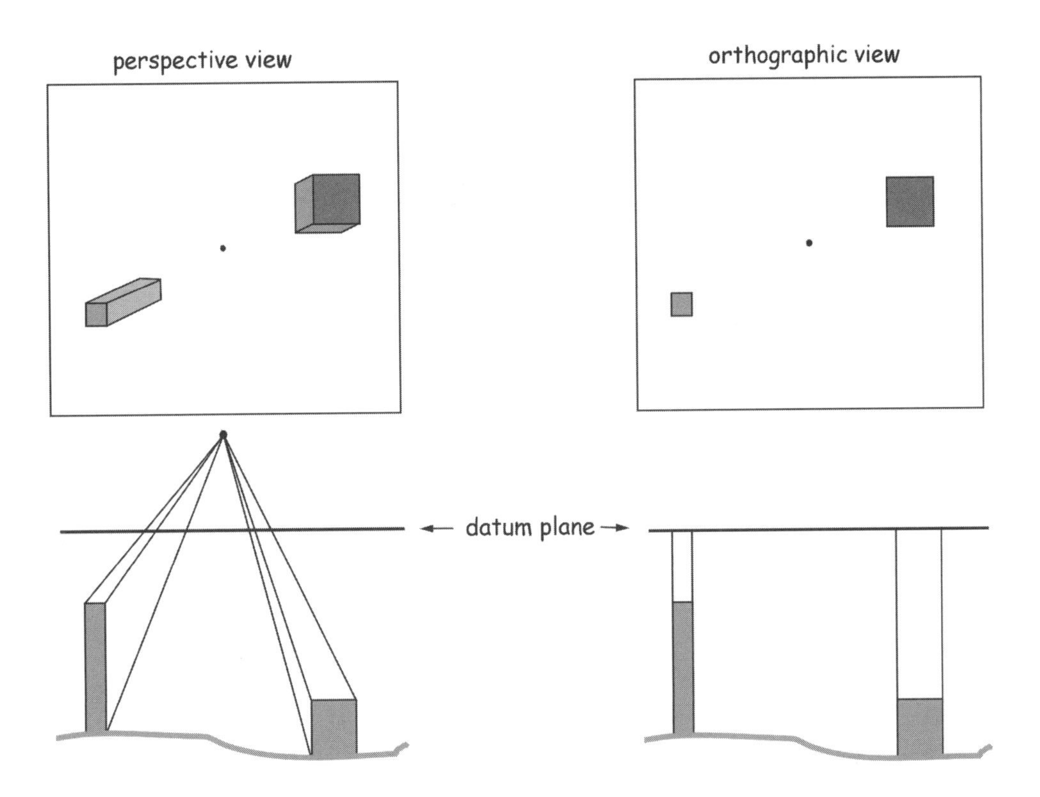

Figure 6-9: Perspective (left) and orthographic (right) views. Perspective views project from the surface onto a datum plane from a fixed viewing location. Orthographic views project at right angles to the datum plane, as if viewing from an infinite distance.

format cameras. Small-format cameras can and have been used for GIS data input, but great care must be taken in ensuring geometric distortion is reduced to acceptable levels when using small-format cameras.

Terrain and Tilt Distortion in Aerial Images

Terrain variation, defined as differences in elevation within the image area, is often the largest source of geometric distortion in aerial photographs. Terrain variation causes *relief displacement*, defined as the radial displacement of objects that are at different elevations. Figure 6-10 illustrates the basic principles of relief displacement. The Figure shows the photographic geometry over an area with substantial differences in terrain. The reference surface (datum plane) in this example is chosen to be at the elevation of the *nadir* point directly below the camera, N on the ground, imaged at n on the photograph. The camera station P is the location of the camera at the time of the photograph. We are assuming a vertical photograph, meaning the optical axis of the lens points vertically below the camera and intersects the reference surface at a right angle at the nadir location. The locations for points A and B are shown on the ground surface. The corresponding locations for these points occur at A′ and B′ on the reference datum surface. These locations are projected onto the film, as they would appear in a photograph taken over this varied terrain. In a real camera the film is behind the lens, however it is easier to visualize the displacement by showing the film in front of the lens, and the geometry is the same. Note that the points a and b are displaced from their reference surface locations, a′ and b′. The point a is displaced radially outward relative to a′, because the elevation at A is higher than the reference surface. The displacement of b is inward relative to b′, because B is lower than the reference datum. Note that any points that have elevations exactly equal to the elevation of the reference datum will not be displaced, because the reference and ground surfaces coincide at those points.

Figure 6-10 illustrates a few key characteristics of terrain distortion in a vertical aerial photograph.

Terrain distortions are radial – that is, relief displacement causes points to be displaced outward or inward from the image center. Elevations higher than the image central point are displaced outward, and elevations lower than the central point are displaced

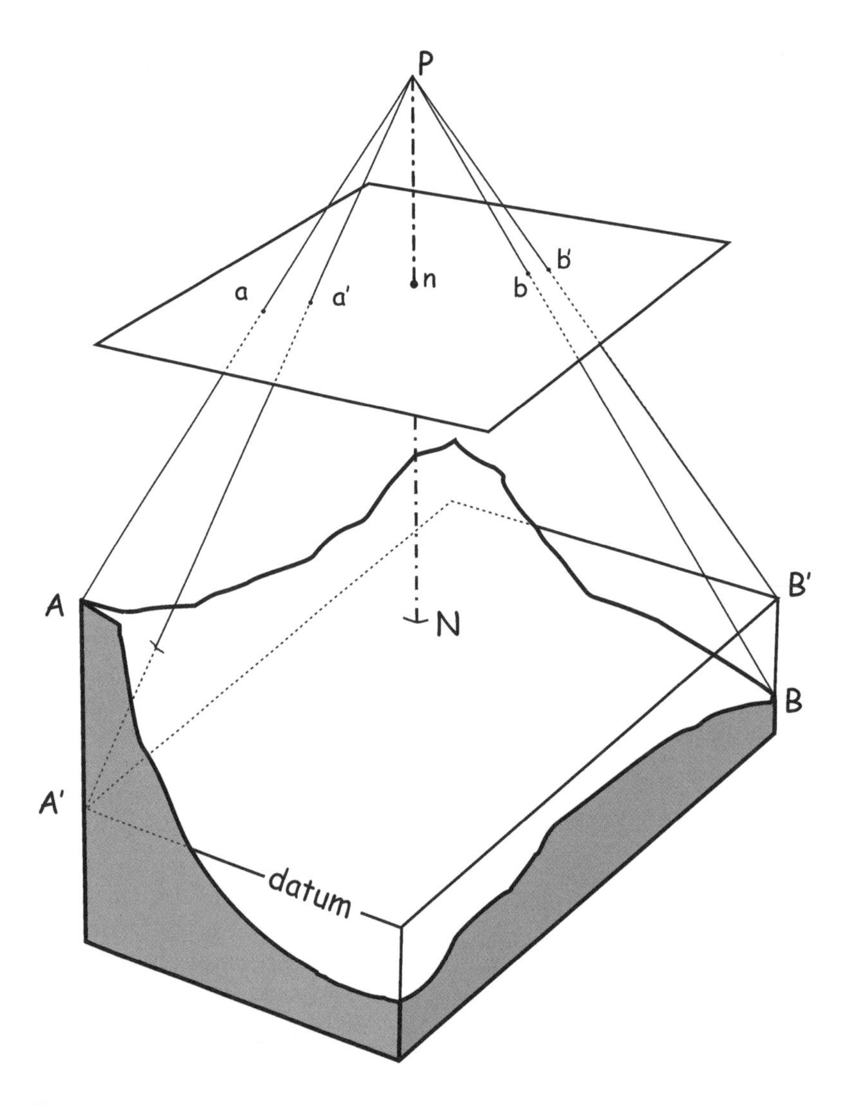

Figure 6-10: Geometric distortion on an aerial photograph due to relief displacement. (adapted from Wolf, 1983, and Lillesand and Kiefer, 1999)

inward relative to the center point. Points at the same elevation as the central point are not displaced.

Relief distortions affect angles and distances on an aerial photograph. Relief distortion changes the distances between points, and may change measured angles. Lines will not be straight, and areas are likely to be too large or too small when measured on uncorrected aerial photographs.

Scale is not constant on aerial photographs. This is particularly true for photographs taken over varied terrain. Scale changes across the photograph and depends on the magnitude of the relief displacement. Thus, we may describe an average or nominal scale for a vertical aerial photograph over varied terrain, but the true scale between any two points may be different than the average scale.

A vertical aerial photograph taken over varied terrain is not orthographic. We cannot expect geographic data digitized from photographs over varied terrain to match orthographic data in a GIS. If the distortions are small relative to digitizing error or other sources of geometric error, then data digitized directly from aerial photographs may appear to match data from orthographic sources. However, if the relief displacement is large, then the relief displacement errors will be significant, and the spatial locations of features digitized from aerial photographs will be in error unless these errors are removed.

Camera tilt may be another large source of positional error in aerial photographs. Camera tilt, in which the optical axis points at a non-vertical angle, results in distortion in aerial photographs (Figure 6-11). Tilt distortion comes primarily from perspective

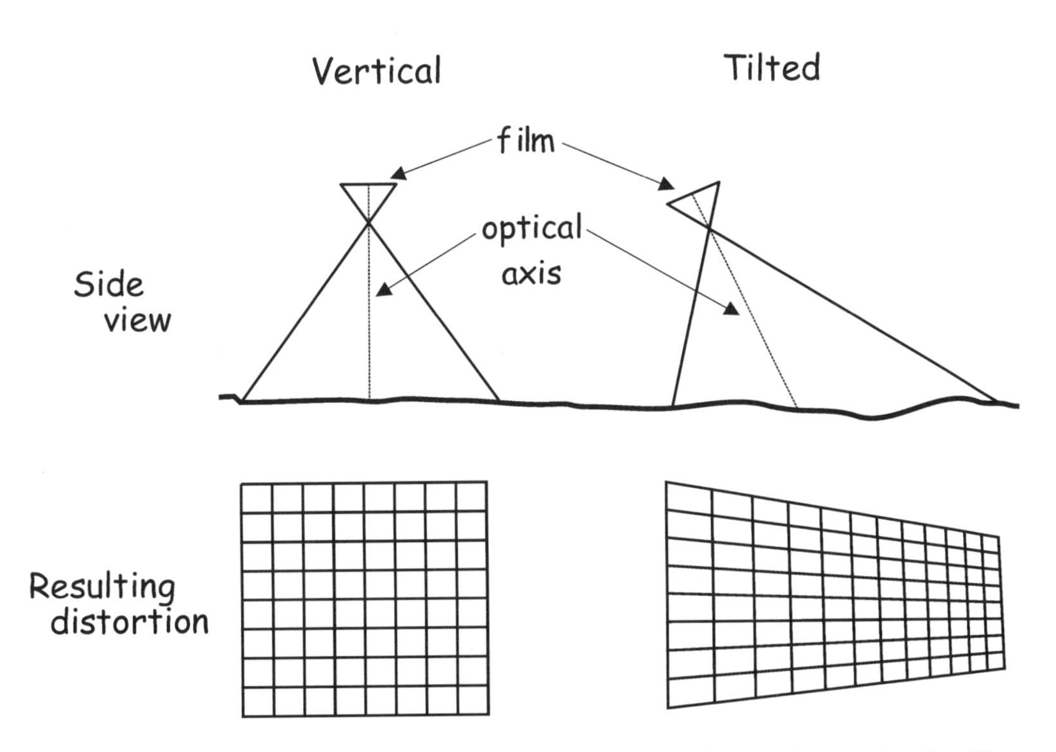

Figure 6-11: Image distortion caused by a tilt in the camera optical axis relative to the ground surface. The perspective distortion, shown at the bottom right, results from changes in the viewing distance across the photograph.

convergence. Objects further away appear to be closer together than equivalently-spaced objects that are nearer the observer. Tilt distortion is zero in vertical photographs, and increases as tilt increases. Contracts for photographic mapping missions typically specify tilt angles of less than three degrees from vertical. Perspective distortion is somewhat difficult to remove, and so efforts are made to minimize tilt distortion by maintaining a vertical optical axis. Great efforts are spent to mount the camera so the optical axis of the lens points directly below, and to keep the aircraft on a smooth and level flight path as much as possible.

Tilt and terrain distortion may both occur on aerial photographs taken over varied terrain. Tilt distortion may occur even on vertical aerial photographs, because tilts up to three degrees are usually allowed. The overall level of distortion depends on the amount of tilt and the variation in terrain, and also on the photographic scale. Not surprisingly, errors increase as tilt or terrain increase, and as photographic scale becomes smaller.

Figure 6-12 illustrates the changes in total distortion with changes in tilt, terrain, and photographic scale. This figure shows the error that would be expected in data digitized from vertical aerial photographs when only applying an affine transformation, a standard procedure used to register orthographic maps (see Chapter 4). The process used to produce these error plots mimics the process of directly digitizing from uncorrected aerial photographs. Note first that there is zero error across all scales when the ground is flat (terrain range is zero) and there is no tilt (bottom line, left panel in Figure 6-12). Errors increase as photographic scale decreases, shown by increasing errors as you move from left to right in both panels. Error also increases as tilt or terrain increases. These errors can be quite large, even for vertical photographs over moderate terrain (Figure 6-12, right side). These graphs clearly indicate that geometric errors will occur when digitizing from vertical aerial photographs, even if the digitizing system is perfect and introduces no error. Thus the magnitude of tilt and terrain errors should be assessed relative to the geometric accuracy required before digitizing from vertical aerial photographs.

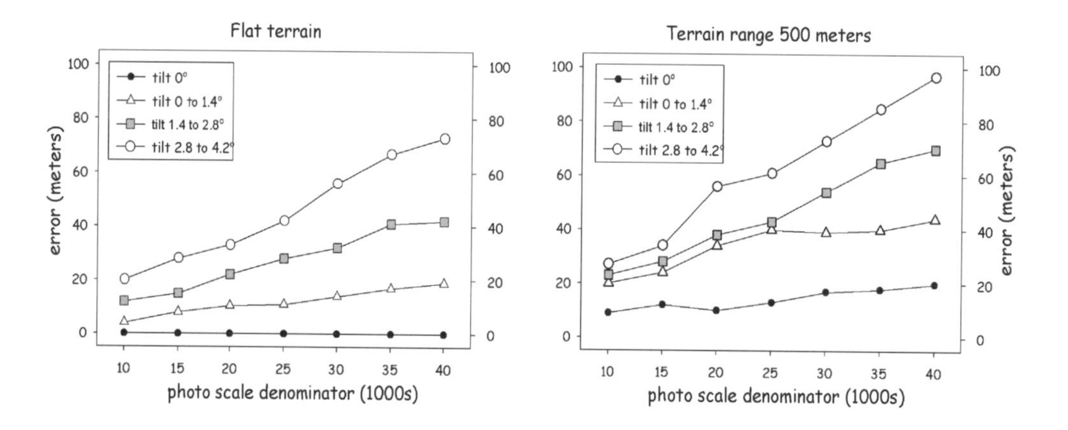

Figure 6-12: Terrain and tilt effects on mean positional error when digitizing from uncorrected aerial photographs. Distortion increases when tilt and terrain increase, and as photo scale decreases. (from Bolstad, 1992)

System Errors: Media, Lens, and Camera Distortion

The film, camera, and lens system may be a significant source of geometric error in aerial photographs. The perfect lens-camera-film system would exactly project the viewing geometry of the target onto the film. The relative locations of features on the film in a perfect camera system would be exactly the same as the relative locations on a viewing plane an arbitrary distance in front of the lens. Real camera systems are not perfect and may distort the image. For example the light from a point may be bent slightly when traveling through the lens, or the film may shrink or swell, both causing a distorted image.

Radial lens displacement is one form of distortion commonly caused by the camera system. Whenever a lens is manufactured there are always some imperfections in the curved shapes of the lens surfaces. These cause a radial displacement, either inward or outward, from the true image location. Radial lens displacement is typically quite small in mapping camera systems, but may be quite large in other systems. A radial displacement curve is often developed for a mapping camera lens, and this curve may be used to correct radial displacement errors when the highest mapping accuracy is required.

Systematic camera distortion also may occur. For example, the *platten* is a flat surface behind the film in a mapping camera. The platten holds the film in a planar shape during exposure. If the camera does not have a platten, or the platten is deformed, then the image will be geometrically distorted. There may be problems with camera mounts, take-up rolls may be mis-aligned, or there may be other imperfections in the camera body that affect the geometric quality of an image.

We wish to emphasize that mapping camera systems are engineered to minimize systematic errors. Lenses are designed and precisely manufactured so that image distortion in minimized. Lens mountings, the platten, and the camera body are optimized to ensure a faithful rendition of image geometry. Films are designed so that there is limited distortion under tension on the camera spools, and films are relatively insensitive to changes in temperature and humidity or other conditions which might cause them to swell or shrink non-uniformly. This optimization leads to extremely high geometric fidelity in the camera/lens system. Thus, camera and lens distortions in mapping cameras are typically much smaller than other errors, e.g., tilt and terrain errors, or errors in converting the image data to forms useful in a GIS.

We also wish to emphasize that camera-caused geometric errors may be quite high when a non-mapping camera is used, e.g., when photographs are taken with small-format 35mm or 70mm camera system. Lens radial distortion may be extreme, typically there is no platten, and the films may stretch non-uniformly from photo to photo or roll to roll. Hence, these systems are likely to have large geometric distortion relative to mapping cameras. That is not to say that mapping cameras must always be used. In some circumstances the distortions inherent in small-format camera systems may be acceptable, or may be reduced relative to other errors, for example, when very large scale photographs are taken, and when qualitative or attribute information are required. However, the geometric quality of any non-mapping camera system should be evaluated prior to use in a mapping project.

Atmospheric distortion is another source of geometric error in aerial images, although it is usually small relative to other errors. Light bends when passing through the atmosphere, and this may result in spatial displacement and hence distortion. Atmospheric distortion increases with increasing atmospheric depth, and is most important when photographs are taken from extremely high altitudes or at oblique angles through the atmosphere. Under most conditions atmospheric distortion is quite small relative to other errors, and is corrected only when the highest spatial accuracies are required.

Stereo Photographic Coverage

As noted above, relief displacement in vertical aerial photographs adds a radial displacement that depends on the relative terrain heights. The bigger the terrain difference, the bigger the radial displacement. This relief displacement may be considered a problem if we take only one photograph and we wish to produce a map from only this single photograph. Photogrammetric methods may be used to remove the distortion. However if two overlapping photographs are taken, called a *stereopair*, then these photographs may be used together to determine the relative elevation differences. Relief displacement in a stereopair may be a positive phenomenon because it may be used to rapidly and inexpensively determine elevation. Many mapping projects collect *stereo photographic coverage* in which subsequent photographs in a flight line overlap, and adjacent lines overlap (Figure 6-13). Overlap in the flight direction is termed *endlap*, and overlap in adjacent flight lines is termed *sidelap*.

Stereopairs reveal elevation due to *parallax*, a shift in relief displacement due to a shift in observer location. Figure 6-14 illustrates parallax. Notice that the block (closer to the viewing locations) apparently shifts more than the sphere when the viewing location is changed. The displacement of any given point is different on the left vs. right ground views because the relative viewing geometry is different. Points are shifted by different amounts, and the magnitude of the shift depends on the distance from the observer (or camera) to the objects. This shift in position with a shift in viewing location is the basis of depth perception.

A *stereomodel* is a three-dimensional perception of terrain or other objects that we see when viewing a stereopair. Each eye looks at a different photograph, and we observe a set of parallax differences, and our brain may convert these to a perception of depth. When we have vertical aerial photographs, the distance from the camera to each point on the ground is determined primarily by the elevations at each point on the ground. We may observe parallax for each

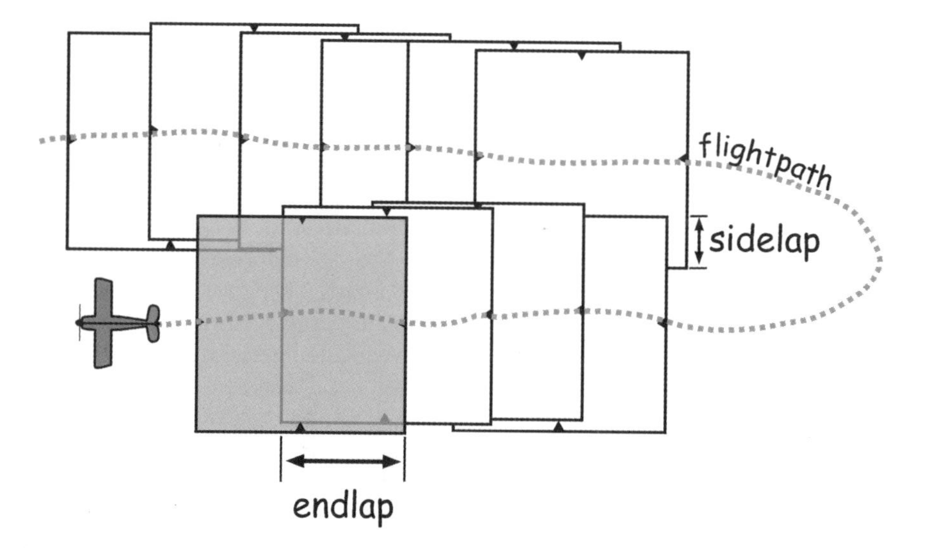

Figure 6-13: Aerial photographs often overlap. This allows three-dimensional measurements and the correction of relief displacement. Overlap is characterized as endlap and sidelap.

point and use this parallax to infer the relative elevation for every point. Stereo viewing creates a three-dimensional stereomodel of terrain heights, with our left eye looking at the left photo and our right eye looking at the right photo. We may project this three-dimensional stereomodel onto a flat surface and interpret a map onto the surface. We may also interpret the relative terrain heights on this three-dimensional surface, and thereby estimate elevation wherever we have stereo coverage. In this way we may draw contour lines or mark spot heights through interpretation of stereopairs. This has historically been the most common method for determining elevation over large areas, e.g., larger than a few hundred hectares.

Geometric Correction of Aerial Photographs

Due to all the geometric distortions described above, it should be quite clear that aerial photographs should not be used directly as a basis for spatial data collection

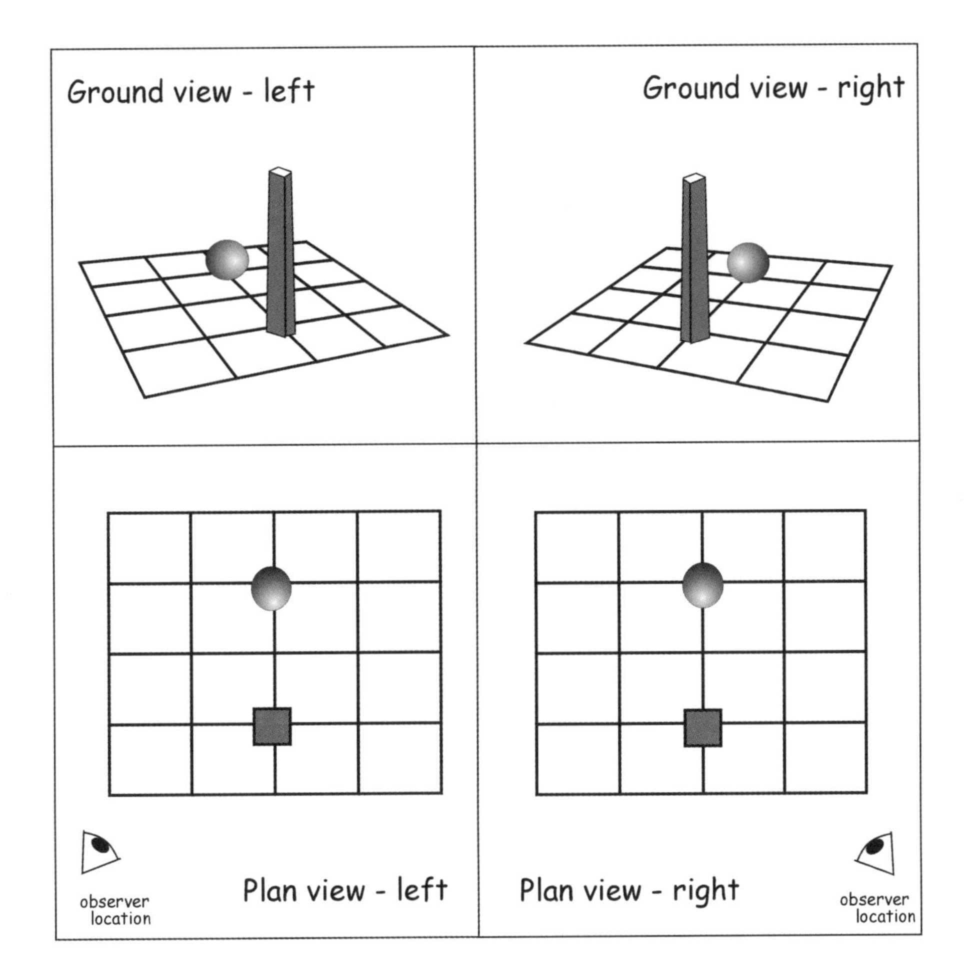

Figure 6-14: Parallax is the relative shift in the position of objects with a change in the observer location. The top panels show a ground view and the bottom panels show the plan view (from above), of observer locations. The relative position of the sphere and the tower changes when the observer location is changed.

under most circumstances. Points, lines, and area boundaries may not occur in their correct relative positions, so length and area measurements may be incorrect. Worse, when spatial data derived from uncorrected photographs are combined with other sources of geographic information, features may not fall in their correct locations. A river may fall on the wrong side of a road or a city may be located in a lake. Given all the positive characteristics of aerial photographs, how do we best use this rich source of information? Fortunately, photogrammetry provides the tools needed to remove geometric distortions from photographs.

Geometric correction of aerial photographs involves calculating the distortion at each point, and then shifting the image location accordingly. Consider the image in Figure 6-15. This image shows a tower located in a flat area. The bottom of the tower at B is imaged on the photograph at point b, and the top of the tower at point A is imaged on the photograph at point a. Point A will occur on top of point B on an orthographic map. If we consider the flat plane at the base of the tower as the datum, we can use simple geometry to calculate the displacement from a to b on the image. We'll call this displacement d, and go through an explanation of the geometry used to calculate the displacement.

Observe the two similar triangles in Figure 6-15, one defined by the points S-N-C, and one defined by the points a-n-C. C is the

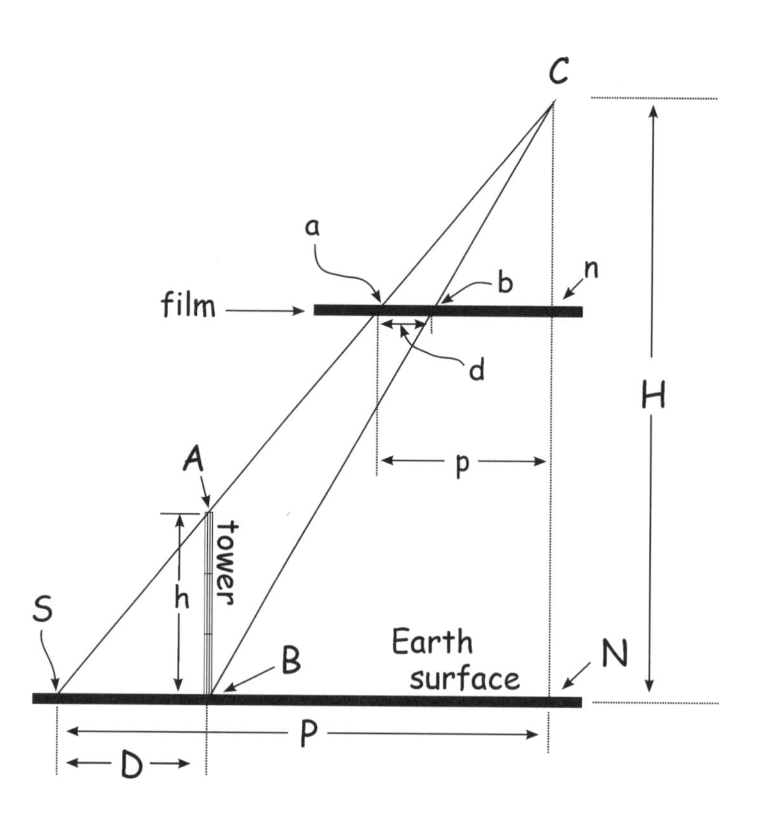

Figure 6-15: Relief displacement may be calculated based on geometric measurement. Similar triangles relate heights and distances in the photograph and on the ground. We usually know flying height, H, and can measure d and p on the photograph.

focal center of the camera lens, and may be considered the location through which all light passes. The film in a camera is placed behind the focal center; however, as in previous figures, the film is shown here in front of the focal center for clarity. Note that the following ratios hold for the similar triangles:

$$D/P = h/H \qquad (6.1)$$

and also

$$d/p = D/P \qquad (6.2)$$

so

$$d/p = h/H \qquad (6.3)$$

rearranging

$$d = p*h/H \qquad (6.4)$$

where:

d = displacement distance

p = distance from the nadir point, n, on the vertical photo to the imaged point

H = flying height

h = height of the imaged point

We usually know the flying height, and can measure the distance p. If we can get h, the height of the imaged point above the datum, then we can calculate the displacement. We might climb or survey the tower to measure its height, h, and then calculate the photo displacement by Equation (6.4). Relief displacement for any elevated location may be calculated provided we know the height.

Equation 6.4 applies to vertical aerial photographs. When photographs are tilted they may contain both tilt and terrain displacement. The geometry of the image distortion is much more complicated, and so are the equations used to calculate tilt and elevation displacement. Equations may be derived that describe the three-dimensional projection from the terrain surface to the two-dimensional film plane. These equations and the methods for applying them are part of the science of photogrammetry, and will not be discussed here. A more complete discussion is included in several photogrammetry textbooks, including the book by P.R. Wolf listed at the end of this chapter. Photogrammetric methods combine ground-surveyed measurements with measurements taken on aerial images to provide precise, orthographically correct coordinate locations.

The process requires the measurement of the photo coordinates and their combination with ground x, y, and z coordinates. Photo coordinates may be measured using a physical ruler or calipers, however they are most often measured using digital methods. Typically the photographs are scanned using a precise image scanning device. Measurements of image x and y are then determined relative to some photo-specific coordinate system. These measurements are obtained from one or many photographs. Ground x, y, and z coordinates come from precise ground surveys. A set of equations is written that relates photo x and y coordinates to ground x, y, and z coordinates. The set of equations is solved, and the displacement calculated for each point on the photograph. The displacement may then be removed and an orthographic image or map produced.

How do we use displacement equations to correct data collected from aerial photographs? Several methods have been developed. With one method, points and vectors may be digitized from the photographs using a common coordinate digitizer. Flying height, camera focal length, elevation, and other measurements are used to calculate the displacement for every digitized x and y coordinate. This displacement is then reversed for each digitized point, adjusting the location outward or inward as required. Note that the elevation must be provided for every digitized coordinate, e.g., the h in

Figure 6-16: A softcopy photogrammetric workstation, used in the production of orthophotographs. (courtesy Z/I Imaging Systems)

Equation (6.4) above, to calculate the displacement. To apply these corrections we would need to determine the elevation of every point, node, or vertex location that we wish to enter into our GIS. These could be obtained from field surveys, or more likely, from digital elevation data for the study area.

An alternate process applies the corrections to photographs prior to digitization. The photograph is typically scanned to produce a digital raster image. The displacement is then calculated for each point in the raster image, and the distortion removed for each raster cell. The raster image is *orthographic*, defined as an image with terrain and perspective distortions removed. Distances, angles, and areas can be measured from the image. These orthographic images, also known as *orthophotographs* or *digital orthographic images*, have the positive attributes of photographs, with their rich detail and timely coverage, and some of the positive attributes of cartometric maps, such as uniform scale and true geometry.

Digital orthophotographs are most often produced using a *softcopy* photogrammetric workstation (Figure 6-16). This method uses softcopy (scanned) images, digital versions of aerial photographs. *Softcopy photogrammetry* uses mathematical models of photogeometry to remove tilt, terrain, camera, atmospheric, and other distortions from digital images. Photographs are scanned, control points identified on sets of photographs, stereomodels developed, and geometric distortions estimated. These distortions are then removed, creating an orthophotograph. Multiple photographs may be analyzed, corrected, and mosaiced at a time (Figure 6-17). This process of developing photomodels of multiple photographs at once is often referred to as a *bundle adjustment*.

The USGS has undertaken production of orthophotographs for the United States. These *Digital Orthophoto Quadrangles* (DOQs) are available for much of the United States, and the USGS plans on complete national coverage. In addition, digital orthophotographs of higher precision or for specific time periods are regularly produced by photogrammetric engineers in public and private organizations.

Photo Interpretation

Photographs and other images are useful primarily because we may identify the position and properties of interesting features. Once we have determined that the film and camera system meet our spatial accuracy and information requirements, we need to collect the photographs and interpret them. Photo (or image) interpretation is the process of converting images into information. Photo interpretation is a well-developed discipline, with many specialized techniques. We will provide a very brief description of the process. A more complete description may be found in several of the sources listed at the end of this chapter.

Interpreters use the size, shape, color, brightness, texture, and relative and absolute location of features to interpret images. Differences in these diagnostic characteristics allow the interpreter to distinguish among features. Different vegetation types may show distinct color or texture variations,

Figure 6-17: Multiple photographs (left) may be scanned, corrected, and mosaiced (right) using a softcopy photogrammetric workstation. This process yields digital image with the orthographic view of maps and the detailed content of photographs. (courtesy Z/I Imaging Systems)

road types may be distinguished by width or the occurrence of a median strip, and building types may be defined by size or shape.

The proper use of all the diagnostic characteristics requires that the photo-interpreter develop some familiarity with the features of interest. For example, it is difficult to distinguish the differences between many crop types until the interpreter has spent time in field, photos in hand, comparing what appears on the photographs with what is found on the ground. This "ground-truth" is invaluable in developing the local knowledge required for accurate image interpretation. When possible, ground visits should take place contemporaneously with the photographs. This is often not possible, and sites may only be visited months or years after the photographs were collected. The affects of changes through time on the ground-to-image comparison must then be considered.

Photointerpretation requires we establish a target set of categories for interpreted features. If we are mapping roads, we must decide what classes to use, e.g., all roads will be categorized into one of the following classes: unpaved, paved single lane, paved undivided multi-lane, and paved divided multi-lane. These categories need to be interpretable, e.g., we must be able to distinguish among the road types on the images, and they must be inclusive, e.g., in our photos there must be no roads that are multi-lane and unpaved. If so, we must ignore them, or create a category for them.

Photointerpretation also requires we establish a minimum mapping unit. This is the smallest area, line length, or point item that we will interpret. A minimum mapping unit in one sense defines the lower limit on what we consider a significant feature. We may not be interested in forest patches smaller than 0.5 hectares, or road segments shorter than 50 meters long. Although they may be visible on the image, features smaller than the minimum mapping unit are not delineated and transferred into the digital data layer.

Finally, photointerpretation to create spatial data requires a method for entering the interpreted data into a digital form. Perhaps the most common method involves on-screen digitizing. Point, line, and area features interpreted on the image may be manually drawn in an editing mode, and captured directly to a data layer. Another common method consists of interpretation directly from the hardcopy image. The image may be treated as a map and attached to a digitizing board. Features are then directly interpreted from the image during digitizing. A third method, commonly applied, involves fixing a clear drafting sheet over the image. Features are drawn on the sheet as they are interpreted. Data are then digitized as with a hardcopy map.

Satellite Images

In previous sections we described the basic principles of remote sensing and the specifics of image collection and correction using aerial photographs. In many respects satellite images are similar to aerial photographs when used in a GIS. The primary motivation is to collect information regarding the location and characteristics of features. However, there are important differences between photographic and satellite-based scanning systems used for image collection, and these differences affect the characteristics and hence uses of satellite images.

Satellite scanners have several advantages and disadvantages relative to aerial photographic systems. Satellite scanners may have an extended spectral range beyond wavelengths that are detectable when using aerial photographs. These differences are particularly important in the longer mid- and far-infrared wavelengths, where differences in vegetation and mineral reflectance properties make some scanners particularly useful for landcover and geological mapping. Satellite scanners also have a very high perspective which significantly reduces terrain-caused distortion. Equation 6.4 shows the terrain displacement (d) on an image is inversely related to the flying height (H). Satellites have large values for H, typically 600 kilometers (360 miles) or more above the Earth surface, so relief displacements are correspondingly small. Because satellites are flying above the atmosphere, their pointing direction (attitude control) is very precise, and so they may be maintained in a near perfect vertical orientation.

There may be a number of disadvantages in choosing satellite images instead of aerial photographs. Satellite images typically cover larger areas, so if the area of interest is small, costs may be needlessly high. Photographs are often available in hardcopy prints, and may be interpreted and entered using a coordinate digitizer. The use of satellite images may require specialized image processing software. Acquisition of aerial photographs may be more flexible, in that a pilot may wait for a clear day at a specific time of year, and decide to fly on the morning of the photo acquisition date. Satellite image acquisition is often scheduled days or weeks in advance, and follow fixed flight schedules. The effective resolution is better for many photographs than satellite images. Finally, aerial photographs are often available at reduced costs from government sources. Many of these differences between aerial photographs and satellite images are diminishing as more, higher-resolution, pointable scanners are placed in orbit.

Basic Principles of Satellite Image Scanners

Satellite scanners differ from aerial camera and film systems in that reflected energy is detected and stored electronically, rather than physically. Scanners use photosensitive *detectors* that respond electronically to incoming energy. Much as with photographic film, the stronger the energy, the higher the response. Scanners differ from photographs in that an electronic signal is recorded digitally, rather than with a chemical change in a film emulsion.

Scanners operate by pointing the detectors at the area to be imaged. Each detector has an *instantaneous field of view*, or IFOV, that corresponds to the size of the area viewed by each detector (Figure 6-18). Although the IFOV may not be square and a raster cell typically is square, this IFOV may be thought of as approximately equal to the raster cell size for the acquired image.

The scanner builds a two-dimensional image of the surface by pointing a detector or detectors at each cell and recording the reflected energy. Data are typically collected in the across-track direction, perpendicular to the flight path of the satellite, and in the along-track direction, parallel to the direction of travel (Figure 6-18). Several scanner designs achieve this across- and along-track scanning. Some designs use a *spot detector* and a system of mirrors and lenses to sweep the spot across track. The forward motion of the satellite positions the scanner for the next swath in the along-track direction. Other designs have a *linear array* of detectors – a line of detectors in the across-track direction. The across-track line is sampled at once, and the forward motion of the satellite positions the array for the next line in the along-track direction. Finally, a *two-dimensional array* may be used, consisting of a rectangular array of detectors. Reflectance is collected in a patch in both the cross-track and the along-track directions.

A remote sensing satellite also contains a number of other subsystems to support image data collection. A power supply is required, typically consisting of solar panels and batteries. Precise altitude and orbital

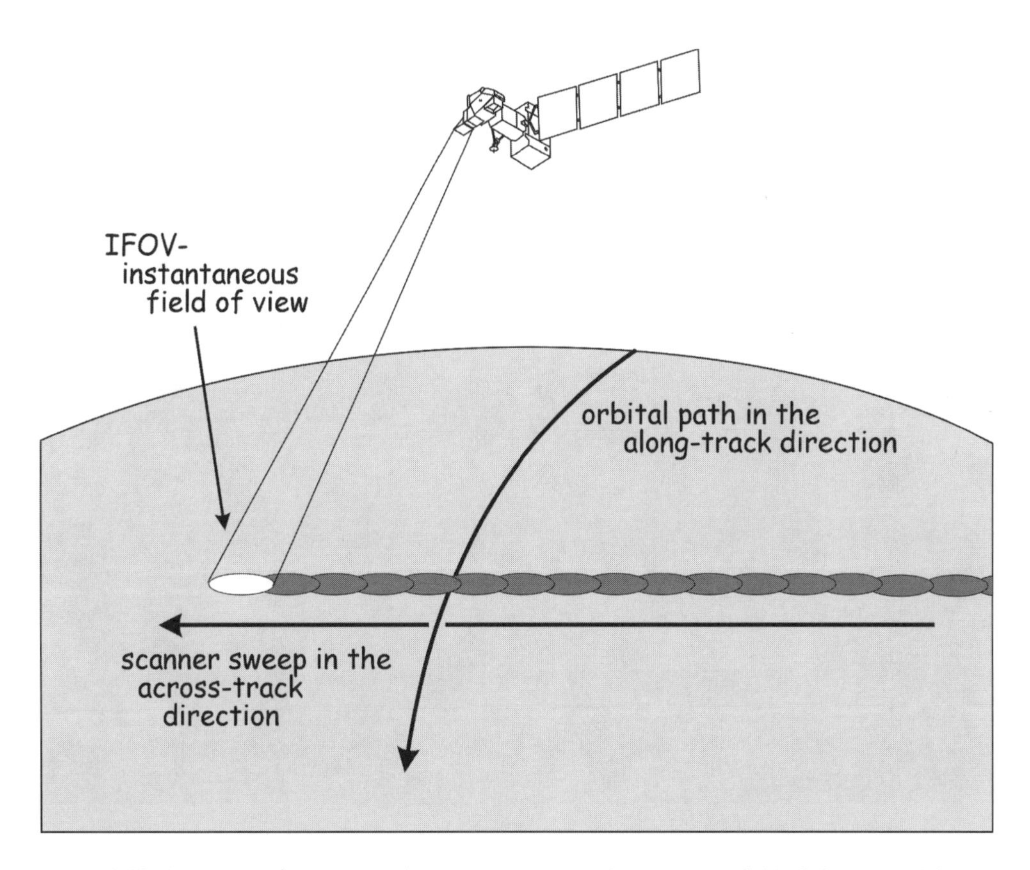

Figure 6-18: A spot scanning system. The scanner sweeps an instantaneous field of view (IFOV) in an across-track direction to record a multispectral response. Subsequent sweeps in an along-track direction are captured as the satellite moves forward along the orbital path.

control are required, so satellites carry navigation and positioning subsystems. Sensors evaluate satellite position and pointing direction, and thrusters and other control components orient the satellite. There is a data storage subsystem, and a communications subsystem for transmitting the data back to Earth and for receiving control and other information. All of these activities are coordinated by an on-board computing system.

Several remote sensing satellite systems have been built, and data have been available for land surface applications since the early 1970s. The detail, frequency, and quality of satellite images have been improving steadily, and there are several satellite remote sensing systems currently in operation. The following sections describe the satellite systems most often used as data sources for GIS-based analyses.

Landsat

The Landsat satellite program began with the launch of the first Landsat satellite on July 23, 1972. This satellite was the first system specifically designed to gather data about the Earth's land surface. Previous satellite systems had focused on observing the weather, oceans, or other features. While earlier systems hinted at the exciting possibilities of land surface remote sensing, these possibilities were not realized until Landsat (Figure 6-19). The success of the first satellite led to the development and launch of six more, and Landsat satellites have been oper-

Figure 6-19: A portion of a Landsat ETM+ image for an area in South Africa. The image was collected in May 2000 in the Northern Province. Vegetation is a mix of savanna and woodland on abandoned grazing land, and agriculture. The image spans approximately 40 kilometers east to west and 30 kilometers north to south.

ating continuously since the launch of the first – more than three decades of coverage.

Landsat satellites have carried three primary imaging scanners. The first three Landsat satellites also carried video cameras, but these have been little-used. The *Multispectral Scanner* (or MSS) was the first satellite-based land scanner, and has been carried on board Landsat satellites one through five. The original MSS provided four *bands* of data. Each band recorded energy for a portion of the electromagnetic spectrum. The MSS recorded a green, a red, and two infrared bands. Landsat 4 and 5 added a fifth band to the MSS. MSS data provide raster cells at a resolution of approximately 80 meters.

Landsat satellites 4 and 5 also carried the Thematic Mapper (TM). TM data include more and different bands than MSS, a higher spatial resolution, better geometric quality, and more exact radiometric calibration. TM data contain seven spectral bands (three visible, a near-infrared, two mid-infrared, and a thermal band), and provide data at approximately a 30 meter grid-cell resolution for the first six bands. The thermal band on the TM has a 120 meter resolution. Landsat 6 was lost shortly after launch.

Landsat 7 carries an *Enhanced Thematic Mapper* (ETM+), an improved version of the TM. The ETM+ adds a 15 meter resolution panchromatic band covering the visible wavelengths, and improves the thermal band resolution to 60 meters. The ETM+ also adds a number of other improvements, including increased *radiometric fidelity*, wherein the recorded energy is more accurately and precisely measured.

All Landsat satellites have been placed in polar, inclined orbits, meaning they travel approximately north-south, but do not pass directly over the North and South Poles. The orbital plane is inclined approximately 9 degrees, and this inclined orbit results in trapezoidal images. Adjacent images do not overlap along a satellite path, but image edges overlap slightly at the equator. Because all satellite paths intersect at the poles, image edge overlap increases at higher latitudes, with more than a 50% image overlap for adjacent rows in the far north and south. The inclined orbits are sun-synchronous, meaning the satellite passes overhead at approximately the same local time.

Return intervals have varied somewhat among systems. Landsat 1 through 3 repeated each orbital path every 18 days, while Landsat 4 and 5 had a return interval of 16 days, as does Landsat 7. Approximately 20 to 22 images may be obtained for most areas on Earth each year. This return interval does not guarantee a usable image, however, as cloud cover may obscure an area for much of the year. Some desert locations routinely have 20 cloud free images a year, while some tropical forest locations have only a few cloud-free images collected during the past 20 years. Mean frequency of cloudy days over most of the inhabited world varies from 10 to 60%, so the probability of getting one cloud-free image in a year is quite high.

Landsat data have been used in a broad array of disciplines. MSS data formed the foundation of continental and global crop forecasting systems. They were widely used to document tropical deforestation, in the detection of other land use change, and in general landcover mapping. TM data were an improvement in these applications, and the increased spectral and spatial resolution allowed the identification of finer land use categories. The new bands on the TM sensor were particularly useful for geologic applications, spurring the development of space-based mineral exploration and prospecting. A wide range of urban and regional planning analyses have been supported using TM, and more recently ETM+ data. Operational crop forecasting systems use TM and other data to predict harvest months in advance. News gathering, geographic education, and even art are based on satellite data from the current generation of Landsat-borne scanners.

SPOT

Following the technological success of the Landsat satellite system, the French Government led the development of the *Systeme Pour l'Observation de la Terre*, or SPOT (Figure 6-20). In February 1986, SPOT-1 was launched, and there have since been four additional SPOT satellites placed in orbit. SPOT differed from Landsat in that it was designed as an operational system. Landsat was ground-breaking, but was initially designed as an experimental program. As such, the early satellites and ground systems were not optimized nor operated in a manner to specifically serve commercial interests, or to operate in a high-volume, production mode. SPOT incorporated many changes to serve this end.

The SPOT scanner provides for both more detailed and more frequent data collec-

Figure 6-20: An example of an image from the SPOT satellite system. The active volcano Popocatepetl, in Mexico, is visible in the center of the image. A bright thermal glow is visible in the center of the crater. This image demonstrates the broad area coverage and fine detail available from the SPOT system. (courtesy SPOT Image Corp.)

tion than Landsat. First, the scanner provides two imaging modes, a panchromatic mode and a high-resolution visible (HRV) mode. The panchromatic mode contains one image band in the visible wavelengths, 0.51 to 0.73 μm. The HRV mode has three bands, one band each in the green, red, and near-infrared portions of the spectrum. The panchromatic mode has a spatial resolution of 10 meters, and the HRV mode has a spectral resolution of 20 meters, both of which are substantial improvements over the initial Landsat TM.

Scanner design was substantially different between SPOT and Landsat systems. Landsat scanners were a point scanning design and used a mirror to sweep the IFOV across the track (Figure 6-18). SPOT employed a *push-broom* scanner design. This push-broom design uses a linear array of sensors that are oriented at right angles to the travel direction of the satellite. An entire cross-track line is sampled at once. This may allow each sensor to remain fixed on a location for a longer period of time, thus providing radiance measurements with a stronger signal and less radiometric "noise".

Finally, the SPOT scanners were developed with pointable optics, meaning they may collect data from directly beneath the satellite or they may be pointed to areas up to 27° to either side of the satellite path. The effective revisit time is reduced to between one and five days. Pointable optics also allow for the collection of satellite stereopairs, and hence SPOT data may be used to map elevation, much as with photographic stereopairs.

The increased resolution of the SPOT system comes at some cost, most notably in image size. Each SPOT image covers an area approximately 60 kilometers (36 miles) on a side, while Landsat images cover an area approximately 185 kilometers (110 miles) on a side. Thus, each Landsat scene provides information for approximately 10 times the area of a SPOT scene. Also, the first three SPOT scanners only provided visible and near-infrared data and omitted the mid-infrared region. Landsat mid-infrared bands have proven most useful, particularly in discriminating among vegetation types and in other landcover mapping. This shortfall was remedied with SPOT 4 and SPOT 5, both of which include a mid-infrared band.

AVHRR

Images from the *Advanced Very High Resolution Radiometer* (AVHRR) have been used as a source of spatial data for GIS, particularly when relatively coarse-resolution data are required over large areas. Despite its name, AVHRR images have a maximum resolution of approximately 1.1 kilometers. This resolution is high for a meteorological satellite. However AVHRR data are low resolution for many spatial analysis needs, although they cover large areas (Figure 6-21). AVHRR are most appropriate for continental and global scale analyses. Images from a single satellite cover the entire globe once each day, and typically there are two satellites in orbit, thus providing coverage approximately every 12 hours.

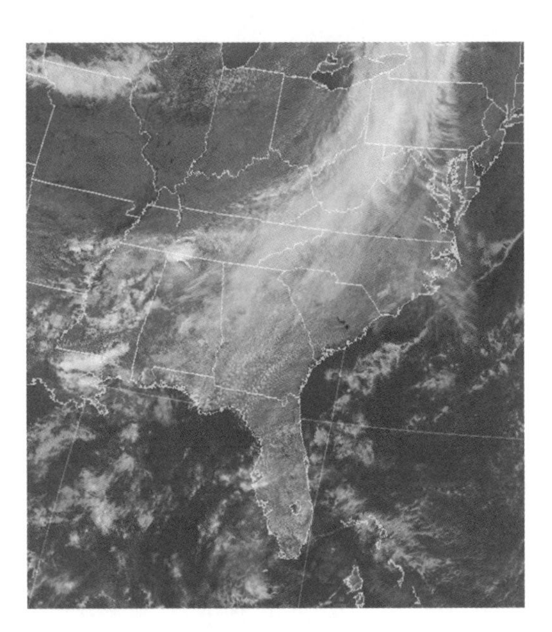

Figure 6-21: This image demonstrates the broad-area coverage provided by the AVHRR. (courtesy NOAA)

AVHRR images are particularly useful when high frequency data are required over large areas, for example when monitoring short-lived or rapidly changing phenomena. AVHRR are commonly used to track weather-related events, and have been applied to monitor storms, fires, and floods.

Other Systems

There are several other airborne and satellite remote sensing systems that are operational or under development. Although some are quite specialized, each may serve as important sources of data. Some may introduce entirely new technologies, while others replace or provide incremental upgrades to existing systems. Space prevents our offering more than a brief description of these satellite systems here.

High Resolution Satellite Systems – There are a number of planned or recently launched high-resolution satellite systems. These systems provide resolutions as high as one meter, and also offer both panchromatic and multispectral data (Figure 6-22). These resolutions begin to blur the distinction between satellite and photo-based images. While the inherent resolution of most photographs is much greater than one meter at typical scales, this resolution is often unneeded. Most features in many applications may be accurately identified at a one-meter resolution. Spectral range, price, availability, and ease of use may become more important factors in selecting between aerial photographs and satellite images.

As of this writing there are two operational high-resolution systems, Ikonos and QuickBird. Both systems are commercial, for-profit enterprises, funded and operated by businesses. The Ikonos satellite was built and the system is run by the Lockheed Martin Corporation. The satellite was launched in September, 1999, and after a brief test period, began providing commercial images. The satellite has an orbital altitude of 680 kilometers and an orbital period of 98 minutes. The Ikonos systems provides one-meter panchromatic and four-meter visible/infrared images. The swath width is 13 kilometers directly below the satellite, but because the Ikonos system is pointable off nadir, large areas are visible from any satellite track. Revisit times are typically one to three days.

Data from the Ikonos archive demonstrate the high quality and utility of high resolution satellite images. These data have been applied by government and industry in a range of data acquisition and development contexts, including urban planning, disaster management, agricultural production, real estate, and transportation.

QuickBird was also developed by a group of companies, including ITT Industries and Hitachi. QuickBird is similar to Ikonos in that it provides both panchromatic and multispectral data. It differs mainly in the resolution available, down to 0.65 meters panchromatic and 2.44 meters multispectral, and in a larger image footprint of 16.5 kilometers vs. 13 kilometers.

Figure 6-22: An example of a one-meter panchromatic image from the Ikonos satellite system. (courtesy Space Imaging Corp.)

Radar Systems – a number of radar-based satellite systems have been used as a source of spatial data for GIS. Radar stands for radio detection and ranging, and is an active system wherein the imaging system is both the source and receiver of electromagnetic energy. Radar wavelengths are much longer than optical remote-sensing systems, from approximately one to tens of centimeters, and may be used day or night, through most weather conditions. Radar images are panchromatic, because they provide information on the strength of the reflected energy at one wavelength. Radar systems have been successfully used for topographic mapping and some landcover mapping, particularly when large differences in surface texture occur, such as between water and land, or forest and recently clearcut areas. Operational systems include the ERS-1, operated by the European Space Agency, the JERS-1, by the National Space Development Agency of Japan, and the Radarsat system, developed and managed by the Canadian Space Agency.

Laser Systems - A number of laser-based systems have been developed and are in the early stages of deployment. Lasers are pointed at the Earth surface, pulses of laser light emitted, and the reflected energy is recorded (Figure 6-23). Like radar, laser systems are active because they provide the energy that is sensed. Unlike radar, lasers have limited ability to penetrate clouds, smoke, or haze.

Experimental systems have been used to gather data about topography, vegetation, and water quality. A series of studies by NASA have demonstrated that lasers may provide information on vegetation density, height, and structure. Laser pulses are scattered from the canopy and the ground, and the strength and timing of the return may be used to estimate canopy height (Figure 6-23).

Commercial laser mapping systems are a recent phenomena, and have been used primarily for collecting elevation data. A laser pulse may be directed at the ground. Some energy is reflected from vegetation, but the signal is usually relatively weak. The time interval between laser pulse generation and the strong ground return may be used to calculate aircraft height above the terrain. If flying height is known, then the terrain elevation may be calculated for each pulse. These pulses may be sent several thousand times a second, so a trace of ground heights may be measured from every few centimeters to few meters along the ground. Pointable lasers allow the measurement of a raster grid of elevations.

Most laser systems have been carried aboard aircraft. The power required and the laser ground footprint both increase as the flying height increases, therefore, it is more difficult to build a laser for a satellite platform. Satellite-based systems have been flown experimentally, and an operational system is currently planned by NASA.

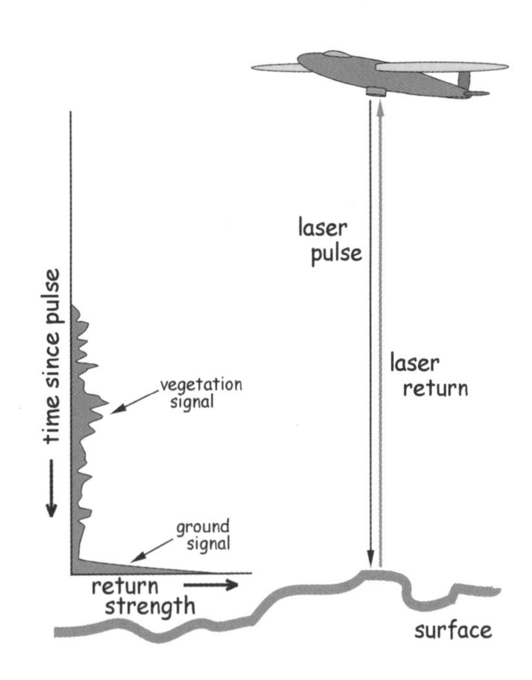

Figure 6-23: Laser mapping systems operate by generating and then sensing light pulses. The return strength may be used to distinguish between vegetation and the ground, and the travel time may be used to determine heights. Mapping systems take measurements in a grid pattern.

Satellite Images in GIS

Satellite images have two primary uses in GIS. First, satellite images are often used to create or update landcover data layers. Satellite images are particularly appropriate for landcover classification by virtue of their uniform data collection over large areas. Landcover classes often correspond to specific combinations of spectral reflectance values. For example, forests often exhibit a distinct spectral signature that distinguishes them from other landcover classes (Figure 6-24). Other landcover classes might correspond to different combinations of spectral reflectance. Satellite image classification involves the identification of the reflectance patterns associated with each landcover class, and then applying this knowledge to classify all areas of a satellite image. A number of techniques have been developed to facilitate landcover mapping using satellite data, as well as techniques for testing the classification accuracy of these landcover data. Regional and statewide classifications are commonly performed, and these data are key inputs in a number of resource planning and management analyses using GIS.

Satellite images are also used to detect and monitor change. The extent and intensity of disasters such as flooding, fires, or

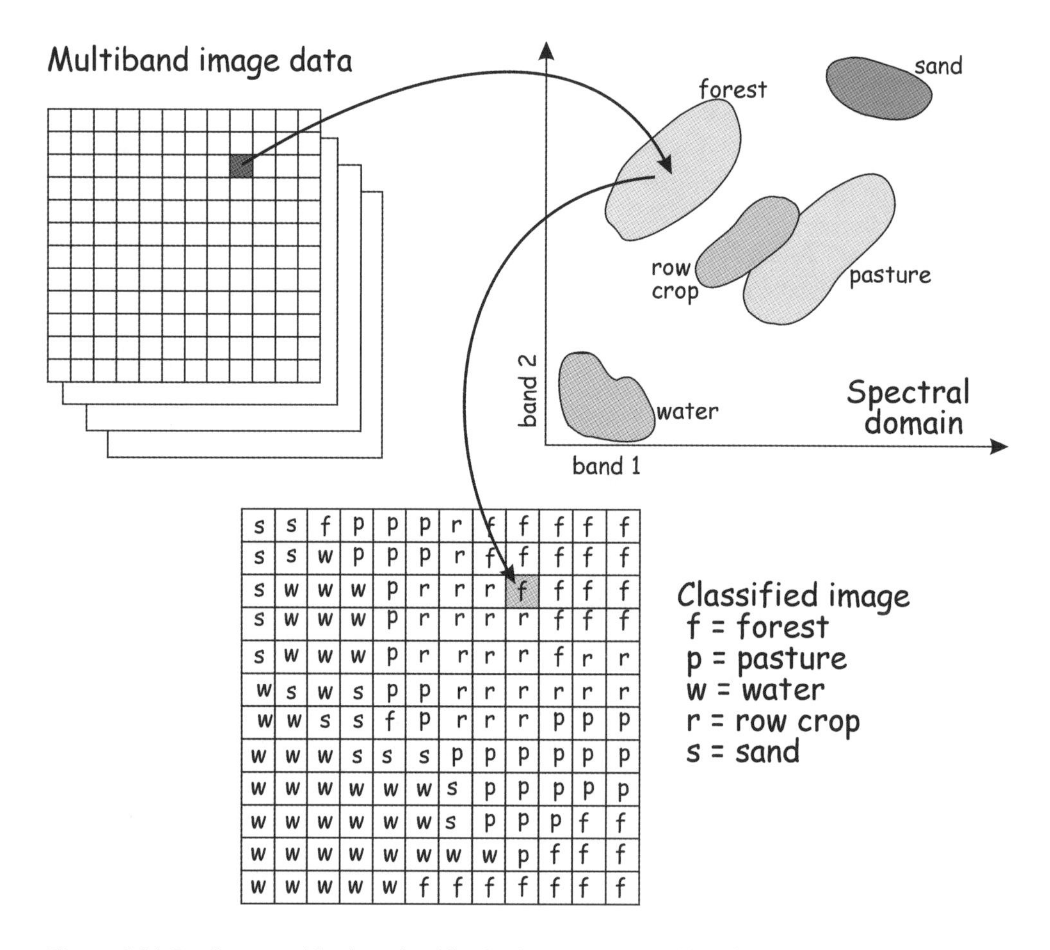

Figure 6-24: Landcover and landuse classification is a common application of satellite images. The spectral reflectance patterns of each cover type are used to assign a unique landcover class to each cell. These data may then be imported into a GIS as a raster data layer.

hurricane damage may be determined using satellite images. Urbanization, forest cutting, agricultural change, or other changes in land use or condition have all been successfully monitored and analyzed based on satellite data. Change detection often involves the combination of new images with previous landcover, infrastructure, or other information in spatial analyses to determine the extent of damage, to direct appropriate responses, and for long-range planning.

Aerial Photographs or Satellite Images in GIS: Which to Use?

The utility of both satellite images and aerial photographs as data sources for GIS should be clear by now. Several sources are often available or potentially available for a given study area. An obvious question is "Which should I use?" A number of factors should be considered when selecting an image source.

First, spatial detail required in the analysis should be considered. Acquired image data should provide the necessary spatial resolution. The resolving power of a system depends in part on the difference in color between two adjacent objects, but resolution is generally defined by the smallest high-contrast object that may be detected. Current high-resolution satellite systems have effective spatial resolutions of from one to several meters. Photographs from mapping cameras, when taken at typical scales and with commonly used film, have maximum resolutions as fine as a few centimeters. Thus the resolving power of photographs is typically much greater than that of satellite systems. Although the gap is narrowing as higher resolution satellite systems are deployed, aerial photographs are currently selected in many spatial analyses that require the highest spatial resolution.

Second, the size of the analysis area should be considered. Aerial photographs are typically less expensive in small-area analyses. Large-scale aerial photographs are often available from government sources at low cost. Each photograph covers from tens to 100's of square kilometers, and cost per square kilometer may be quite low. As the size of the study area increases, the costs of using photographs may increase. Multiple photographs must often be placed in a mosaic, and this is typically a time-consuming operation. Terrain distortion may be significant in mountainous areas, and distortion removal may be more costly with photographs than with satellite images. Satellites, because of their high view and large area coverage, are less likely to require the creation of mosaics and may have significantly less terrain distortion. Therefore, satellite images may be less expensive for larger study areas.

Satellite scanners provide a broader spectral range and narrower bands relative to aerial photographs. As noted earlier, satellite scanners may be designed to detect well beyond the visible and near-infrared portions of the spectrum to which aerial photographs are limited. If important features are best detected using these portions of the spectrum, then satellite data are preferred. Specialized sensors may be particularly important under specific conditions, for example when clouds prevent the use of aerial photographs. In such cases radar remote sensing may be used.

The inherent digital format is another slight advantage for images from satellite scanners. Aerial photographs are analog, and must be converted if digital images are required. While desktop scanners are inexpensive, the scanning process may not be, depending on the time required for scanning and processing each photograph. Scanners may also introduce non-linear distortions into the scanned image, and these distortions may be difficult to remove. Unlike tilt and terrain distortion, there is unlikely to be an analytical model that will allow the removal of scanning distortion. Expensive scanners may be required to ensure geometric fidelity.

Finally, satellite images may require specialized software and training to be effectively utilized. Satellite images are often quite useless for analysis in a printed form. Single images printed on standard or large-sized media result in quite small scales. Most features of interest are not visible at these small scales, and satellite images must be viewed and manipulated with a computer. Image display and manipulation systems may be quite sophisticated and require specialized training. In contrast, under many conditions printed photographs may show all the features of interest with little or no magnification. Tilt and terrain distortion may be quite low on vertical photographs taken in relatively flat portions of the Earth. If geometric distortion is within acceptable limits, features may be digitized directly from aerial photographs. Data may be converted to digital formats using standard digitizing software and hardware, and require no additional training or equipment.

Sources of Images

National, state, provincial, or local governments are common sources of aerial photographs. These photographs are often provided at a reduced cost. For example, the National Aerial Photography Program (NAPP) provides full coverage of the lower 48 United States on at least a ten-year return interval. State cost-sharing may reduce the interval to as short as five years. NAPP provides 1:40,000-scale color infrared aerial photographs with stereographic coverage. Photographs may be acquired during the leaf-off or leaf-on season. The NAPP program is coordinated through the USGS, and online and hardcopy indexes are available to aid in identifying appropriate photographs.

Aerial photographs may also be purchased from other government agencies or from private organizations. The U.S. Natural Resource Conservation Service (NRCS) and the U.S. Forest Service (USFS) both routinely take aerial photographs for specialized purposes. The USFS uses aerial photographs to map forest type and condition, and often requires images at a higher spatial resolution and different time of year than those provided by NAPP. The NRCS uses aerial photographs in the development of county soil surveys, and the Farm Services Agency uses them to monitor crops and agricultural program compliance. These organizations are also excellent sources of historical aerial photographs. Many government agencies contribute to a national archive of aerial photographs, which may be accessed at the internet addresses listed in Appendix B.

Satellite images may be obtained from various sources. Current Landsat data are available through the NASA and the USGS. New acquisitions may be scheduled or recent collections browsed and ordered. Ordering older scenes may be somewhat more complicated, as Landsat 4 and 5 were for a time partly administered by a private organization. As of this writing, data availability and cost are scene and time-specific. However data available from the U.S. federal government have been catalogued, and may be obtained through the USGS. SPOT and other satellite system data may be obtained directly from the managing sources, listed in Appendix B.

Summary

Aerial photographs and satellite images are valuable sources of spatial data. Photos and images provide large-area coverage, geometric accuracy, and a permanent record of spatial and attribute data, and techniques have been well-developed for their use as a data source.

Remote sensing is based on differences among features in the amount of reflected electromagnetic energy. Chemical or electronic sensors record the amount of energy reflected from objects. Reflectance differences are the basis for images, which may in turn be interpreted to provide information on the type and location of important features.

Aerial photographs are a primary source of coordinate and attribute data. Camera-based mapping systems are well-developed, and are the basis for most large-scale topographic maps currently in use. Camera tilt and terrain variation may cause large errors on aerial photographs; however, methods have been developed for the removal of these errors. Terrain-caused photographic displacement is the basis for stereophotographic determination of elevations.

Satellite images are available from a range of sources and for a number of specific purposes. Landsat, the first land remote sensing system, has been in operation for nearly 30 years, and has demonstrated the utility of satellite images. SPOT, AVHRR, Ikonos, and other satellite systems have been developed that provide a range of spatial, spectral, and temporal resolutions.

Aerial photographs and satellite images often must be interpreted to provide useful spatial information. Aerial photographs are typically interpreted manually. An analyst identifies features based on their shape, size, texture, location, color, and brightness, and draws boundaries or locations, either on a hardcopy overlay, or on a scanned image. Satellite images are often interpreted using automated or semi-automated methods. Classification is a common interpretation technique that involves specifying spectral and perhaps spatial characteristics common to each feature type.

The choice of photographs or satellite imagery depends on the needs and budgets of the user. Aerial photographs often provide more detail, are less expensive, and are easily and inexpensively interpreted for small areas. Satellite images cover large areas in a uniform manner, sense energy across a broader range of wavelengths, and are provided in a digital format.

Suggested Reading

American Society of Photogrammetry, Manual of Remote Sensing, 2nd Edition. ASP, Falls Church, 1983.

Atkinson, P. and Tate, N., Advances in Remote Sensing and GIS Analysis, Wiley, New York, 1999.

Avery, T. E., Interpretation of Aerial Photographs, Burgess, Minneapolis, 1973.

Befort, W., Large-scale sampling photography for forest habitat-type identification, *Photogrammetric Engineering and Remote Sensing*, 1986, 52:101-108.

Campbell, J. B., Introduction to Remote Sensing, 2nd Edition,Guilford, New York, 1996.

Bolstad, P.V., Geometric errors in natural resource GIS data: the effects of tilt and terrain on aerial photographs, *Forest Science*, 1992, 38:367-380.

Drury, S. A., A Guide to Remote Sensing: Interpreting Images of the Earth, Oxford University Press, New York, 1990.

Elachi, C., Introduction to the Physics and Techniques of Remote Sensing, Wiley, New York, 1987.

Kodak, Kodak Data for Aerial Photography, Eastman Kodak Company, Rochester, 1982.

Light, D., The National Aerial Photography Program as a geographic information systems resource, *Photogrammetric Engineering and Remote Sensing*, 1993, 59:61-65.

Lillesand, T. M. and Kiefer, R. W., Remote Sensing and Image Interpretation, 4th Edition. Wiley, New York, 1999.

Meyer, M. P., Place of small-format aerial photography in resource surveys, *Journal of Forestry*, 1982, 80:15-17.

Nelson, R. and Holben, B., Identifying deforestation in Brazil using multiresolution satellite data, *International Journal of Remote Sensing*, 1986, 7:429-448.

Teng, W. L., AVHRR monitoring of U.S. crops during the 1988 drought, *Photogrammetric Engineering and Remote Sensing*, 1990, 56:1143-1146.

Warner, W., Accuracy and small-format surveys: the influence of scale and object definition on photo measurements, *ITC Journal*, 1990, 1:24-28.

Welch, R., Integration of photogrammetric, remote sensing and database technologies for mapping applications, *Photogrammetric Record*, 1987, 12:409-428.

Wolf, P. R., Elements of Photogrammetry, McGraw-Hill, New York, 1983.

Study Questions

Can you describe several positive attributes of images as data sources?

What is the electromagnetic spectrum, and what are the principle wavelength regions?

What is a spectral reflectance curve? Can you draw typical curves for vegetation and soil through the visible and infrared portions of the spectrum?

What is photogrammetry? Why is it important for GIS?

Can you describe the structure and properties of the main types of photographic film, including their spectral sensitivity curves?

What are the basic components of a camera used for taking aerial photographs?

Can you describe the most commonly used camera formats for aerial photography, and their relative advantages?

What are the major sources of geometric distortion in aerial photographs, and why? What are other, usually minor, sources of geometric distortion in aerial photographs?

What are typical magnitudes of geometric errors in uncorrected aerial photographs? How might these be reduced?

Can you describe stereo photographic coverage, and why it is obtained?

What is parallax, and why is it useful?

Can you describe the basic process of terrain distortion removal?

What is photointerpretation and what are the main photographic characteristics used during interpretation?

How are images from satellite scanners different from photographs? How are they similar?

What are some of the basic types and principles of operation of satellite imaging scanners?

Can you describe and contrast the Landsat MSS, Landsat TM, SPOT, AVHRR, and Ikonos satellite imaging systems?

What are some of the criteria used in selecting the type of images for spatial data development?

7 Digital Data

Introduction

Many spatial data currently exist in digital forms. Roads, political boundaries, water bodies, land cover, soils, elevation, and a host of other features have been mapped and converted to digital spatial data for much of the World. Because these data are often distributed at low or no cost, these existing digital data are often the easiest, quickest, and least expensive source for many spatial data layers.

Digital data are often developed by governments because these data help provide basic public services such as safety, health, transportation, water, and energy. Spatial data are required for disaster planning and management, national defense, infrastructure development and maintenance, and other governmental functions. Many national, regional, and local governments have realized that once these data have been converted to digital formats for use within government, they may also be quite valuable for use outside government. Business, nonprofit, education, science, as well as governmental bodies may draw benefit from the digital spatial data, as these organizations benefited in prior times from government-produced paper maps. Although digital data are also available from private sources, governmental sources provide a diverse set of data over broader areas, often at low costs. Some data commonly available throughout the United States are described in this chapter.

Digital Raster Graphics

A Digital Raster Graphic (DRG) is a georeferenced raster image of a scanned USGS map (Figure 7-1). DRGs are available

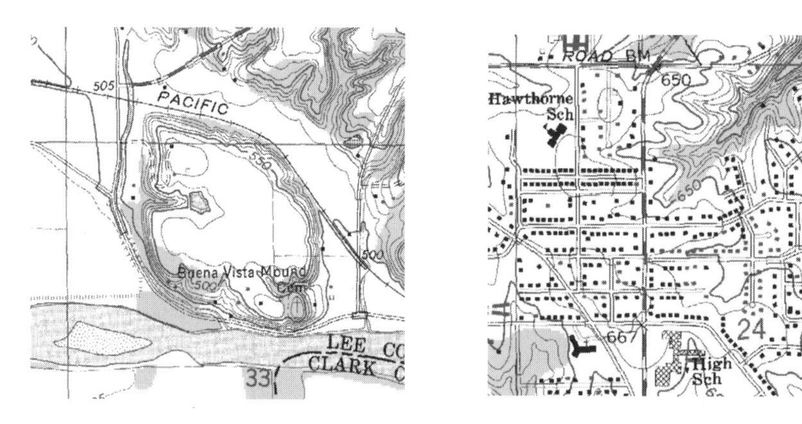

Figure 7-1: Examples from USGS 1:24,000-scale digital raster graphics (DRG) files.

for most of the 1:24,000-scale USGS quad maps, and 1:100,000 and 1:250,000-scale maps. Color maps from the standard map series have been scanned at a high spatial resolution and referenced to a standard map projection. These data may then be used as a source for on-screen digitizing, as a backdrop for other data collection or viewing, or as digital maps. DRGs are particularly useful as a quick check for consistency in new data collection, and for navigation when displaying new data.

DRGs are created by scanning a paper or other appropriate map. The map is placed on a high resolution scanner and an image captured. Colors are modified to enhance viewing, and the scanned maps are georeferenced to an appropriate coordinate system, e.g., a UTM coordinate system for most 1:24,000-scale USGS maps. The image is resampled to just under 100 dots per millimeter (250 dots per inch), and converted to a compressed GeoTIFF image format.

Digital Line Graphs

Digital line graph data (DLGs) are vector representations of most features portrayed on USGS national series maps (Figure 7-2). Most DLGs have been produced from USGS series maps, and most features on these maps have been recorded in DLGs, so there is a close correspondence between DLGs and USGS series maps. DLGs are available by map series designation, e.g., there are 1:2 million DLGs that correspond to the data included on 1:2 million scale maps. DLGs for 1:100,000 and 1:24,000-scale maps are also available. The extent of an individual DLG typically corresponds to the extent of the map series. For example, a 1:24,000-scale map in the USGS national series covers a 7.5 by 7.5 minute area. The corresponding DLG contains digital data from the same area. Thus the DLGs are typically tiled just as the corresponding map series.

Data recorded on most DLGs have been edge matched to those recorded on the adjacent DLGs (Figure 7-3). Edge matching is performed at the time of DLG creation. In most cases the original paper maps were edge matched, so DLG data match across boundaries by virtue of accurate digitizing. If the map on one side of the boundary has been revised, or the original maps were not

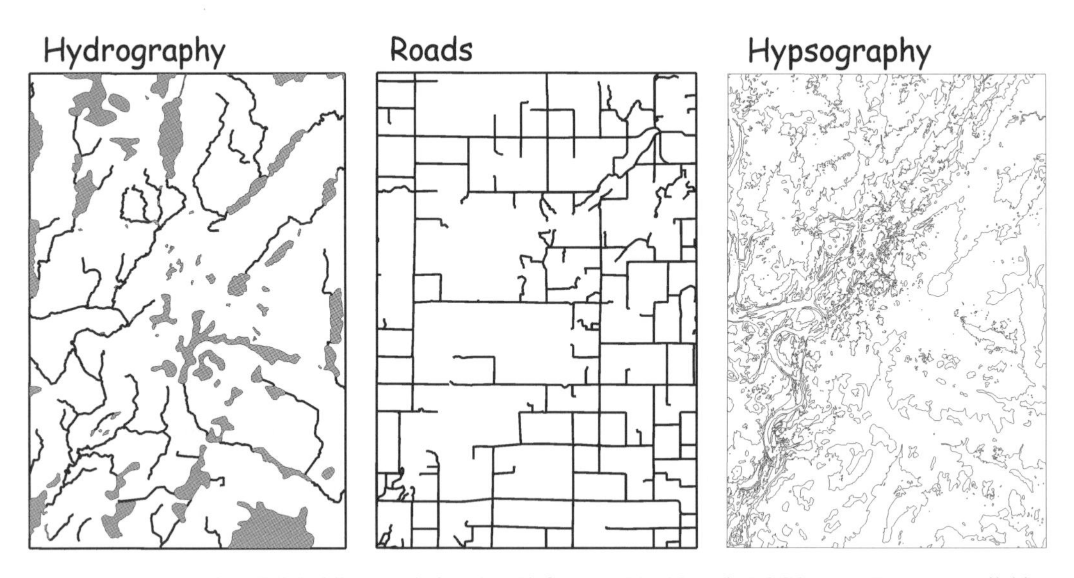

Figure 7-2: Examples of digital line graph data (DLG) from 1:100,000-scale USGS maps. Data are available at this and other scales for most of the United States.

edge matched or the series map has not been produced for one of the sides, edge matching may be enforced manually or through customized software at the time of DLG creation. This edge matching ensures smooth transitions and consistent features when joining multiple DLGs.

DLGs are provided in separate themes, and these themes vary by map series. For example, there are typically four DLGs for the 1:100,000-scale map series: boundaries, hydrography (water bodies), roads, and hypsography (elevation contours). There are commonly nine separate DLGs for the 1:24,000-scale map series: transportation, hypsography (contours and spot elevations), hydrography, (lakes, rivers, glaciers, etc.), boundaries (political and administrative), vegetation features, non-vegetation feature (sand, gravel, lava, etc.), monuments and control points, Public Land Survey System, and man-made features. Each of these is provided in a separate data layer. Not all data layers may be available for every mapped area, as some features may not exist in a map sheet, or were not recorded at the time the map was produced. For example the Public Land Survey System (PLSS) was used in the original surveys of many western territories, and is included on most 1:24,000-scale maps and corresponding DLGs for western states. PLSS was not defined or used in many eastern states, and so is not a DLG layer on most maps for that region.

Figure 7-3 Data from DLGs are often edge matched along the map seams. Here the vertical line visible near the center of the figure shows the edge between DLGs.

DLGs are provided in a number of standardized formats. Originally there were three formats, DLG-1 through DLG-3, with differences in positional accuracy, attribute coding, and other spatial information. Among these formats only the DLG-3 is commonly used, and so most references to DLGs are to the DLG-3 format. A format that complies with the Spatial Data Transfer Standard (SDTS) is increasingly available. Larger scale DLGs, those derived from 1:24,000-scale USGS topographic maps, were originally provided in a standard coordinate system, in inch coordinates relative to an arbitrary lower left or map center origin. The "optional" DLG format using a map projected coordinate system has proven more popular, and is most commonly encountered. For example, 1:24,000-scale DLGs are most often provided in a Universal Transverse Mercator coordinate system. Data are typically delivered in text files. Information on the source, theme, coordinate system, and features is provided in the text files. Most GIS packages are able to ingest DLG-3 and SDTS formats.

DLG data provide a limited set of attribute information, and are highly structured to convey topological relationships. Numeric codes are associated with point, line, and area features. These codes may be used to categorize area, line, or point features, e.g., road type. Categories are typically similar to those represented on hardcopy USGS maps, resulting in a few broad classes, e.g., major road, minor road, or unpaved road. DLG data are topological in the sense that the full set of adjacency and connectivity information are provided.

Digital Elevation Models

Digital elevation models (DEMs) provide elevation data in a raster format and are available for most of the Earth. The U.S. and other governments have created DEMs at a range of scales, and coarse-scale data are available for every continent (Figure 7-4). Worldwide coverage currently exists at relatively coarse resolutions of a few kilometers or more, although DEMs for small areas are available at very high resolutions. These are most often produced for specific projects, e.g., large developments such as dams or roads, and may be created via standard photogrammetric methods, or using optical satellite measurements (Chapter 6). New technologies promise improved global data sets, such as those based on space-based radar topographic measurement. At present high resolution DEMs are common in developed countries.

Ground and aerial surveys are the primary source of original elevation measurements for most digital elevation models. Traditional distance and angle measurements with surveying equipment may provide precise elevations at specified locations. Because these methods are relatively slow they can only provide a dense network over small areas. Large-area surveys typically can only afford a sparse network of points. Global positioning system technologies have greatly improved survey speed and accuracy, but even these technologies are too slow to be the sole elevation data collection method over all but the smallest areas. Aerial photographs complement field surveying by increasing the number and density of measured elevations. Accurate elevation data may be collected over broad areas with the appropriate selection of cameras, survey points, photo locations, and subsequent anal-

Figure 7-4: Digital elevation models (DEMs) are available at various resolutions and coverage areas for most of the World. This figure of North America was produced using data that have been derived from 1:2,000,000-scale maps and are available from the USGS.

Figure 7-5: An example of contour lines depicted on a 1:24,000-scale USGS topographic map. The contour interval is 10 feet, with the 950 and 1000 foot contour labels visible near the left center and center of the figure.

ysis and processing. These photogrammetric methods are discussed in Chapter 6.

Much of the currently available elevation data are derived from contour lines that were produced using photogrammetric techniques. Photogrammetry has been operationally applied since the 1930's to map elevation, and data have been developed for most of the Earth's surface using these techniques. These lines are printed on cartometric-quality maps, typically published by national mapping agencies such as the U.S. Geological Survey or the British Ordnance Survey (Figure 7-5). Most currently available DEMs are derived from manual or automatic digitization of these mapped contour lines.

Digital elevation data may also be derived more directly from the photographic source. Early photogrammetric equipment required contours be drafted onto hardcopy materials; these hardcopy maps could subsequently be digitized. Later equipment was outfitted with three-axis digital recorders. These digital recorders bypassed the paper compilation step and allowed the interpreted contours or profiles to be directly entered in a digital form. Hybrid optical-mechanical-electronic devices such as the Gestalt Photo Mapper II were developed which automated the identification of contours and elevation data.

Completely digital, "softcopy" photogrammetric workstations are now standard equipment, used in most large projects today (see Chapters 4 and 6). The basic principle of parallax is still applied. Photographs are scanned to produce digital images, and the camera locations, parallax measurement, and contour generation all take place in a digital computer environment. These systems may support full or semi-automated DEM and contour generation, techniques that greatly improve the efficiency of DEM creation in new surveys. Softcopy systems are often the best choice when new or improved digital elevation data are required, although they may at some point in the future be replaced by airborne or satellite-based laser or radar systems. While the bulk of the DEM data available by the mid 1990's was developed from manual or automated digitizing of mapped contour lines, softcopy photogrammetric workstations are currently used to produce most large-area digital elevation data.

Laser-based elevation mapping and DEM generation are becoming more common. These systems are based on a downward-pointing laser placed on an aircraft (see Chapter 6). The laser measures height above terrain, and may include pointing optics that can measure the heights of locations both directly below and to either side of the aircraft. A precise GPS system measures flying heights, and these may be combined with laser measurements to produce a set of terrain heights from across the landscape. Height measurements may be combined to create a TIN or grid-based DEM. Laser systems are most often used when precise elevation measurements are required over small areas, but they will be more widely applied as the technology develops.

DEMs with a 30 meter horizontal sampling frequency are available for most of the

United States, excluding Alaska. These high-resolution DEMs are delivered in tiles 7.5 minutes on a side, corresponding to the 1:24,000-scale map sheets published by the USGS. These DEMs have been produced using a number of methods, including contour digitizing and conversion, optical-mechanical instruments specialized for DEM generation, and directly from scanned and parallax-matched aerial photographs. Data are provided in a south to north sequence of elevation postings, arrayed from west to east across the 7.5 minute quadrangle (Figure 7-6). Data are projected in a UTM coordinate system, clipped by the straight lines between the 7.5 minute coordinate corners. Because of this, there is an occasional "dropped" posting between tiles. A dropped posting occurs when the first posting in one tile is approximately 60 meters further north than the last posting in the corresponding line of the adjacent tile. Data are also provided with each DEM that describe the quadrangle name, corner coordinates, format, method of data production, and other important information.

DEM data for the United States are also available with various other sampling frequencies and area coverage:

National elevation dataset (NED): these are a seamless combination of the highest-resolution DEM data for the United States. Elevation data from 7.5 minute are used for the conterminous U.S., and 15 minute data for Alaska. Coarser-resolution sources are used

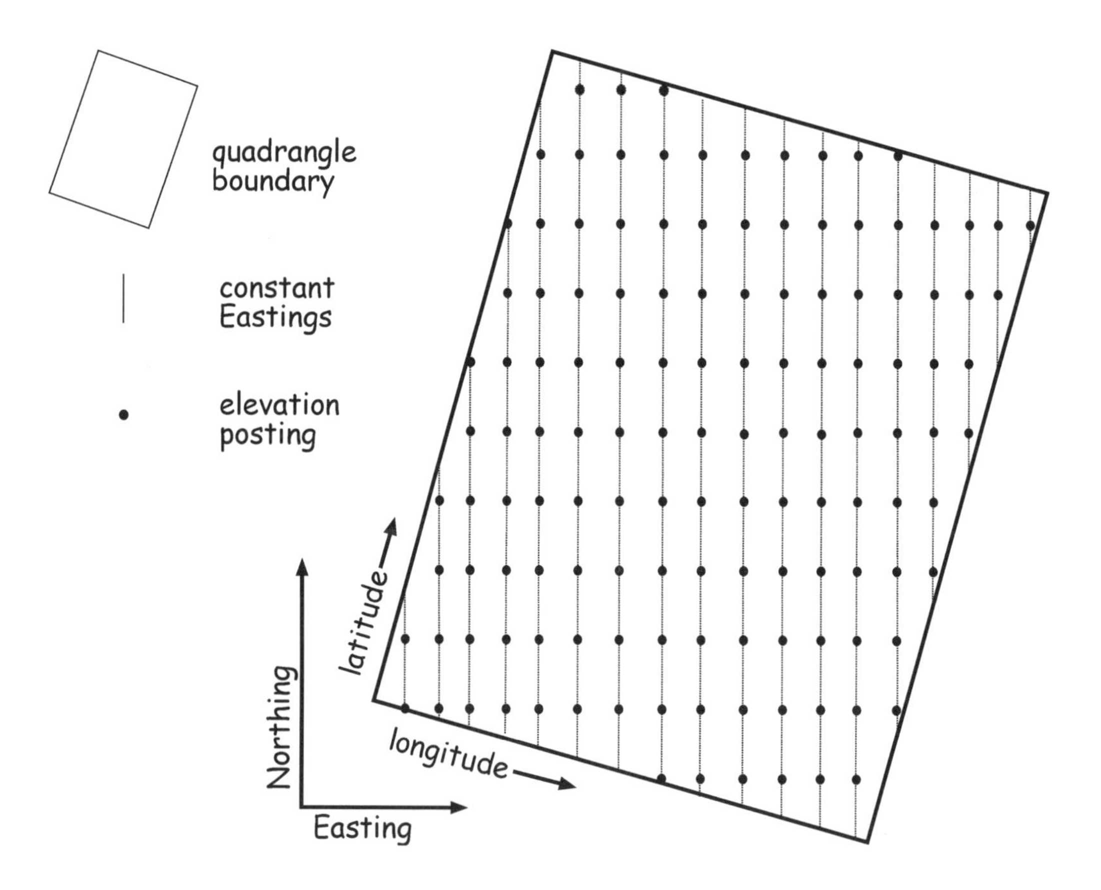

Figure 7-6: DEMs provided by the USGS are arranged as a series of postings in lines parallel to the local Northing direction. Data are provided at a fixed sampling frequency that is usually constant along and between posting lines.

in the rare instances where these data are lacking. NED elevation values are provided as decimal meters in the NAD83 datum, and all the data are recast in a geographic projection. Source DEMs from the 7.5 and 15 minute quads were produced by methods that at times produced artifacts such as pits, spikes, or linear drainage systems. The NED assembly process is designed to remove most of these artifacts. This improves the quality of the slope, aspect, shaded-relief, and drainage information that may be derived from the elevation data.

2 arc-second DEMs: these data have an elevation posting frequency every two seconds of arc. Columns and rows are in approximately northerly and easterly directions. Data are provided in tiles covering 30 minutes of latitude and longitude. Tile edges correspond to the edges of USGS 1:24,000-scale, 7.5 minute quadrangle maps. Data are available for the conterminous United States and Hawaii, and were produced by the USGS.

3 arc-second DEMs: these data are provided with an elevation posting frequency in both northerly and easterly directions at 3 seconds of arc. This corresponds to approximately 80 meters over most of the United States. These data are available for the entire U.S., and were produced by the Defense Mapping Agency (now NIMA). These data are distributed by the USGS in 1 degree by 1 degree blocks.

Alaska DEMs: there are two DEM series available for Alaska, one at a 1 by 2 arc-second spacing corresponding to 7.5 minute quadrangles, and one at 2 by 3 arc-second spacing in tiles covering 15 minutes of latitude to between 20 and 36 minutes of longitude. Variation in size is due to longitudinal convergence from south to north. Spacing in meters varies with location, also due to longitudinal convergence.

Digital elevation data sets that cover most of the world are available from the US government, although current world-wide coverage is provided at rather coarse scales. Data are available at a 30 arc-second sampling frequency, approximately 1 kilometer (0.6 miles) for all continents. These data are tiled, e.g., in a 40 by 60 degree block, and provided in latitude/longitude coordinates.

Higher resolution terrain data will soon be available, based on radar measurements from NASA's Shuttle Radar Topography Mission (SRTM). Collected in the year 2000, SRTM data were not processed for much of the World at the time of this writing. SRTM DEMs will provide elevation data at an approximate 100 meter resolution for most of the Earth surface.

Many developed countries have higher-resolution elevation data specific to their areas of control. For example, elevation data on a 20 meter contour or 3-second grid are available for Australia through the Australian National Mapping Division, and data for the United Kingdom are provided at a 50 meter (160 foot) raster cell size by the Ordnance Survey of Britain.

Digital Orthophoto Quadrangles

Digital orthophoto quadrangles (DOQs) are scanned photographic images that have been corrected for distortions due to camera tilt, terrain displacement, and other factors. These corrections yield photographs that are planimetrically correct, similar to large-scale topographic maps. The DOQs are registered to an Earth coordinate system. These photographs may be a valuable source of spatial information, because they may be used as the basis for creating or updating thematic data layers such as roads, vegetation, or buildings (Figure 7-7).

DOQs are available for much of the United States. The USGS leads production of DOQs for use by a number of agencies and non-government organizations. The U.S. Natural Resource Conservation Service uses DOQs in farm tracking and regulation compliance, the U.S. Forest Service uses them in forest management, and many county and city governments and private firms use DOQs for digital data development.

Figure 7-7: An example of a digital orthophotoquad (DOQ) for an area in eastern Minnesota. DOQs are available for much of the United States, and provide geometrically corrected images at a 1 meter resolution. The USGS plans on providing nationwide coverage based on a tiling scheme compatible with the 1:24,000-scale map series.

DOQs are produced by first scanning a transparency, known as a diapositive, of an aerial photograph. Transparencies are scanned at a high resolution with a precision scanner. The scanned image is combined with ground coordinate information to remove positional errors. Geometric distortion due to camera tilt, camera systematic errors, and terrain displacement are removed through the application of projection geometry, as described in Chapter 6.

This projection geometry should not be confused with a map projection. Although similar terms are used, projection geometry is used here to describe the photogrammetric projection from the distorted location on a photograph onto a corrected location on a planar, mapped surface. A photogrammetric process called a space resection combines elevation data, the image data, and precise ground survey information to calculate where each point on the image would fall in an undistorted (orthographic) projection. Each point in an image is projected onto the orthographic plane. Images are typically recorded in a compressed format known as a JPEG to ease distribution and reduce storage requirements.

DOQs have many uses in GIS. One primary use is the identification of additional geographic control points. We often need to register maps or imagery that lack adequate control information. Most features larger than one meter in size are visible on DOQs. Any small, well-defined feature on both the map to be registered and the DOQ may serve as a control point because the DOQ is a registered image. Because DOQs are usually provided in a known UTM coordinate system we may obtain the coordinates of many potential control points, including road intersections, small buildings, or other well-defined features.

DOQs are also useful as a source of data for creating new or updating old data sets. Urban expansion, road construction, or other changes in land use have occurred over many areas since hardcopy maps were produced or updated. While the original data were correct for their time period, the data derived from them may lead to erroneous conclusions because of changes through time. Imagery such as DOQs provide detailed spatial and categorical information about these changes, and may be used as a backdrop in on-screen digitizing to update important data sets.

National Wetlands Inventory

Data on the location and condition of wetlands are available for much of the United States through the National Wetlands Inventory (NWI) program. NWI data are produced by the US Fish and Wildlife Service. NWI data portray the extent and characteristics of wetlands, including open water (Figure 7-8), and as of the year 2000 were available for approximately 90% of the conterminous United States. About 40% of the conterminous United States is available in digital formats. NWI data were produced through the 1970s and 1980s, with an update in the 1990s. Decadal updates are planned.

NWI data are produced through a combination of field visitation and the interpretation of aerial photographs. Spring photographs at a range of scales and types are used. Color infrared photographs at a scale of 1:40,000 are commonly used, however black and white photographs and scales ranging between 1:20,000 and 1:62,500 have been employed. Spring photographs typically record times of highest water tables and are most likely to record ephemeral wetlands. There is substantial year-to-year variability in surface water levels, and hence there may be substantial wetland omission when photographs are acquired during a dry year.

NWI data provide information on wetland type through a hierarchical classification scheme, with modifiers (Figure 7-9). Wetlands are categorized as part of a lacustrine (lake), palustrine (pond), or riverine system. Subsystem designators then specify further attributes, e.g., is the wetland perennial, intermittent, littoral or deep water. Further class and subclass designators and modifiers provide additional information on

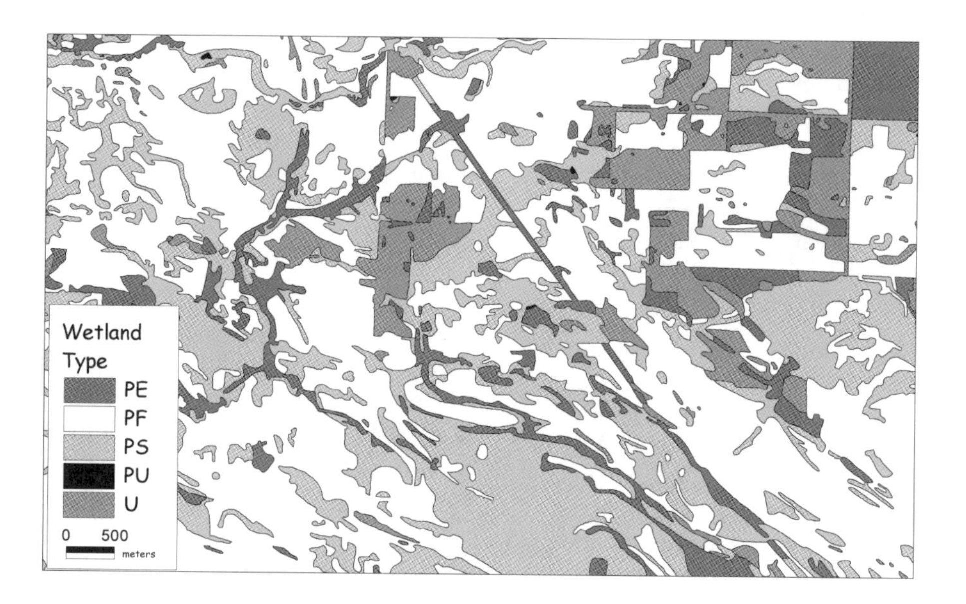

Figure 7-8: An example of national wetlands inventory (NWI) data. Digital NWI data are available for most of the United States, and provide information on the location and characteristics of wetlands.

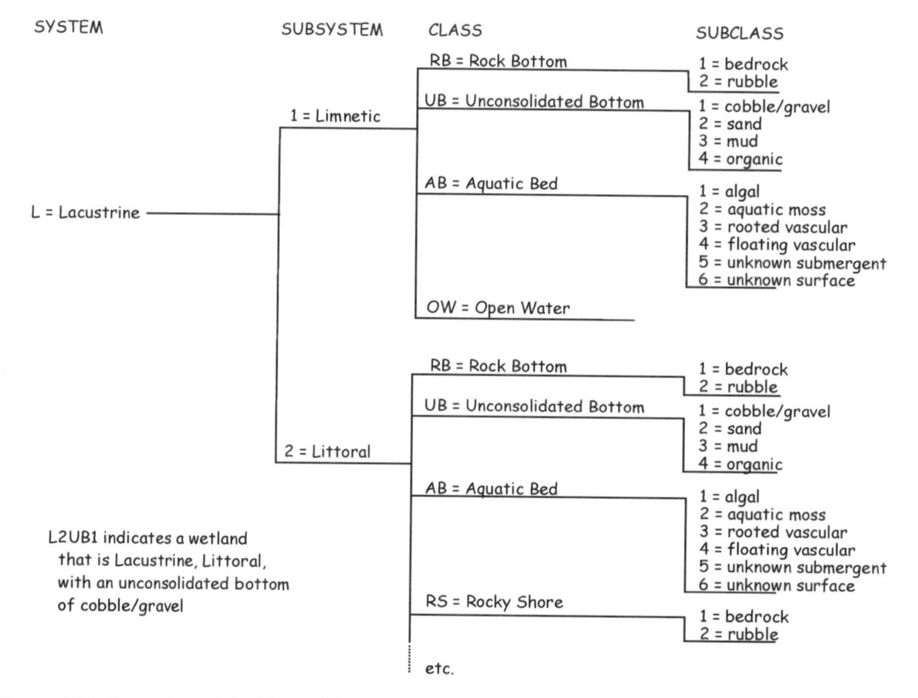

Figure 7-9: A portion of the hierarchical classification system provided in NWI wetlands designation. Attribute data on the wetland system, subsystem, class, modifiers, and other characteristics are provided.

wetland characteristics. A shorthand designator is often used to specify the wetland class, e.g., a wetland may be designated L1UB2G, as system = lacustrine (L), subsystem = limnetic (1), class = unconsolidated bottom (UB), with subclass = sand (2), and a modifier indicating the wetland is intermittently exposed (G). Wetland data are often provided in a DLG format.

The *minimum mapping unit* (MMU) is the target size of the smallest feature captured. Features smaller than the MMU are not recorded in these data. NWI data typically specify MMUs of between 0.5 and 2 hectares. MMUs vary by vegetation type, film source, region, and time period. MMUs are typically largest in forested and smallest in agricultural or developed areas, as it is more difficult to detect many forested wetlands. MMUs also tend to be larger on smaller scale photographs. The MMU, scale, and other characteristics of the wetlands data are available in map-specific metadata.

NWI data do not exhaustively define the location of wetlands in an area. Because of the photo scales and methods used, many wetlands are not included. Statutory wetland definition typically includes not only surface water, but also characteristic vegetation or evidence on the surface or in the soils that indicates a period of saturation. Since this saturation may be transient or the evidence may not be visible on aerial photographs, many wetlands may be omitted from the NWI. Nonetheless, NWI data are an effective tool for identifying the location and extent of large wetlands, the type of wetland, and for directing further, more detailed ground surveys.

Digital Soils Data

The Natural Resource Conservation Service (NRCS) of the United States Department of Agriculture has developed three digital soils data sets. These data sets differ in the scale of the source maps or data, and hence the spatial detail and extent of coverage. The National Soil Geography (NATSGO) data set is a highly generalized, national coverage soils map, developed from small scale maps. NATSGO data have limited use for most regional or more detailed analyses and will not be further discussed here. State Soil Geographic (STATSGO) data are intermediate in scale and resolution, and Soil Survey Geographic (SSURGO) data provide the most spatial and categorical detail.

SSURGO data are intended for use by land owners, farmers, and planners at the large farm to county level. SSURGO maps indicate the geographic location and extent of the soil map units within the soil survey area (Figure 7-10). Soil map units typically correspond to general grouping, called phases, of detailed soil mapping types. These detailed mapping types are called soil series. There are approximately 18,000 soil series in the United States, and several phases for most series, so there are potentially a large number of map units. Only a small subset is likely to occur in a mapped area, typically fewer than a few hundred soil series or series phases. A few to thousands of distinct polygons may occur.

SSURGO data are not intended for use at a site-specific level, such as crop yield predictions for an individual field or septic system location within a specific parcel. SSURGO data are more appropriate for broader-scale application, e.g., to identify areas most sensitive to erosion, or to plan land use and development. SSURGO data and the soil surveys on which they are based are the most detailed soils information available over most of the United States.

SSURGO data are developed from soil surveys. These surveys are produced using a combination of field and photo-based measurements. Trained soil surveyors conduct a series of field transects in an area to determine relationships between soil mapping units and terrain, vegetation, and land use. Aerial photographs at scales of 1:12,000 to 1:40,000 are used in the field to aid in location and navigation through the landscape. Soil map unit boundaries are then interpreted onto aerial photographs or corresponding orthophotographs or maps. Typical photo scales are 1:15,840, 1:20,000, or 1:24,000. These maps are then digitized in a manner that does not appreciably affect positional accuracy. Soil linework is joined and archived in tiles corresponding to USGS quadrangle maps. Soil surveys are often conducted on a county basis, so county mosaics of SSURGO data are common.

SSURGO data are reported to have positional accuracy no worse than that for 1:24,000-scale quadrangle maps. This corresponds to a positional error of less than 13 meters (40 feet) for approximately 90% of the well-defined points when SSURGO data are compiled at 1:24,000-scale. SSURGO data are provided using standard DLG formatting.

SSURGO data are linked to a Map Unit Interpretations Record (MUIR) attribute data base (Figure 7-11). Key fields are provided with the SSURGO data, including a unique identifier most often related to a soil map unit, known as the map unit identifier (muid). Tables in the MUIR data base are linked via the muid, and other key fields. Most tables contain the muid field, so a link may be created between the muid value for a polygon and the muid value in another table, e.g, the Compyld table (Figure 7-11). This creates an expanded table that may be further linked through cropname, clascode, or other key fields. These table structures and linkages are discussed more generally in Chapter 8.

Tables are linked via keys, so that the specific variables of interest may be accessed for each soil type. Variables include an extensive set of soil physical and chemical properties. Data are reported for water capacity, soil pH, salinity, depth to bedrock, building suitability, and most appropriate crops or other uses. Most MUIR data report a range of values for each soil property. Ranges are determined from representative field-collected samples for each map unit, or from data collected from similar

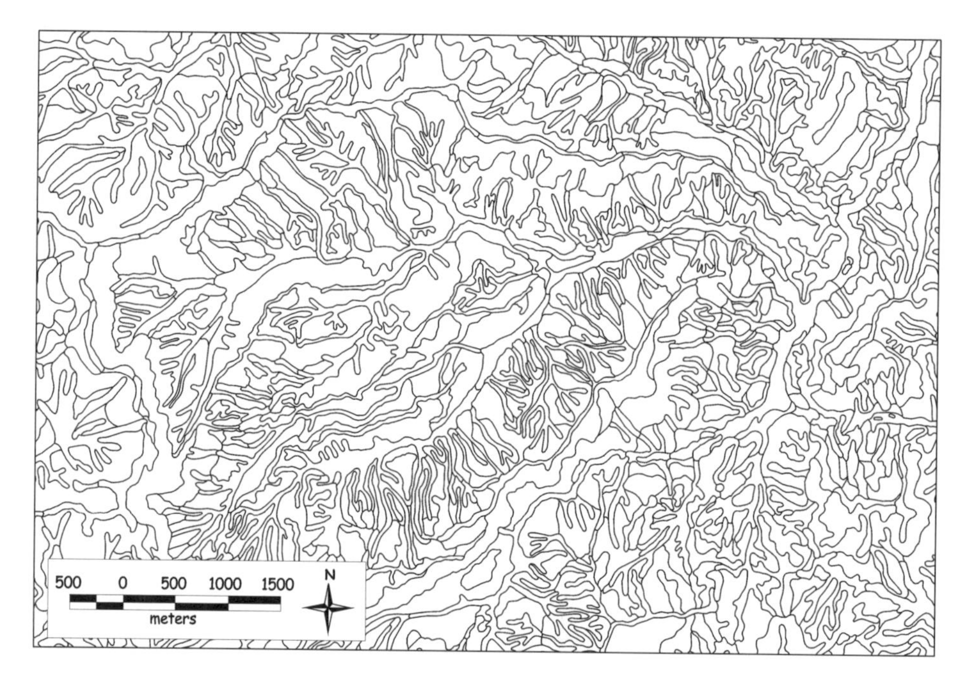

Figure 7-10: An example of SSURGO digital soils data available from the NRCS.

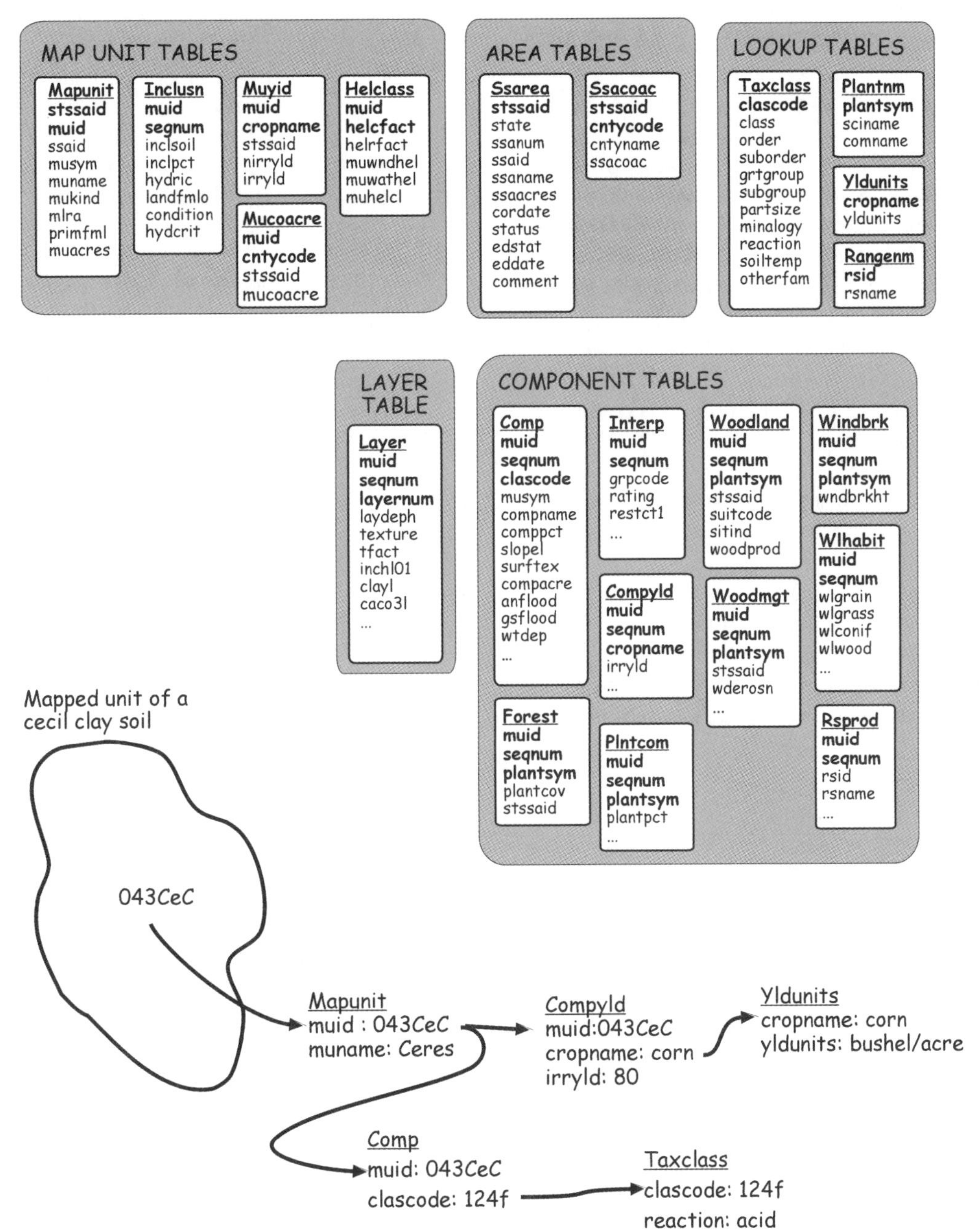

Figure 7-11: The database schema associated with the SSURGO digital soils data. Variables describing soil characteristics are provided in a set of relatable tables. Keys in each table, shown in bold, provide access to items of interest. Codes provided with the digital geographic data, e.g., the map unit identifier (muid), provide a link to these data tables. The relation of a mapped soil polygon to attribute data is shown in the example at the bottom. The muid is related from the MAPUNIT and COMP table, which in turn are used to access other variables through additional keys.

map units. Samples are analyzed using standardized chemical and physical methods.

STATSGO digital soil maps are smaller scale and cover broader areas than SSURGO soil data. STATSGO data are typically created by generalizing SSURGO data. If SSURGO data are not available, STASGO data may be generated from a combination of topographic, geologic, vegetation, land use, and climate data. Relationships between these factors and general soil groups are used to create STATSGO maps. STASTGO map units are larger, more generalized, and do not necessarily follow the same boundaries as SSURGO map units (Figure 7-12). In addition, STATSGO polygons contain from one to over 20 different SSURGO detailed map units. A SSURGO map unit type is a standard soil type used in mapping. It is often a phase of a map series. Each STATSGO map unit may be made up of thousands of these more detailed SSURGO polygons, with over 20 different SSURGO map unit types represented within a STATSGO polygon. STATSGO data provide information on some of this variability. Data and properties on multiple components are preserved for each STATSGO map unit.

STATSGO data are developed via a redrafting of more detailed data onto 1:250,000-scale base maps. These data may be existing county soil surveys that form the basis for SSURGO data, or a combination of these and other materials such as previously published statewide soil maps, satellite imagery, or other statewide resource maps. Soil map units are drafted onto mylar sheets overlain on 1:250,000-scale maps. A minimum mapping unit of 625 hectares is specified, and map unit boundaries are edgematched across adjoining 1:250,000-scale maps. These features are then digitized to produce digital soils data for each 1:250,000-scale map. Data may then be joined into a statewide soils layer. Geographic data are then attached to appropriate attributes. Data are most often digitized in a local zone UTM coordinate system, and converted to a common Alber's equal area projection. STATSGO data are provided in a USGS DLG format.

Figure 7-12: An example of STATSGO data for Colorado, USA.

Digital Census Data

The United States Census Bureau developed and maintains a database system to support the national census. This system, known as the Census TIGER system (Topologically Integrated Geographic Encoding and Referencing) has been created to help collect and maintain digital geographic data for the U.S. and its territories. The TIGER system is used to organize areas by state, county, census tract, and other geographic units for data collection and reporting. It also allows the assignment of individual addresses to geographic entities. The census TIGER system links geographic entities to census statistical data on population size, age, income, health, and other factors (Figure 7-13). These entities are typically polygons defined by roads, streams, political boundaries, or other features. The TIGER system is a key tool of government in the collection of census data. TIGER also aids in the application of census data during the apportionment of federal government funds, in congressional re-districting, in transportation management and planning, and in other federal government activities.

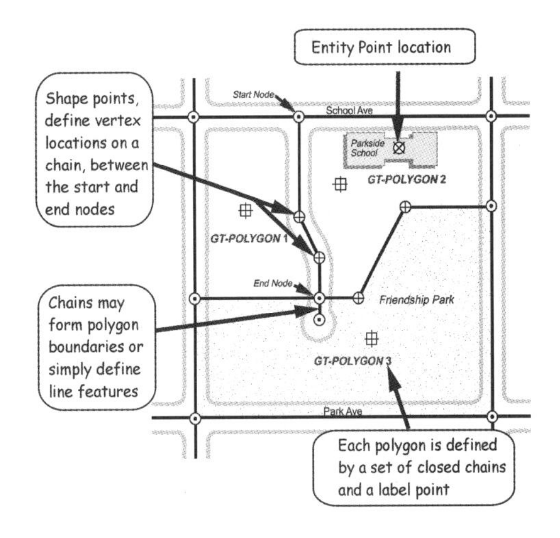

Figure 7-14: TIGER data provide topological encoding of points, lines (chains), and polygons. (from U.S. Dept. of Commerce)

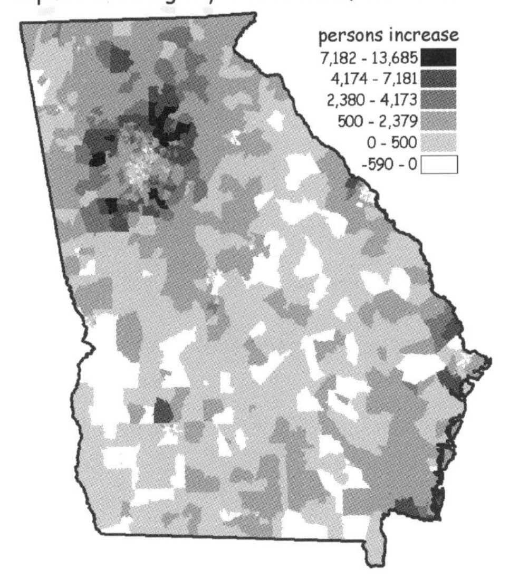

Figure 7-13: Digital census data provide spatially referenced demographic and other data for the U.S. These data are quite useful in a number of disciplines, for example in this analysis of population change in Georgia.

TIGER/Line files are at the heart of the system. They define line, landmark, and polygon features in a topologically integrated fashion. Lines (called chains) most often represent roads, hydrography, and political boundaries, although railroads, power lines, and pipelines are also represented. Polygon features include census tabulation areas such as census block groups and tracts, and area landmarks such as parks and cemeteries. Point landmarks such as schools and churches may also be represented. Points, lines, and polygons are used to define these features (Figure 7-14).

Nodes and vertices are used to identify line segments. A set of topological attributes are attached to the nodes and lines, such as the polygons on either side of the line segment, or the line segments that connect to the node. Point landmarks and polygon interior points are other topological elements of TIGER/Line files.

TIGER/Line files contain information to identify street address labels. Starting and ending address numbers are recorded corresponding to starting and ending nodes (Figure 7-15). Addresses may then be assigned within the address range. The system does not allow specific addresses to be assigned to specific buildings. However it does restrict the addresses on a city block to a limited range of numbers, something of great use to field workers responsible for collecting census information.

TIGER/Line files are organized in 17 different record types. A collection of records reports the location and attribute information about a set of census features, e.g., the location, shape, addresses and other census attributes for a county. There is an identifier based on the U.S. Federal Information Processing Standards Code (FIPS) that is used to identify the file and record type. The record types for a county are organized in files that include the unique FIPS code, e.g., all records of type 1 for Pulaski County, Georgia, FIPS code 13235, are delivered in a file named tgr13235.rt1. All or part of the 17 files are used to describe geographic features within the county. Separate records define the end nodes for lines, the vertices for lines, line identifiers, landmark features, names, internal points, ZIP codes, and index and linkage attributes. The index and linkage attributes are used to combine the data across record types. The linkages and indices allow groups of points, lines, or areas to represent features. Identifiers are also included to link areas to Summary Tape (STF) files, Congressional District files, County files, Census Tract/Block Numbering files, and other data files produced by the U.S. Census Bureau.

Specialized software packages are available to convert TIGER/Line and related census files to data layers in specific GIS formats, and most GIS provide utilities to ingest TIGER/Line and census data. These products are often bundled with census data and provide for the selection of areas based

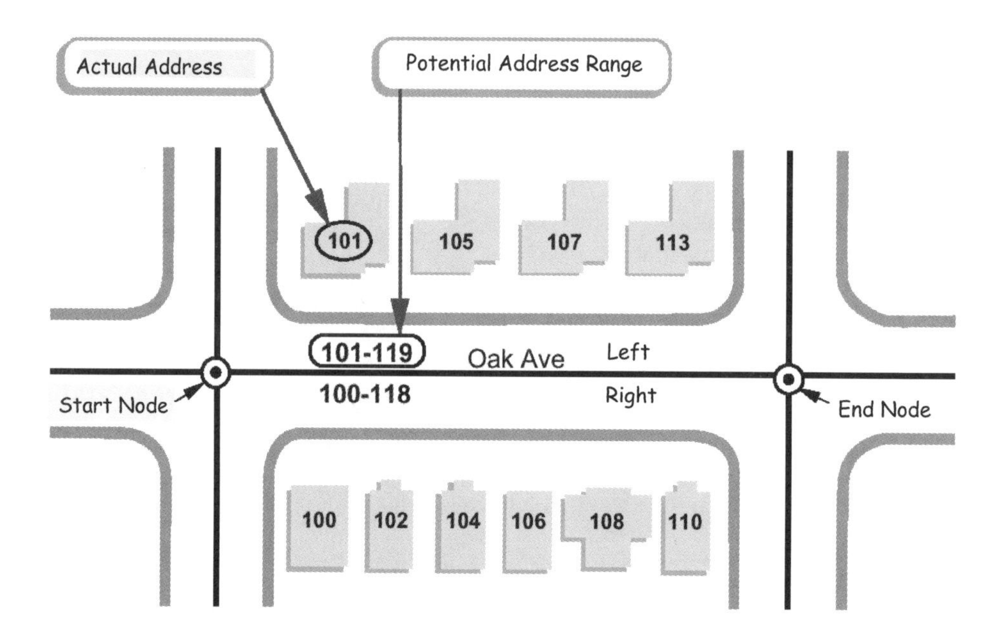

Figure 7-15: TIGER data provide address ranges for line (chain) segments. These ranges may be distributed across the line, giving approximate building locations on a street. (from U.S. Dept. of Commerce)

on states, counties, zip codes, or other area units, and allow selection of available attribute data. TIGER/Line files contain a wealth of attributes for line and area features, including state and county boundaries, school district, city, township, and other minor civil divisions, place names, park locations, road and other infrastructure attributes, as well as summary population data. These data may be extracted in a customized manner.

National Land Cover Data

While land cover is important when managing many spatially distributed resources, data on land cover are quite expensive to obtain over large areas. These data are often scarce, at low categorical or spatial resolution, and rarely available over broad areas. While individual states, counties, metropolitan areas, or private landholders have developed detailed land cover maps, there have been few national efforts to map land cover in a consistent manner. There are two national data sets available, the Land Use Land Cover (LULC) data from the 1970s and early 1980s, and the National Land Cover Data (NLCD) from the 1990s (Figure 7-16).

LULC data were developed by the USGS by the interpretation of 1970s and early 1980s aerial photographs. These photographs were taken at a range of scales, although scales between 1:40,000 and 1:60,000 were most commonly used. Land use was manually photointerpreted. Photointerpretation involved viewing the photographs and assigning land use and land cover based on differences in the color, tone, shape, and other information contained in the photograph. Technicians were trained to become familiar with the land cover and land uses common to an area, and how these land uses appeared on photographs. The knowledge of image/land use relationships was most often developed through extensive field surveys in areas for which photographs were available. Interpreters visited a number of sites of each land use type. These interpreters noted how variation in land use, terrain, and other factors interact with the film type and other photo characteristics and this

Figure 7-16: An example of National Land Cover Data (NLCD) for the State of Virginia. NLCD categorizes land cover into 21 classes and are provided in a 30 meter raster cell format. In this example forests are shown in darker tones, with urban and agricultural areas in lighter tones of gray.

allowed interpreters to classify land use and land cover for unvisited locations.

LULC data were interpreted onto hard-copy maps. Land use/land cover polygons were drafted from the photographs onto paper or plastic material using monocular or stereo viewing equipment (see Chapter 6). Minimum mapping units of 4 to 16 hectares were applied, depending on the feature type. Maps were then manually digitized. LULC data are available in either vector or raster formats. Data are tiled by 1:100,000 or 1:250,000 USGS map series boundaries.

NLCD are a more recent and detailed source of national land cover information. NLCD are produced in a cooperative effort by a number of U.S. federal government agencies. Their goal is a consistent, current land cover data record for the conterminous United States. Land cover classifications are based primarily on 30 meter Landsat Thematic Mapper data. Land cover is assigned to one of 21 classes (Table 7-1). Full coverage is obtained from adjacent or overlapping, cloud-free Landsat images. Multiple dates are often acquired in order to improve accuracy and categorical detail through phenologically-driven changes. For example, evergreen vs. deciduous forests are more easily distinguished when both leaf-off and leaf-on imagery are used. Other spatial data sets are used to improve the accuracy and categorical detail possible through spectral data alone. These data include digital elevation, slope, aspect, Bureau of Census population and housing density data, USGS

Table 7-1: NLCD Land cover classes.

Water
11 open water
12 perennial ice/snow

Developed
21 low intensity residential
22 high intensity residential
23 commercial/industrial/transportation

Barren
31 bare rock/sand/clay
32 quarries/strip mines/gravel pits
33 transitional

Forested Upland
41 deciduous forest
42 evergreen forests
43 mixed forests

Shrubland
51 shrubland

Non-natural Woody
61 orchard/vineyard/other

Herbaceous Upland Natural
71 grassland/herbaceous

Herbaceous Planted/Cultivated
81 pasture/hay
82 row crop
83 small grains
84 fallow
85 urban/recreational grasses

Wetlands
91 woody wetlands
92 emergent herbaceous wetlands

LULC data, National Wetlands Inventory data, and STATSGO soils data.

Data are processed in a uniform manner within each state or region, and a national set of categories and protocols followed. All classifications were subjected to a standardized accuracy assessment, and reported and delivered in a standard format. Accuracy assessments were based on NAPP or other medium to high resolution aerial photographs. Areas were stratified based on the photographs, and sampling units defined. Photointerpretations of land cover were assumed true, and compared to NLCD classification assignments. Errors were noted and reported using standard methods.

Statewide Spatial Data

Many state governments in the U.S.A. distribute statewide spatial data.These data are often edited and improved versions of federally-developed data, with additional error checks applied. Statewide datasets developed wholly by state governments are also available.These data are often provided at low or no cost.

Statewide data are often accessed via a web browser, with selectable layers in a graphical user interface. The user may select both by layer type, and by geographic area, e.g., by county, mapsheet boundary, or watershed (Figure 7-17). Data are available for most states, e.g., for Georgia at http://gis.state.ga.us/, for North Carolina at http://cgia.cgia.state.nc.us/ncgdc/, or from Minnesota at http://deli.dnr.state.mn.us/. Users interested in a specific state are encouraged to contact statewide GIS offices, or transportation, environmental conservation, or other divisions of state government that manage spatially-distributed resources.

Figure 7-17: Spatial data are often available from state governments. This example from Wisconsin illustrates the browse capabilities that are commonly provided. (courtesy, State of Wisconsin)

Summary

Digital data are available from a number of sources, and provide a means for rapidly and inexpensively populating a GIS database. Most of these data have been produced by government organizations and are available at little or no cost, often via the internet. Data for elevation, transportation, water resources, soils, population, landcover, and imagery are available, and should be evaluated when creating and using a GIS.

Suggested Reading

Broome, F. R. and Meixler, D. B., The TIGER database structure, *Cartography and Geographic Information Systems*, 1990, 17:39-47.

Carter, J. R., Digital representations of topographic surfaces, *Photogrammetric Engineering and Remote Sensing*, 1988, 54:1577-1580.

Chrisman, N. R., Deficiencies of sheets and tiles: Building sheetless databases, *International Journal of Geographical Information Systems*, 1990, 4:157-168.

Decker, D., GIS Data Sources, Wiley, New York, 2001.

Dubayah, R. and Rich, P. M., Topographic solar radiation models for GIS, *International Journal of Geographical Information Systems*, 1995, 9:405-419

Elassal, A. A. and Caruso, V. M., Digital Elevation Models. Geological Survey Circular 895-B, United States Geological Survey, Reston, 1984.

Lytle, D. J., Bliss, N. B., and Waltman, S. W., Interpreting the State Soil Geographic Database (STATSGO), (in) GIS and Environmental Modeling: Progress and Research Issues, Goodchild, M. F., Steyaert, L. T., Parks, B. O., Johnston, C., Maidment, D., Crane, M., and Glendinning, S., eds., GIS World, Fort Collins, 1996.

Marx R.W., The TIGER system: automating the geographic structure of the United States Census, *Government Publications Review*, 1986, 13:181-201.

Study Questions

What are some advantages and disadvantages of using digital spatial data?

Can you describe each of the following data sets, i.e., who produces them, what are the source materials, what do they contain, how are they delivered: digital raster graphics (DRGs), digital line graphs (DLGs), digital elevation models (DEMs), digital orthophotoquads (DOQs), National Wetlands Inventory data (NWI), SSURGO and STATSGO soils data, TIGER census data, and national land cover data (NLCD) sets?

What is edge matching and why is it important?

8 Attribute Data and Tables

Introduction

We have described how spatial data in a GIS are often split into two components, the coordinate information describing object geometry, and the attribute information, describing non-spatial properties of objects (Figure 8-1). Because these non-spatial data are frequently presented to the user in tables, these attribute data are often referred to as tabular data. Tabular data summarize the most important non-spatial characteristics of each cartographic object. For example, we may define the location and boundaries of all the counties in a study region, and store the coordinate data defining the county boundaries in appropriately structured files. We also may define the non-spatial attributes characterizing counties. These attributes might include the county name, Federal Information Processing Standards (FIPS) code, population, area, and population density. Values for each of these attributes may be collected for each county and organized into tables.

Attribute information in a GIS are typically entered, analyzed, and reported using a *database management system* (DBMS). A DBMS is a specialized computer program

Name	FIPS	Pop90	Area	PopDn
Whatcom	53073	128	2170	59
Skagit	53057	80	1765	45
Clallam	53009	56	1779	32
Snohomish	53061	466	2102	222
Island	53029	60	231	261
Jefferson	53031	20	1773	11
Kitsap	53035	190	391	485
King	53033	1507	2164	696
Mason	53045	38	904	42
Gray Harbor	53027	64	1917	33
Pierce	53053	586	1651	355
Thurston	53067	161	698	231
Pacific	53049	19	945	20
Lewis	53041	59	2479	24

Figure 8-1: Data in a GIS include both spatial (left) and attribute (right) components.

for organizing and manipulating data. The DBMS stores the properties of geographic objects and the relationships among the objects. A DBMS incorporates a specialized set of software tools for managing tabular data, most often specialized computer programs for efficient data storage, retrieval, indexing, and reporting. DBMS were initially developed in the 1960s, and continual refinements since then have led to robust, sophisticated systems employed by government, businesses, and other organizations. A somewhat standard set of DBMS tools and methods have been developed and are provided by many vendors.

Note that the terms DBMS and database are sometimes used interchangeably. In most cases this is incorrect and in all cases imprecise. A DBMS is a computer program that allows you to enter, organize, and manipulate data. A database is an organized collection of data, often created or manipulated with the help of a DBMS. The database may have a specific form dictated by a DBMS, but it is not the system.

Database Components and Characteristics

The basic components of a traditional database are *data items* or *attributes*, the indivisible named units of data (Figure 8-2). These items can be identifiers, sizes, colors, or any other suitable characteristic used to describe the features of interest. Attributes may be simple, for example one word or number, or they may be compound, for example an address data item that consists of a house number, a street name, a city, and a zip code.

A collection of related data items that are treated as a unit represents an *entity*. In a GIS, the database entities are typically roads, counties, lakes, or other types of geographic features. A specific entity, e.g., a specific county, is an *instance* of that entity. Entities are defined by a set of attributes. In our

Attribute or Item

Record

Name	FIPS	Pop90	Area	PopDn
Whatcom	53073	128	2170	59
Skagit	53057	80	1765	45
Clallam	53009	56	1779	32
Snohomish	53061	466	2102	222
Island	53029	60	231	261
Jefferson	53031	20	1773	11
Kitsap	53035	190	391	485

Figure 8-2: Components of an attribute data table.

example in Figure 8-2, the attributes about a county include the name, a FIPS code identifier, the 1990 population in thousands of persons, area, and population density. These related data items are often organized as a row or line in a table, called a *record*. A *file* may then contain a collection of records, and a group of files may define the database. Specific database systems may have refined the terms differently for each of these parts. For example in the relational database model, the record is called a *row* or a *n-tuple*, and the records are typically organized into a file called a *relational table*.

You should note that the concept of an entity when referring to a database may be slightly different than an entity in a GIS data model. This difference stems from two different groups, geographers and computer scientists, using a word for slightly different but related concepts. An entity in a geographic data model is often used for the real-world item or phenomenon we are trying to represent with a cartographic object. These entities are typically a physical object, e.g., a lake, city, or building, but they may also be a conceptual or defined phenomenon, e.g., a property boundary. In contrast, computer scientists and database managers often define an entity somewhat differently. The database manager often defines an entity as the principal data objects about which information will be collected. In the DBMS literature the entity is the data object which denotes a physical thing, and not the thing itself. Thus properties of entities, and relationships among entities refer to the structure of the DBMS. These properties and relationships are used to represent the real-world phenomenon, and it is a subtle distinction in terminology, but these different definitions can lead to confusion unless the meaning is noted. For the remainder of this chapter we will use the definition of an entity as a data object.

A database typically contains complex structures, primarily to provide data security, stability, and to allow multiple users or programs to access the same data simultaneously. Database users often demand shared access, that is, multiple users or programs may be allowed to open, view, or modify a data set simultaneously. However if each program or user has direct file access, then multiple copies of a database may be open for modification at the same time (Figure 8-3, top). With direct file access, multiple users may try to write to the data file simultaneously, with unforeseen results. The data saved may be the most recent, the first updates, or some mix in between. Because data may be lost or modified in unforeseen ways with direct file access, a DBMS may be designed to manage multi-user access (Figure 8-3, bottom). Some DBMS manage shared files and data, and enforce a pre-determined precedence in simultaneously accessed files. The DBMS may act as an intermediary between the files and the application programs or user. The DBMS may prevent errors due to simultaneous access. Other DBMS programs do not manage simultaneous access, and users of such systems generally must avoid opening multiple copies of the database at one time.

A database provides many useful functions. A database may provide *data independence*, a valuable characteristic when working with large data sets. Data independence is the ability to make changes in the structure of the database that are transparent to any user or application program. This means restructuring the database does not require a user or application programmer to modify their procedures. Data independence ensures that the specific file structures of the DBMS do not need to be known by the applications programmer or user.

Databases may also provide for *multiple user views*. Different users may require different information from the database, or the same information delivered in a different format or arrangement. Profiles or forms can be developed which change the way data are provided to each program or user. Database management systems are able to automatically reformat data to meet the specific input requirements of various users. The DBMS

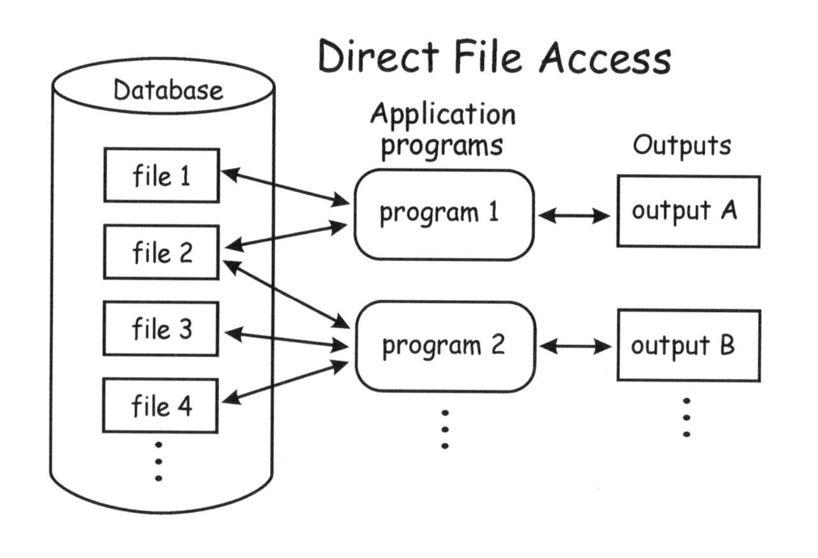

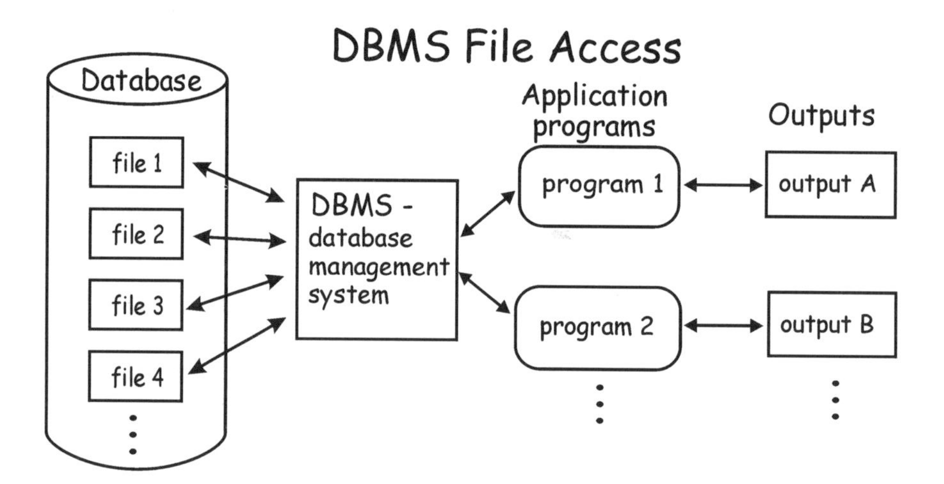

Figure 8-3: Direct and database management system file access. (adapted from Aronoff, 1989)

avoids the need for multiple copies of the same data, one for each application or user, by changing the presentation to meet each specific need. This simplifies maintenance and updates of the database, because there is only one copy to manage.

A database also allows *centralized control and maintenance* of important data. One "standard" copy of the data may be maintained, and updated on a regular, known basis. These data may be time stamped or provided with a version number to aid in management. These data are then distributed to the various users. A single person or group may be charged with maintaining data currency, quality, and completeness, and with resolving contradictions or differences between various versions of the database.

Adopting a database may come at some cost. Specialized training may be required to develop, use, and maintain a database. Defining the entities and relationships among entities may be a complex task that may require database specialists. Structuring the database for efficient access or creating customized forms will often require a specialist. The software itself may be quite expensive, particularly when multiple users will access the database simultaneously.

Physical, Logical, and Conceptual Structures

A database may be viewed as having conceptual, logical, and physical structures. These structures define the entities and their relationships and specify how the data files or tables are referred one to another.

The conceptual structure is often represented in *schema.* This structure may be succinctly described in standard shorthand notations, e.g., using *entity-relationship* diagrams, also known as E-R diagrams. We will not describe E-R diagrams or other conceptual methods here.

A schema is a compact graphical representation of the entities and the relationships among them. The schema represents the conceptual models of the entities and their relationships. The relationships may be one-to-one, between one entity and another, or they may be one-to-many, or many-to-many, connecting several objects. These relationships are represented by lines connecting the entities, and may indicate if the relationships are between one or many entities.

Databases also have a *logical database design*. Most databases are developed using commercial programs that use a specific database model or type. This constrains the specific way a conceptual model may be implemented, and so the conceptual model must be converted to a specific set of elements that define the structure and interaction of database components. The logical structure influences the specific physical structures that are adopted.

Physical referencing may by achieved in many ways. One common method uses file pointers to connect records in one file with those in other files. Much of this structure is designed to speed access, aid updates, and provide data integrity. This structuring is part of the *physical design* of the database. The design typically strives to physically cluster or link data used together in processes so that these processes may be performed quickly and efficiently.

Consider the data in Figure 8-4. These data describe a set of forests, trails in the forests, and types of recreational features found on the trails. These data may be physically

Forests

Forest Name	Location	Size
Nantahala	North Carolina	184,447
Cherokee	North Carolina	92,271

Recreation features

Feature	Description	Activities
Wfall	Waterfall	Photography, Swimming
Ogrth	Old-Growth Forest	Photography, Hiking
Vista	Scenic overlook	Photography, viewing
Wlife	Wildlife Viewing	Photography, Birding
Cmp	Camping	Camping

Trail features

Trail Name	Difficulty	Forest	Feature
Bryson's Knob	E, M	Nantahala	Vista, Ogrth
Slickrock Falls	M	Cherokee	Wfall, Ogrth
North Fork	M	Nantahala	-
Cade's Cove	E	Cherokee, Nantahala	Ogrth, Wlife
Appalachian	M, D	Nantahala, Cherokee	Wfall, Ogrth, Vista, Wlife, Cmp

Figure 8-4: Forest data to be used in following examples of different database organization schemes. Data are provided for forests, trails within the forests, and recreational features.

structured in a database to respond to frequent requests, for example, "what trails pass by waterfalls?". This structuring might involve placing direct links between the forest and waterfalls data, or physically placing the data in adjacent locations on storage media to increase access speed.

Hierarchical Databases

A database may be organized in several different ways to improve access or provide a logical structure. The data in Figure 8-4 may be placed in a hierarchical structure, as shown in Figure 8-5. Forest data is the root and other entities are related in a hierarchical tree. The root is the base of the structure, and is considered the *parent* of other elements in the hierarchy. The elements one step down the hierarchy are termed *children* of the root. Parent-child relationships are repeated successively down the hierarchy. Each parent may have many children, however each child has only one parent. In our current example forests are children of the Forest Data entity, trails are children at the next level down the hierarchy (Bryson's Knob, North Fork, etc.), and recreational features in turn are the children of trails (Vista, Ogrth, etc.).

Data are accessed in a hierarchical structure by traversing down the branches of the hierarchy. The links from parents to children may be one-to-one or one-to-many, but they are never many-to-many. Thus, all children of a given parent may be quickly accessed. All the trails in the Cherokee forest are directly connected to the parent entity (Figure 8-5) and so may be directly accessed from that branch in the tree. The structure ensures the retrieval of data on a single branch is efficient, because there is a direct link between a parent and all directly related child entities. If a small set of queries are repeated frequently, a database may be structured in a hierarchy such that it responds very quickly to those queries.

This simple example also shows some of the limitations of hierarchical database designs. Attribute information may be stored many times in the database, leading to significant redundancy. The Wlife recreational feature occurs for several trails, meaning multiple copies are represented in the data-

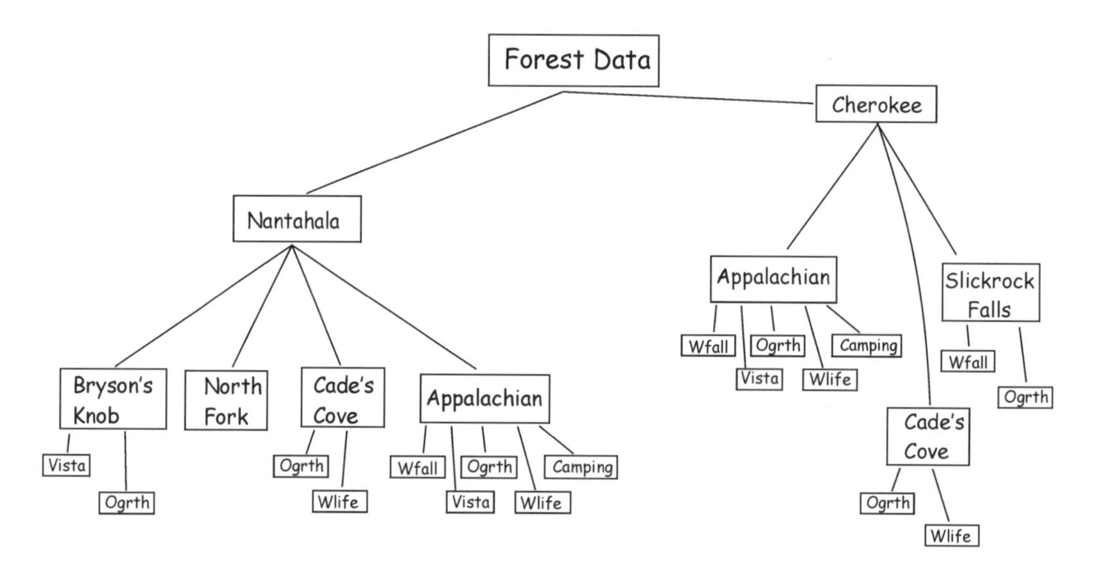

Figure 8-5: Forest data in a hierarchical database system.

base. These copies take up storage space redundantly, and also complicate updates or modifications. Any changes to a recreational feature must be made to all copies. In our example, if we wish to change the description or attributes associated with the Wlife recreational feature, we must find all copies of the feature in our database and change them in each instance.

Searching across lower-level attributes is also a limiting factor in hierarchical database designs. For example, searching for all trails which support photography is time-consuming because the entire network must be traversed, and the data are not indexed by more than one attribute at each level in the hierarchy.

Network Databases

Network database designs differ from hierarchical designs chiefly in that a child may have more than one parent. Having multiple parents allows for many-to-many relationships as well as one-to-one and one-to-many relationships while structuring the database (Figure 8-6). Each entity may be a node in the network and is connected as appropriate to other entities in the network.

Network database designs have less redundant data than hierarchical designs by virtue of their additional relationships. Only one copy of a specific entity may be required. In Figure 8-6 the Wlife recreational feature is represented only once and is connected to the appropriate trails in the database.

Network designs may be powerful and flexible. However, designing and implementing the network structure can be quite complex, and network databases are often difficult to modify. A great deal of information is stored in the complex network relationships, and updates require modifications of these linkages. Significant amounts of time, money, and expertise are usually required. While network database designs have a large and important installed user base in some sectors, they have not been widely adopted for use with GIS, perhaps because of these limitations.

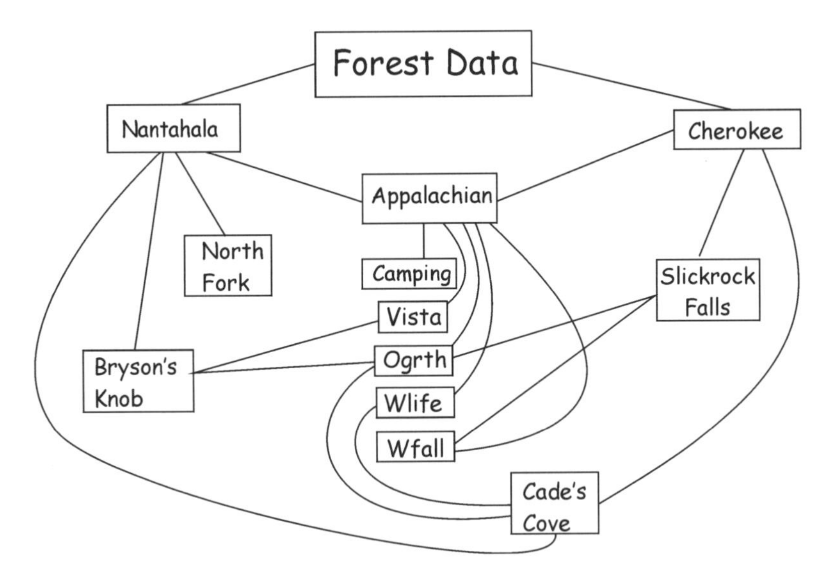

Figure 8-6: The forest data in a network database structure.

Relational Databases

Relational databases are among the most common database designs. There are a number of reasons for their popularity. The relational model is more flexible than hierarchical or network designs, flexible enough to represent these structures. The tables structure does not restrict processing or queries as with the other designs. The organization is simple to understand and hence learn, and is easy to implement relative to other designs. It can accommodate a wide range of data types, and it is not necessary to know in advance the kind of queries, sorting, and searching that will be performed on the database. Because of their lack of structure, processing a relational database may be slow relative to other database designs. Performance may be improved when information on the types of processes is used to optimize the relational database structure. In addition, improvements in computer speed are allowing acceptable access times in extremely large relational databases, so these performance differences are becoming less important in most applications.

A relational database design represents data in tables (Figure 8-7). Each instance of an entity is represented by a row in a table. In our forest data example there may be a forest table, with a row for each forest, and other tables representing the trails, trail features, and recreational opportunities. The

Forests

Forest Name	Forest-ID	Location	Size
Nantahala	1	N. Carolina	184,447
Cherokee	2	N. Carolina	92,271

Trails

Trail Name	Forest-ID
Bryson's Knob	1
Slickrock Falls	2
North Fork	1
Cade's Cove	1
Cade's Cove	2
Appalachian	1
Appalachian	2

Characteristics

Trail Name	Feature	Difficulty
Bryson's Knob	Vista	E,M
Bryson's Knob	Ogrth	E,M
Slickrock Falls	Ogrth	M
Slickrock Falls	Wfall	M
North Fork	-	M
Cade's Cove	Ogrth	E
Cade's Cove	Wlife	E
Appalachian	Wfall	M,D
Appalachian	Ogrth	M,D
Appalachian	Vista	M,D
Appalachian	Wlife	M,D
Appalachian	Cmp	M,D

Recreational features

Feature	Description	Activity1	Activity2
Wfall	Waterfall	Photography	Swimming
Ogrth	Old-Growth Forest	Photography	Hiking
Vista	Scenic Overlook	Photography	Viewing
Wlife	Wildlife Viewing	Photography	Birding
Cmp	Camping	Camping	-

Figure 8-7: Forest data in a relational database structure.

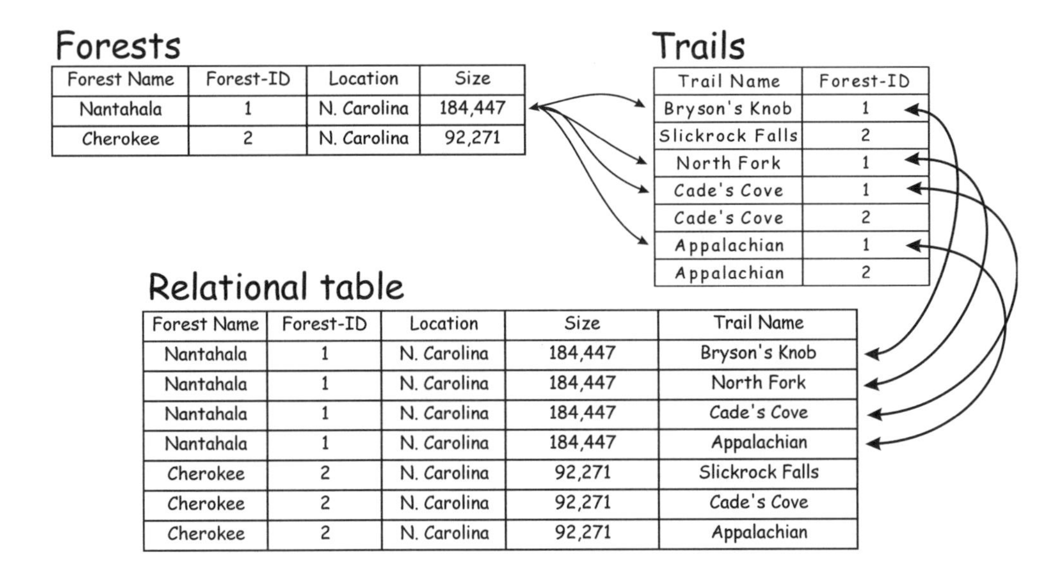

Forests

Forest Name	Forest-ID	Location	Size
Nantahala	1	N. Carolina	184,447
Cherokee	2	N. Carolina	92,271

Trails

Trail Name	Forest-ID
Bryson's Knob	1
Slickrock Falls	2
North Fork	1
Cade's Cove	1
Cade's Cove	2
Appalachian	1
Appalachian	2

Relational table

Forest Name	Forest-ID	Location	Size	Trail Name
Nantahala	1	N. Carolina	184,447	Bryson's Knob
Nantahala	1	N. Carolina	184,447	North Fork
Nantahala	1	N. Carolina	184,447	Cade's Cove
Nantahala	1	N. Carolina	184,447	Appalachian
Cherokee	2	N. Carolina	92,271	Slickrock Falls
Cherokee	2	N. Carolina	92,271	Cade's Cove
Cherokee	2	N. Carolina	92,271	Appalachian

Figure 8-8: Forest and trails data in a relational data structure. Rows hold records associated with an entity, and columns hold items. A key, here Forest-ID, is used to join tables.

rows are also called *records* or *tuples*. Columns contain *attributes* that describe entities. Attributes may also be referred to as *items* or *variables*. Tables are related through *keys*, one or more fields that meet certain requirements and may be used to index the records. A database with fewer key fields is generally simpler, has smaller data files, and may be searched more quickly. Some keys are used to index and add flexibility in selecting data. Too few keys may result in difficulties searching or sorting the database. Figure 8-8 shows our forest and trails data in a relational data structure, with a primary key Forest-ID in the Forests table. Note that not all data need be represented in one table, and thus a mechanism is required to link between tables.

Tables are related through values found in the keys. Values in one or more keys are matched across tables, and the information is combined based on the matching. For example our Forest data table may be linked to our trails data table through the Forest-ID key field (Figure 8-8). Each Forest-ID entry in the Trails table is matched to the Forest-ID value in the Forests table, and the data joined or related through the values of these items. The forest records are related to the Trails records using the Forest-ID attribute, creating a joined table that combines the attributes of both tables.

Hybrid Database Designs in GIS

Data in a GIS are often stored using hybrid designs. Hybrid designs often store coordinate data using specialized database structures, and attribute data in a relational database. Thousands to millions of coordinate pairs or cells are often required to represent the location and shape of objects in a GIS. Even with modern, high-speed computers, retrieval of coordinate data stored in a relational database design is often too slow. Therefore, the coordinate data are often stored using structures specifically designed for rapid retrieval. This structuring typically involves grouping coordinates for carto-

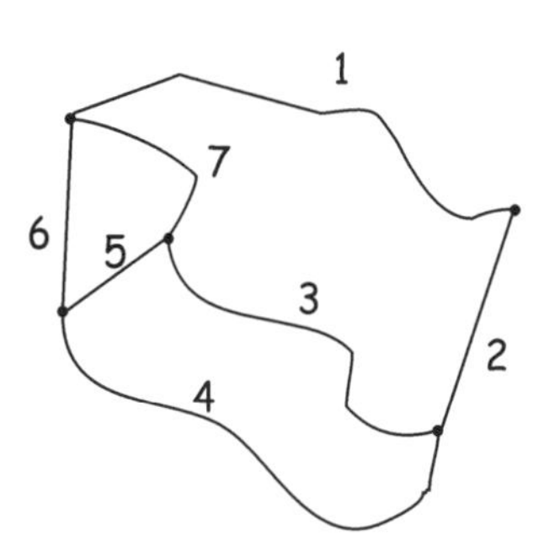

Attribute data

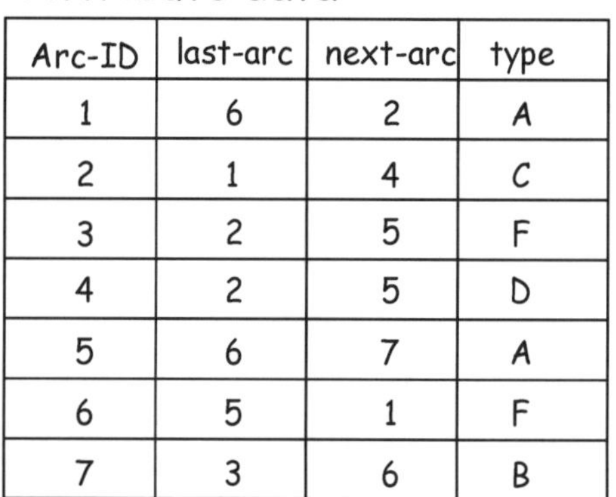

Arc-ID	last-arc	next-arc	type
1	6	2	A
2	1	4	C
3	2	5	F
4	2	5	D
5	6	7	A
6	5	1	F
7	3	6	B

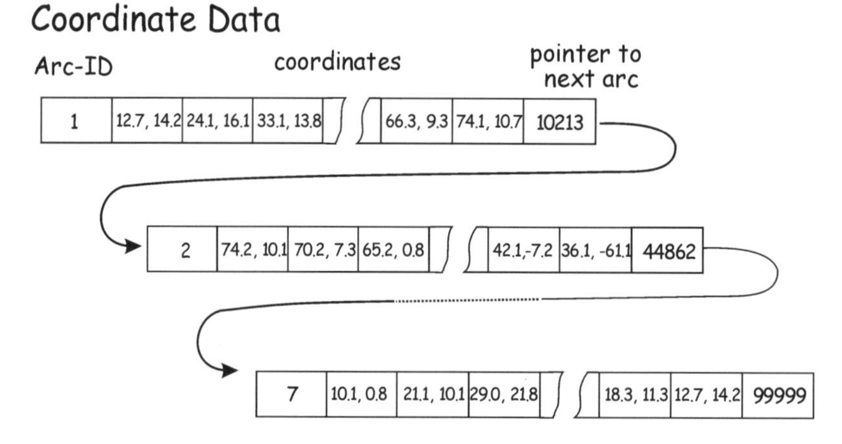

Figure 8-9: A small example of a hybrid database system for spatial data. Attribute data are stored in a relational table, while coordinate data are stored in a network or other structure.

graphic objects, e.g., storing ordered lists of coordinate pairs to define lines, and indexing or grouping lines to identify polygons. Pointers are used to link related lines or polygons, and unique identifiers link the geographic features (points, lines, or polygons) to corresponding attribute data (Figure 8-9).

Topological relationships may be explicitly encoded to improve analyses or to increase access speed. Addresses to the previous and next data are explicitly stored in an indexing table, and pointers are used to connect coordinate strings. Explicitly recording the topological elements of all geographic objects in a data layer may improve geographic manipulations, including determination of adjacencies, line intersection and hence polygon and line overlay, and network definition. Coordinates for a given feature or part of a feature may be grouped and these groups linked or indexed to speed manipulation or display.

Hybrid data designs typically store attribute data in a DBMS. These data are linked to the geographic data through unique identifiers or labels which are an attribute in the DBMS. Data may be stored in a manner that facilitates the use of more than one brand of DBMS, and allows easy transport of data from one DBMS to another.

Selection Based on Attributes

Table Queries

Queries are among the most common operations in a DBMS. A query may be viewed as the selection of a subset of records based on the values of specified attributes. Queries may be simple, using one variable, or they may be compound, using conditions on more than one variable, or using multiple conditions on one variable. In concept, queries are quite simple, but basic query operations may be combined to produce quite complex selections.

The left side of Figure 8-10 demonstrates the selection from a simple query. A single condition is specified, Area > 20. The set of selected records is empty at the start of the query. Each record is inspected and added to the selected set if the attribute Area meets the specified criteria. Each record in the selected set is shown in gray in Figure 8-10.

The right side of Figure 8-10 demonstrates a compound query based on two attributes. This query uses the AND condition to select records that meet two criteria. Records are selected that have a Landuse value equal to Urban AND a Municip value equal to City. All records that meet both of these requirements are placed in the selected set. All records that fail to comply with either of these requirements are in the unselected set. The boolean operations AND, OR, and NOT may be applied in combination to select records that meet multiple criteria.

Simple selection:

records with Area > 20.0

ID	Area	Landuse	Municip
1	10.5	Urban	City
2	330.3	Farm	County
3	2.4	Suburban	Township
4	96.0	Suburban	County
5	22.1	Urban	City
6	30.2	Farm	Township
7	4.4	Urban	County

AND selection:

records with (Landuse = Urban) **and** (Municip = City)

ID	Area	Landuse	Municip
1	10.5	Urban	City
2	330.3	Farm	County
3	2.4	Suburban	Township
4	96.0	Suburban	County
5	22.1	Urban	City
6	30.2	Farm	Township
7	4.4	Urban	County

Figure 8-10: Simple selection - applying one criterion to select records (left), and compound selection, applying multiple requirements (right).

OR selection:

records with (Area > 20.0) OR (Municip = City)

ID	Area	Landuse	Municip
1	10.5	Urban	City
2	330.3	Farm	County
3	2.4	Suburban	Township
4	96.0	Suburban	County
5	22.1	Urban	City
6	30.2	Farm	Township
7	4.4	Urban	County

NOT selection:

records with Landuse NOT Urban

ID	Area	Landuse	Municip
1	10.5	Urban	City
2	330.3	Farm	County
3	2.4	Suburban	Township
4	96.0	Suburban	County
5	22.1	Urban	City
6	30.2	Farm	Township
7	4.4	Urban	County

Figure 8-11: OR and NOT compound selections.

AND combinations typically decrease the number of records in a selected set. They add restrictive criteria, and provide a more strenuous set of conditions that must be met for selection. In the example in Figure 8-10 the record with ID = 7 meets the first criterion, Landuse = Urban, but does not meet the second criterion specified in the AND, Municipality = City. Thus, the record with ID = 7 is not selected. ANDs add restrictions, and winnow the selected set.

OR combinations typically increase or add to a selected set in compound queries. OR conditions may be considered as inclusive criteria. The OR adds records that meet a criterion to a set of records defined by previous criteria. In the query on the left side of Figure 8-11, the first criterion, Area > 20, results in the selection of records 2, 4, 5, and 6. The OR condition adds any records that satisfy the criterion Municip = City, in this case the record with ID = 1.

The NOT is the negation operation, and may be interpreted as meaning select those records that do not meet the condition following the NOT. The right side of Figure 8-11 demonstrates the negation operation. The operation may be viewed as first substituting equals for the NOT, and identifying all records. Then the remaining records are placed in the selected set, and the identified records placed in the unselected set. On the right side of Figure 8-11 all records that have a value of Landuse equal to Urban are identified. These are placed in the unselected set. The remaining records, shown in gray, are placed in the selected set (Figure 8-11, right side).

ANDs, ORs, and NOTs can have complex effects when used in compound conditions, and the order or precedence is important in the query. Combinations of these three operations may be used to perform very complex selections.

Figure 8-12 demonstrates that queries are not generally distributive. For example, if OP1 and OP2 are operations, such as AND or NOT, then,

NOT [(Landuse = Urban) AND (Municip = County)]

ID	Area	Landuse	Municip
1	10.5	Urban	City
2	330.3	Farm	County
3	2.4	Suburban	Township
4	96.0	Suburban	County
5	22.1	Urban	City
6	30.2	Farm	Township
7	4.4	Urban	County

[NOT (Landuse = Urban)] AND [NOT (Municip = County)]

ID	Area	Landuse	Municip
1	10.5	Urban	City
2	330.3	Farm	County
3	2.4	Suburban	Township
4	96.0	Suburban	County
5	22.1	Urban	City
6	30.2	Farm	Township
7	4.4	Urban	County

Figure 8-12: Selection operations may not be distributed, and order in application is very important. When the NOT operation is applied after the AND (left) a different set of records is selected than when the NOT operation is applied before the AND (right side). Order of operation is important, and ambiguity should be removed by using parentheses or other delimiters.

OP1 (ConditionA OP2 ConditionB) (8.1)

is not always the same as

(OP1 ConditionA)OP2(OP1 ConditionB) (8.2)

For example,

NOT [(Landuse = Urban) AND (Municipality = County)] (8.3)

does not yield the same set of records as the expression

[NOT (Landuse = Urban)] AND [NOT (Municipality = County)] (8.4)

Parentheses or other delimiters should be used to ensure unambiguous queries.

Relational databases may support a *structured query language* known as SQL. SQL was initially developed by the International Business Machines Corporation but is supported by a number of software vendors. SQL is a non-procedural query language in that the specification of queries does not depend on the structure of the data. The language can be powerful, general, and transferable across systems, and so has become widely adopted. Because SQL as initially defined has limitations for spatial data processing, many spatial operations are not easily represented in SQL. Many more selections may be specified only with complex queries, so various SQL extensions appropriate for spatial data have been developed.

Normal Forms in Relational Databases

Keys and Functional Dependencies

Although relational tables are among the most commonly-used database designs, they can suffer from serious problems in performance, consistency, redundancy, and maintenance. If all data are stored in one large table there may be large amounts of redundant data or wasted space, and long searches may be needed to select a small set of records. Updates on a large table may be slow, and the deletion of a record may result in the unintended deletion of valuable data from the database.

Consider the data in Figure 8-13, in which building records are stored in a single table. Attributes include Parcel-ID, Alderman, Tship-ID, Tship_name, Thall_add, Own-ID, Own_name, and Own_add. Some information is stored redundantly, for example, changing the address for the Alderman named Johnson would require changing many rows, and identifying all apartment buildings would require a search of the entire table. This storage redundancy is costly both because it takes up disk space and because each extra record adds to the search and access times. A second problem comes with changes in the data. For example, if Devlin, Yamane, and Prestovic sell the parcel they jointly own (first data row), deleting the parcel record for Devlin would purge the database of her address and tax payment history. If these data on Devlin were required later, they would have to be re-entered from an external source.

We may place relational databases in *normal forms* to avoid many of these problems. Data are structured in sequentially higher normal forms to improve correctness, consistency, simplicity, non-redundancy, and stability. There are several levels in the hierarchy of normal forms, but the first three levels, known as the first through third normal forms, are most common. Data are usually structured sequentially, that is, first all tables are converted to first normal forms, then converted to 2nd and then 3rd normal forms as needed. Prior to describing normal forms we must introduce some terminology and properties of relational tables.

Land Records table, unnormalized form

parcel-ID	Alderman	Tship-ID	Tship_name	Thall-add	Own-ID	Own_name	Own_add	Own-ID	Own_name	Own_add	Own-ID	Own_name	Own_add
2303	Johnson	12	Birch	15W	122	Devlin	123_pine	337	Yamane	72_lotus	890	Prestovic	12_clayton
618	DeSilva	14	Grant	35E	457	Suarez	453_highland	890	Prestovic	12_clayton	231	Sherman	64_richmond
9473	Johnson	12	Birch	15W	337	Yamane	72_lotus	-	-	-	-	-	-

Figure 8-13: Land records data used in future examples, in unnormalized form. The table is shown in two parts because it is too wide to fit across the page.

As noted earlier, relational tables use keys to index data. There are different kinds of keys. A *super key* is one or more attributes that may be used to uniquely identify every record (row) for a table. A subset of attributes of a super key may also be a super key, and is called a *candidate key*. The *primary key* for indexing a table is chosen from the set of candidate keys. There may be many potential primary keys for a given table, however it is usual to use only one primary key per table. The Parcel-ID is a primary key for the table in Figure 8-13, because it uniquely identifies each row in the table.

Functional dependency is another important concept. Attributes are functionally dependent if at a given point in time each value of the dependent attribute is determined by a value of another attribute. In our example in Figure 8-13, we may know that Own_add is functionally dependent on Own_name, and for each Tship_name there is only one Thall-add. Each owner can only have one resident address, so for a given Own_name, e.g., Jordan Pratt, the Own_add is determined. In a similar manner, there is only one Township name, Tship_name, for each Town Hall address, Thall-add. Functional dependency is often denoted by

Own_name -> Own_add

Tship_name -> Thall-add

These indicate that Own_add is functionally dependent on Own_name, and Thall-add is functionally dependent on Tship_name.

Functional dependencies are transitive, so if A -> B, and B -> C, then A -> C. This notation means that if B is functionally dependent on A, and C is functionally dependent on B, then C is functionally dependent on A.

While relational database designs are flexible, the use of keys and functional dependencies places restrictions on relational tables:

•There cannot be repeated records, that is, there can be no two or more rows where all attributes are equal.

•There must be at least one candidate key in a table that may be used as the primary key for the table. This key allows each record to be uniquely identified.

•No member of a column that forms part of the primary key can have a null value. This would allow multiple records which could not be uniquely identified by the primary key.

The First and Second Normal Forms

We begin by gathering all our data, perhaps in a single table. Normal forms typically result in many compact, linked tables, so it is quite common to split tables as the database is *normalized*, that is, as the data table or tables are placed in normal forms. After normalization the tables have an indexing system that speeds searches and isolates values for updating.

Tables with repeat groupings, for example the table at the top of Figure 8-14, are *unnormalized*. A repeating group exists in a relational table when an attribute is allowed to have more than one value represented within a row. Owner-ID repeats itself for dwellings with multiple owners.

A table is in first normal form when there are no repeat columns. The Land Records table at the bottom of Figure 8-14 has been *normalized* by placing each owner into a separate row. This is a table in the first normal form (also 1NF), because each column appears only once in the table definition. This is the most basic level of table normalization, however this 1NF table structure still suffers from excessive storage redundancy, inefficient searches, and potential loss of data on updating. First normal forms have an advantage over unnormalized tables because queries are easier to code and implement. However, 1NF tables are usually converted to higher-order normal forms. Most data tables are converted to at least third normal form, 3NF, however it is useful to understand second normal forms before describing 3NF tables.

A table is in second normal form (2NF) if it is in first normal form and every non-key attribute is functionally dependent on the primary key. Remember that functional dependency means that knowing the value for one attribute of a record automatically specifies the value for the functionally dependent attribute. The non-key attributes may be directly dependent on the primary key through some functional dependency, or they may be dependent through a transitive dependency. The Land Records table at the bottom of Figure 8-14 has only one possible primary key, the composite of **Parcel-ID** and **Own-ID**. However this table is not in second normal form because it has non-key attributes which are not functionally dependent on the primary key attributes. For example **Tship_name** and **Thall_add** are not functionally dependent on the composite key made up by **Parcel-ID** and **Own-ID**.

The **Land Records** table at the bottom of Figure 8-14 is repeated at the top of Figure 8-15. This table exhibits the primary disadvantages of the first normal form. **Parcel-ID**, **Alderman**, and **Tship-ID** are duplicated when there are multiple owners of a parcel, causing burdensome data redundancy. Each time these records are updated, for example when a new **Alderman** is elected, data must be changed for each duplicate record. If a parcel changes hands and the seller does not own another parcel represented in the table, then information on the seller is lost.

Land Records table, unnormalized form

parcel-ID	Alderman	Tship-ID	Tship_name	Thall-add	Own-ID	Own_name	Own_add	Own-ID	Own_name	Own_add	Own-ID	Own_name	Own_add
2303	Johnson	12	Birch	15W	122	Devlin	123_pine	337	Yamane	72_lotus	890	Prestovic	12_clayton
618	DeSilva	14	Grant	35E	457	Suarez	453_highland	890	Prestovic	12_clayton	231	Sherman	64_richmond
9473	Johnson	12	Birch	15W	337	Yamane	72_lotus	-	-	-	-	-	-

Land Records table, first normal form (1NF)

Parcel-ID	Alderman	Tship-ID	Tship_name	Thall_add	Own-ID	Own_name	Own_add
2303	Johnson	12	Birch	15W	122	Devlin	123_pine
2303	Johnson	12	Birch	15W	337	Yamane	72_lotus
2303	Johnson	12	Birch	15W	890	Prestovic	12_clayton
618	DeSilva	14	Grant	35E	457	Suarez	453_highland
618	DeSilva	14	Grant	35E	890	Prestovic	12_clayton
618	DeSilva	14	Grant	35E	231	Sherman	64_richmond
9473	Johnson	12	Birch	15W	337	Yamane	72_lotus

Figure 8-14: Relational tables in unnormalized (top) and first normal forms (bottom).

Some of these disadvantages can be removed by converting the first normal form table to a group of second normal form tables. To create second normal form tables we make every non-key attribute fully dependent on a primary key in the new tables. Note that the 1NF table will often be split into two or more tables when converting to 2NF, and each new table will have its own key. Any non-key attributes in the new tables will be dependent on the primary keys. The bottom of Figure 8-15 shows our Land Records converted to second normal form. Each of the three tables in second normal form isolates an observed functional dependency, so each table and dependency will be described in turn.

We know that parcels occur in only one township, and that each township has a unique Tship-ID, a unique Tship_name, a unique Thall_add, and one Alderman. This means that if we have identified a parcel by its Parcel-ID, the Alderman, Tship-ID, Tship_name, and Thall_add are known. We assign a unique identifier to each parcel of land, and the Alderman, Tship_name, and Thall_add are all dependent on this identifier. This means if we know the identifier, we know these remaining values. This is the definition of functional dependency. We represent these functional dependencies by:

Parcel-ID -> Alderman

Parcel-ID -> Tship-ID

Parcel-ID -> Tship_name

Parcel-ID -> Thall_add

This functional dependency is incorporated in the table named Land Records 1 in Figure 8-15.

Second, note that once Own-ID is specified, the Own_name and Own_add are determined. Each owner has a unique identifier and only one name (aliases not allowed). Also, each owner has only one permanent home address. Own_name and Own_add are functionally dependent on Own-ID. The functional dependencies are:

Own-ID -> Own_name

Own-ID -> Own_add

These functional dependencies are represented in the table Land Records 2 in Figure 8-15.

Finally, note that we need to tie the owners to the parcels. These relationships are presented in the table Land Records 3 in Figure 8-15.

The three tables Land Records 1 through 3 satisfies the conditions of a second normal form. This form eliminates some of the redundancies associated with the first normal form. Note that the redundancy in storing the information on Alderman, Tship-ID, Tship_name, and Thall_add have been significantly reduced, and the minor redundancy in Own_name has also been removed. Editing the tables becomes easier, for example, changes in Alderman entail modifying fewer records. Finally, deletion of a parcel does not have the side effect of deleting the information on the owner, Own-ID, Own_name, and Own_add.

The Third Normal Form

The second normal form still contains problems, although they are small compared to a table in the first normal form. They can still suffer from transitive functional dependencies. If a transitive functional dependency exists in a table, then there is a chain of dependencies. A transitive dependency occurs in our example table named Land Records 1 (Figure 8-15). Note that Parcel-ID specifies Tship-ID, and Tship-ID speci-

Land records table, first normal form (1NF)

Parcel-ID	Alderman	Tship-ID	Tship_name	Thall_add	Own-ID	Own_name	Own_add
2303	Johnson	12	Birch	15W	122	Devlin	123_pine
2303	Johnson	12	Birch	15W	337	Yamane	72_lotus
2303	Johnson	12	Birch	15W	890	Prestovic	12_clayton
618	DeSilva	14	Grant	35E	457	Suarez	453_highland
618	DeSilva	14	Grant	35E	890	Prestovic	12_clayton
618	DeSilva	14	Grant	35E	231	Sherman	64_richmond
9473	Johnson	12	Birch	15W	337	Yamane	72_lotus

Given functional dependencies:

Parcel-ID → Alderman, Tship-ID

Tship-ID → Tship_name, Thall_add

Own-ID → Own_name, Own_add

Land records tables, second normal form (2NF)

Land Records 1

Parcel-ID	Alderman	Tship-ID	Tship_name	Thall_add
2303	Johnson	12	Birch	15W
618	DeSilva	14	Grant	35E
9473	Johnson	12	Birch	15W

Land Records 2

Own-ID	Own_name	Own_add
122	Devlin	123_pine
337	Yamane	72_lotus
890	Prestovic	12_clayton
457	Suarez	453_highland
231	Sherman	64_richmond

Land Records 3

Parcel-ID	Own-ID
2303	122
2303	337
2303	890
618	457
618	890
618	231
9473	337

Figure 8-15: Ownership data, converted to second normal forms.

fies Tship_name and Thall_add. In our notation of functional dependencies:

Parcel-ID - > Tship-ID, Alderman

and

Tship-ID -> Tship_name, Thall_add

This causes a problem when we delete a parcel from the database. To delete a parcel we remove the parcel from tables Land Records 1 and Land Records 3. In so doing we also lose the relationship between Tship-ID, Tship_name, and Thall_add. To avoid these problems we need to convert the tables to the third normal form.

A table is in the third normal form if and only if for every functional dependency A -> B, A is a super key, or B is a member of a candidate key. This requirement means we must identify transitive functional dependencies and remove them, typically by splitting the table that contains them. The tables Land Records 2 and Land Records 3 in Figure 8-15 are already in the third normal form. However the table Land Records 1 in Figure 8-15 is not in third normal form because the right hand side for the functional dependencies for table Land Records 1 are:

Tship-ID -> Tship_name, Thall_add

Tship-ID is not a super key for the table, nor are Tship_name and Thall_add members of a primary candidate key for that table. Removing the transitive functional dependency by splitting the table will create two new tables, each of which satisfies the criteria for the third normal form. Figure 8-16 contains the tables Land Records 1a and Land Records 1b, both of which now satisfy the third normal form criteria, and preserve

Land records, third normal form

Land Records 1a

FD: Parcel-ID→ Alderman, Tship-ID

Parcel-ID	Alderman	Tship-ID
2303	Johnson	12
618	DeSilva	14
9473	Johnson	12

Land Records 1b

FD: Tship-ID →Tship_name, Thall_add

Tship-ID	Tship_name	Thall_add
12	Birch	15W
14	Grant	35E

Land Records 2

FD: Own-ID → Own_name, Own_add

Own-ID	Own_name	Own_add
122	Devlin	123_pine
337	Yamane	72_lotus
890	Prestovic	12_clayton
457	Suarez	453_highland
231	Sherman	64_richmond

Land Records 3

No Functional Dependencies

Parcel-ID	Own-ID
2303	122
2303	337
2303	890
618	457
618	890
618	231
9473	337

Figure 8-16: Ownership data in third normal form.

the information contained in the first normal form table in Figure 8-14. Note that Parcel-ID is now a superkey for Table 1a and Tship-ID is a superkey for Table 1b, so the third normal form criteria are satisfied.

A general goal in defining a relational database structure is to have the fewest tables possible that contain the important relationships, and have all tables in at least a third normal form. Higher order normal forms have been described and provide further advantages, however these are often specialized and depend on the intended use of the database. Placing the tables in normal forms and identifying the functional dependencies is a valuable exercise. It significantly improves database maintenance, integrity, and significantly reduces redundancy.

While relational tables in normal forms have certain useful characteristics, they may suffer from relatively long access times for specific queries. These structures are based on usage, or the most common processes expected for the database. Tables may be organized in non-normal forms to improve access. Relational databases may purposely be *denormalized* when access speed is of primary importance. These denormalizations typically add extra columns or permanent joins to the database structure. This may add redundancy or move a table to a lower normal form, but these disadvantages often allow significant gains in processing speed. The need to denormalize tables has diminished in many instances with improvements in computing power, however denormalization may be required for extremely large databases, or where access speed is of primary importance.

Trends in Spatial DBMS

Two database designs are currently active areas of research and may find significant application in the future. First, both spatial and attribute data may be integrated into relational DBMS. Coordinate data are stored in tables that are linked to represent higher-order structures and connect them with attribute data. Data access is admittedly slower than with the hybrid vector-topological systems, however increasing hardware capabilities may render this a minor penalty for all but the largest databases. Several integrated model systems have been developed, including those that maintain normal forms but overcome some of the speed-of-access problems.

Object oriented database models are a second area that has received significant research and commercial interest in recent times. Object orientation has been defined in various ways at various times, so there is some ambiguity in the term. However most definitions include the ability to define classes of objects with values that are both private (restricted access) and public, and define procedures or functions with the same name and general outcome, but which differ for each class of objects. Sub-classes may be derived from super-classes, and inherit some or all of the properties of the super-class. Object oriented designs have been described as more flexible and more easily understood and maintained, although there are relatively few successful examples of object-oriented spatial database management systems.

Summary

Attribute data are an important component of spatial data in a GIS. These data may be organized in several ways, but data structures that use relational tables have become the most common method for organizing and manipulating attribute data in GIS.

Selections, or queries, are among the most common analyses conducted on attribute data. Queries mark a subset of records in a table, often as a precursor to subsequent analyses. Queries may use AND, OR, and NOT operations, among others, alone or in combination.

Relational tables are often placed in normal forms to improve correctness and consistency, to remove redundancy, and to ease updates. Normal forms seek to break large tables into small tables that contain simple functional dependencies. This significantly improves maintenance and integrity of the database. Normal forms may cause some cost in speed of access, although this is a receding problem as computer hardware improves.

Suggested Reading

Adam, N. and Gangopadhyay, A., Database Issues in Geographic Information Systems, Kluwer Academic Publishers, Dordrecht, 1997.

Bhalla, N., Object-oriented data models: a perspective and comparative review, *Journal of Information Science*, 1991, 17:145-160.

Date, C. J., An Introduction to Database Systems, 7th Edition. Addison-Wesley, New York, 2000.

Frank, A. U., Requirements for a database management system for a GIS, *Photogrammetric Engineering and Remote Sensing*, 1988, 54:1557-1564.

Lorie, R. A. and Meier, A., Using a relational DBMS for geographical databases, *Geoprocessing*, 1984, 2:243-257.

Milne, P., Milton, S., and Smith, J. L., Geographical object-oriented databases: a case study, *International Journal of Geographical Information Systems*, 1993, 7:39-55.

Teorey, T. J., Database Modeling and Design, 3rd Edition. Morgan Kaufmann, San Francisco, 1999.

Study Questions

What are the main components of a database management system?

Can you describe the difference between single and multiple user views?

What are the main differences between hierarchical, network, and relational databases?

Why have relational database structures proven so popular?

Can you describe and perform basic selection operations on a database table, including simple selections and compound selections using combinations of ANDs, ORs, and NOTs?

What are normal forms in relational databases? Why are they used, that is, what are the advantages of putting data in higher normal forms?

Can you define the basic differences between first, second, and third normal forms?

Can you give an example of a functional dependency?

9 Basic Spatial Analyses

Introduction

Spatial data analysis involves the application of operations to coordinate and related attribute data. Spatial analyses are most often applied to solve a problem, for example, to identify high crime areas, to generate a list of road segments that need repaving, or to select the best location for a new business. There are hundreds of *spatial operations* or *spatial functions* that involve the manipulation or calculation of coordinates or attribute variables.

We note that the terms spatial operations and spatial functions are often used interchangeably. Some insist an operation doesn't necessarily produce any output, while a function does. However, in keeping with many authors, GIS practitioners, and software vendors, we will use the terms function and operation interchangeably.

Spatial operations may be applied sequentially to solve a problem. Each spatial operation may create output, and the output may serve as input to other spatial operations. A chain of spatial operations is often specified, with the output of each spatial operation serving as the input of the next (Figure 9-1). Part of the challenge of geographic analysis is the selection of the appropriate spatial operations, applied in the appropriate order.

The table manipulations we described in Chapter 8 are included by our definition of a spatial operation. Indeed, selection and modification of attribute data in spatial data layers are included at some time in nearly all complex spatial analyses. Many operations incorporate both the attribute and coordinate data, and the attributes must be further selected and modified in the course of a spatial analysis. Some might take issue with our inclu-

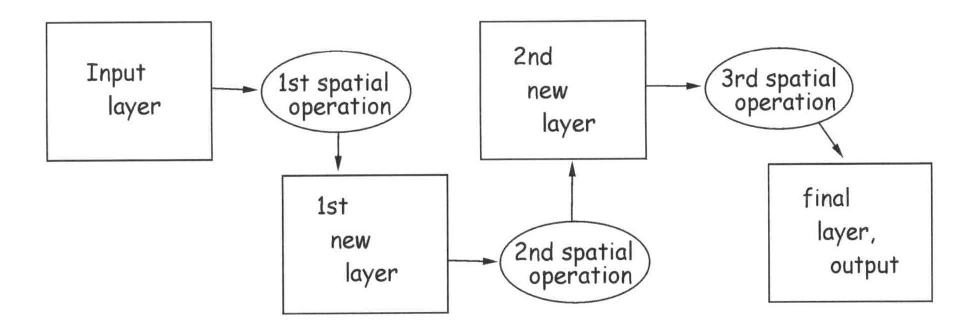

Figure 9-1: A sequence of spatial operations are often applied to obtain a desired final data layer.

sive definition of a spatial operation, in that an operation might be applied only to the "non-spatial" attribute data stored in the database tables. Attribute data are part of the definition of spatial objects, and it seems artificial to separate operations on attribute data from operations that act on only the coordinate portion of spatial data.

The discussion in the present chapter will expand on rather than repeat the selection operations treated in Chapter 8. This chapter describes spatial data analyses that involve sort, selection, classification, and spatial operations that are applied to both coordinate and associated attribute data.

Input, Operations, and Output

Spatial data analysis typically involves using data from one or more layers to create output (Figure 9-2). The analysis may consist of a single operation applied to a data layer, or many operations that integrate input data from many layers to create the desired output.

As shown in Figure 9-2, there may be single or multiple inputs or outputs from a spatial data operation. Many operations require a single data layer as input and generate a single output data layer. Vector to raster conversion is an example of an operation that typically takes a single input data layer and generates a single output data layer (Figure 9-3). There is a one-to-one correspondence between input and output layers.

There are also operations that generate several output data layers from an input, or that require several inputs to generate a single output data layer. Terrain analysis functions may take a raster grid of elevations as an input data layer and produce both slope (how steep each cell is) and aspect (the slope direction). Two outputs are generated for each input elevation data set. A layer average is an example of the use of multiple input layers to produce a single output layer. Mean annual grain production might be stored for 10 separate years in raster data layers. To calculate the average grain production over the 10 year period, the annual layers are combined and averaged on a cell-by-cell basis. This results in a sin-

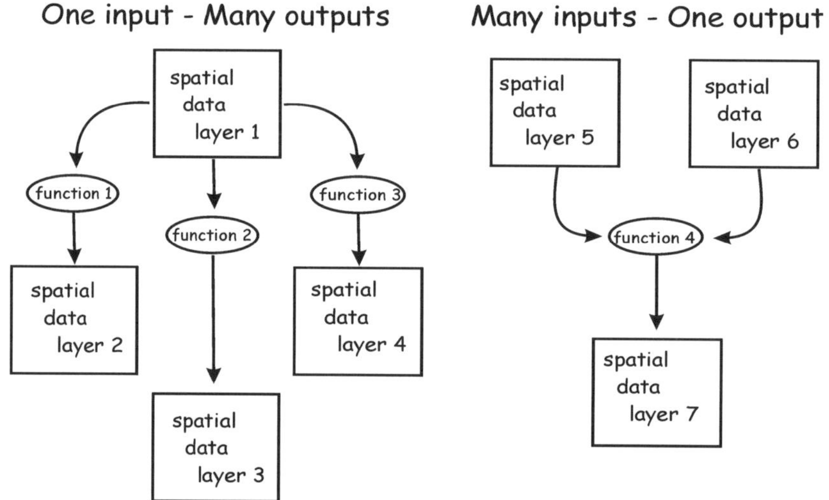

Figure 9-2: Much basic spatial data analysis consists largely of spatial operations. These operations are applied to one or more input data layers to produce one or more output data layers.

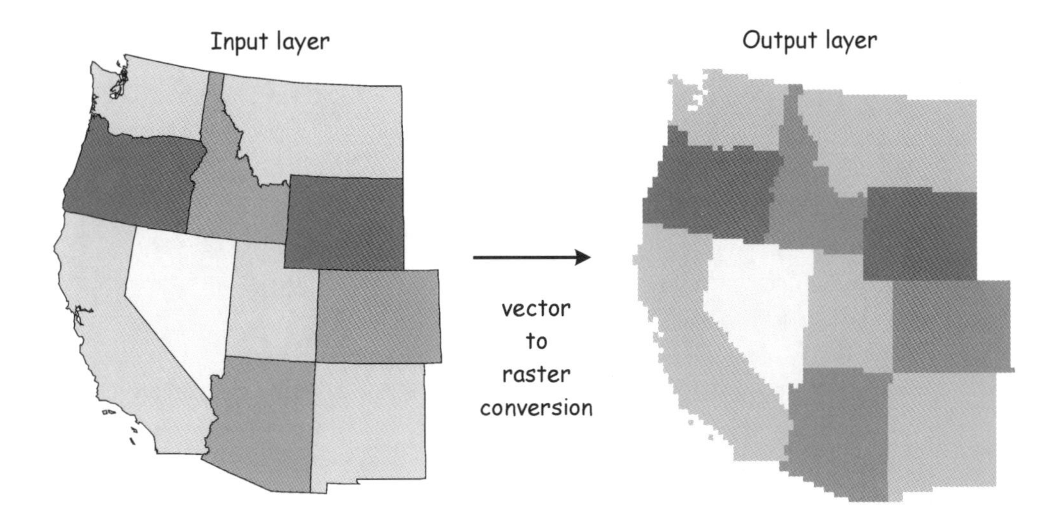

Figure 9-3: Vector to raster conversion, an example of a spatial data operation.

gle output data layer. Finally, there are some spatial operations that require many input layers and generate many output layers.

Spatial operations may be considered the basic components of spatial data analyses, and several operations are often used in quite long sequences when problem solving with spatial data. A single operation alone is usually not sufficient to solve a significant problem, and a series of operations is performed in most spatial data analyses. Most significant spatial analyses require multiple primary data inputs. Tens of spatial operations are applied, many of which generate new data layers, all of which may be used in creating output layers or scalars (single numbers).

The output from a spatial operation may be spatial, in that a new data layer is produced, or the output may be non-spatial, in that the spatial operation may produce a scalar value, a list, or a table, with no explicit geometric data attached. A layer mean function may simply calculate the mean cell value found in a raster data layer. The input is a spatial data layer, but the output is a scalar, the single number representing the mean raster value.

Other operations create aspatial output, for example, a list of all landcover types may be extracted from a data layer. The list indicates all the different types of landcover that can be found in the layer, but is non-spatial, because it does not attach each landcover to specific locations on the surface of the Earth. One might argue that the list is referenced in a general way to the area covered by the spatial data layer, but each landcover class in the list is not explicitly related to any specific polygons or points of interest in the data layer. Thus, a spatial input is passed through this "occurrence" operator and provides a list output.

Scope

Spatial data operations may be characterized by their *spatial scope*, the extent or area of the input data that are used in determining the values at output locations (Figure 9-4). There is often a direct

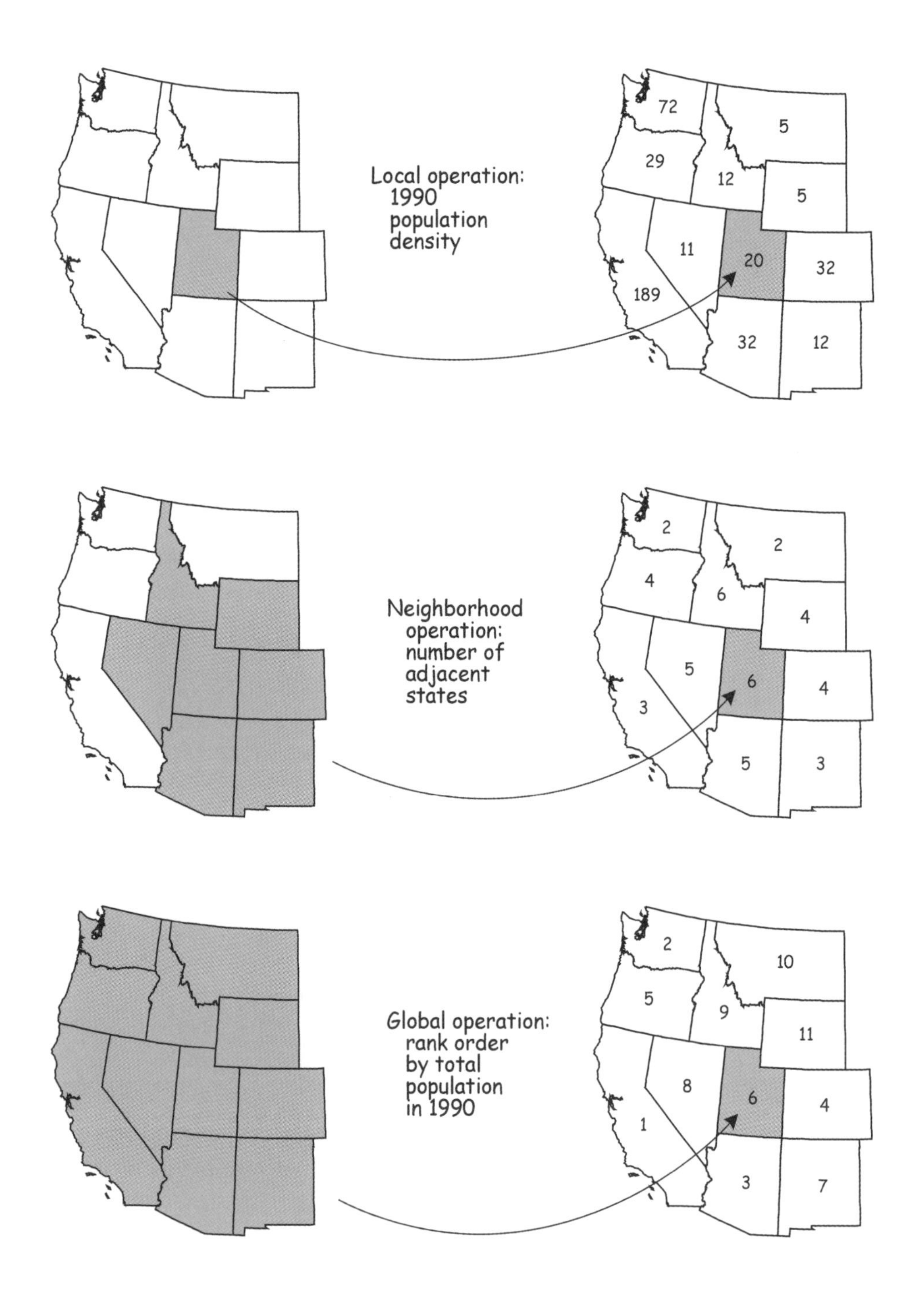

Figure 9-4: Local, neighborhood, and global operations. Specific input and output regions are shown for Utah, the shaded area on the right side of the figure. Shaded areas on the left contribute to the values shown in the shaded area on the right. Local operation output (top right) depends only on data at the corresponding input location (top left). Neighborhood operation output (middle right) depends on input from the local and surrounding areas (middle left). Global operation output (bottom right) depends on all features in the input data layer (bottom left).

correspondence between input data at a location and output data at that same location. The geographic or attribute data corresponding to an input location are acted on by a spatial operation, and the result is placed in a corresponding location in the output data layer. Spatial operations may be characterized as local, neighborhood, or global, to reflect the extent of the source area used to determine the value at a given output location.

Local operations use only the data at one input location to determine the value at a corresponding output location (Figure 9-4). Attributes or values at adjacent locations do not affect the operation or contribute to the value placed in an output location.

Neighborhood operations use data from both an input location plus nearby locations to determine the output value (Figure 9-4). The extent and relative importance of values in the nearby region may vary, but the value at an output location is influenced by more than just the value of data found at the corresponding input location.

Global operations use data values from the entire input layer to determine each output value. The value at each location depends in part on the values at all input locations (Figure 9-4).

The set of available spatial operations depends on the data model and type of spatial data used as input. Some operations, e.g., buffering or proximity operations (described later in this chapter) may be easily applied to raster or vector data, and to point, linear, or areal features. While the details of the specific implementation may change, the concept of the operation does not. Other operations may be possible in only one data model.

Characteristics of a data model will determine how a concept is applied. Analyses will vary depending on the model and the specific operations the GIS software developers chooses to include in their computer programs. Operation scope provides a good example. A local operation in a raster data layer typically matches input and output for a specific raster cell. The cell is uniformly defined in size, shape, and location. A local operation for a vector polygon data set is likely to have a variable size, shape, and location. In Figure 9-4 the local operation follows a state boundary. Therefore, the operation applies to a different size and shape for each state. Neighborhood analyses are affected by the shape of adjacent states in a similar manner. Summary values such as populations of adjacent states may be greatly influenced by changes in neighborhood size, so great care must be taken when interpreting the results of a spatial operation. Knowledge of the algorithm behind the operation is the best aid to interpreting the results.

While most operations might be conceptually compatible with most spatial data models, there are often significant differences among models in the ease with which a spatial operation may be implemented. Some operations are quite easy to program when using raster data models, and quite difficult when using vector data models. The reverse is true for other spatial data operations. In many instances it is more efficient to convert the data between data models and apply the desired operations, and if necessary, convert the results back to the original data model.

Selection and Classification

Selection operations involve identifying features that meet one to several conditions or criteria. The attributes or geometry of features are checked against criteria. Those that satisfy the conditions defined by the criteria are selected. These features may then be written to a new output data layer, or the geometry or attribute data may be manipulated in some manner.

Figure 9-5 contains an example of a selection operation that involves the attributes of a spatial data set. Two conditions are applied and the features that satisfy both conditions are included in the selected set. This example shows the selection of those states in the "lower 48" states of the United States of America that are a) entirely north of Arkansas, and b) have an area greater than 84,000 square kilometers. The complete set of features that will be considered is shown at the top of the figure. This set is comprised of the lower 48 states, with the state of Arkansas indicated by shading. The next two maps of Figure 9-5 shows those states that match the individual criteria. The second map from the top shows those states that are entirely north of Arkansas, while the third map shows all those states that are greater than 84,000 square kilometers. We are required to identify states that meet both conditions, so we need to identify states that are shaded in both the second and third maps. The bottom part of Figure 9-5 show those states that satisfy both conditions. This figure illustrates two basic characteristics of selection operations. First, there is a set of features that are candidates for selection, and second, these features are selected based singly or on some combination of the geographic and attribute data.

The simplest form of selection is an *on-screen query*. A data layer is displayed, and features are selected by a human operator. The operator uses a pointing device to locate a cursor over a feature of interest and sends a command to select, often via a mouse click or keyboard entry. On-screen (or interactive) query is used to gather information about specific features, and is often used for interactive updates of attribute or spatial data. For example, it is common to set up a process such that when a feature is selected the attribute information for the feature is displayed. These attribute data may then be edited, and changes saved.

Queries may also be specified by applying conditions solely to the aspatial components of spatial data. These selections are most often based on the attribute data tables for a layer or layers. These selection operations are applied to a set of features in a data layer or layers. The attributes for each feature are compared against a set of conditions. If the attributes match the conditions, they are selected; if the features fail to match the conditions, they are placed in an unselected set. In this manner selection splits the data into two sets: a selected set and an unselected set. Selected data are typically then acted on in some way. Selected data are often saved to a separate file, deleted, or changed in some manner.

Selection operations on tables were described in general in Chapter 8. The description here expands on that information and draws attention to specific characteristics of selections applied to spatially-related data. Table selections have spatial relevance because each record in a table is associated with a geographic features. Selecting a record in a table may also be used to select cells, points, lines, or areas. Spatial selections may be combined with table selections to identify a set of selected geographic features.

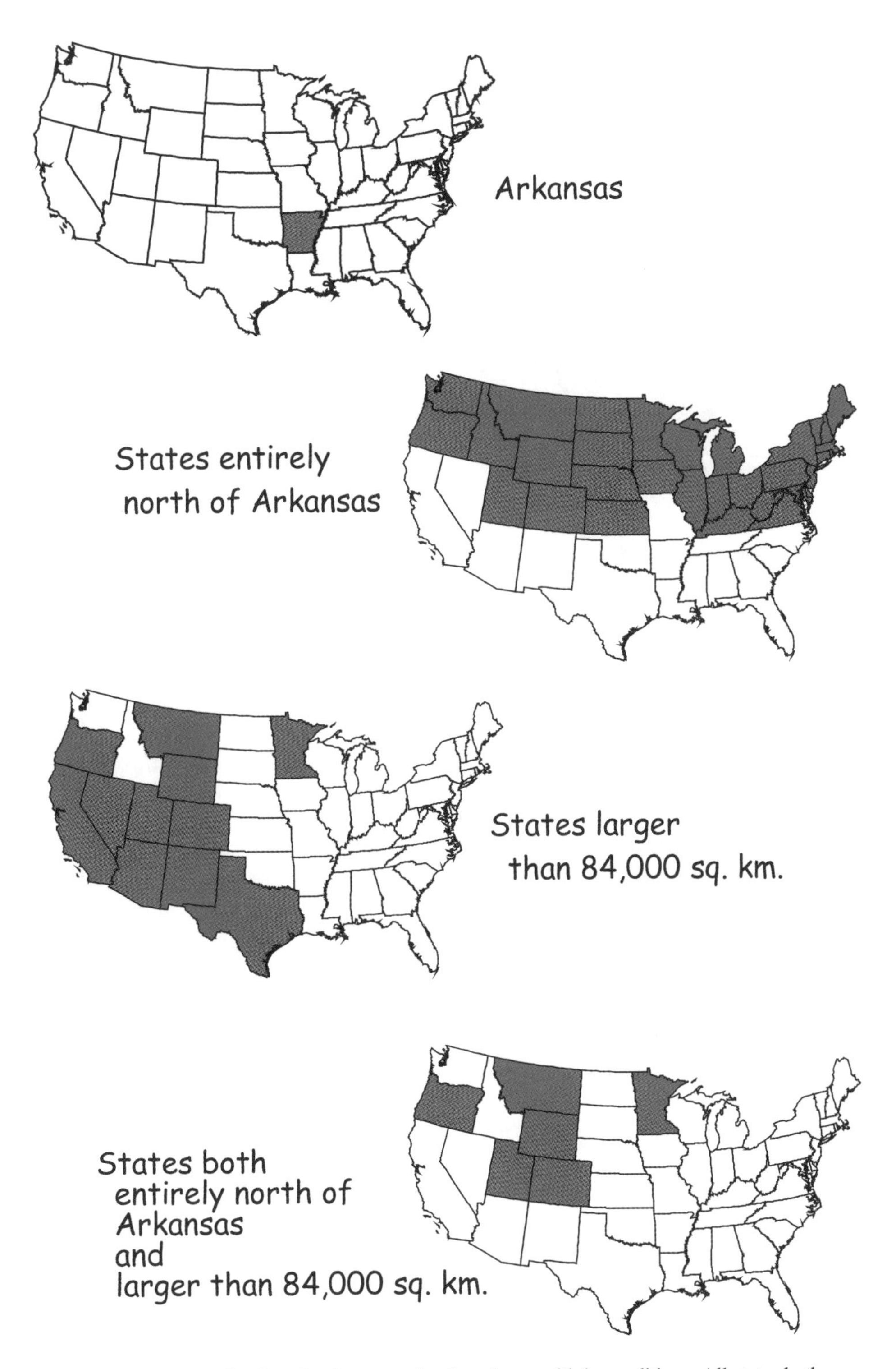

Figure 9-5: An example of a selection operation based on multiple conditions. All states both entirely north of Arkansas and larger than 84,000 square kilometers are selected. Arkansas is shown in the upper panel, followed by states matching each condition. States meeting both conditions are shown at the bottom of the figure.

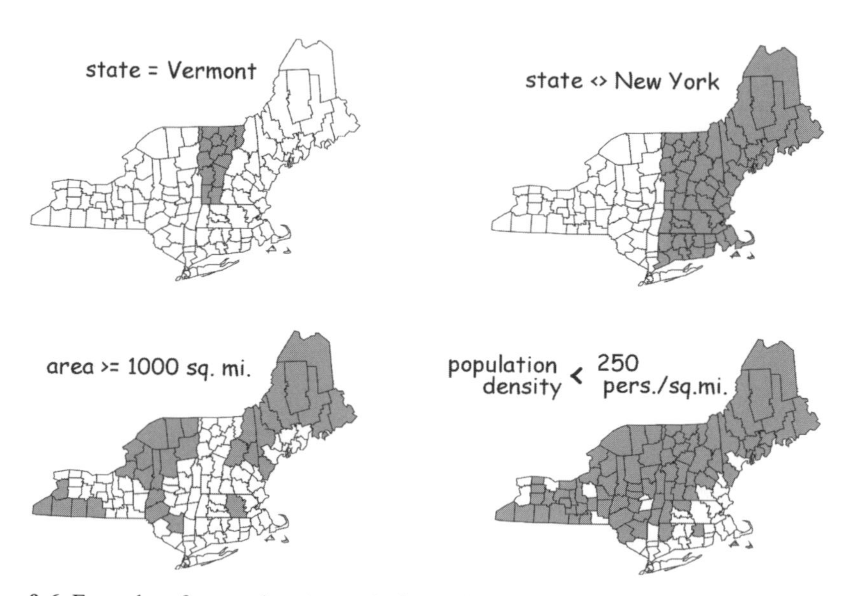

Figure 9-6: Examples of expressions in set algebra and their outcome. Selected features are shaded.

Set Algebra

Selection conditions are often formalized using *set algebra*. Set algebra uses the operations less than (<), greater than (>), equal to (=), and not equal to (< >). These selection conditions may be applied either alone or in combination to select features from a set.

Figure 9-6 shows four set algebraic expressions and the selection results for a set of counties in New England and New York. The upper two selections illustrated in Figure 9-6 show an equal to (=) and a not equal to (< >) selection. The upper-left shows all counties with a value for the attribute state that equals Vermont, while the upper right shows all counties with a value for state that are not equal to New York. The lower selections in Figure 9-6 show examples of ordinal comparisons. The left figure shows all counties with a size greater than or equal to (> =) 1000 square miles, while the right side shows all counties with a population density less than (<) 250 persons per square mile.

Set algebra operations greater than (>) or less than (<) may not be applied to nominal data, because there is no implied order in nominal data. Green is not greater than yellow, and red is not less than blue. Only the set algebra operations equal to (=) and not equal to (< >) apply to these or other nominal variables. All set algebra operations may be applied to ordinal data, and all are often applied to interval/ratio data.

Boolean Algebra

Boolean algebra uses the conditions OR, AND, and NOT to select features. Boolean expressions are most often used to combine set algebra conditions and create compound spatial selections. The Boolean expression consists of a set of Boolean operators, variables, and perhaps constants or scalar values.

Boolean expressions are evaluated by assigning an outcome, true or false, to each condition. Figure 9-7 shows three example Boolean expressions. The first is an expression using a Boolean AND. There are two arguments for the expression. The first argument specifies a condition on a vari-

Boolean expressions

```
1. (area > 100,000)
   AND
   (farm_income < 10 billion)

2. NOT ( state = Texas )

3. [ ( rainfall > 1000 )
     AND
     ( taxes = low) ]
    OR
   [ ( house_cost < 65,000)
     AND
     NOT (crime = high) ]
```

Figure 9-7: Examples of Boolean expressions

able named area, and the second argument a condition on a variable named farm_income. Features are selected if they satisfy both arguments, that is, if their area is larger than 100,000 AND farm_income is less than 10 billion. Features meeting both conditions will become part of the selected set.

Expression 2 in Figure 9-7 illustrates a Boolean NOT expression. This conditions specifies that all features with a variable state which is not equal to Texas will return a true value, and hence be selected. NOT is also often known as the negation operator. This is because we might interpret the application of a NOT operation as exchanging the selected set for the unselected set. The argument of expression 2 in Figure 9-7 is itself a set algebra expression. When applied to a set of features, this expression will select all features for which the variable state is equal to the value Texas. The NOT operation reverses this, and selects all features for which the variable state is not equal to Texas, e.g., the other 49 states of the United States.

The third expression in Figure 9-7 shows a more complex Boolean expression. This is a compound expression that specifies multiple conditions. Boolean operators AND, OR, and NOT are used to combine four set algebra expressions. This example shows what might be a naive attempt to select areas for retirement. Our grandparent is interested in selecting areas that have high rainfall and low taxes (a gardener on a fixed income), or low housing cost and low crime.

The spatial outcomes of specific Boolean expressions are shown in Figure 9-8. The figure shows three overlapping circular regions, labeled A, B, and C. Areas may fall in more than one region, e.g., the center, where all three regions overlap, is in A, B, and C. As shown in the figure, Boolean AND, OR, or NOT may be used to select any combination or portions of these regions.

OR conditions return a value of true if either argument is true. Areas in either region A or region B are selected at the top center of Figure 9-8. AND requires the conditions on both sides of the operation be met; an AND operation results in a reduced selection set (top right, Figure 9-8). NOT is the negation operator, and flips the effect of the previous operations, i.e., it turns true to false and false to true. The NOT shown in the lower left portion of Figure 9-8 returns the area that is only in region C. Note that this is the converse, or opposite set returned by the comparative AND, shown in the top center of Figure 9-8. The NOT operation is often applied in combination with the AND Boolean operator, as shown at the bottom center of Figure 9-8. Again, this selects the converse (or complement) of the corresponding AND. Compare the bottom center selection to the top right selection in Figure 9-8. NOTs, ANDs, and ORs may be further combined to select specific combinations of areas, as shown in the lower right of Figure 9-8.

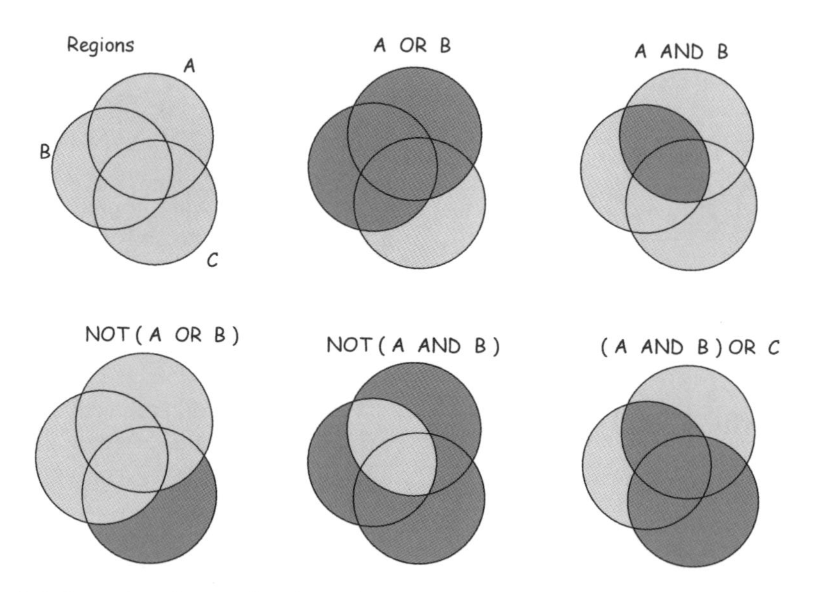

Figure 9-8: Examples of expressions in Boolean algebra, and their outcomes. Sub-areas of three regions are selected by combining AND, OR, and NOT conditions in Boolean expressions. Any sub-area or group of sub-areas may be selected by the correct Boolean combination.

Note that as with table selection discussed in Chapter 8, the order of application of these Boolean operations is important. In most cases you will not select the same set when applying the operations in a different order. Therefore, parentheses, brackets, or other delimiters should be used to specify the order of application. The expression A AND B OR C will give different results when interpreted as (A AND B) OR C, as shown in Figure 9-8, than when interpreted as A AND (B OR C). Verify this as an exercise. Which areas does the second Boolean expression select?

Spatial Selection Operations

Many spatial operations select sets of features. These operations are applied to a spatial data layer and return a set of features that meet a specified condition. Adjacency and containment are commonly used spatial selection operations.

Adjacency operations are used to identify those features that "touch" other features. Features are typically considered to touch when they share a boundary, as when two polygons share an edge or bounding line. A target or key set of polygon features is identified, and all features that share a boundary with the target features are placed in the selected set.

Figure 9-9a shows an example of a selection based on polygon adjacency. The state of Missouri is shaded on the left side of Figure 9-9a, and states adjacent to Missouri are shaded on the right portion of Figure 9-9a. States are selected because they include a common border with Missouri.

There are many ways the shared border may be detected. With a raster data layer an exhaustive cell-by-cell comparison may be conducted to identify adjacent pairs with different state values. Vector adjacency may be identified by observing the topological relationships (see Chapter 2 for a discussion of topology). Line and polygon topology typically records the polygon

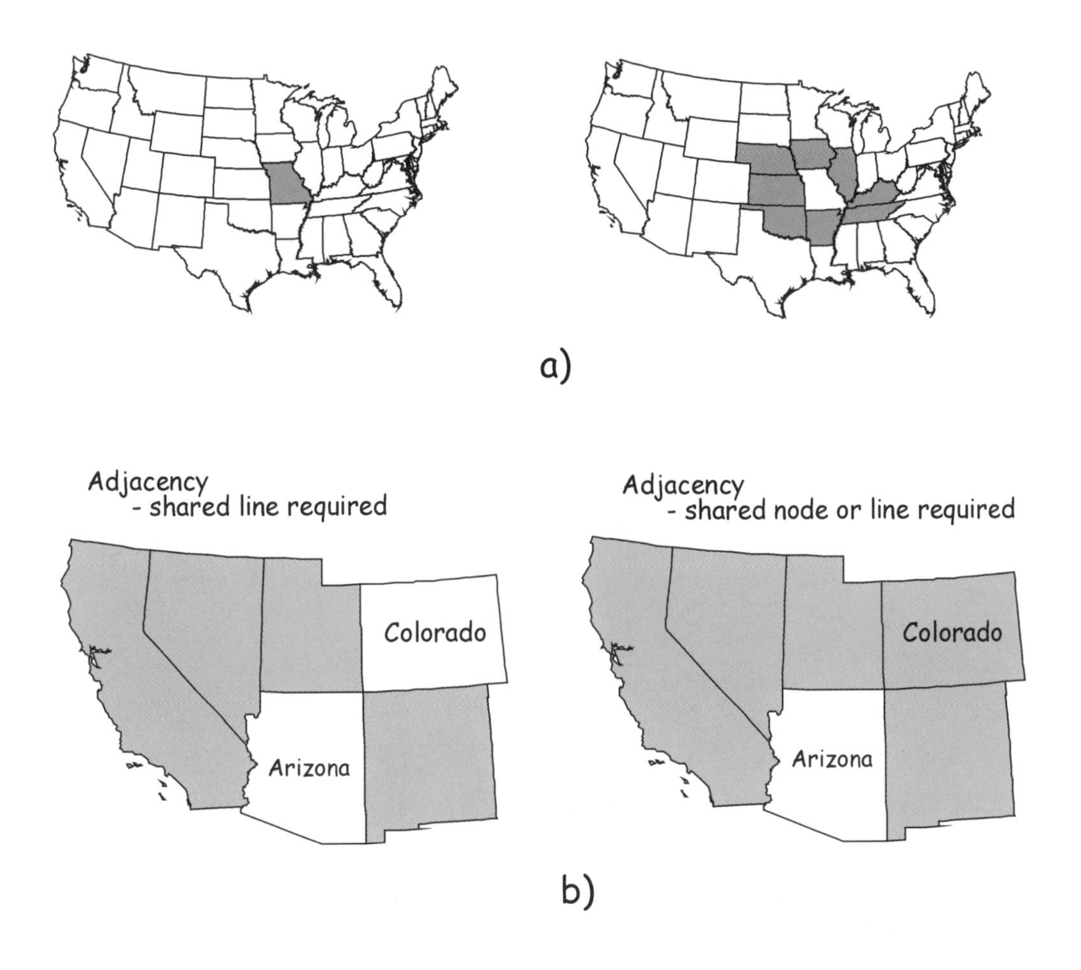

Figure 9-9: Examples of selections based on adjacency. a) Missouri, USA is shown on the left and all states adjacent to Missouri shown on the right. b) Different definitions of adjacency result in different selections. Colorado is not adjacent to Arizona when line adjacency is required (left), but is when node adjacency is accepted (right).

identifiers on each side of a line. All lines with Missouri on one side and a different state on the other side may be flagged, and the list of states adjacent to Missouri extracted.

Adjacency is defined in Figure 9-9a as sharing a boundary for some distance greater than zero. Figure 9-9b shows how a different definition of adjacency may affect selection. The left of Figure 9-9b shows the state of Arizona and a set of adjacent (shaded) western states. By the definition of adjacency used in Figure 9-9a, Arizona and Colorado are not adjacent, because they do not share a boundary along a line segment. Arizona and Colorado share a border at a point, called Four Corners, where they join with Utah and New Mexico. When a different definition of adjacency is used, with a shared node qualifying as adjacent, then Colorado is added to the selected adjacent set (right, Figure 9-9b). This is another illustration of an observation made earlier; there are often several variations of any single spatial

Figure 9-10: An example of a selection based on containment. All states containing a portion of the Mississippi River or its tributaries are selected.

operation. Care must be taken to test the operation under controlled conditions until the specific implementation of a spatial operation is well understood.

Containment is another spatial selection operation. Containment selection identifies all features that contain or surround a set of target features. For example, the California Department of Transportation may wish to identify all counties, cities, or other governmental bodies that contain some portion of Highway 99, because they wish to consider improving road safety. A spatial selection may be used to identify these governmental bodies.

Figure 9-10 illustrates a containment selection based on the Mississippi River in North America. We wish to identify states that contain some portion of the River and its tributaries. A query is placed, identifying the features that are contained, here the Mississippi River network, and the target features that may potentially be selected. The target set in this example consists of the lower 48 states of the United States. All states that contain a portion of the Mississippi River or its tributaries are shaded as part of the selected set.

Classification

Classification is a spatial data operation that is often used in conjunction with selection. A classification, also known as a *reclassification* or *recoding*, will categorize geographic objects based on a set of conditions. For example, all the polygons larger than one square mile may be assigned a size value equal to Large, all polygons from 0.1 to one square mile may be assigned a size equal to Mid, and all polygons smaller than 0.1 square miles may be assigned a size equal to Small (Figure 9-11). Classifications may add to or modify the attribute data for each geographic object. These modified attributes may in turn be used in further analyses, e.g., for more complex combinations in additional classification.

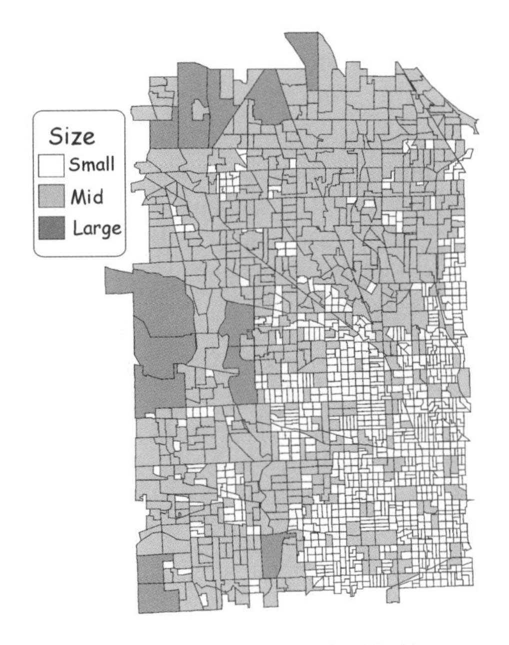

Figure 9-11: Land parcels re-classified by area.

Classification may be used for many other purposes. One common end is to group objects for display or map production. These objects have a common property, and the goal is to display them with a uniform color or symbol so the similar objects are identified as a group. The display color and/or pattern is typically assigned based upon the values of an attribute or attributes. A range of display shades may be chosen, and corresponding values for a specific attribute assigned. The map is then displayed based on this classification.

A classification may be viewed as an assignment from an existing set of classes to a new set of classes. The assignment from input attribute values to new class values may be defined manually, or the assignment may be defined automatically. For manual classifications, the class transitions are specified entirely by the human analyst.

Classifications are often specified by a table or array. The table identifies the input class, and the output class for each input class. Figure 9-12 illustrates the use of a classification table to specify class assignment. Input values of A or B lead to an output class value of 1, an input value of E leads to an output value of 2, and an input value of I leads to an output value of 3. The table provides a complete specification for each classification assignment.

Figure 9-12 illustrates a classification based on a manually defined table. The table is manually defined in that each col-

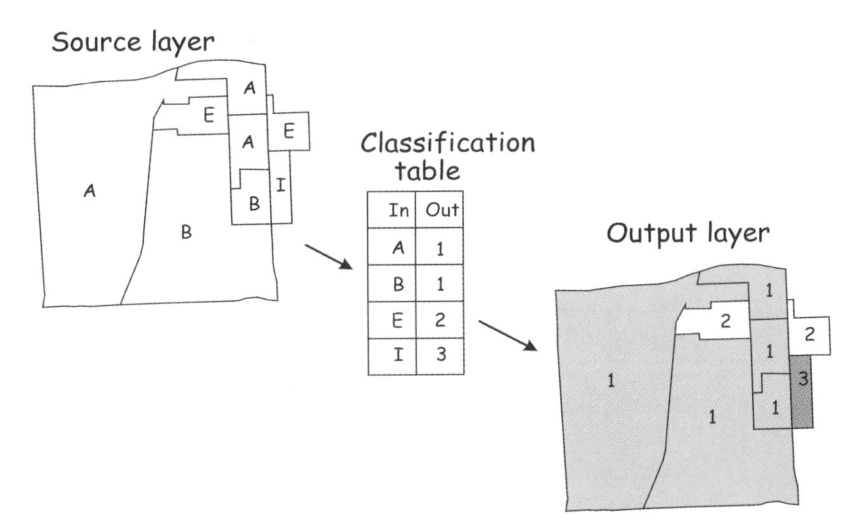

Figure 9-12: The classification of a thematic layer. Values for a specific attribute are used with a classification table to assign classes in an output layer.

umn entry is specifically assigned and entered by the human analyst. The analyst specifies the **In** items for the source data layer. She enters these values directly into the classification table, as well as the corresponding output value for each **In** variable. **Out** values must be specified for each input value or there will be undefined features in the output layer. Manual definition of the classification table provides the greatest control over class assignment. Alternatively, classification tables may be automatically assigned, in that a number of classes may be specified and some rule used to assign output classes to input classes.

A *binary classification* is perhaps the simplest form of classification. A binary classification places objects into two classes – 0 and 1, true and false, A and B, or some other two-level classification. A set of features is selected and assigned a value, e.g., one. The complement of the set, all remaining features in the data layer, is assigned the alternate binary value, e.g., zero.

A binary classification is often used to store the results of a complex selection operation. A large number of Boolean and set algebra expressions may be applied, resulting in a selected set. The selected set of features is assigned a unique value, and all features in the unselected set a different value. We may wish to record membership in the selected set, and so we create a new variable, and assign all selected records a value for this new variable, and all unselected records an alternate value.

For example, we may wish to select states west of the Mississippi River as an intermediate step in an analysis (Figure 9-13). We may be using this classification in many subsequent spatial operations. Thus, we wish to store this characteristic,

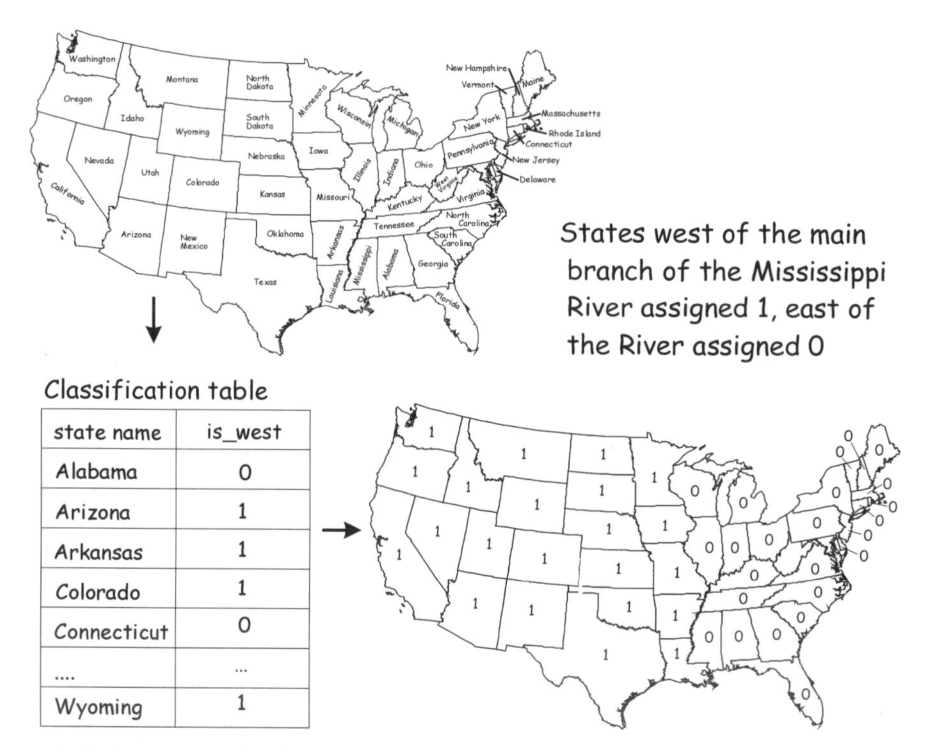

state name	is_west
Alabama	0
Arizona	1
Arkansas	1
Colorado	1
Connecticut	0
....	...
Wyoming	1

Figure 9-13: An example of a binary classification. Features are placed into two classes in a binary classification, west (1) and east (0) of the Mississippi River. The classification table codifies the assignment.

whether the state is west or east of the Mississippi River. States are selected based on location and reclassified. We record this classification by creating a new attribute and assigning a binary value to this attribute, e.g., one for those parcels that satisfy the criteria, and zero for those that do not (Figure 9-13). The variable `is_west` records the state location relative to the Mississippi River. Additional selection operations may be applied, and the created binary variable preserves the information generated in the initial selection.

The manual definition of the classification table may not always be necessary, and may be tedious or complex. Suppose we wish to assign a set of display colors to a set of elevation values. There may be thousands of distinct elevation values in the data layer, and it would be inconvenient at best to assign each color manually. Automatic classification methods are often used in these instances.

An automatic classification uses some rule to specify the input class to output class assignments. The input and output class boundaries are often based on a set of parameters used to guide class definition.

A potential drawback from an automated class assignment stems from our inability to precisely specify class boundaries. A mathematical formula or algorithm defines the class boundaries, and so specific classes of interest may be split. The analyst sacrifices precise control over class specification when an automated classification is used. Considerable time may be saved by automatic class assignment, but we may have to manually change some class boundaries as the only way to achieve a desired classification.

Figure 9-14 describes a data layer we will use to illustrate automatic class assignment. The figure shows a set of "neighborhoods" with populations that range from 0 to 5133. We wish to display the neighborhoods and populations in three distinct

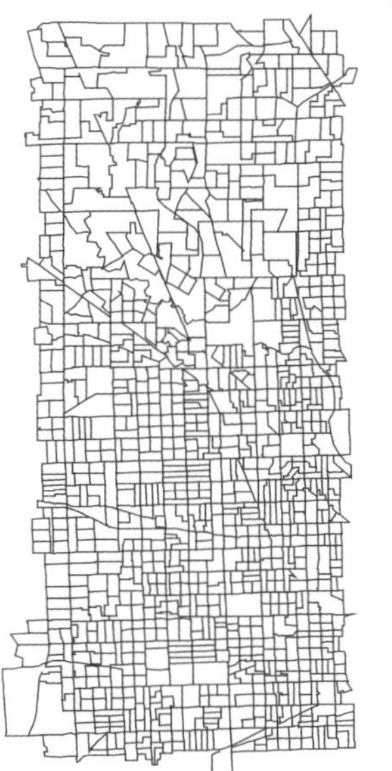

Neighborhoods
- 1074 polygons
- population for neighborhoods ranges from 0 to 5133 (3 outliers > 3300)

A bar graph shows the frequency of neighborhood population, e.g., 8.1 % of the neighborhoods have a population between 3000 and 3100

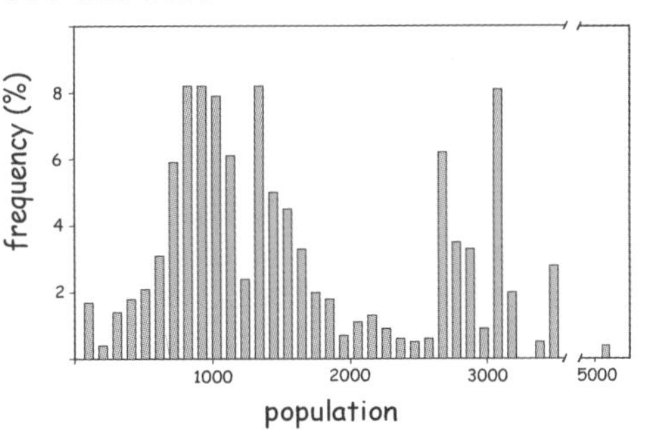

Figure 9-14: Neighborhood polygons and population levels used in subsequent examples of classification assignment. The populations for these 1074 neighborhoods ranges from 0 to 5133. The histogram at the lower right shows the frequency distribution. Note that there is a break in the chart between 3500 and 5000 to show the three "outlier" neighborhoods with populations above 3300.

classes, high, medium, and low population. High will be shown in black, medium in gray, and low in white. We must decide how to assign the categories - what population levels define high, medium, and low? In many applications the classification levels are previously defined. There may be an agreed-upon standard for high population, and we would simply use this level. However, in many instances the classes are not defined, and we must choose them.

Figure 9-14 includes a bar graph depicting the population frequency distribution, commonly called a histogram. The frequency histogram shows the number of neighborhoods that are found in each bin of a set of very narrow population categories. For example, we may count the number of neighborhoods that have a population between 3000 to 3100. If a neighborhood has 3037, 3004, 3088, or any other number between 3000 and 3100, we add one to our frequency sum for the category covering 3000 and 3100. We review all neighborhoods in our area, and plot the percentage of neighborhoods that have a population between 3000 and 3100 as a vertical bar on the histogram. Approximately 8.4 percent of the neighborhoods have a population in this range, so a vertical bar corresponding to 8.4 units high is plotted. We count and plot the historgram values for each of our narrow categories, e.g., the number from 0 to 100, from 100 to 200, from 200 to 300, until the highest population value is plotted (Figure 9-14).

We may view our class assignment problem as deciding where to place the class boundaries. Should we place the boundary between the low and medium population classes at 1000, or 1200? Where should the boundary between medium and high population classes be placed? The location of the class boundaries will change the appearance of the map, and also the resulting classification.

One common method for automatic classification specifies the number of output classes and requests equal interval classes over the range of input values. This *equal-interval* classification simply subtracts the lowest value of the classification variable from the highest value, and defines equal-width boundaries to fit the desired number of classes into the range.

Figure 9-15 illustrates an equal-interval classification for the population variable. Three classes assigned over the range of 0 to 5133 are specified. Each interval is approximately one-third of this range. This range is evenly divided by 1711. The small class extends from 0 to 1711, the medium class from 1712 to 3422, and the large class from 3423 to 5133. Population categories are shown colored accordingly on the map and the bar graph, with the small (white), medium (gray) and large (black) classes shown.

Note that the low population class shown in white dominates the map; most of the neighborhoods fall in this population class. This often happens when there are features that have values much higher than the norm. There are a few neighborhoods with populations above 5000 (to the right of the break in the population axis of the bar graph), while most neighborhoods have populations below 3000. The outliers shift the class boundaries to higher values, 1711 and 3422, resulting in most neighborhoods falling in the small population category.

Another common method for class assignment results in an *equal-area* classification (Figure 9-16). Class boundaries are defined to place an equal proportion of the study area into each of a specified number of classes. This usually leads to a visually-balanced map in that all classes have approximately equal extents. Equal-area classes are often desirable for reasons other than generating a balanced map. Resources may need to be distributed over equal areas, or equally-sized overlapping sales territories may be specified.

Note that the class width may change considerably with an equal-area classification. An equal-area classification sets class boundaries so that each class covers approximately the same area. A class may

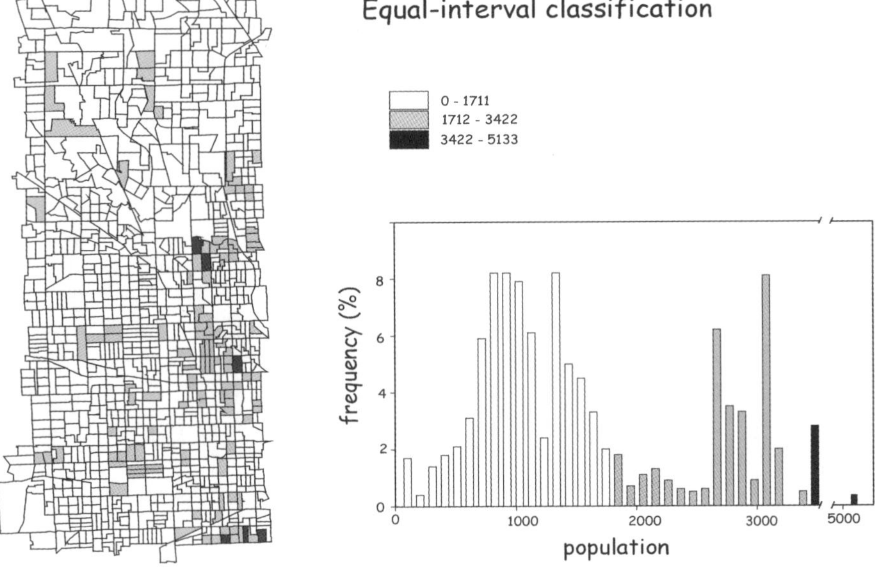

Figure 9-15: An equal interval classification. The range 0 - 5133 is split into three equal parts. Colors are assigned as shown in the map of the layer (left), and in the frequency plot (right). Note the relatively few polygons assigned to the high classes in black. A few neighborhoods with populations near 5000 shift the class boundaries upward.

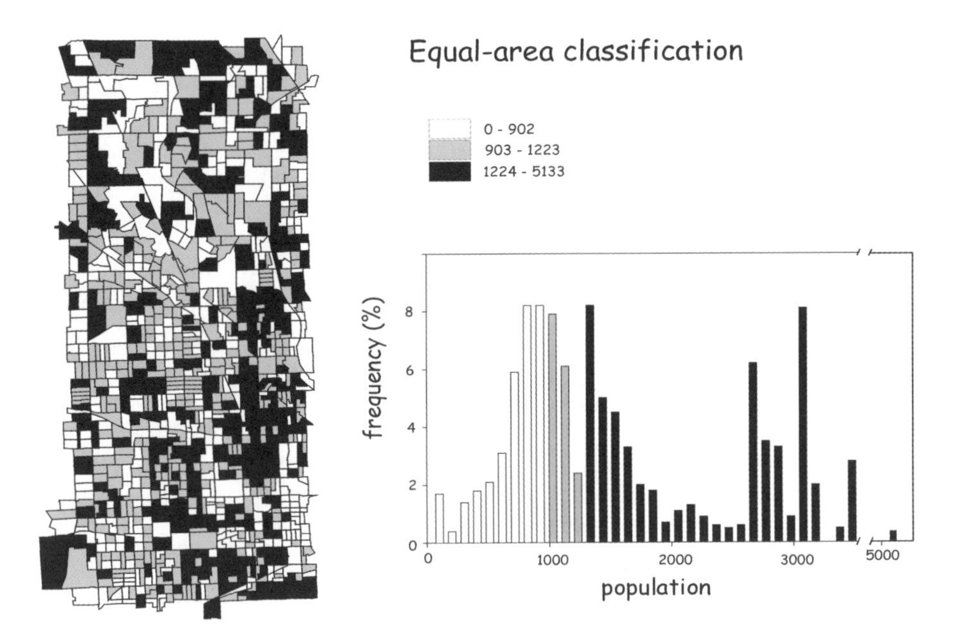

Figure 9-16: Equal-area classification. Class boundaries are set such that each class has approximately the same total area. This often leads to a smaller range when groups of frequent classes are found. In this instance the medium class spans a small range, from 903 to 1223, while the high population class spans a range that is almost 10 times broader, from 1224 to 5133.

consist of a few or even one large polygon. This results in a small range for the large-polygon classes. Classes also tend to have a narrow range of values near the peaks in the histogram. Many polygons are represented at the histogram peaks, and so these may correspond to large areas. Both of these effects are illustrated in Figure 9-16. The middle class of the equal-area classification occurs at population values between 903 and 1223. This range of populations is near the peak in the frequency histogram, and these population levels are associated with larger polygons. This middle class spans a range of approximately 300 population units, while the small and large classes span near 900 and 4000 population units, respectively.

Note that equal-area assignments may be highly skewed when there are a few polygons with large areas, and these polygons have similar values. Although not occurring in our example, there may be a relationship between the population and area for a few neighborhoods. Suppose in a data set similar to ours there is one very large neighborhood dominated by large parks. This neighborhood has both the lowest populations and largest area. An equal-area classification may place this neighborhood in its own class. If a large parcel also occurs with high population levels, we may get three classes: one parcel in the small class, one parcel in the high population class, and all the remaining parcels in the medium population class. While most equal-area classifications are not this extreme, unique parcels may strongly affect class ranges in an equal-area classification.

We will cover a final method for automated classification, a method based on *natural breaks*, or gaps, in the data. Natural breaks classification looks for "obvious" breaks. It attempts to identify naturally occurring clusters of data, not clusters based on the spatial relationships, but rather clusters based on an ordering variable.

There are various algorithms used to identify natural breaks. Large gaps in an ordered list of values are one common method. Baring gaps, low points in the frequency histogram may be identified. There is usually an effort to balance the need for relatively wide and evenly distributed classes and the search for natural gaps. Many narrow classes and one large class may not be acceptable in many instances, and there may be cases where the specified number of gaps does not occur in the data histogram. More classes may be requested than obvious gaps, so some natural break methods include an alternative method, e.g., equally-spaced intervals, for portions of the histogram where no natural gaps occur.

Figure 9-17 illustrates a natural break classification. Two breaks are evident in the histogram, one near 1300 and one between 1900 and 2600 persons per neighborhood. Small, medium, and large populations are assigned at these junctures.

Figure 9-15 through Figure 9-17 strongly illustrate an important point: you must be careful when producing or interpreting class maps, as the apparent relative importance of categories may be manipulated by altering the starting and ending values that define each class. Figure 9-15 suggests most neighborhoods are low population, Figure 9-16 that high population neighborhoods cover the largest areas and these are well mixed with areas of low and medium population, while Figure 9-17 indicates the area is dominated by low and medium population neighborhoods. Precisely because there are no objectively defined population boundaries we have great flexibility in manipulating the impression we create. The legend in class maps should be scrutinized, and the range between class boundaries noted. A histogram, as shown in these figures, and an indication of the highest and lowest data values are valuable when interpreting the legend.

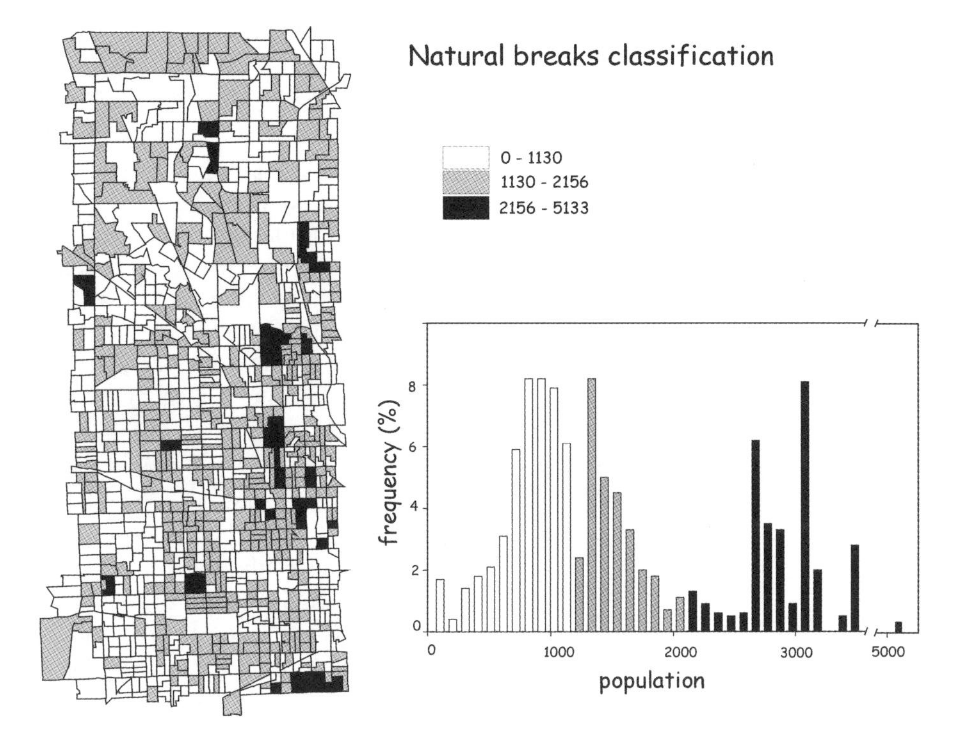

Figure 9-17: Natural breaks classification. Boundaries between classes are placed where natural gaps or low points occur in the frequency distribution.

Dissolve

A *dissolve* function has the primary purpose of combining like features within a data layer. Adjacent polygons may have identical values for an attribute. For example, a wetlands data layer may specify polygons with several sub-classes, e.g., wooded wetlands, herbaceous wetlands, or open water. If an analysis requires we identify only the wetland areas vs. the upland areas, then we may wish to dissolve all boundaries between adjacent wetlands. We are only interested in preserving the wetland/upland boundaries.

Dissolve operations are usually applied based on a specific "dissolve" attribute associated with each feature. A value or set of values is identified that belong in the same grouping. Each line that serves as a boundary between two polygon features is assessed. The values for the dissolve attribute are compared across the boundary line. If the values are the same, the boundary line is removed, or dissolved away. If the values for the dissolve attribute differ across the boundary, the boundary line is left intact.

Figure 9-18 illustrates the dissolve operation that produces a binary classification. This classification places each of the contiguous United States into one of two categories, those entirely west of the Mississippi River (1) and those east of the Mississippi River (0). The attribute named `is_west` contains values indicating location. A dissolve operation applied on the variable is_west removes all state boundaries between similar states. This has the result of reducing the set from 48 polygons to two polygons.

Dissolve operations are often needed prior to applying an area-based selection in spatial analysis. For example, we may wish to select areas from the natural breaks classification shown on the left of Figure 9-19. We seek polygons that are greater than three square miles in area and have a medium population. The polygons may be composed of multiple neighborhoods. We typically must dissolve the boundaries between adjacent, medium-sized neighborhoods prior to applying the size test. Otherwise two adjacent, medium population neighborhoods may be discarded because both cover approximately two square miles. Their total area is four square miles, above the specified threshold, yet they will not be selected unless a dissolve is applied first.

Dissolves are also helpful in removing unneeded information. After the classification into small, medium, and large size classes, many boundaries may become redundant. Unneeded boundaries may inflate storage and slow processing. A dissolve has the primary advantage of removing unneeded geographic and tabular data, thereby improving processing speed and reducing data volumes and complexity.

Figure 9-19 illustrates a dissolve of the natural breaks classification described previously. Each line that separates two polygons is inspected. If the same size class occurs on both sides of the line, the line is removed. Otherwise, the line is retained. New spatial and attribute data must be gen-

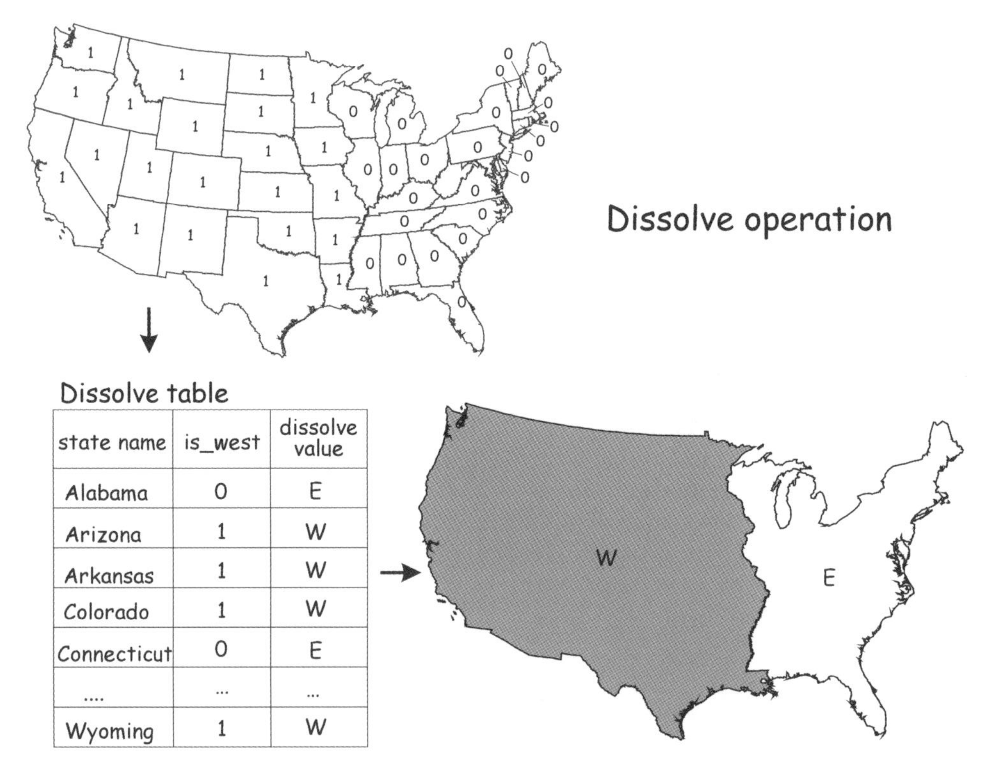

state name	is_west	dissolve value
Alabama	0	E
Arizona	1	W
Arkansas	1	W
Colorado	1	W
Connecticut	0	E
....	...	...
Wyoming	1	W

Figure 9-18: An illustration of a dissolve operation. Boundaries are removed when they separate states with the same value for the dissolve attribute is_west.

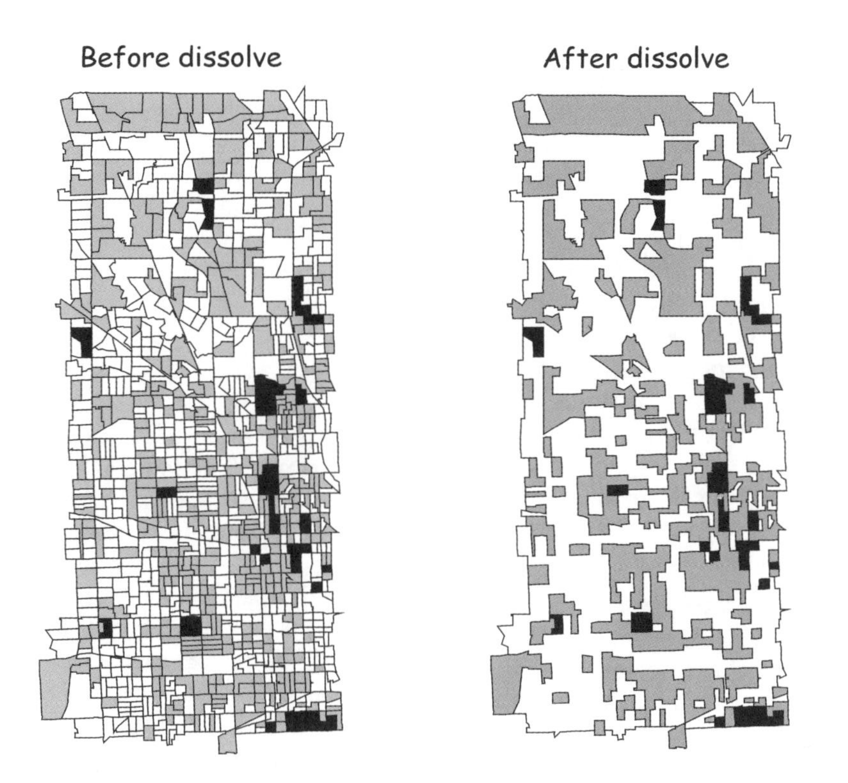

Figure 9-19: An example of a dissolve operation. Note the removal of lines separating polygons of the same size class. This greatly reduces the number of polygons.

erated for the layer once all appropriate lines have been dissolved. This example illustrates the space saving and complexity reduction common when applying a dissolve function. The number of polygons is reduced approximately nine-fold by the dissolve, from 1074 on the left to 119 polygons on the right of Figure 9-19.

Proximity Functions and Buffering

Proximity functions or *operations* are among the most powerful and common spatial analysis tools. Many important questions hinge on proximity, the distance between features of interest. How close are schools to an oil refinery, what neighborhoods are far from convenience stores, and which homes will be affected by an increase in freeway noise? These and other questions regarding proximity are often answered through analyses in a GIS.

Proximity functions modify existing features or create new features that depend in some way on the distance from existing features. For example, one simple proximity function creates a raster of the minimum distance to a set of features (Figure 9-20). The figure shows a distance function applied to water holes in a wildlife reserve. Water is a crucial resource for many animals, and the reserve managers may wish to ensure that most of the area is within a short distance of water. In this instance point features are entered. The point features represent the location of permanent water. Water holes are represented by individual points, and rivers by a group of points set along the river course. The distance function calculates the distance to all water points for each raster cell. The minimum distance is selected and placed in an output raster data layer. The distance layer is illustrated in Figure 9-20 along with water locations. The short distances are shown in white and longest distances in black. The distance function creates a mosaic of what appear to be overlapping circles. Although the shading scheme shows apparently abrupt transitions, the raster cells contain a smooth gradient in distance away from each water feature.

Distance values are calculated based on the Pythagorean formula (Figure 9-21). These values are typically calculated from cell center to cell center when applied to a raster data set. Although any distance is possible, the distances between adjacent cells change in discrete intervals related to the cell size. Note that distances are not restricted to even multiples of the cell size, because distances measured on diagonal angles are not even multiples of the cell dimension. There may be no cells that are exactly some fixed distance away from the target features, however there may be many cells less than or greater than that fixed distance.

Buffers

Buffering is one of the most commonly used proximity functions. A *buffer* is a region that is less than or equal to a specified distance from one or more features (Figure 9-22). Buffers may be determined for point, line, or area features, and for raster or vector data. Buffering is the process of creating buffers. Buffers typically identify areas that are "outside" some given threshold distance vs. those "inside" some threshold distance.

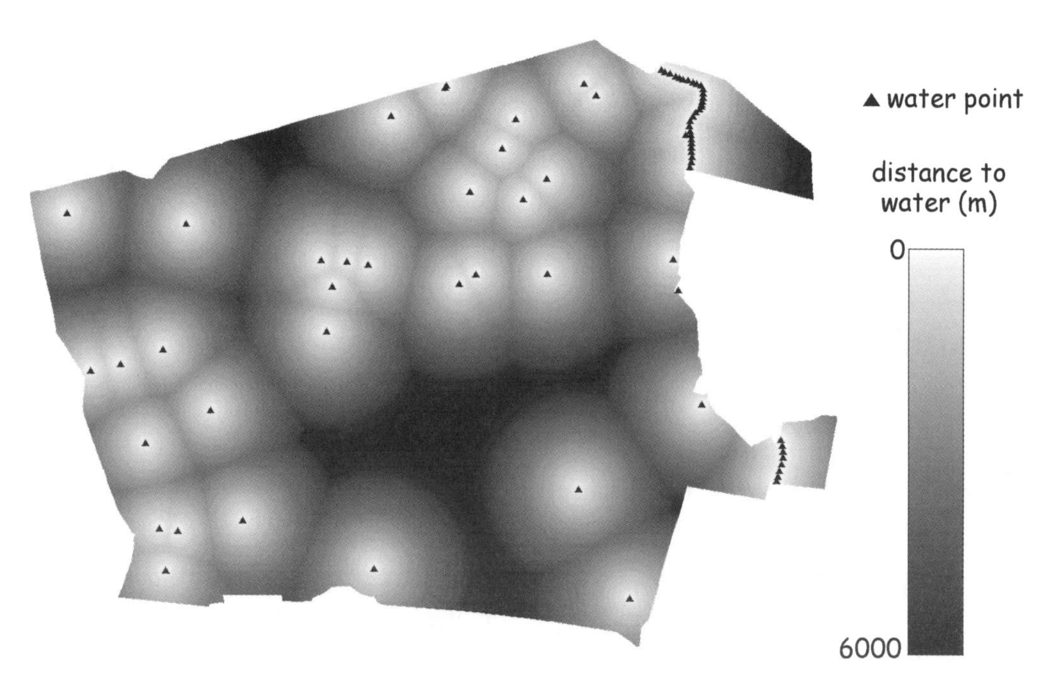

Figure 9-20: An example of a distance function. This distance function is applied to a point data layer and creates a raster data layer. The raster layer contains the distance to the nearest water feature.

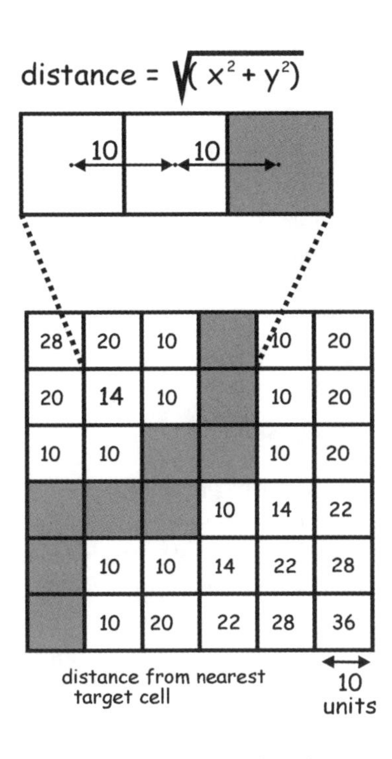

Figure 9-21: A distance function applied to a raster data set.

Buffers are used often because many spatial analyses are concerned with distance constraints. For example, emergency planners might wish to know which schools are within 1.5 kilometers of an earthquake fault, a park planner may wish to identify all lands more than 10 kilometers from the nearest highway, or a business owner may wish to identify all potential customers within a given radius of her store. All these questions may be answered with the appropriate use of buffering.

Raster Buffers

Buffer operations on raster data entail calculating the distance from each source cell center to all other cell centers. Output cells are assigned an in value whenever the cell-to-cell distance is less than the specified buffer distance. Those cells that are further than the buffer distance are assigned an out value (Figure 9-23).

Raster buffers may be viewed as a combination of a minimum distance function and a binary classification function. A minimum distance function calculates the

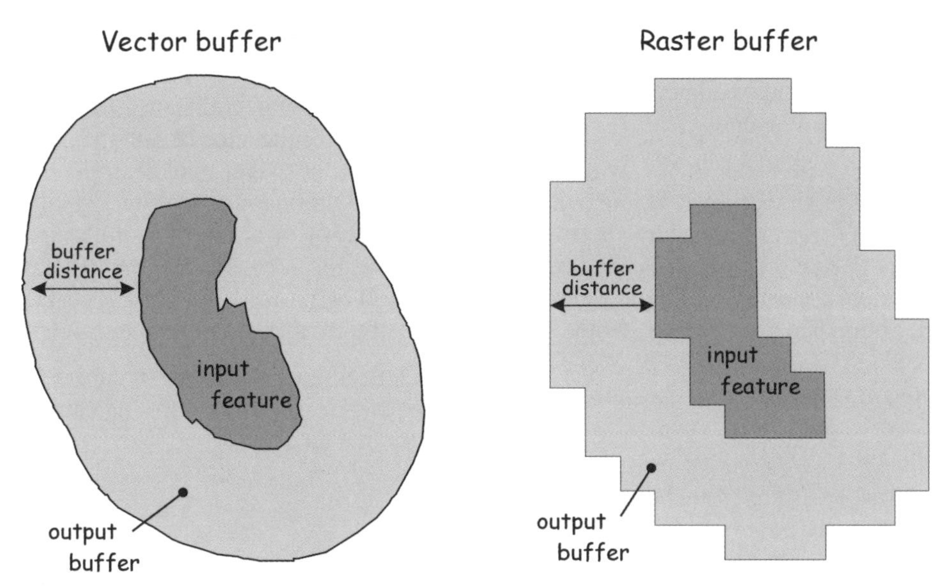

Figure 9-22: Examples of vector and raster buffers derived from polygonal features. A buffer is defined by those areas that are within some buffer distance from the input features.

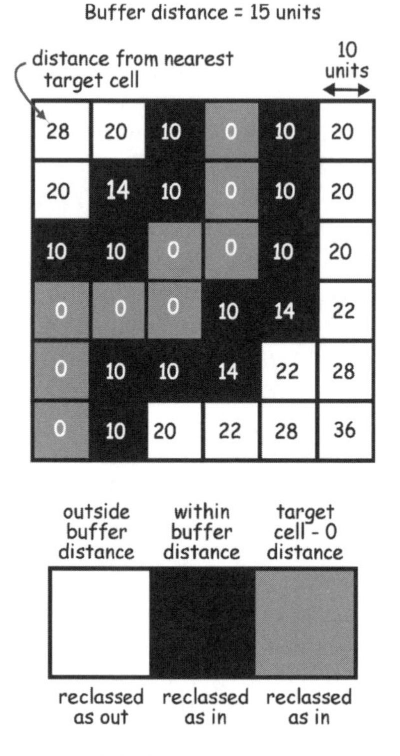

Figure 9-23: Raster buffering as a combination of distance and classification.

shortest distance from a set of target features and stores this distance in a raster data layer. The binary classification function splits the raster cells into two classes: those with a distance greater than the threshold value, and those with a distance less than or equal to a threshold value.

Buffering with raster data may produce a "stair-step" boundary, because the distance from features is measured between cell centers. When the buffer distance runs parallel and near a set of cell boundaries, the buffer boundary may "jump" from one row of cells to the next (Figure 9-23). This phenomenon is most often a problem when the raster cell size is large relative to the buffer distance. A buffer distance of 100 meters may be approximated when applied to a raster with a cell size of 30 meters. A smaller cell size relative to the buffer distance results in less obvious "stair-stepping". The cell size should be small relative to the spatial accuracy of the data, and small relative to the buffer distance. If this rule is followed, then stairs-stepping should not be a problem, because buffer sizes should be many times greater than the uncertainty inherent in the data. Buffer distances on the same order of magnitude as the spatial accuracy should be avoided.

Vector Buffers

Vector buffering may be applied to point, line, or area features, but regardless of input, buffering always produces an output set of area features (Figure 9-24). There are many variations in vector buffering. *Fixed distance buffering*, the simplest and most common form of vector buffering delineates areas a fixed distance from the input features (Figure 9-24). Fixed distance, or simple buffering separates the areas farther from any input features than the specified buffer distance from those regions that are closer than the buffer distance from an input feature. Simple buffering does not distinguish between regions that are close to one, two, three, or more features. A location is either near any feature, or farther away. Simple buffering also uses a uniform buffer distance for all features. A buffer distance of 100 meters specified for a roads layer may be applied to every road in the layer, irrespective of road size, shape, or location. In a similar manner, buffer distances for all points in a point layer will be uniform, and buffer distances for all area features will be fixed.

Buffering on vector point data is based on the creation of circles around each point feature in a data set.

The equation for a circle with an origin at $x=0$, $y=0$ is:

$$r = \sqrt{x^2 + y^2} \tag{9.1}$$

where r is the buffer distance. The more general equation of a circle with a center at x_1, y_1, is:

$$r = \sqrt{(x - x_1)^2 + (y - y_1)^2} \tag{9.2}$$

Equation (9.2) reduces to equation (9.1) at the origin, where $x_1 = 0$, and $y_1 = 0$. The general equation creates a circle centered on the coordinates x_1, y_1, with a buffer distance equal to the radius, r. Point buffers are created by applying this equation of a circle successively to each point feature in a data layer. The x and y coordinate locations of each point feature are used for x_1 and y_1, placing the point feature at the center of a circle (Figure 9-25). This generates a set of circles, one for each point feature.

Circles may overlap. In simple buffering, the circle boundaries that occur in overlap areas are removed. For example, areas within 10 kilometers of hazardous waste sites may be identified with the creation of a buffer layer. We may have a data layer in which hazardous waste sites are represented as points (Figure 9-26a). A circle with a 10 kilometer radius is drawn around each point. When two or more circles overlap, internal boundaries are dissolved, resulting in non-circular polygons (Figure 9-26b).

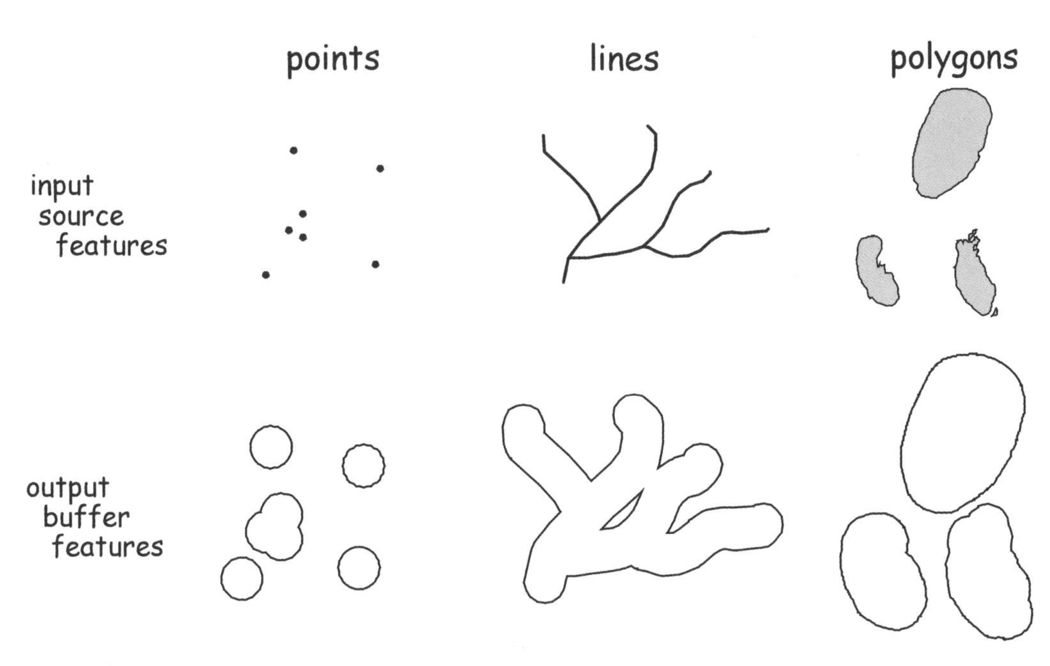

Figure 9-24: Vector buffers produced from point, line, or polygon input features. In all cases the output is a set of polygon features.

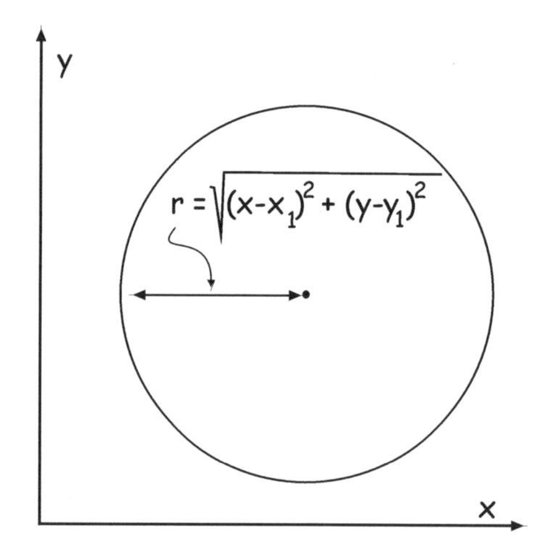

Figure 9-25: The formation of a point buffer boundary.

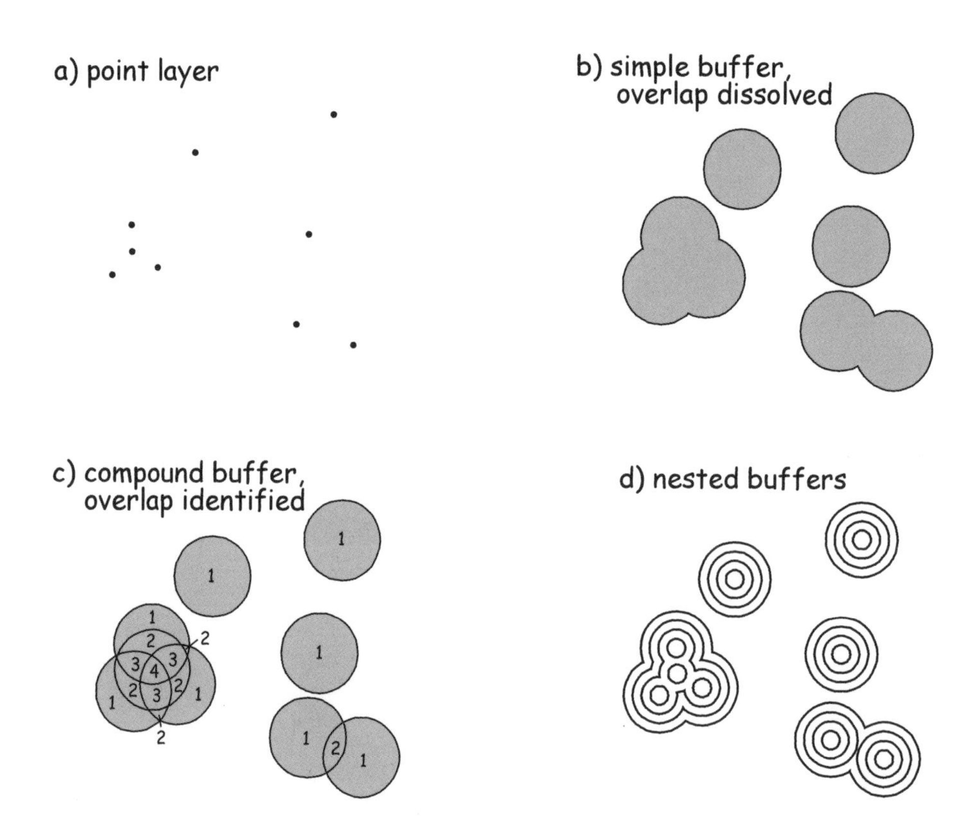

Figure 9-26: Various types of point buffers. Simple buffers dissolve areas near multiple features, more complex buffers do not. Multi-ring buffers provide distance-defined zones around each feature.

More complex buffering may be applied. These may identify buffer areas by the number of features within the given buffer distance, or apply variable buffer distances depending on the characteristics of the input features. We may be interested in areas that are near multiple hazardous waste sites. These zones may entail added risk and therefore require special monitoring or treatment. This in turn mandates the identification of all areas within the buffer distance of a hazardous waste site, and the number of sites. In most applications the majority of areas will be close to one site, but some will be close to two, three, or more sites. The simple buffer, described above, will not provide the required information. This simple buffering discards the boundaries specified by overlapping buffers.

A buffering variant, referred to here as *compound buffering*, provides the needed information. Compound buffers maintain all overlapping boundaries (Figure 9-26c). All circles defined by the fixed-radius buffer distance are generated. These circles are then intersected to form a planar graph. An attribute is created for each area that records the number of features within the specified buffer distance.

Nested (or multi-ring) buffering is another common buffering variant (Figure 9-26d). We may require buffers at multiple distances. In our hazardous waste site example, suppose threshold levels have been established with various actions required for each threshold. Areas very close to hazardous waste sites require evacuation, intermediate distance require remediation, and areas further away require monitoring. These zones may be defined by nested buffers.

Buffering on vector line data is also quite common. The formation of line buffers may be envisioned as a sequence of steps. First, circles are created that are centered at each node or vertex (Figure 9-27). Tangent lines are then generated. These lines are parallel to the input feature lines

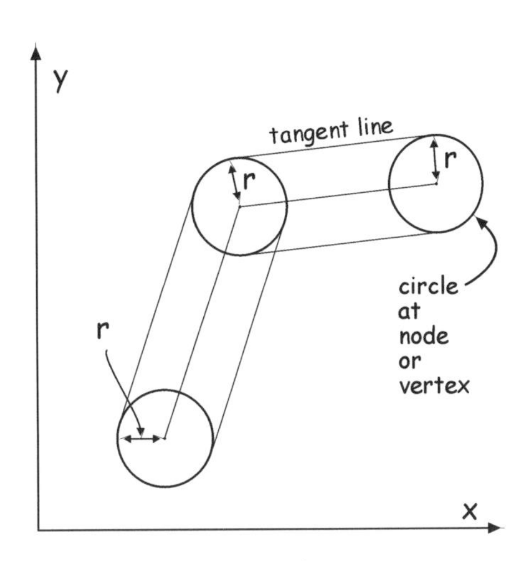

Figure 9-27: The creation of a line buffer at a fixed distance r.

and tangent to the circles that are centered at each node or vertex. The tangent lines and circles are joined and interior circle segments dissolved.

The location of the tangent lines and their intersections with the circles are based on a complicated set of rules and algebra. These operations identify the buffer segments that define the boundaries between in and out areas. These segments are saved to create a buffer for an individual line. Buffers for separate lines may overlap. With simple line buffering the internal boundaries in the overlap areas are dissolved.

Note that three different types of areas may be found when creating simple line buffers (Figure 9-28). The first type of area is within the buffer distance of the line features. An example of this area is labeled inside buffer in Figure 9-28. The second type of area is completely outside the buffer. This area is labeled outside buffer in Figure 9-28. The third type of area is labeled enclosed area in Figure 9-28. This type of area is further than the buffer distance from the input line data, but completely enclosed within a surrounding buffer polygon. These enclosed areas occur occasionally when buffering points and polygons, but enclosed areas are most frequent when buffering line features.

Area buffers are developed in a manner similar to line buffers. Each polygon is defined by a set of lines. Circles are calculated for each vertex and node, and tangent lines placed and intersected. A bounding polygon is defined by these geometric figures, and assembled for each polygon. Polygon buffering may be simple, in which case overlapping lines between separate polygon buffers are dissolved. Area buffers may also be more complex, maintaining location and number of overlaps.

Variable-distance buffers are another common variant of vector buffering. As indicated by the name, the buffer distance is variable, and may change among features. The buffer distance may increase in steps, for example, we may have one buffer distance for a given set of features, and a different buffer distance for the remaining features. In contrast, the buffer distance may vary smoothly, for example, the buffer distance around a city may be a function of the population density in the city.

There are many instances for which we may require a variable-distance buffer. Public safety requires different zones of protection that are dependant on the magnitude of the hazard. We may wish to specify a larger buffer zone around large fuel storage facilities when compared to smaller fuel storage facilities. We often require more stringent protections further away from large rivers compared to small rivers, and give large landfills a wider bearth than small landfills.

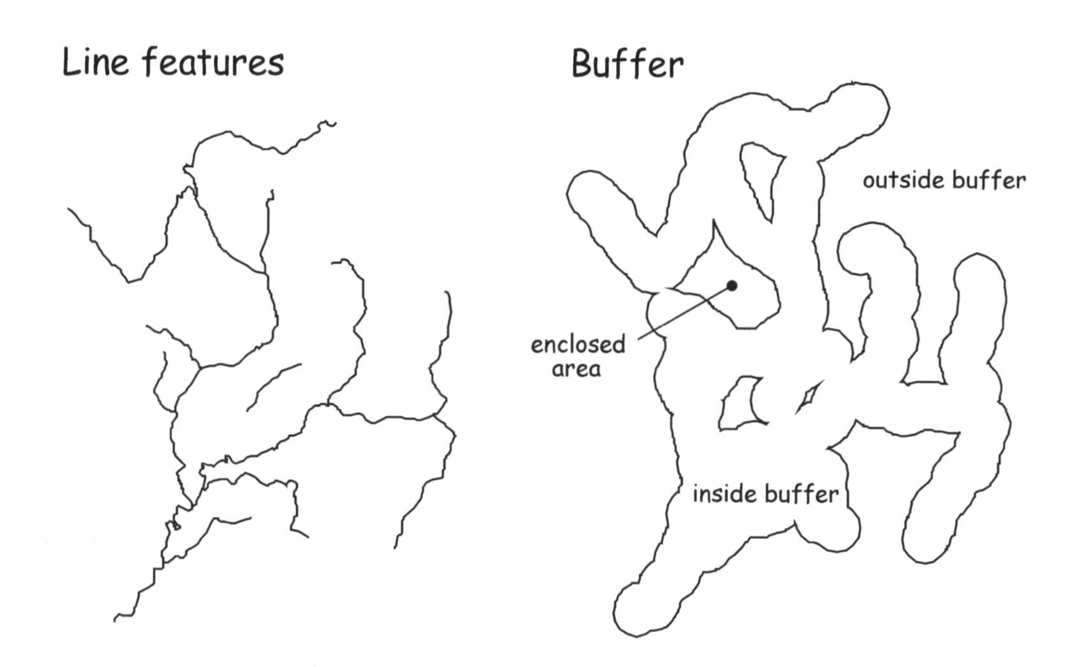

Figure 9-28: Simple buffers that are based on line features may result in three types of areas. Areas are defined inside the buffer distance. Those areas outside the buffer area may be either completely outside the buffer distance, or outside the buffer distance, but enclosed within surrounding buffer areas.

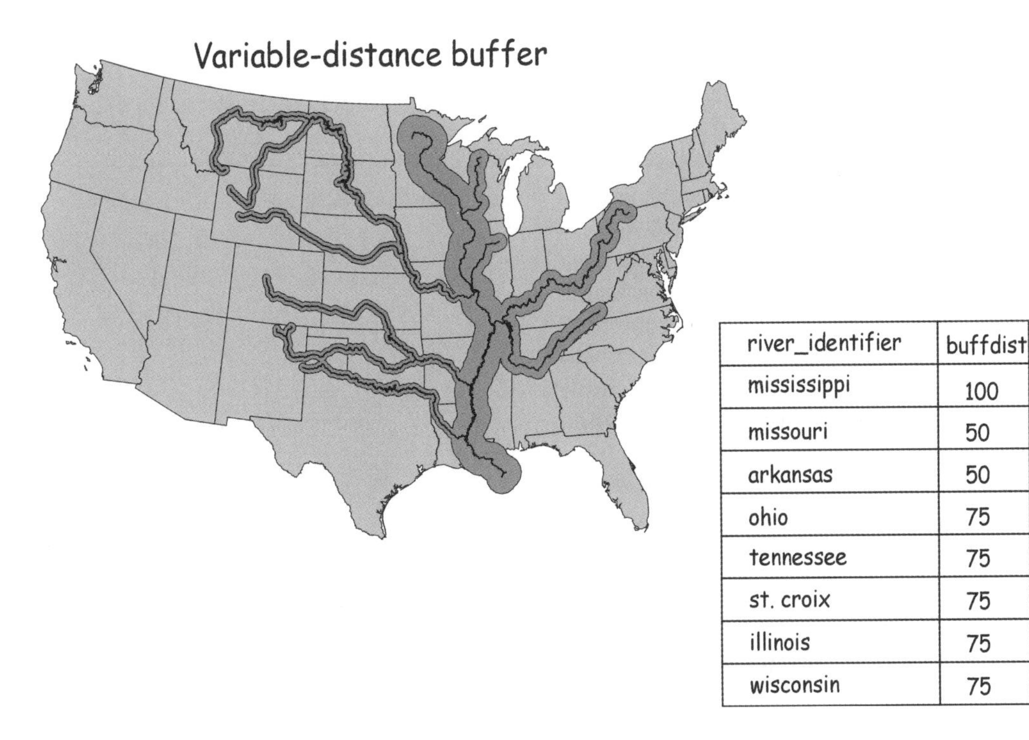

river_identifier	buffdist
mississippi	100
missouri	50
arkansas	50
ohio	75
tennessee	75
st. croix	75
illinois	75
wisconsin	75

Figure 9-29: An illustration of a variable-distance buffer. A line buffer is shown with a variable buffer distance based on a river_identifier. A variable buffer distance, buffdist, is specified in a table and applied for each river segment.

Figure 9-29 illustrates the creation of buffers around a river network. These buffers may be used to analyze or restrict land use near rivers. We may wish to increase the buffer distance for larger rivers. The increase in distance may be motivated by an increased likelihood of flooding downstream, or an increased sensitivity to pollution, or a higher chance of bank erosion as river size increases. We may specify a buffer distance of 50 kilometers for small rivers, 75 kilometers for intermediate size rivers, and 100 kilometers for large rivers. There are many other instances when variable distance buffers are required, e.g., larger distances from noisier roads, smaller areas where travel is difficult, or bigger buffers around larger landfills.

The variable buffer distance is often specified by an attribute in the input data layer. This is illustrated in Figure 9-29. A portion of the attribute table for the river data layer is shown. The attribute table contains the river name in river_identifier and the buffer distance stored in buffdist. The attribute buffdist is accessed during buffer creation, and the size of the buffer adjusted automatically for each line segment. Note how the buffer size depends on the value in buffdist.

Overlay

Overlay operations are powerful spatial analysis tools, and were an important driving force behind the development of GIS technologies. Overlays involve combining spatial and attribute data from two or more spatial data layers, and are among the most common and powerful spatial data operations (Figure 9-30). Many problems require the overlay of thematically different data. For example, we may wish to know where there are inexpensive houses in good school districts, where whale feeding grounds overlap with proposed oil drilling areas, or the location of farm fields that are on highly erodible soils. In the later example a soils data layer may be used to identify highly erodible soils, and a current land use layer used to identify the locations of farm fields. The boundaries of erodible soils will not coincide with the boundaries of the farm fields in most instances, so these soils and land use data must somehow be combined. Overlay is the primary means of providing this combination.

An overlay operation requires that data layers use a common coordinate system. Overlay uses the coordinates that define each spatial feature to combine the data from the input data layers. The coordinates for any point on Earth depend on the coordinate system used (Chapter 3). If the coordinate systems used in the various layers are not exactly the same, the features in the data layers will not align correctly.

Overlay may be viewed as the vertical stacking and merger of spatial data (Figure 9-30). Features in each data layer are set one "on top" another, and the points, lines, and area feature boundaries merged into a

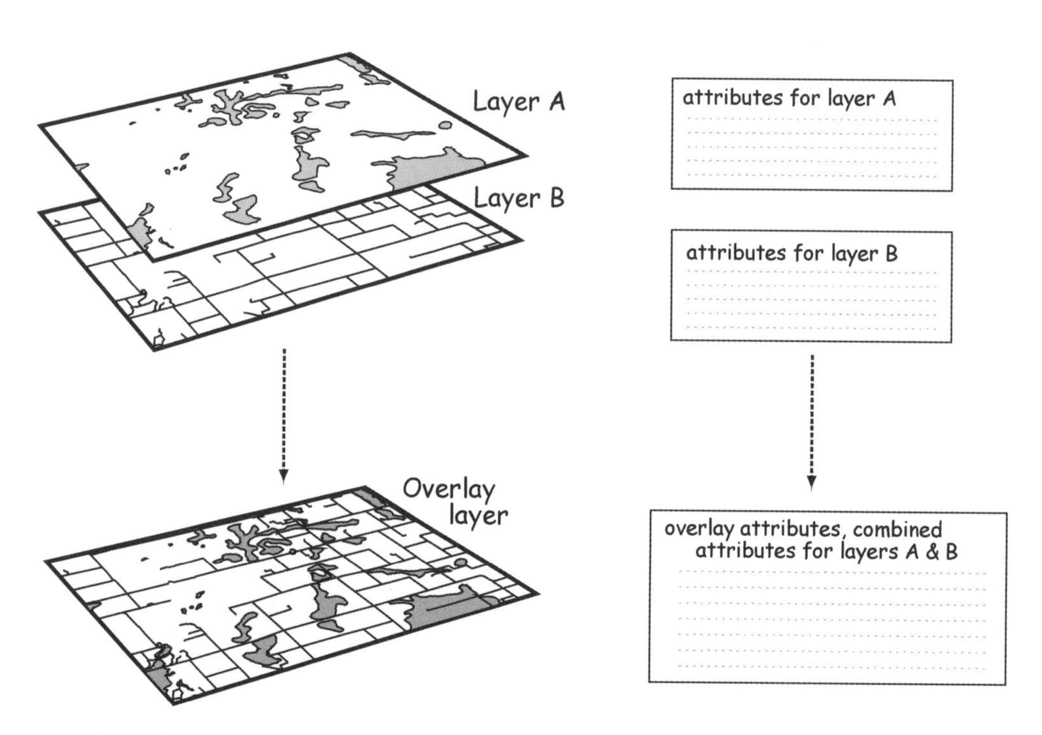

Figure 9-30: Spatial data overlay. Overlay combines both the coordinate information and the attribute information across different data layers.

single data layer. The attribute data are also combined, so that the new data layer includes information contained in each input data layer.

Raster Overlay

Raster overlay involves the cell-by-cell combination of two or more data layers. Data from one layer in one cell location correspond to a cell in another data layer. The cell values are combined in some manner and an output value assigned to a corresponding cell in an output layer.

Raster overlay is typically applied to nominal or ordinal data. A number or character stored in each raster cell represents a nominal or ordinal category. Each cell value corresponds to a category for a raster variable. This is illustrated in the input data sets shown at the left and center of Figure 9-31. Input Layer A represents soils data. Each raster cell value corresponds to a specific soil value. In a similar manner input Layer B records land use, with values 1, 2, and 3 corresponding to particular land uses. These data may be combined to create areas with combinations of the two input layers, cells with values for both soil type and land use.

There are as many potential categories as there are possible combinations of input layer values. In Figure 9-31 there are two soil types in Layer A, and three land use types in Layer B. There are potentially six different combinations in the output layer. Not all combinations will necessarily occur in the overlay, as shown in Figure 9-31. In this example only four of the six overlay combinations occur. Unique identifiers must be generated for each observed combination, and placed in the appropriate cell of the output raster layer.

The number of possible combinations is important to note because it may change the number of binary digits or bytes required to represent the output raster data layer. Raster cells typically hold a number or character, and may be one-byte integer, two-byte integer, or some other size. Raster data sets typically use the smallest required data size. As discussed in Chapter 2, one unsigned byte may store up to 256 different

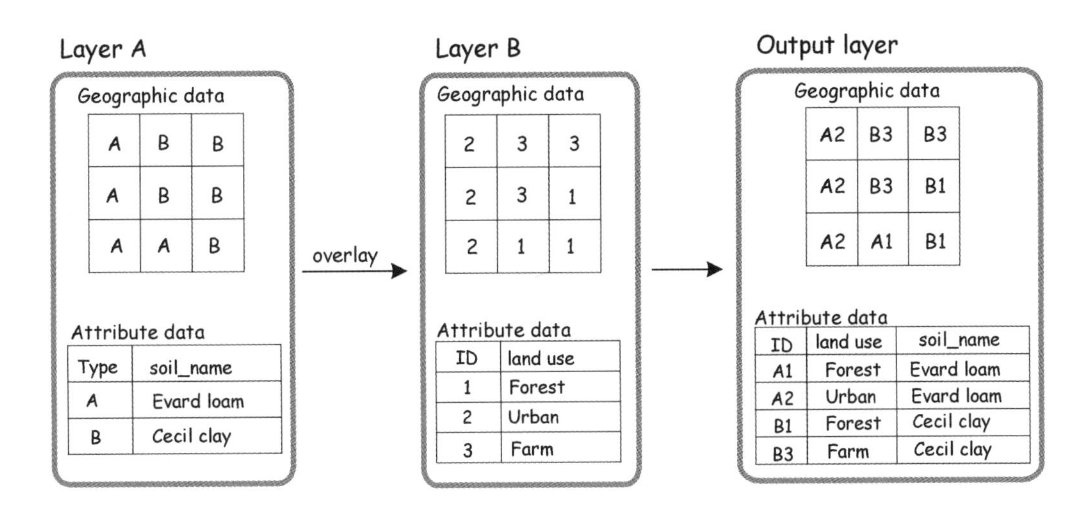

Figure 9-31: Cell-by-cell combination in raster overlay. Two input layers are combined in raster overlay. Nominal variables for corresponding cells are joined, creating a new output layer. In this example a soils layer (Layer A) is combined with a land use layer (Layer B) to create a composite Output layer.

values. Raster overlay may result in an output data layer that requires a higher number of bytes per cell. Consider the overlay between two raster data layers, one layer which contains 20 different nominal classes, and a second layer with 27 different nominal classes. There is a total of 20 times 27, or 540 possible output combinations. If more than 256 combinations occur the output data will require more than one byte for each cell. Typically two bytes will be used. This causes a doubling in the output file size. Two bytes will hold more than 65,500 unique combinations; if more categories are required then four bytes per cell are often used.

Raster overlay requires the input raster systems be compatible. This typically means they should have the same cell dimension and coordinate system, including the same origin for x and y coordinates. If the cell sizes differ, there will likely be cells in one layer that match parts of several cells in the second input layer (Figure 9-32). This may result in ambiguity when defining the input attribute value to use.

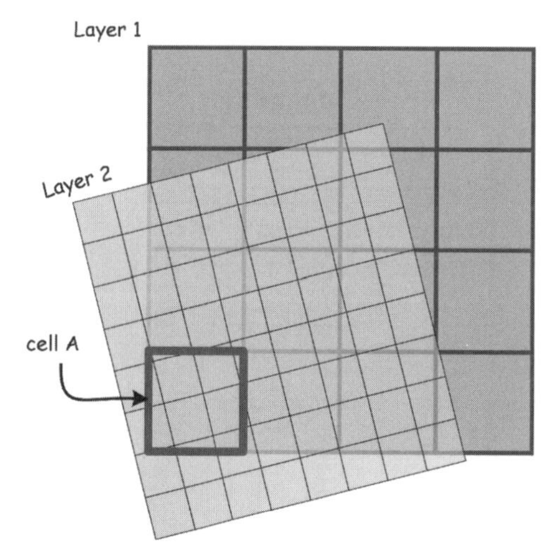

Figure 9-32: Overlain raster layers should be compatible to ensure unambiguous overlay. Cell orientation should be coincident and cell size should be compatible. In the overlay depicted here it is not clear which cells from Layer 2 should be combined with cell A in Layer 1.

Overlay may work if the cells are integer multiples with the same origin, e.g., the boundaries of a 1 by 1 meter raster layer may be set to coincide with a 3 by 3 raster layer, however this rarely happens. Data are normally converted to compatible raster layers before overlay. This is most often done using a coordinate transformation and/or resampling, as described in Chapter 4. In our example, we might choose to resample Layer 2 to match Layer 1 in cell size and orientation. Values for cells in Layer 2 would be combined through a nearest neighbor, bilinear interpolation, cubic convolultion, or some other resampling formula to create a new layer based on Layer 2 but compatible with Layer 1.

Vector Overlay

Overlay when using a vector data model involves combining the point, line, and polygon geometry and associated attribute data. This overlay creates new geometry. Overlay involves the merger of both the coordinate and attribute data from two vector layers into a new data layer. The coordinate merger may require the intersection and splitting of lines or areas and the creation of new features.

Figure 9-33 illustrates the overlay of two vector polygon data layers. This overlay requires the intersection of polygon boundaries to create new polygons. The overlay also entails the combination of attribute data during polygon overlay. The data layer on the left is comprised of two polygons. There are only two attributes for Layer 1, one an identifier (ID), and the other specifying values for a variable named class. The second input data layer, Layer 2, also contains two polygons, and two attributes, ID and cost. Note that the two tables have an attribute with the same name, ID. These two ID attributes serve the same function in their respective data layers, but they are not related. A value of 1 for the ID attribute in Layer 1 has nothing to do with the ID value of 1 in Layer 2.

Vector overlay of these two polygon data layers results in four new polygons. Each new polygon contains the attribute information from the corresponding area in the input data layers. For example, note that the polygon in the output data layer with the ID of 1 has a class attribute with a value of 0 and a cost attribute with a value of 10. These values come from the values found in the corresponding input layers. The boundary for the polygon with an ID value of 1 in the output data layer is a composite of the boundaries found in the two input data layers. The same holds true for the other three polygons in the output data layer. These polygons are a composite of geographic and attribute data in the input data layers.

The topology of vector overlay output will likely be different from that of the input data layers. Vector overlay functions typically identify line intersection points during overlay. Intersecting lines are split and a node placed at the intersection point. Thus topology must be re-created if it is needed in further processing.

Any type of vector feature may be overlain with any other type of vector feature, although some overlay operations rarely provide useful information and are performed infrequently. Point-in-polygon overlays are quite common, point in line much less so, and point-on-point nonsensical in most cases. In theory points may be overlain on point, line, or polygon feature layers, lines on all three types, and polygons on all three types. Point-on-point or point-on-line overlay rarely results in intersecting features, and so are rarely applied. Line-on-line overlay are sometimes required, for example when we wish to identify the intersections of two networks such as road and railroads. Overlays involving polygons are the most common by far.

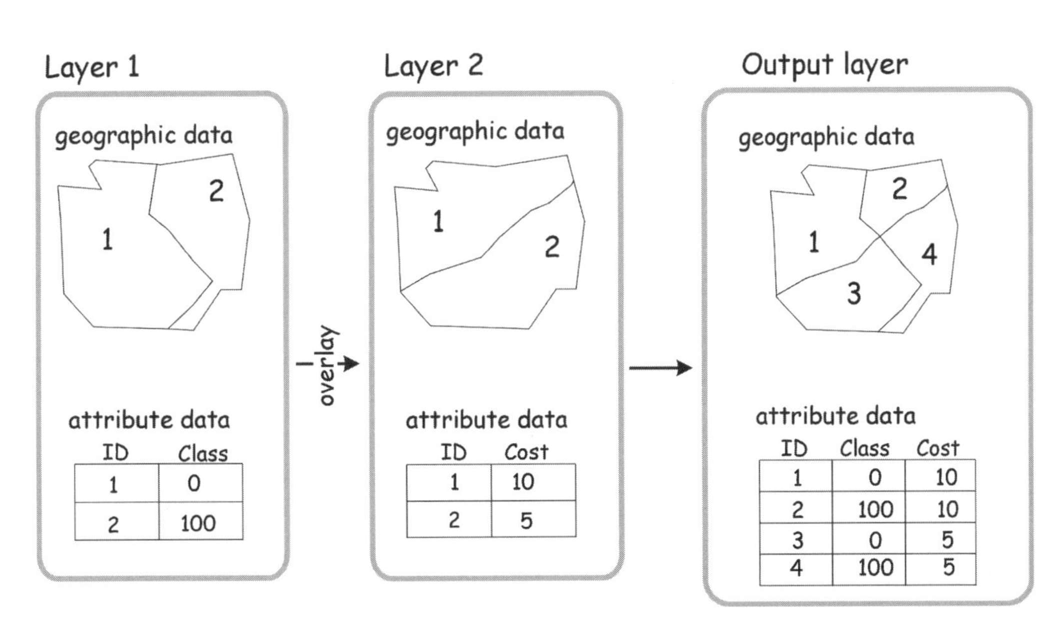

Figure 9-33: An example of vector polygon overlay. In this example output data contain a combination of the geographic (coordinate) data and the attribute data of the input data layers. New features may be created with topological relationships distinct from those found in the input data layers.

Overlay output typically takes the lowest dimension of the inputs. This means point-in-polygon overlay results in point output, and line-in-polygon overlay results in line output. This avoids problems when multiple lower-dimension features intersect with higher-dimension features.

Figure 9-34 illustrates an instance where multiple points in one layer fall within a single polygon in an overlay layer. Output attribute data for a feature are a combination of the input data attributes. If polygons are output (Figure 9-34, right, top) there is ambiguity regarding which point attribute data to record. Each point feature has a value for an attribute named class. It is not clear which value should be recorded in the output polygon, the class value from point A, point B, or point C. When a point layer is output (Figure 9-34, right, bottom), there is no ambiguity. Each output point feature contains the original point attribute information, plus the input polygon feature attributes.

One method for creating polygon output from point-in-polygon overlay involves recording the attributes for one point selected arbitrarily from the points that fall within a polygon. This is usually not satisfactory because important information may

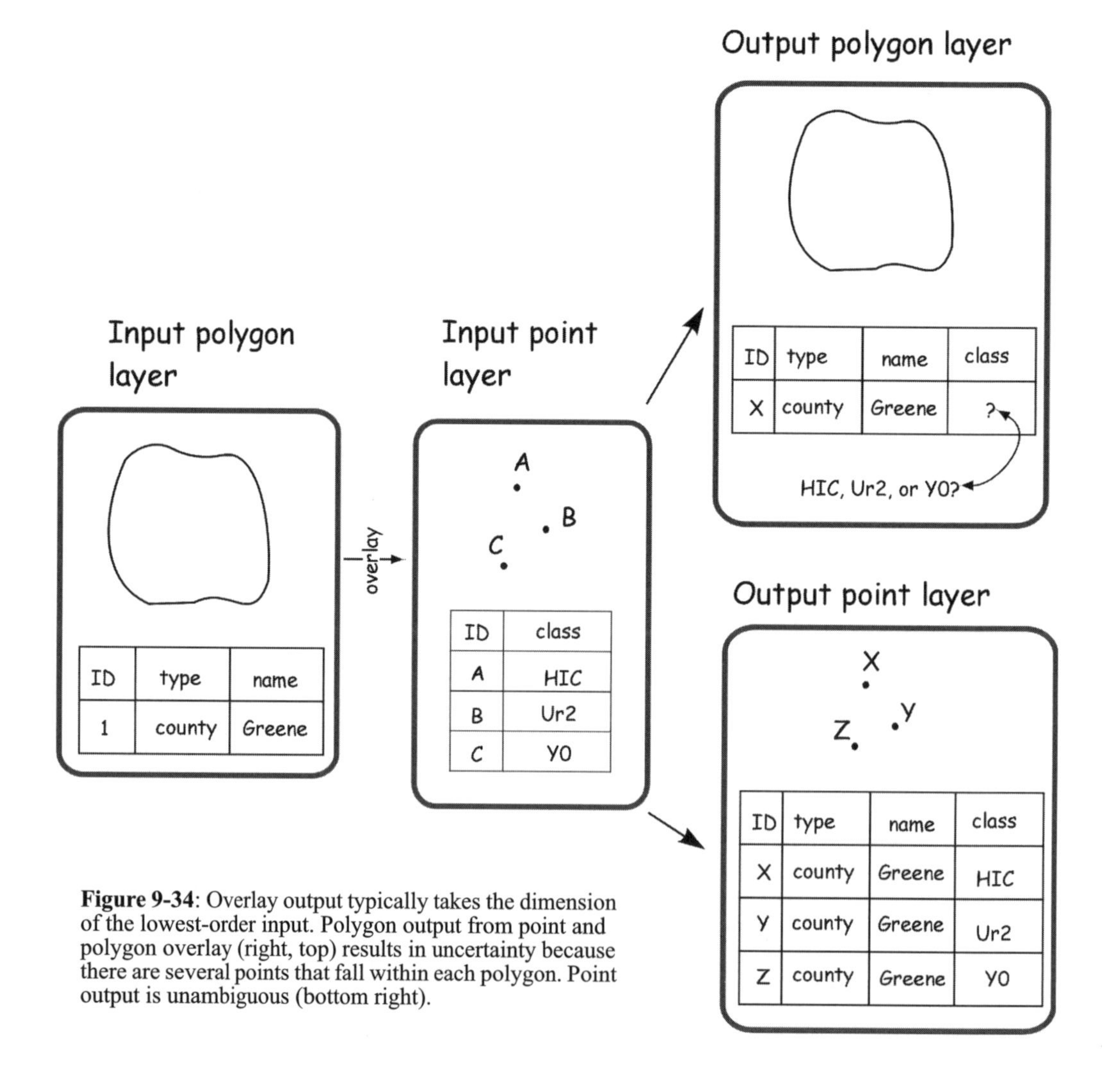

Figure 9-34: Overlay output typically takes the dimension of the lowest-order input. Polygon output from point and polygon overlay (right, top) results in uncertainty because there are several points that fall within each polygon. Point output is unambiguous (bottom right).

be lost. An alternative solution involves adding columns to the output polygon to preserve multiple points per polygon. However this would still result in some ambiguity, e.g., what should be the order of duplicate attributes? It may also add a substantial number of sparsely used items, thus increasing file size inefficiently. Forcing the lower order output during overlay avoids these problems, as shown in the lower right of Figure 9-34.

Note that the number of attributes in the output layer increases after each overlay. This is illustrated in Figure 9-34, with the combination of a point and polygon layer in an overlay. The output point attribute table shown in the lower right portion of the figure contains four items. This output attribute table is a composite of the input attribute tables.

Large attribute tables may result if overlay operations are used to combine many data layers. When the output from an overlay process is in turn used as an input for a subsequent overlay, the number of attributes in the next output layer will usually increase. In rare instances the number of attributes in an output layer will be the same as the larger of the two input layers. As the number of attributes grows, tables may become unwieldy, and there may be a need to delete redundant or superfluous attributes. Processes that require the overlay of many layers often include the removal of unneeded attributes.

Figure 9-35 illustrates the most common types of vector overlay. Each row in the figure represents an overlay operation with point, line or polygon input. The input data layers are arranged in the left and middle columns and the output data layer in the right-hand column. Rather than show the complete attribute tables, labels are used to represent the combination of attributes. The bottom row illustrates this for polygon overlay. The two polygons in the first input data layer are labeled 100 and 200. The two polygons in the second input layer (middle panel) are labeled R and S. The resultant composite polygons in the output layer are shown on the left with labels 100R, 100S, 200R, and 200S to represent the combination of attributes from both input data layers.

Common characteristics of various types of overlay are apparent in Figure 9-35. Point-on-line overlay is shown in the top row. This results in joining line data for only one point, 2, which in the overlay layer is labeled 2B. The other points do not intersect a line, as indicated by the minus in the labels denoting attributes, e.g., 3-. This is common with point-in-line overlay. Points have no dimension, and lines have zero width. The likelihood of points and lines intersecting are quite low for most data sets. Only in special circumstances such as when point data are restricted to fall on a linear network (e.g., points representing accident locations) are there likely to be frequent point and line intersections.

Point features result from point-in-polygon overlay, as shown in the second row of Figure 9-35. Points take the attributes of the coincident polygon. Point location is not changed nor are any geographic data from the polygon features typically incorporated into output point features.

Vector line-on-line overlay results in a line data layer (middle row, Figure 9-35). A planar graph is produced, meaning nodes (open circles) are placed at each line intersection. Each respective line segment maintains the attributes from the source data layer. Node attribute tables, if they exist, may contain information that originated from lines in each of the input data layers.

Line-on-polygon overlay is shown in the fourth row of Figure 9-35. This type of overlay typically produces a vector line output layer. Each line in the output data layer contains attributes from both the original input line data layer and the coincident polygon attribute layer. Line segments are split where they cross a polygon boundary, e.g., the line segment labeled 10 in Input layer 1 is split into two segments in the

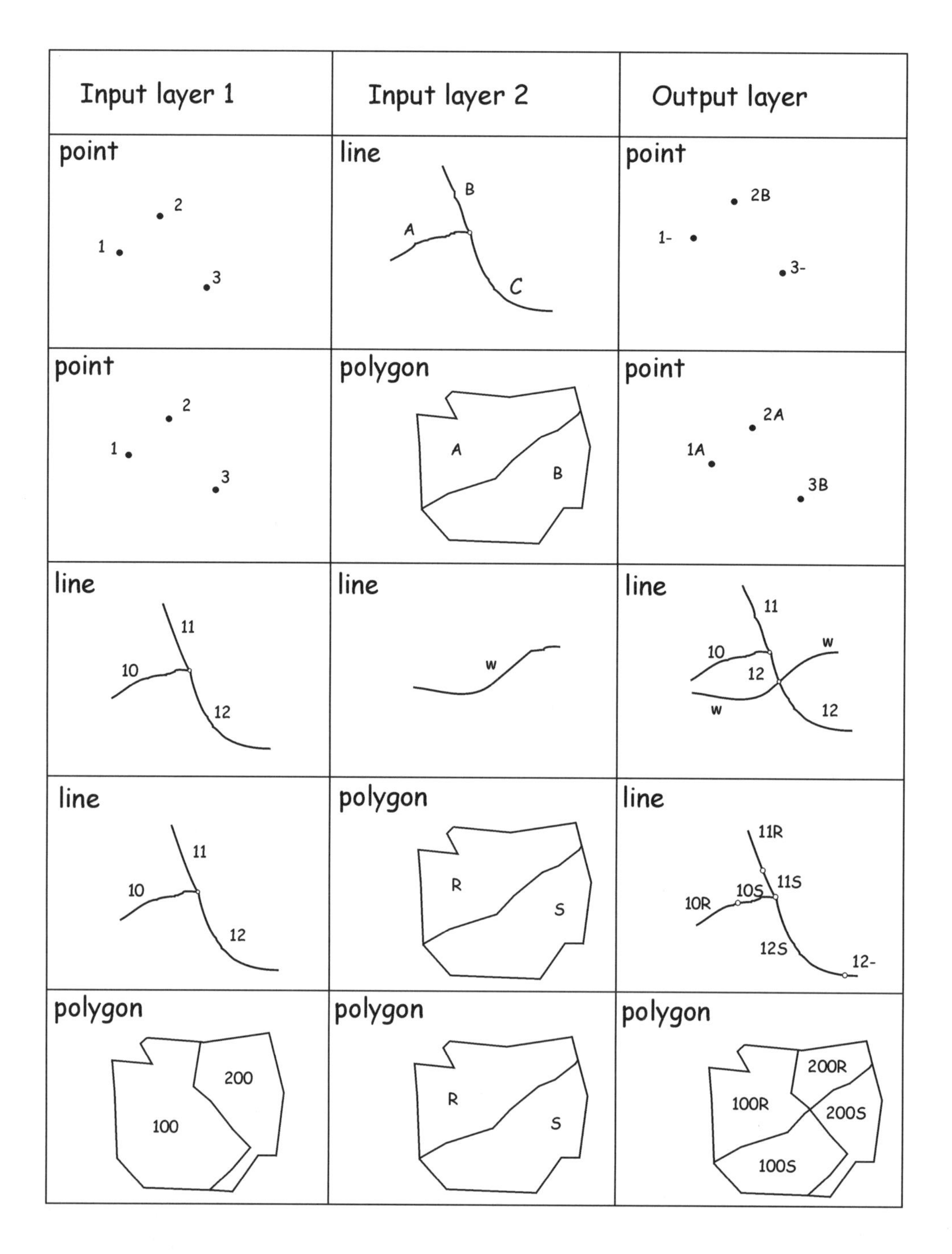

Figure 9-35: Examples of vector overlay. In this example, point, line, or polygon layers are combined in an inclusive overlay. The combination results in an output layer that contains the attribute and geographic data from input layers. The output data are typically the minimum order of the input, e.g., if point data are used as input, the output will be point. If line and line or line and polygon data are used, the output will be a line data layer.

Output layer. Each segment of this line exhibits a different set of attributes: 10R and 10S. Note that not all line segments may contain a complete set of attributes derived from polygons. Line segments falling outside all polygons contain a null value for polygon attributes, such as the segment at the lower right of line-on-line output panel. This segment is labeled 12-. The minus (-) denotes that polygon attributes are not recorded. These "outside" line segments typically contain null or flag values in the attribute table for the polygon items.

A polygon-on-polygon overlay is shown in the bottom row of Figure 9-35. The combination of polygon features has been discussed previously, and will not be described further here.

Overlays that include a polygon layer are most common. We are often interested in combination of polygon features with other polygons, or in finding the coincidence of point or line features with polygons. What counties include hazardous waste sites? Which neighborhoods does one pass through on E Street? Where are there shallow aquifers below cornfields? All these examples involve the overlay of area features, either with other area features, or with point or line features.

Clip, Intersect, and Union: Special Cases of Overlay

There are three common ways overlay operations are applied: as a *clip*, an *intersection*, or a *union* (Figure 9-36). The basic layer-on-layer combination is the same for

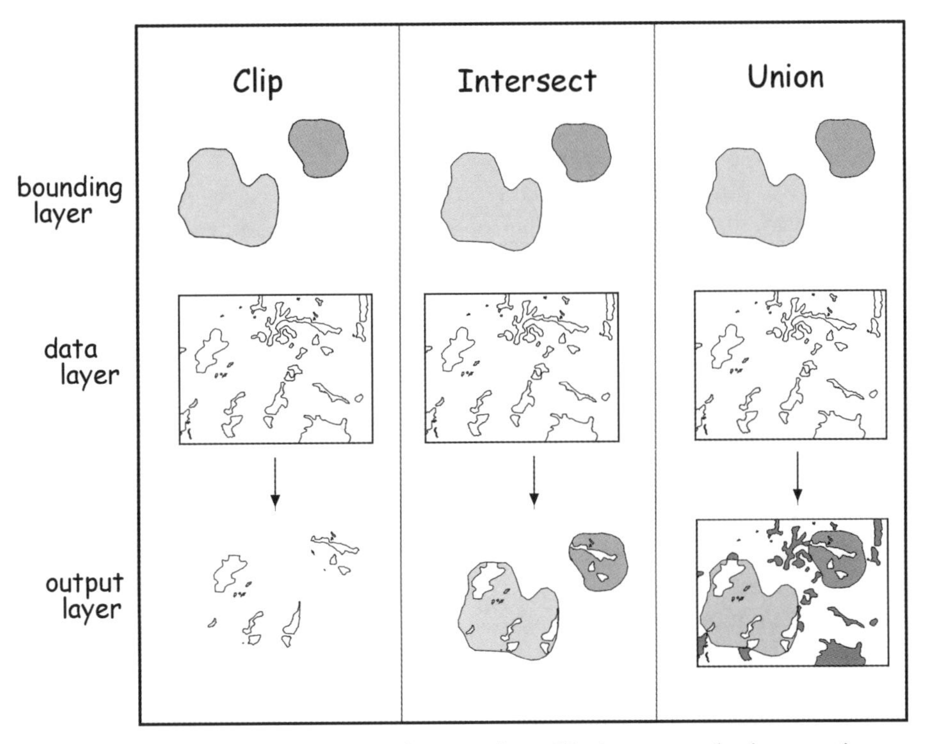

Figure 9-36: Common variations of overlay operations. Clip, intersect, and union operations are used to combine data from separate layers. The clip preserves information only from the data layer and only for the area of the bounding layer. An intersect is restricted to the area of the bounding layer, but combines data from both input layers. The union combines data from both input layers over the combined extent of both layers.

all three. They differ in the geographic extent for which vector data are recorded, and in how data from the attribute layers are combined. Intersection and union are derived from general set theory operations. The intersection operation may be considered in some ways to be a spatial AND, while the union operation is related to a spatial OR. The clip operation may be considered a combination of an intersection and an elimination. All three are common and supported in some manner as stand-alone functions by most GIS software packages.

A *clip* may be considered a "cookie-cutter" overlay. A bounding polygon layer is used to define the areas for which features will be output. This bounding polygon layer defines the clipping region. Point, line, or polygon data in a second layer are "clipped" with the bounding layer. In most versions of the clip function the attributes for the clipping layer are not included in the output data layer. Only features in the second input data layer are contained in the output data layer.

An example of a clip is shown on the left side of Figure 9-36. The bounding data layer consists of two polygons, and the data layer contains many small wetland boundaries. The presence of polygon attributes in the bounding layer is indicated by the different shades for the two polygons in that layer. The output from the clip consists of those portions of wetlands in the area contained by the bounding layer polygons. Note that the polygon boundaries defining the bounding layer are not included in the output data layer.

An *intersection* may be defined as an overlay that combines data from both layers but only for the region where both layers contain data. This is illustrated in the central panel of Figure 9-36. Data from the polygons of the bounding layer and data from the data layer are combined at the bottom of the central panel. Note that all or parts of polygons in the data layer that are outside the bounding layer are clipped and discarded. A spatial intersection operation may differ from a clip in that data from the bounding layer are also included in the output layer. Each polygon in the output layer may include attributes defined in either the bounding layer or the data layer.

A *union* is an overlay that includes all data from both the bounding and data layers. A union for our example is shown on the right of Figure 9-36. No geographic data are discarded in the union operation, and corresponding attribute data are saved for all regions. New polygons are formed by the combination of coordinate data from each data layer.

Many software packages support additional variants of overlay operations. Some support a complement to the clip function, in which the area covered by the input layer are "cut out" or erased from the bounding layer. Other software packages support other variants on unions or intersections. Most of these specialized overlay operations may be created from the application of union or overlay operations in combination with selection operations.

A Problem in Vector Overlay

Polygon overlays often suffer when there are common features that are represented in both input data layers. We define a common feature as a different representation of the same phenomenon. Figure 9-37 illustrates this problem. A parcel boundary may coincide with a change in a vegetation type. However the parcel boundary and vegetation data layers were digitized from different source materials, at different times, and using different systems. Thus, these two representations may differ even though they identify the same boundary on the Earth surface.

In most data layers the differences will be quite small, and will not be visible except at very large display scales, i.e., when the on-screen zoom is quite high. The differences have been exaggerated in Figure 9-37. When the vegetation and parcels data layers are overlain, many small polygons are formed along the boundary. These polygons are quite small, but they are often quite numerous.

These "sliver" polygons cause problems because there is an entry in the attribute table for each polygon. One-half or more of the polygons in the output data layer may be these slivers. Slivers are a burden because they take up space in the attribute table but are not of any interest or use. Analyses of large data sets are hindered because all selections, sorts, or other operations must treat all polygons.

There are several methods to reduce the occurrence of these slivers. One method involves identifying all common boundaries across different layers. The boundary with the highest coordinate accuracy is substituted into all other data layers, replacing the less accurate representations. Replacement involves considerable editing, and so is most often used as a strategy when developing new data layers.

Another method involves manually identifying and removing slivers. Small polygons may be selected, or polygons with two bounding arcs, as commonly occurs with sliver polygons. Bounding

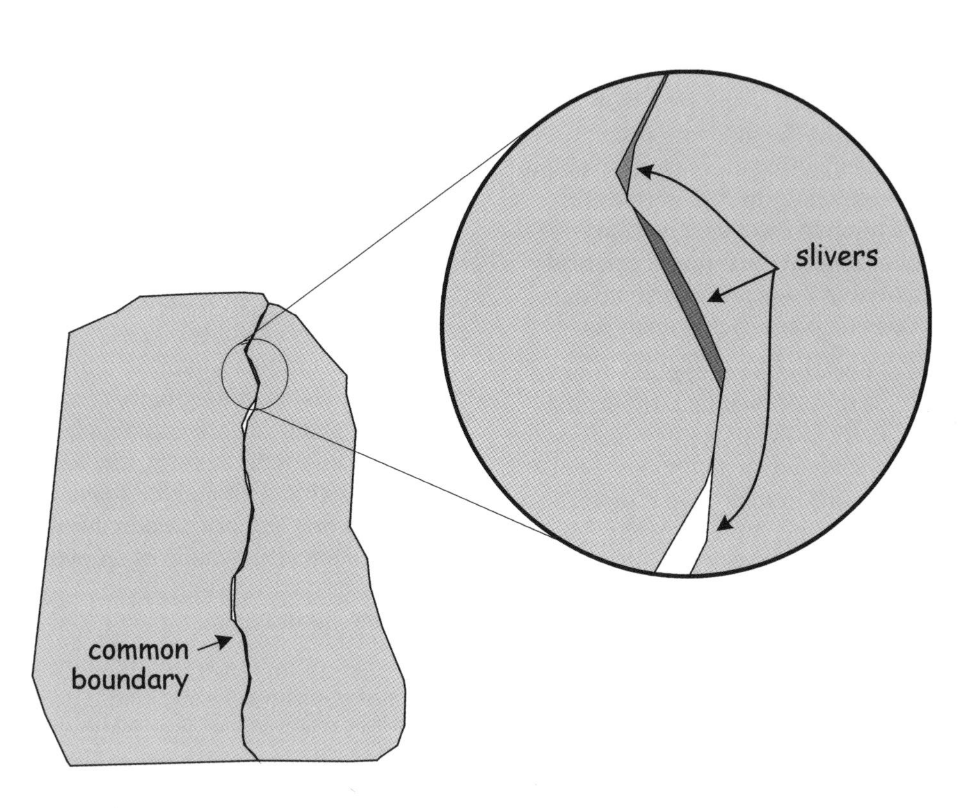

Figure 9-37: Sliver polygons may occur when two representations of a feature are combined. A common boundary between two features has been derived from different sources. The representations differ slightly. This results in small gaps and "sliver" polygons along the margin between these two layers. While slivers may be removed during initial data development, they often occur as a result of overlay operations. Some form of automated or manual sliver removal is often required after multi-layer overlay.

lines may then be adjusted or removed. Manual removal is not practical for many large data sets due to the high number of sliver polygons.

A third method for sliver reduction involves defining a snap distance during overlay. Much as with a snap distance during data development (described in chapter 4), this forces nodes or lines to be coincident if they are within the specified snap distance during overlay. As with data entry, this snap distance should be small relative to the spatial accuracy of the input layers and the required accuracy of the output data layers. If the two representations of a line are within the snap distance then there will be no sliver polygons. In practice, not all sliver polygons are removed, but their numbers are substantially reduced, thereby reducing the time spent on manual editing.

Summary

Spatial analysis, along with map production, is one of the most important uses of GIS. They are often the reason we obtain GIS and invest the substantial time and money required to develop a working system. Any analytical operation we perform on our spatial or associated attribute data may be considered as spatial analysis.

Spatial operations are applied to input data and generate output data. Inputs may be one to many layers of spatial data, as well as non-spatial data. Outputs may also number from one to many. Operations also have a spatial scope, the area of the input data that contributes to output values. Scopes are commonly local, neighborhood, or global.

Selection and classification are among the most often-used spatial data operations. A selection identifies a subset of the features in a spatial database. The selection may be based on attribute data, spatial data, or some combination of the two. Selection may apply set or Boolean algebra, and may combine these with analyses of adjacency, connectivity, or containment. A selected set may be classified in that variables may be changed or new variables added that reflect membership in the selected set.

Classifications may be assigned automatically, but the user should be careful in choosing the assignment. Equal-area, equal-interval, and natural-breaks classifications are often used. The resulting classifications may depend substantially on the frequency histogram of the input data layer, particularly when outliers are present.

A dissolve operation is often used in spatial analysis. Dissolves are routinely applied after a classification, as they remove redundant boundaries that may slow processing.

Proximity functions and buffers are also commonly applied spatial data operations. These functions answer questions regarding distance and separation among features in the same or different data layers. Buffering may be applied to raster or vector data, and may be simple, with a uniform buffer distance, or complex, with multiple nested buffers or variable buffer distances.

Overlay is a final spatial operation described in this chapter. Overlay involves the vertical combination of data from two or more layers. Both geometry (coordinates) and attributes are combined. Any combination of points, lines, and area features is possible, although overlays involving at least one layer of area features are most common. The results of an overlay usually take the lowest geometric dimension of the input layers.

Overlay sometimes results in the generation of gaps and slivers. These occur most often when a common feature occurs in two or more layers. These gaps and slivers may be removed via several techniques.

Suggested Reading

Aronoff, S., Geographic Information Systems, A Management Perspective, WDL Publications, Ottawa, 1989.

Batty, M and Xie, Y., Model structures, exploratory spatial data analysis, and aggregation, *International Journal of Geographical Information Systems*, 1994, 8:291-307.

Bonham-Carter, G. F., Geographic Information Systems for Geoscientists: Modelling with GIS, Pergamon, Ontario, 1996.

Carver, S. J., Integrating multi-criteria evaluation with geographical information systems, *International Journal of Geographical Information Systems*, 1991, 5:321-340.

Chou, Y. H., Exploring Spatial Analysis in Geographic Information Systems, Onword Press, Albuquerque, 1997.

Cliff, A. D and Ord, J. K., Spatial Processes: Models and Applications, Pion, London, 1981.

DeMers, M., Fundamentals of Geographic Information Systems, 2nd Edition. Wiley, New York, 2000.

Heuvelink, G. B. M. and Burrough, P. A, Error propagation in cartographic modelling using Boolean logic and continuous classification, *International Journal of Geographical Information Systems*, 1993, 7:231-246.

Laurini, R. and Thompson, D., Fundamentals of Spatial Information Systems, Academic Press, London, 1992.

Malczewski, J., GIS and Multicriteria Decision Analysis, Wiley, New York, 1999.

Martin, D., Geographical Information Systems and their Socio-economic Applications, 2nd Edition. Routledge, London, 1996.

Monmonier, M., How To Lie With Maps, Chicago Press, Chicago, 1993.

Steinitz, C. P. and Jordan, L., Hand-drawn overlays: their history and prospective uses, *Landscape Architecture*, 1976, 56:146-157.

Worboys, M. F., GIS: A Computing Perspective, Taylor and Francis, London, 1995.

Study Questions

Can you define and give examples of local, neighborhood, and global spatial operations?

What are selection and classification operations?

Can you describe set and Boolean algebra?

Can you describe three different classification methods?

What is a dissolve operation? What are they typically used for?

What is the basic principle behind buffering? What are some of the buffering variants that are applied?

How are raster proximity functions different from vector proximity functions?

How are point, line, and area buffers constructed?

What is the basic concept behind layer overlay?

Can you diagram and contrast raster and vector overlay?

Why are output features in vector overlay typically set to the minimum dimensional order (point, line or polygon) of the input features?

How are clip, intersection, and union overlay different?

What is the sliver problem in vector layer overlay? How might this problem be resolved?

10 Topics in Raster Analysis

Introduction

Raster analyses range from the simple to the complex, largely due to the early invention, simplicity, and flexibility of the raster data model. Raster structures are based on two-dimension arrays, constructs supported by many of the earliest programming languages. The raster row and column format is among the easiest to specify in computer code, and the structure is easily understood, thereby encouraging modification, experimentation, and the creation of new raster operations. Raster cells may store nominal, ordinal, or interval/ratio data, so a wide range of variables may be represented. Complex constructs may be built from raster data, including connected cells to form networks, or groups of cells to form areas.

The flexibility of raster analyses has been amply demonstrated by the wide range of problems they have helped solve. Raster analyses are routinely used to predict the fate of pollutants in the atmosphere, the spread of disease, animal migration, and crop yields. Time varying and wide-area phenomena are routinely analyzed using raster data, particularly when remotely-sensed inputs are available. Raster analyses are routinely used at the small parcel level, for example by the U.S. Environmental Protection Agency in hazard analysis of urban superfund sites, to global-scale estimates of forest growth. Local, state, and regional organizations have used raster analyses at many scales in between.

Numerous research projects have expanded and embellished the basic raster data structure, as well as investigated a general set of raster tools for spatial data analyses. Yale University, the Ohio State University, the Idrisi Project of Clark University, and the Harvard School of Design are among public institutions that developed raster analysis packages for research over the past four decades. Commercial products have been developed by a number of companies.

The long history and level of interest in raster analyses have resulted in a basic set of tools that should be understood by every GIS user. Many of the tools are based on a common conceptual basis, and form a set of generic methods that may be adapted to several types of problems. In addition, specialized methods have been developed for less frequently encountered problems. The GIS user may more effectively apply raster data analysis if she has developed an understanding of the underlying concepts and has become acquainted with a range of specialized raster analysis methods.

Map Algebra

Dana Tomlin and Joseph Berry introduced and developed the concept of *map algebra*. Map algebra is the cell-by-cell combination of raster data layers. The combination entails the application of a set of local and neighborhood functions, and to a lesser extent global functions, to raster data.

Tomlin and Berry's concept of map algebra is based on the simple yet flexible and useful data structure of numbers stored in a raster grid. Each number represents a value at a raster cell location. Simple operations may be applied to each number in a raster. Further, raster layers may be combined through operations such as layer addition, subtraction, and multiplication. Map algebra may be used to solve many spatial problems.

Map algebra is based on operations applied to one or more raster data layers. *Unary* operations apply to one data layer. *Binary* operations apply to two data layers, and higher-order operations may involve many data layers.

A simple unary operation applies a function to each cell in an input raster layer, and outputs a calculated value to the corresponding cell in an output raster. Figure 10-1a illustrates the multiplication of a raster by a scalar. In this example the raster is multiplied by two. This might be denoted by the equation:

$$\text{Outlayer} = \text{Inlayer} * 2 \qquad (10.1)$$

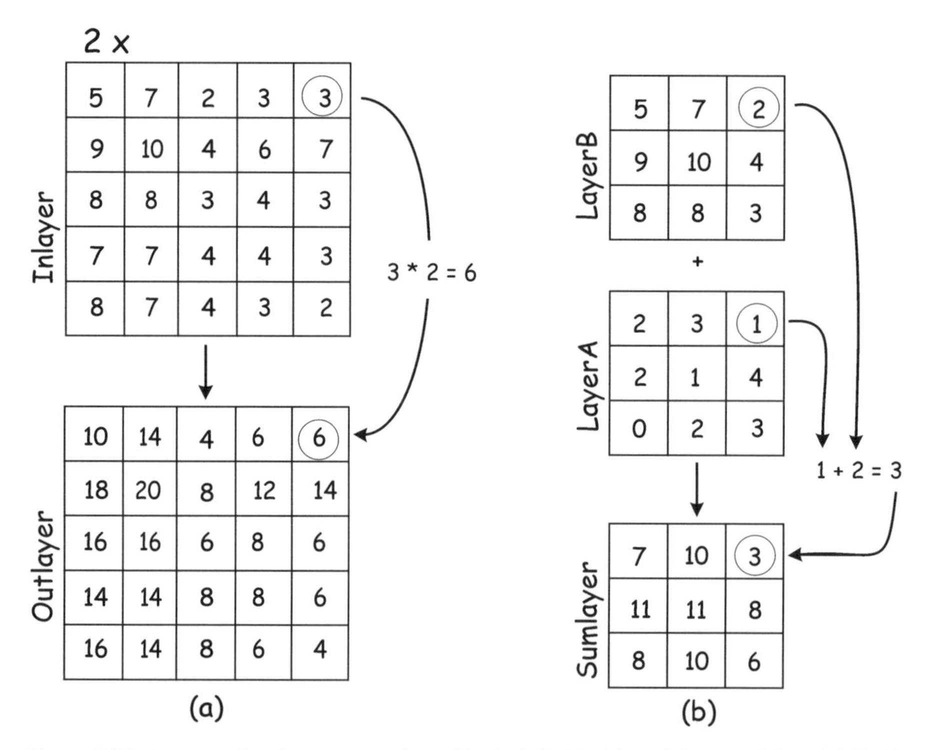

Figure 10-1: An example of raster operations. On the left side (a) each input cell is multiplied by the value 2, and the result stored in the corresponding output location. The right side (b) of the figure illustrates layer addition.

This function involves multiplying each cell value of In_layer by two, and placing the result in the corresponding cell in Outlayer. This multiplication is applied on a cell-by-cell basis, with output placed in the cell location corresponding to each input cell location. Other unary functions are applied in a similar manner, e.g., each cell may be raised to an exponent, divided by a fixed number, or converted to an absolute value.

Binary operations are similar to unary operations in that they involve cell-by-cell application of operations or functions. Binary operations differ from unary operations in that they combine data from two raster layers. Addition of two layers might be specified by:

Sumlayer = LayerA + LayerB (10.2)

Figure 10-1b illustrates this raster addition operation. Each value in LayerA is added to the value found in the corresponding cell in LayerB. These values are then placed in the appropriate raster cell of Sumlayer. The cell-by-cell addition is applied for the area covered by both LayerA and LayerB, and the result placed in Sumlayer. There are many other raster operations that may be applied, including addition, subtraction, maximum, and others.

Note that in our example LayerA and LayerB have the same extent – they cover the same area. This may not always be true. An operation may be specified where different input layers may not have the same extent. When this occurs, most GIS software will either restrict the operation to the area where there are data for both input layers, or assign a "missing data" number to cells where input data are not present. This number acts as a flag, indicating there are no results. It is often a number such as –9999 that will not occur from an operation, but any placeholder may be used, as long as the software and users understand the placeholder indicates no valid data are present.

Incompatible raster cell sizes cause ambiguities when raster layers are combined (Figure 10-2). This problem was described briefly in the previous chapter, and is illustrated with an additional example here. Consider cell A in Figure 10-2 when Layer1 and Layer2 are combined in a raster operation. Several cells in Layer2 correspond to cell A in Layer1. If these two layers are added there are likely to be several different input values for Layer2 corresponding to one input value for Layer1. The problem is compounded in cell B, because a portion of the cell is not defined for Layer2. It falls outside the layer boundary. Which Layer2 value should be used in a raster operation? Is it best to choose only the values in Layer2 from the cells with complete overlap, or to use the median number, the average cell number, or some weighted average? This ambiguity will arise whenever raster data sets are not aligned or have incompatible cell sizes. While the GIS software may have a default method for choosing the "best" input when cells are different sizes or do not align, these decisions are best controlled by the human analyst prior to the application of the raster operation. The analyst may resample the data to a compatible coordinate system, using transformation and resampling methods described in Chapter 4.

As with vector operations, raster operations may be categorized as local, neighborhood, or global operations (Figure 10-3). Local operations use only the data in a single cell to calculate an output value. Neighborhood operations use data from a set of cells, and global operations use all data from a raster data layer.

The concepts of local and neighborhood operations are more uniformly specified in raster data analysis than in vector data analysis. Cells within a layer have uniform size, so a local operation has a uniform input area. Vector areas represented by polygons may have vastly different areas. The local area defined by Alaska is different than the local area defined by Rhode Island. A local operation in a given raster is uniform in that it specifies a particular cell size and dimension.

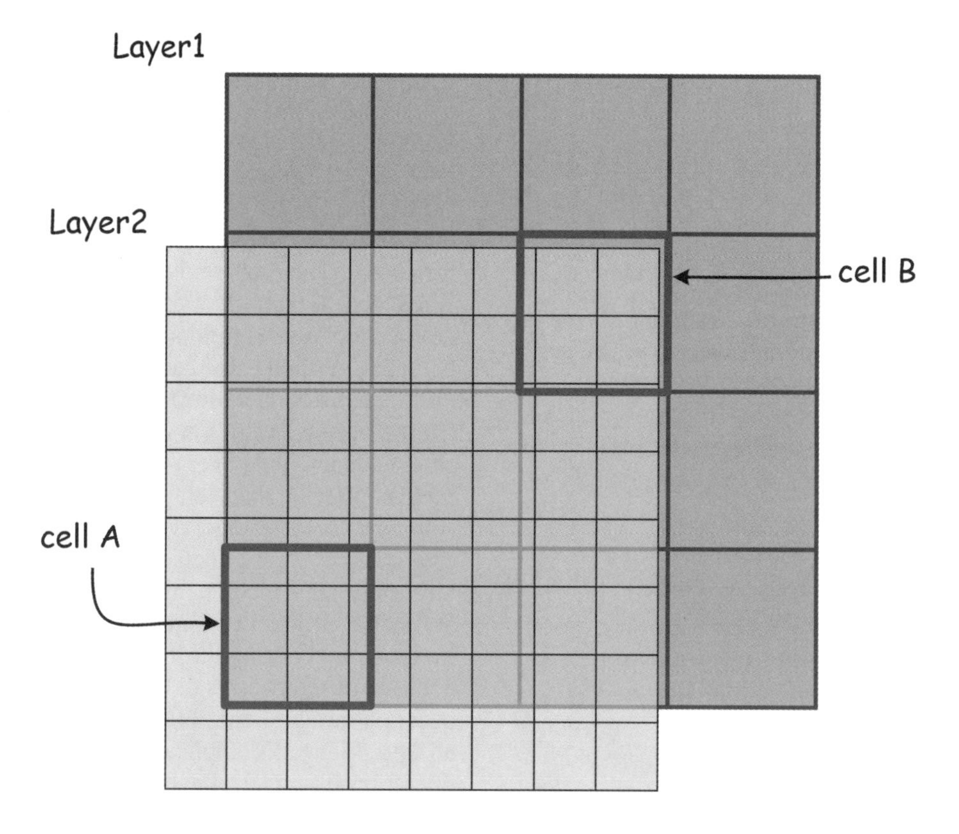

Figure 10-2: Incompatible cell sizes and boundaries confound multi-layer raster operations. This figure illustrates ambiguities in selecting input cell values from Layer2 in combination with Layer1. Multiple full and partial cells may contribute values for an operation applied to cell A. A portion of cell B is undefined in Layer2. These ambiguities are best resolved by resampling prior to layer combination.

Neighborhood operations in raster data sets are also more uniformly defined than in vector data sets. A neighborhood may be defined by a fixed number of cells in a specific arrangement, e.g., the neighborhood might be defined as a cell plus the eight surrounding cells. This neighborhood has a uniform area and dimension for most of the raster, with some minor adjustments needed near the edges of the raster data layer. Vector neighborhoods may depend not only on the shape and size of the target feature, but the shape and sizes of adjacent vector features.

Global operations in map algebra may produce uniform output, or they may produce different values for each raster cell. Global operations that return a uniform value are in effect returning a single number that is placed in every cell of the output layer. For example, the global maximum function for a layer might be specified as:

```
Out_num=globalmax(In_layer)          (10.3)
```

This would assign a single value to Out_num. The value would be the largest number found when searching all the cells of In_layer. This "collapsing" of data from a two-dimensional raster may reduce the map algebra to scalar algebra. Many other functions return a single global value placed in every cell for a layer, e.g., the global mean, maximum, or minimum.

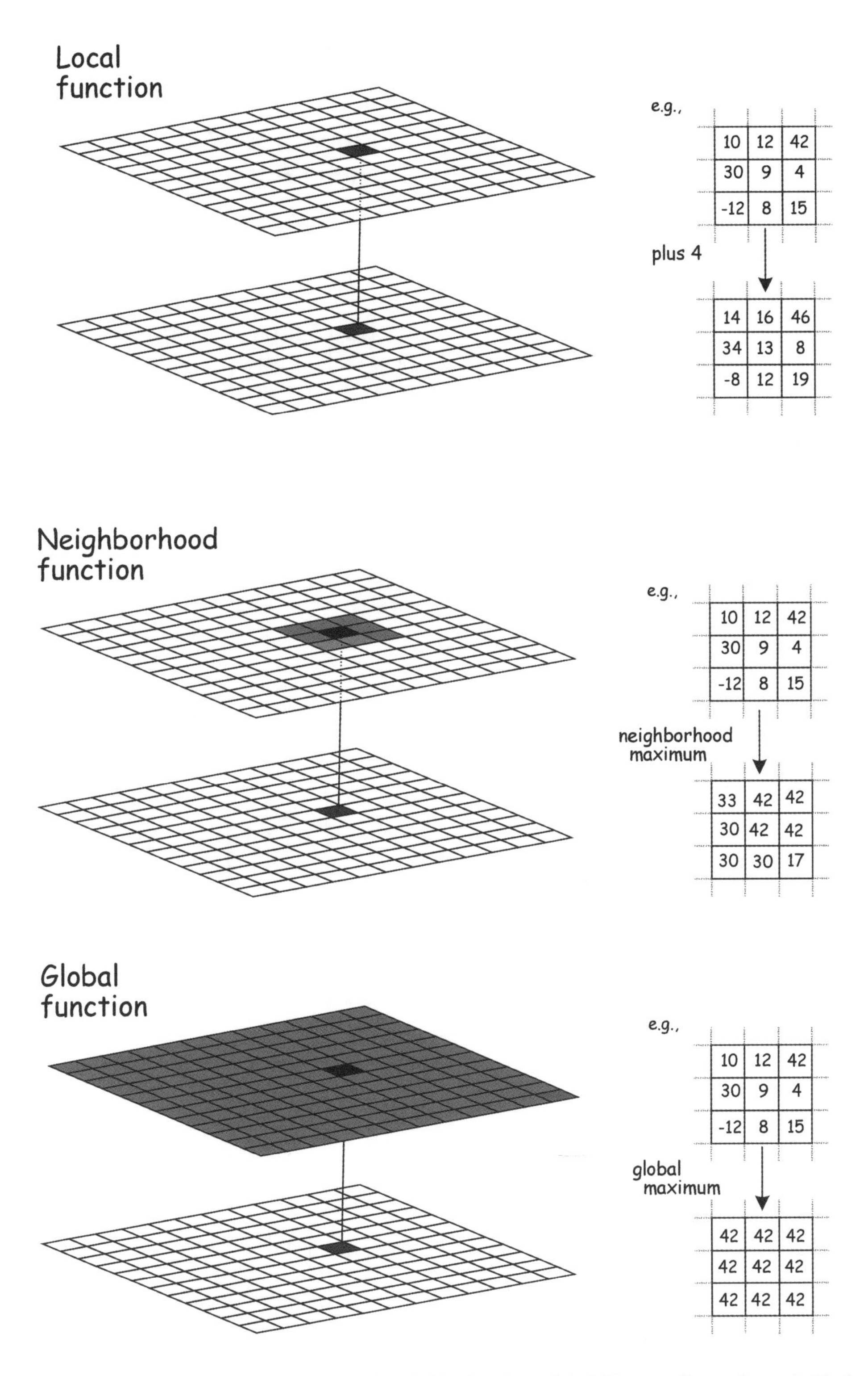

Figure 10-3: Raster operations may be local, neighborhood, or global. Target cells are shown in black, and source cells contributing to the target cell value in the output are shown in gray. Local operations show a cell-to-cell correspondence between input and output. Neighborhood operations include a group of nearby cells as input. Global operations use all cells in an input layer to determine values in an output layer.

Global operations are at times quite useful. Consider an analysis of regional temperature. We may wish to identify the areas where daily maximum temperatures were warmer this year than the highest regional temperature ever recorded. This analysis might help us to identify the extent of a warming trend. We would first apply a maximum function to all previous yearly weather records. This would provide a scalar value, a single number representing the regional maximum temperature. We would then compare a raster data set of maximum temperature for each day in the current year against the "highest ever" scalar. If the value for a day were higher than the regional maximum we would output a flag to a cell location. If it were not, we would output a different value. The final output raster would provide a map of the cells exceeding the previous regional maximum. Here we use a global operation first to create our single scalar value (highest regional temperature). This scalar is then used in subsequent operations that output raster data layers.

Neighborhood Operations and Moving Windows

Neighborhood operations (or functions) in raster analyses deserve an extended discussion because they offer substantial analytical power and flexibility. Neighborhood operations are applied more often than local or global operations. Slope, aspect, and many other raster operations use neighborhoods.

Neighborhood operations most often depend on the concept of a *moving window*. A "window" is a configuration of raster cells used to specify the input values for an operation (Figure 10-4). The window is positioned on a given location over the input raster, and an operation applied that involves the cells contained in the window. The result of the operation is usually associated with the cell

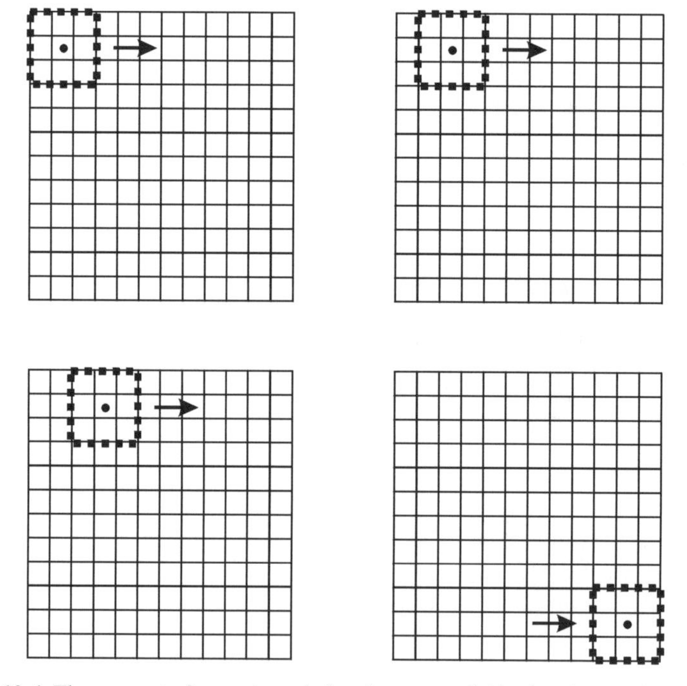

Figure 10-4: The concept of a moving window in raster neighborhood operations. Here a 3 by 3 window is swept from left to right and from top to bottom across a raster layer. The window at each location defines the input cells used in a raster operation.

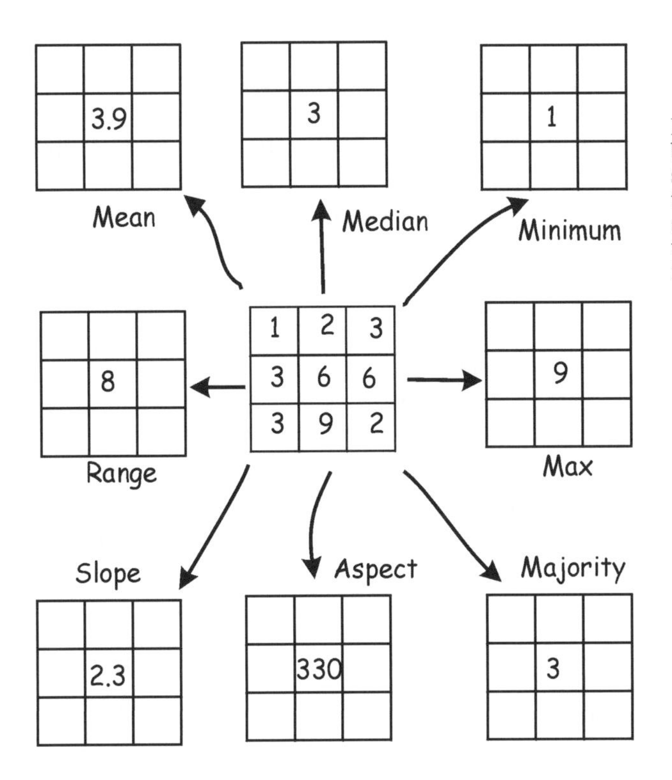

Figure 10-5: A given raster neighborhood may define the input for several raster neighborhood operations. Here a 3 by 3 neighborhood is specified. These nine cells may be used as input for mean, median, minimum, or a range of other functions.

at the center of the window position. The result of the operation is saved to an output layer at the center cell location. The window is then "moved" to be centered over the adjacent cell and the computation repeated (Figure 10-4). The window is swept across a raster data layer, usually from left to right in successive rows from top to bottom. At each window location the moving window function is calculated and the result output to the new data layer.

Moving windows are defined in part by their dimensions. For example, a 3 by 3 moving window has an edge length of three cells in the x and y directions, for a total of nine cells. Moving windows may be any size and shape, however they are typically odd-numbered in both the x and y directions to provide a natural center cell, and they are typically square. A 3 by 3 cell window is the most common size, although windows may also be rectangular. Windows may also have irregular shapes, e.g., L-shaped, circular, or wedge-shaped moving windows are sometimes specified.

There are many neighborhood functions that use a moving window. These include simple functions such as mean, maximum, minimum, or range (Figure 10-5). Neighborhood functions may be quite complicated, e.g., the statistical standard deviation, and they may be non-arithmetic, e.g., a count of the number of unique values, the mode, or a Boolean occurrence. Any function that combines information from a consistently shaped group of raster cells may be implemented with a moving window.

Figure 10-6 shows an example of a mean calculation using a moving window. The function scans all nine values in the window. It sums them and divides the sum by the number of cells in the window, thus calculating the mean cell value for the input window. The multiplication may be represented by a 3 by 3 grid containing the value one-ninth (1/9). The mean value is then stored in an output data layer in the location corresponding to the center cell of the moving window. The window is then shifted to the right and the process repeated. When the end of a row is reached the window is returned to the left-most columns, shifted down one row, and the process repeated until all rows have been included.

The moving window for many simple mathematical functions may be defined by a *kernel*. A kernel for a moving window function is the set of constants for each cell in a given window size and shape. These constants are applied with a function for every moving window location. The kernel in (Figure 10-6) specifies a mean. As the figure shows, each cell value for the Input layer at a given window position is multiplied by the corresponding kernel constant. The result is placed in the Output layer.

Note that when the edge of the moving window is placed on the margin of the original raster grid, we are left with at least one border row or column for which output values are undefined. This is illustrated in Figure 10-6. The moving window is shown placed in the upper right corner of the input raster. The window is as near the top and to the right side of the raster as can be without placing input cell locations in the undefined region, outside the boundaries of the raster layer. The center cell for the window is one cell to the left and down from the corner of the input raster. Output values are not defined for the cells along the top, bottom, and side margins of the output raster. Each successive operation applied to a layer may erode the margin further.

There are several common methods of addressing this margin erosion. One is to define a study area larger than the area of interest. Data may be lost at the margins, but these data are outside the area of interest. A second common approach defines a different kernel for margin windows (Figure 10-7). Margin kernels are similar to the main kernels, but modified as needed by the change

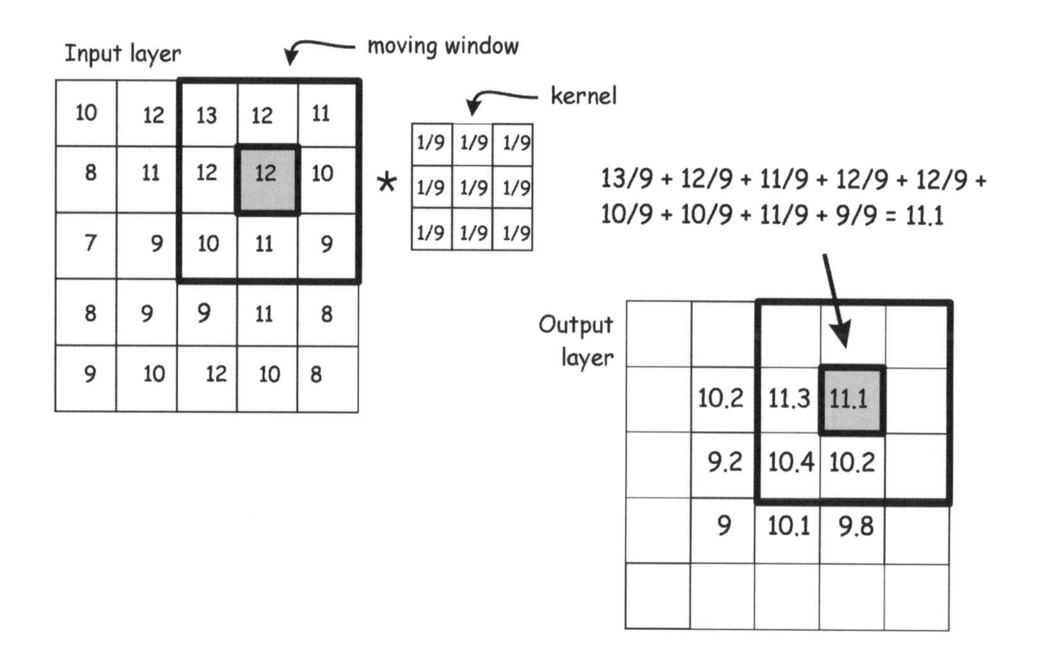

Figure 10-6: An example of a mean function applied via a moving window. Input layer cell values from a 3 by 3 moving window (upper left) are multiplied by a kernel to specify corresponding output cell values. This process is repeated for each cell in the input layer.

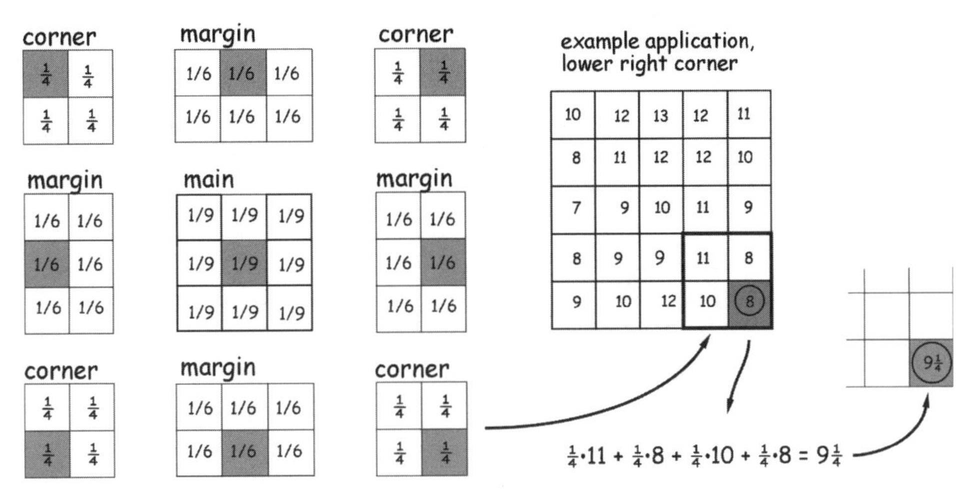

Figure 10-7: kernels may be modified to fill values for a moving window function at the margin of a raster data set. Here the margin and corner kernels are defined differently than the "main" kernel. Output values are placed in the shaded cell for each kernel. The margin and corner kernels are similar to the main kernel, but are adjusted in shape and value to ignore areas "outside" the raster.

in shape and size. Figure 10-7 illustrates a 3 by 3 kernel for the bulk of the raster. Corner values may be determined using a 2 by 2 kernel, and edges with 2 by 3 kernels. Outputs from these kernels are placed in the appropriate edge cells.

Different moving windows and kernels may be specified to implement many different neighborhood functions. For example, kernels may be used to detect edges within a raster layer. We might be interested in the difference in a variable across a landscape. For example, a railway accident may have caused a chemical spill and seepage through adjacent soils. We may wish to identify the boundary of the spill from a set of soil samples. Suppose there is a soil property, such as a chemical signature, that has a high concentration where the spill occurred, but a low concentration in other areas. We may apply kernels to identify the edges, where there are abrupt changes in the levels of this chemical.

Edge detection is based on comparing differences across a kernel. The values on one side of the kernel are subtracted from the values on the other side. Large differences result in large output values, while small differences result in small output values. Edges are defined as those cells with output values larger than a defined threshold.

Figure 10-8 illustrates the application of an edge-detection operation. The kernel on the left side of Figure 10-8 amplifies differences in the x direction. The values in the left of three adjacent columns are subtracted from the value in the corresponding right-hand row of cells. This process is repeated for each cell in the kernel, and the values summed across all nine cells. Large differences result in large values, either positive or negative, saved in the center cell. Small differences between the left and right rows lead to a small number in the center cell. Thus, if there are large differences between values when moving in the x direction, this differ-

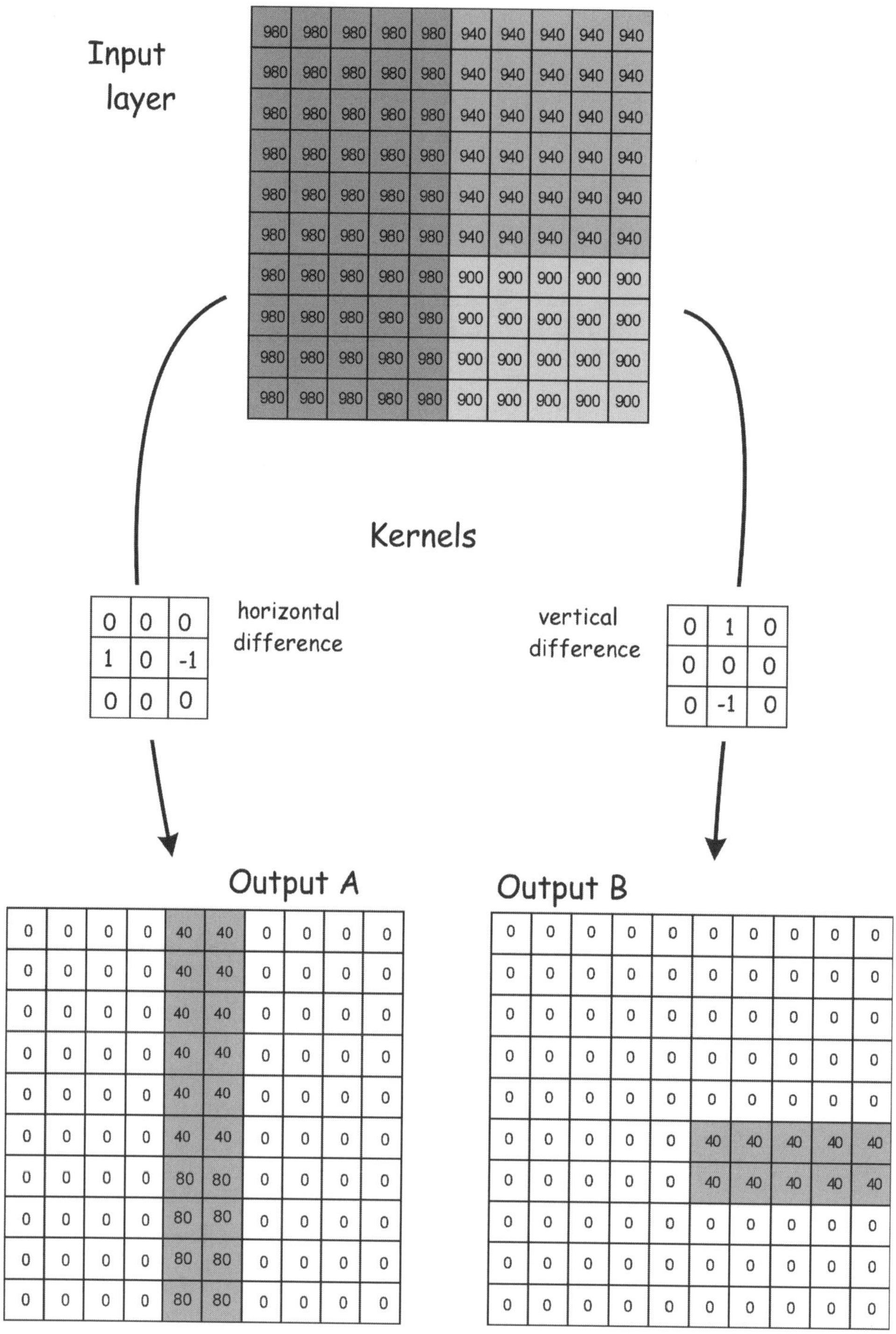

Input layer

980	980	980	980	980	940	940	940	940	940
980	980	980	980	980	940	940	940	940	940
980	980	980	980	980	940	940	940	940	940
980	980	980	980	980	940	940	940	940	940
980	980	980	980	980	940	940	940	940	940
980	980	980	980	980	940	940	940	940	940
980	980	980	980	980	900	900	900	900	900
980	980	980	980	980	900	900	900	900	900
980	980	980	980	980	900	900	900	900	900
980	980	980	980	980	900	900	900	900	900

Kernels

horizontal difference

0	0	0
1	0	-1
0	0	0

vertical difference

0	1	0
0	0	0
0	-1	0

Output A

0	0	0	0	40	40	0	0	0	0
0	0	0	0	40	40	0	0	0	0
0	0	0	0	40	40	0	0	0	0
0	0	0	0	40	40	0	0	0	0
0	0	0	0	40	40	0	0	0	0
0	0	0	0	40	40	0	0	0	0
0	0	0	0	80	80	0	0	0	0
0	0	0	0	80	80	0	0	0	0
0	0	0	0	80	80	0	0	0	0
0	0	0	0	80	80	0	0	0	0

Output B

0	0	0	0	0	0	0	0	0	0
0	0	0	0	0	0	0	0	0	0
0	0	0	0	0	0	0	0	0	0
0	0	0	0	0	0	0	0	0	0
0	0	0	0	0	0	0	0	0	0
0	0	0	0	0	40	40	40	40	40
0	0	0	0	0	40	40	40	40	40
0	0	0	0	0	0	0	0	0	0
0	0	0	0	0	0	0	0	0	0
0	0	0	0	0	0	0	0	0	0

Figure 10-8: There are a large number of kernels used with moving windows. The kernel on the left amplifies differences in the x direction, while the kernel on the right amplifies differences in the y direction. These and other kernels may be used to detect specific features in a data layer.

ence is highlighted. Spatial structure such as an abrupt change in elevation may be detected by this kernel. The kernel in the middle-right of Figure 10-8 may be used to detect differences in the y direction.

Neighborhood functions may also smooth the data. A mean kernel, described above, will reduce the difference between a cell and surrounding cells. This is because windows average across a group of cells, so there is much similarity in the output values calculated from adjacent window placements. These smoothing properties may be used to reduce differences among adjacent raster cells.

Raster data may contain noise. Noise are values that are large or small relative to their spatial context. In an elevation data layer noise might take the form of isolated cells with a very high value in an area of otherwise low values. Noise may come from several sources, including measurement errors, mistakes in recording the original data, miscalculations, or data loss. There is often a need to correct these errors. If it is impossible or expensive to revisit the study area and collect new data, the noisy data may be smoothed using a kernel and moving window.

There are functions known as *high-pass filters* with kernels that accentuate differences between adjacent cells. These high-pass filter kernels may be useful in identifying the spikes or pits that are characteristic of noisy data. Cells identified as spikes or pits may then be evaluated and edited as appropriate, removing the erroneous values. High-pass kernels generally contain both negative and positive values in a pattern that accentuates local differences.

Figure 10-9 demonstrates the use of a high-pass kernel on a data set containing noise. An elevation data set shown in the left-center portion of the figure contains a number of anomalous cells. These cells have extremely high values (spikes, shown in black) or low values (pits, shown in white) relative to nearby cells. If uncorrected they will affect slope, aspect, and other terrain-based calculation, so pits and spikes should be identified and modified.

The high-pass kernel shown contains a value of 9 in the center and -1 in all other cells. Each value is divided by nine to reduce the range of the output variable. The kernel returns a value near the local average in smoothly-changing areas. The positive and negative values balance, returning small numbers in flat areas.

The high-pass kernel generates a large positive value when centered on a spike. The large differences between the center cell and adjacent cells are accentuated. Conversely, a large negative value is generated when a pit is encountered. An example shows the application of the high-pass filter for a cell near the upper left corner of the input data layer (Figure 10-9). Each cell value is multiplied by the corresponding kernel coefficient. These numbers are summed, and divided by nine, and the result placed in the corresponding output location. Calculation results are shown as real numbers, but cell values are shown here recorded as integers. Output values may be real numbers or integers, depending on the programming algorithm and perhaps the specifications set by the user.

Figure 10-10 illustrates the application of a mean filter to a spatial data layer. As with the high-pass kernel, the window is positioned over a set of cells and the mean kernel values multiplied by the cell values and summed. Notice that for the mean kernel this amounts to multiplying each cell value by one-ninth (1/9) and summing these values. The result is placed in the center cell. Note how the mean function smooths the data. Spikes and pits are removed when the mean kernel is applied with a moving window to a data layer.

The mean filter is representative of many moving window functions in that it increases the *spatial covariance* in the output data set. High spatial covariance means values are autocorrelated (introduced in the discussion of kriging in Chapter 13). A large positive spatial covariance means cells near

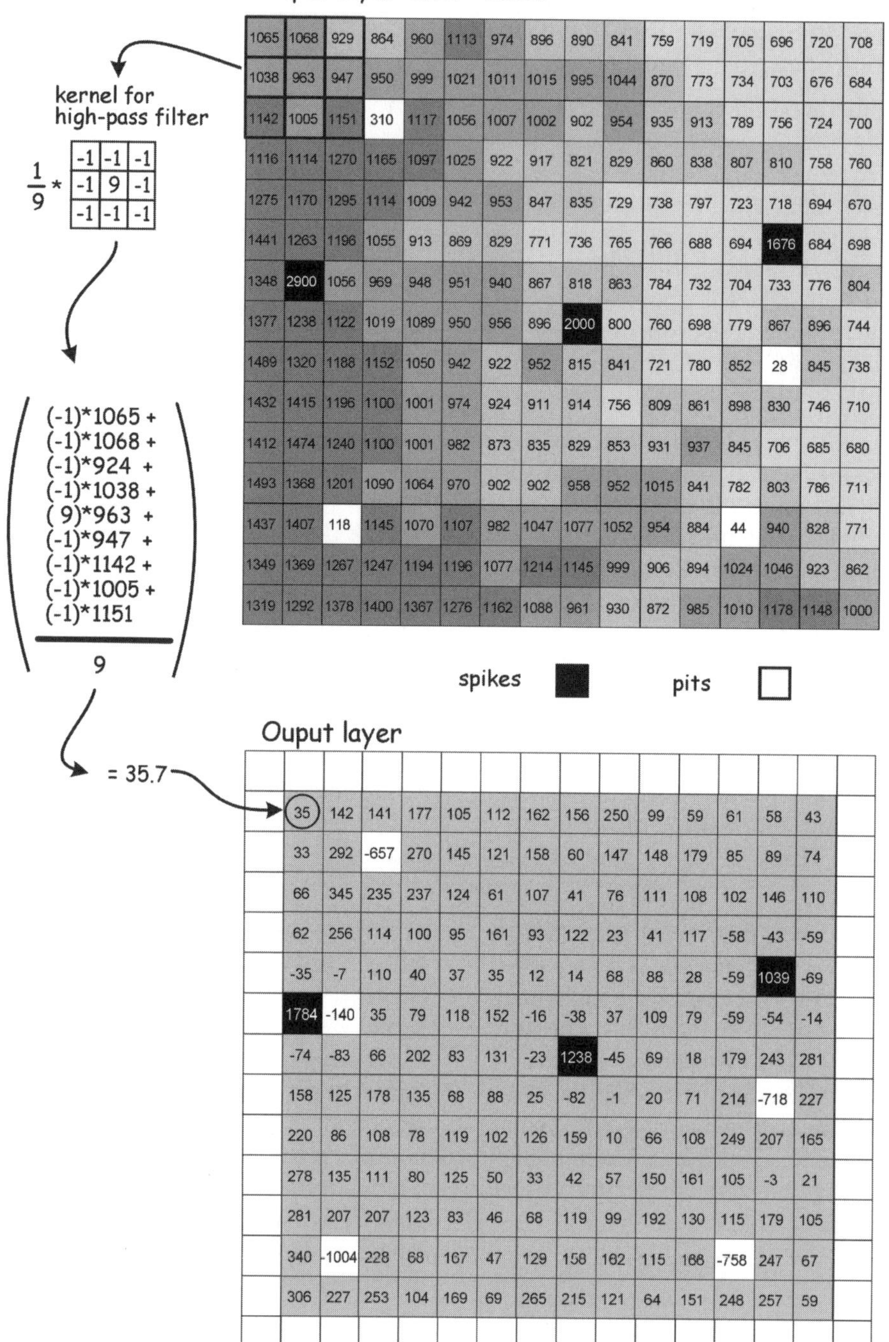

Figure 10-9: An example of a moving window function. Raster data often contain anomalous "noise" (dark and light cells). A high-pass filter and kernel, shown at top left, highlights "noisy" cells. Local differences are amplified so that anomalous cells are easily identified.

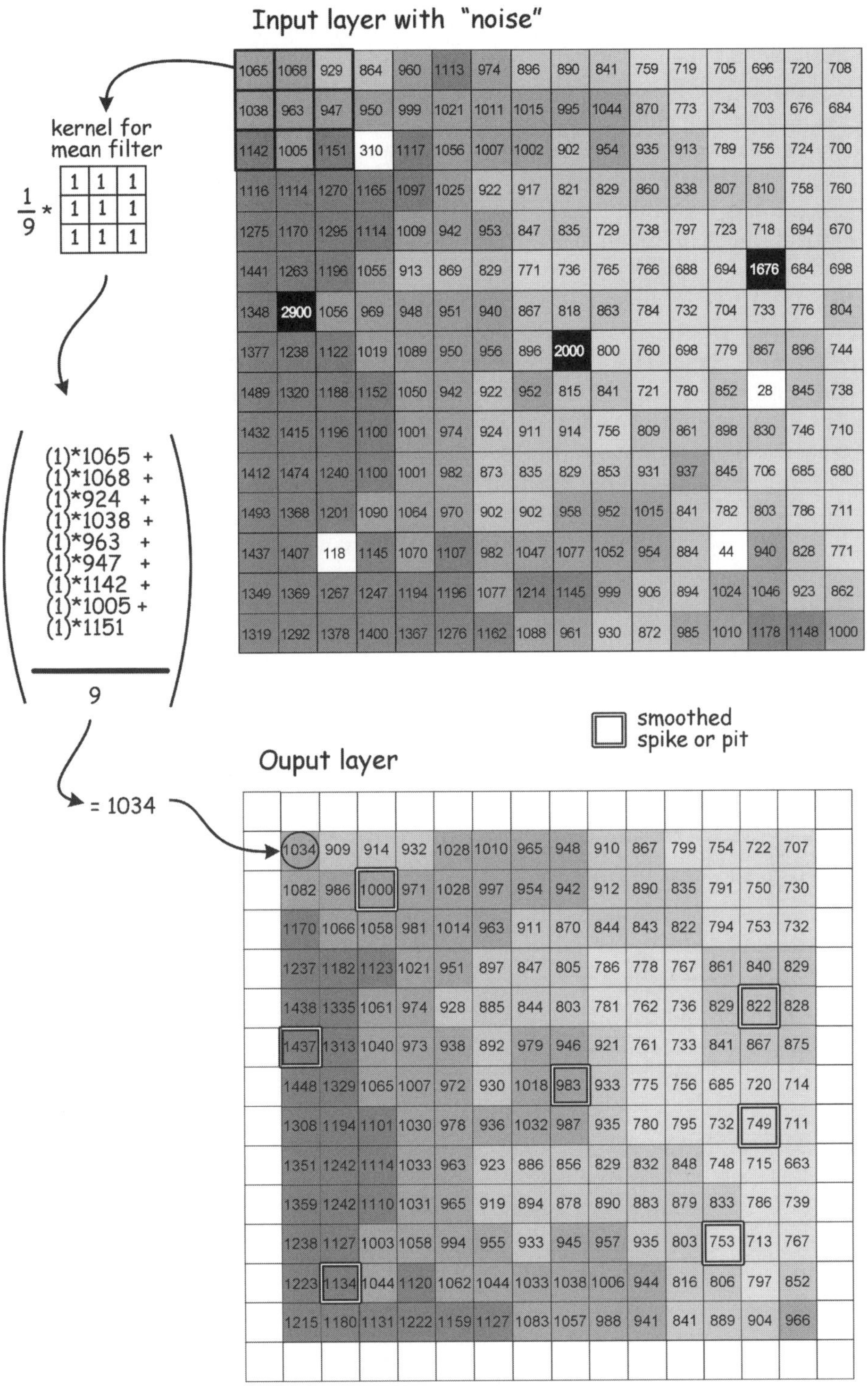

Input layer with "noise"

1065	1068	929	864	960	1113	974	896	890	841	759	719	705	696	720	708
1038	963	947	950	999	1021	1011	1015	995	1044	870	773	734	703	676	684
1142	1005	1151	310	1117	1056	1007	1002	902	954	935	913	789	756	724	700
1116	1114	1270	1165	1097	1025	922	917	821	829	860	838	807	810	758	760
1275	1170	1295	1114	1009	942	953	847	835	729	738	797	723	718	694	670
1441	1263	1196	1055	913	869	829	771	736	765	766	688	694	1676	684	698
1348	2900	1056	969	948	951	940	867	818	863	784	732	704	733	776	804
1377	1238	1122	1019	1089	950	956	896	2000	800	760	698	779	867	896	744
1489	1320	1188	1152	1050	942	922	952	815	841	721	780	852	28	845	738
1432	1415	1196	1100	1001	974	924	911	914	756	809	861	898	830	746	710
1412	1474	1240	1100	1001	982	873	835	829	853	931	937	845	706	685	680
1493	1368	1201	1090	1064	970	902	902	958	952	1015	841	782	803	786	711
1437	1407	118	1145	1070	1107	982	1047	1077	1052	954	884	44	940	828	771
1349	1369	1267	1247	1194	1196	1077	1214	1145	999	906	894	1024	1046	923	862
1319	1292	1378	1400	1367	1276	1162	1088	961	930	872	985	1010	1178	1148	1000

kernel for mean filter

$$\frac{1}{9} * \begin{array}{|c|c|c|} \hline 1 & 1 & 1 \\ \hline 1 & 1 & 1 \\ \hline 1 & 1 & 1 \\ \hline \end{array}$$

$$\left(\frac{(1)*1065 + (1)*1068 + (1)*924 + (1)*1038 + (1)*963 + (1)*947 + (1)*1142 + (1)*1005 + (1)*1151}{9} \right)$$

= 1034

smoothed spike or pit

Ouput layer

1034	909	914	932	1028	1010	965	948	910	867	799	754	722	707
1082	986	1000	971	1028	997	954	942	912	890	835	791	750	730
1170	1066	1058	981	1014	963	911	870	844	843	822	794	753	732
1237	1182	1123	1021	951	897	847	805	786	778	767	861	840	829
1438	1335	1061	974	928	885	844	803	781	762	736	829	822	828
1437	1313	1040	973	938	892	979	946	921	761	733	841	867	875
1448	1329	1065	1007	972	930	1018	983	933	775	756	685	720	714
1308	1194	1101	1030	978	936	1032	987	935	780	795	732	749	711
1351	1242	1114	1033	963	923	886	856	829	832	848	748	715	663
1359	1242	1110	1031	965	919	894	878	890	883	879	833	786	739
1238	1127	1003	1058	994	955	933	945	957	935	803	753	713	767
1223	1134	1044	1120	1062	1044	1033	1038	1006	944	816	806	797	852
1215	1180	1131	1222	1159	1127	1083	1057	988	941	841	889	904	966

Figure 10-10: Example of a mean kernel applied in a moving window function. This function smooths the input data, removing most extreme values.

each other are likely to have similar values. Where you find one cell with a large number, you are likely to find more cells with large numbers. If spatial data have high spatial covariance, then small numbers are also likely to be found near each other. Low spatial covariance means nearby values are unrelated – knowing the value at one cell does not provide much information about the values at nearby cells. High spatial covariance in the "real world" may be a good thing. If we are prospecting for minerals then a sample with a high value indicates we are probably near a larger area of ore-bearing deposits. However if the spatial autocorrelation is increased by the moving window function we may get an overly optimistic impression of our likelihood of striking it rich.

The spatial covariance increases with many moving window functions because these functions share cells in adjacent calculations. Note the average function in Figure 10-11. The left of Figure 10-11 shows sequential positions of a 3 by 3 window. In the first window location the mean is calculated and placed in the output layer. The window center is then shifted one cell to the right, and the mean for this location calculated and placed in the next output cell to the right. Note that there are six cells in common for these two means. Adjacent cells in the output data layer share six of nine cells in the mean calculation. When a particularly low or high cell occurs, it affects the mean of many cells in the output data layer. This causes the outputs to be quite similar, and increases the spatial covariance.

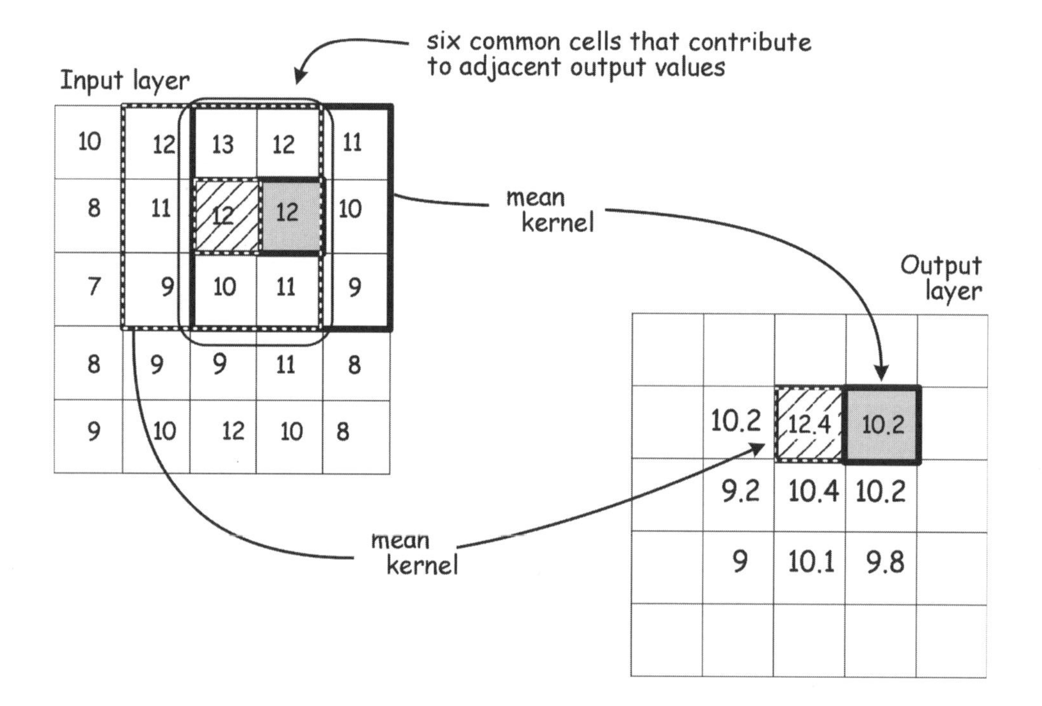

Figure 10-11: A moving window may increase spatial covariance. Adjacent output cells share many input cell values. In the mean function shown here this results in similar output cell values.

Overlay in Raster Map Algebra

Overlay functions in map algebra may be performed through addition and multiplication. Union operations may be performed with layer addition, and clip operations through multiplication. These two raster algebra operations may be used to combine raster data layers in a number of ways.

The union of two raster data layers may be performed through addition. When two layers are added, values are added on a cell-by-cell basis and the results placed in corresponding cell locations for an output data layer. Each output value may be used to identify the combinations of input values.

Figure 10-12 shows the overlay of two data layers through raster addition. Cells in Layer A have values 10 through 40, and cells in Layer B have values 1 through 7. These might correspond to five different soil types in Layer A and six different vegetation types in Layer B. Data layers are combined on a cell-by-cell basis, so each cell value in Layer A is added to the corresponding cell value in Layer B. In the upper right corner cell, the value 20 from Layer A is added to 2 in Layer B, and the resultant value 22 is placed in the Output layer. Cell addition of these two layers will result in a set of numbers between 12 and 47. These correspond to the various combinations of soil types and vegetation.

Attribute data may be matched to the corresponding attribute tables. Each unique value in the input data layer is identified and associated with the appropriate combinations in the output data layer. Attribute assignment is illustrated in Figure 10-12. Cells in Layer A with an ID value of 10 are associated with Type = A and Spec = 22b. Every cell in the Output layer that exhibit an ID = 10 for input Layer A have a Type = A and Spec = 22b in the corresponding output cells.

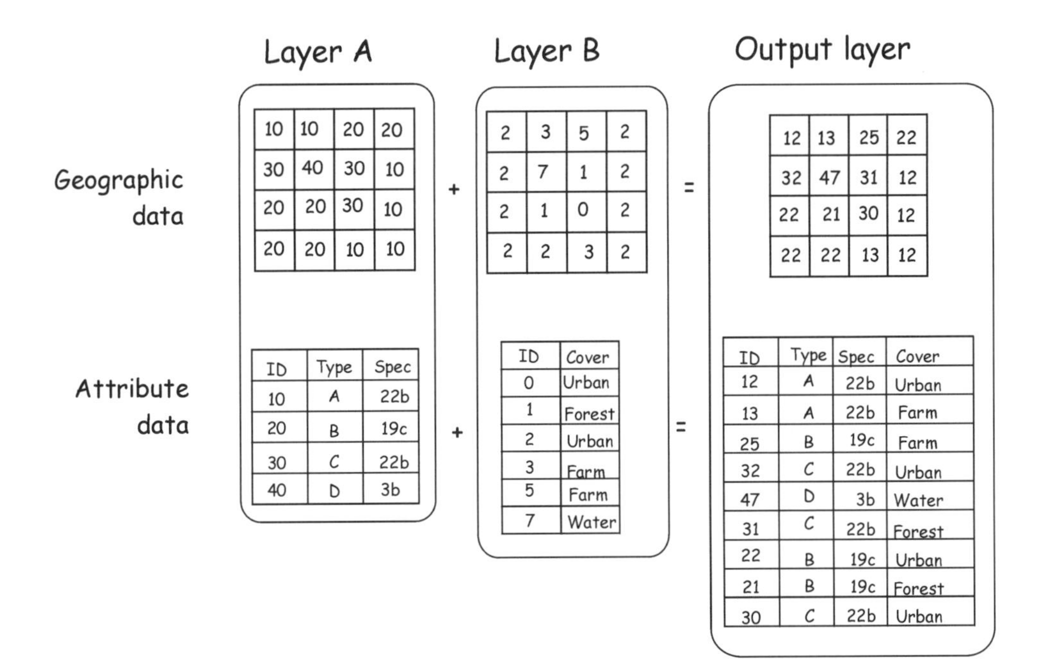

Figure 10-12: Overlay through raster addition. Cell values in Layer A and Layer B are added to yield values in the Output layer. Attribute values associated with each input layer may be combined in an associated table.

Note that there may be some spatial ambiguity in the assignment of output values in that the same input combinations may be disjunct in the output data layer. The ID values of 12 are found in two disjunct sets, one in the upper-left corner and one in the lower-right corner (Figure 10-12). These cells with ID=12 are all referred to by the same entry in the attribute table. This many-to-one relationship may occur quite often, for example, it also exists for ID=22 in the Output layer.

Raster data sets often do not uniquely identify disjunct but otherwise identical areas. Raster data sets often have thousands to millions of unique cell locations. When data layers are combined through raster overlay the average area of contiguous, identical cells decreases; there are often many small areas that have the same combination of input values, but are separated from other cells with the same value. If each group of cells is assigned a unique identifier the attribute table may grow quite large. Large data sets are becoming less of a handicap as computing power and space increases, but many software packages by default implement a one-to-many relationship. This contrasts with the common approach applied by vector overlay software packages, which typically uniquely identify each polygon, whether or not there are other polygons with an identical set of attribute values. The GIS user needs to understand the output form so that output from overlay operations may be properly interpreted and applied.

Care must be taken to avoid ambiguous combinations when using raster layer addition for overlay. Output numbers derived from two different combinations must be avoided. Consider the example shown in Figure 10-13. Two data layers are overlain through addition. A value of 4 may occur in the output layer from multiple combinations: 2 in Layer A and 2 in Layer B, or 1 in Layer A and 3 in Layer B. There are similar problems for output values equal to 3, so our results are ambiguous. We must ensure this cannot occur. We typically do this with a re-classification of a data layer. For example, we could multiply Layer A by 10, thus giving values of 10, 20, 30, and 40. The output values will

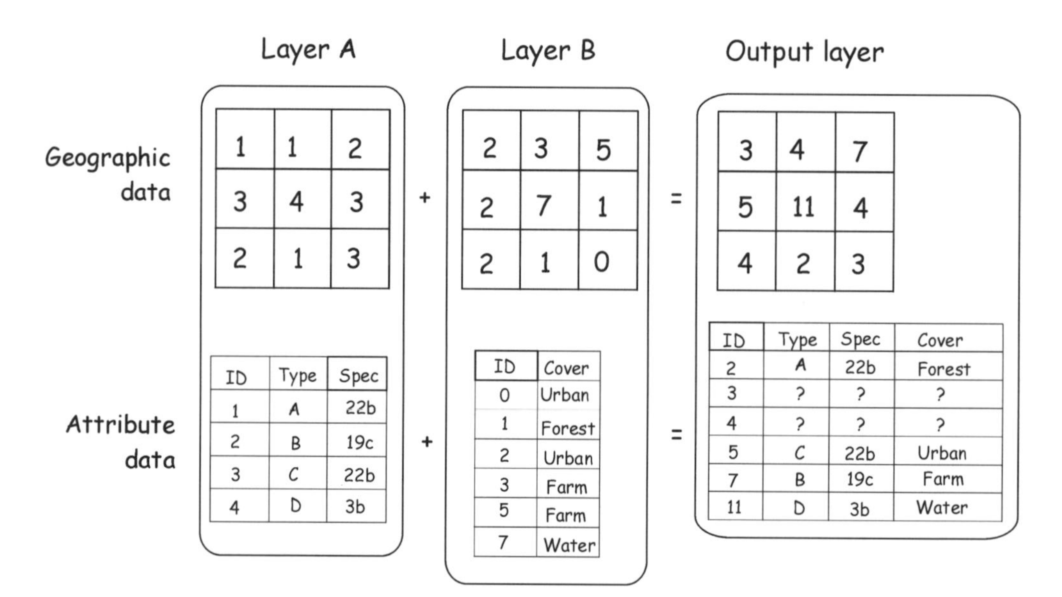

Figure 10-13: Raster addition will lead to ambiguous output when different combinations of inputs lead to identical output combinations. Input layers may be classified or renumbered to ensure unique output combinations.

then uniquely identify the combination of inputs.

Cost Surfaces

Many problems require an analysis of travel costs. These may be monetary costs of travel, e.g., the price one must charge to profitably deliver a package from the nearest distribution center to all points in a region. Travel costs might also be measured in other units, e.g., the time it takes to travel from a school to the nearest hospital, or as a likelihood, e.g., the chance of a noxious foreign weed spreading out from an introduction point. These analyses may be performed with the help of *cost surfaces*. A cost surface contains the minimum cost of reaching cells in a layer from one or more source cells (Figure 10-14).

The simplest cost surface is based on a uniform travel cost. Travel cost depends only on the distance covered, with a fixed cost applied per unit distance traveled. This cost per unit distance does not change from cell to cell. There are no barriers, and so the straight line distance is converted to a cost. First, the distance is calculated to each cell. As illustrated in Figure 10-14, the distance is calculated based on the Pythagorean formula. Distances to each cell in the x and y directions contribute to the total distance from a source cell or cells.

The distance from a source cell is combined with a fixed cost per unit distance to calculate travel cost. As shown in the right side of Figure 10-14, each distance value is multiplied by the fixed cost factor. This results in a cost surface, a raster layer containing the travel cost to each cell. If there are multiple source cells, travels costs are calculated from each source cell, and the lowest cost is typically placed in the output cell.

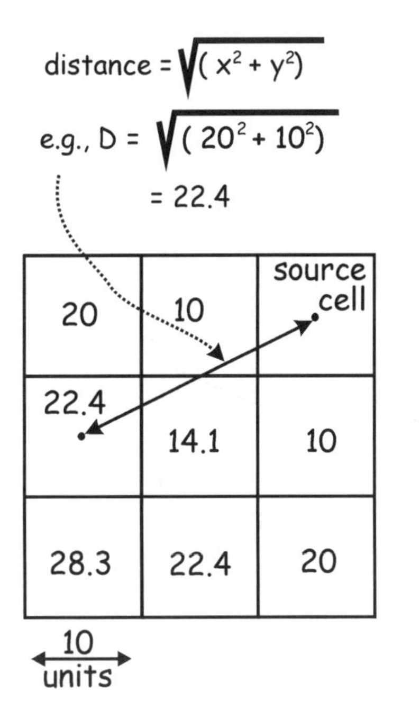

Figure 10-14: A cost surface based on a fixed cost per unit distance. Minimum distance from a set of source cells is multiplied by a fixed cost factor to yield a cost surface.

Travel costs may also be calculated using a *friction surface*. The cell values of a friction surface represent the cost per unit travel distance for crossing each cell. Friction surfaces are used to represent areas with a variable travel cost. Imagine a large military base. Part of the base may include flat, smooth areas such as drill fields, parking lots, or parade grounds. These areas are relatively easy to cross, with corresponding low travel times per unit distance. Other parts of the base may be covered by open grasslands. While the surface may be a bit rougher, travel times are still moderate. Other parts may be comprised of forests. These areas would have correspondingly high travel times, as a vehicle would have to pick a path among the trees. Finally, there may be areas occupied by water, fences, or buildings. These areas would have effectively infinite travel times.

Each cell in the friction surface contains the cost required to traverse a portion of the cell. A value of 3 indicates it costs three units (of time, money, or other factor) per unit distance in the cell. If a cell is 10 wide and costs 3 units per unit distance, and the cell is crossed along the width, then the cost for traversing the cell is 10 times 3, or 30 units.

The actual cost for traversing the cell depends on the distance traveled through the cell. When a cell is traversed parallel to the row or column edge then the distance is simply the cell dimension. When a cell is traversed at any other angle the distance will vary. It may be greater or less than the cell dimension, depending on the angle and location of the path.

The travel cost required to reach each cell is the minimum accumulated total of the cost times the distance to a source cell. We specify a minimum accumulated cost because if there is more than one source cell, there are a large number of potential paths to each of these source cells. Distance across each cell is multiplied by the friction surface

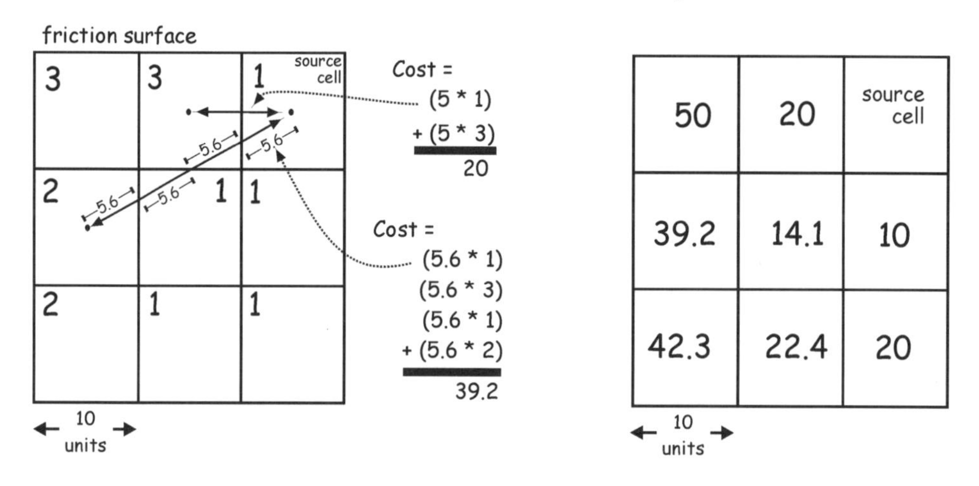

Figure 10-15: A cost surface based on spatially-variable travel costs. A friction surface specifies the spatially varying cost of traveling through raster cells. The distance traversed through each cell is multiplied by the cost in the friction surface. The values are summed for each path to yield a total cost.

cost for that cell and summed for a path to accumulate the total travel cost. The lowest cost path from a source location to a cell is usually assigned as the travel cost to that cell.

Figure 10-15 shows an example of calculations for the friction cost along a set of paths. These are straight-line paths that travel either parallel to the cell boundaries (purely in an x or y direction) or at some angle across cells.

Sample calculation of the friction costs for a path parallel to the x axis is shown at the top middle and on the left side of Figure 10-15. Note that when travelling parallel to a cell boundary, one half-cell width is traversed in the starting and ending cells. Intermediate cells are crossed at a full cell width. When moving from the starting cell to the adjacent left cell, a friction surface value of 1 is encountered, then a friction surface value of 3. One-half the distance, 5 units, is through the top-right cell at a per-unit friction cost of 1. One-half the distance is through the adjacent cell to the left, at a per-unit friction cost of 3. The total cost is then the distance traveled in each cell multiplied by the per-unit friction cost of the cell:

$$5 * 1 + 5 * 3 = 20 \qquad (10.4)$$

The friction cost when traversing cells at an angle is illustrated at the bottom left and bottom center of Figure 10-15. The friction cost is the sum of the cell cost per unit distance multiplied by the distance traveled in each cell. The path begins at the source cell and ends two cells to the left and one cell down. Each intervening cell is traversed for a distance of 5.6 cell units. The distance traversed in each cell is multiplied by the friction value for each cell. The total cost for this leg is:

$$5.6*1 + 5.6*3 + 5.6*1 + 5.6*2 = 39.2 \qquad (10.5)$$

In general, the cost of any path is expressed as:

$$Totalcost = d1*c1 + d2*c2 + \ldots dn*cn \qquad (10.6)$$

where d_i is the distance and c_i is the cost across each cell of a path.

Most implementations calculate the minimum cost to travel to a cell from a set of source cells. There are many routes from any source cell to any destination cell, thousands of distinct routes in most instances if repeated cell crossings are not allowed. Software typically implements some optimization algorithm to eliminate routes early on and reduce search time, thereby arriving at the cost surface in some acceptable time period.

Note that barriers may be placed on a cost surface to preclude travel across portions of the surface. These barriers may be specified by setting the cost so high that no path will include them. Any circuitous route will be less expensive than traveling over the barriers. Some software allows the specification of a unique code to identify barriers, and this code precludes movement across the cell.

Summary

Raster analyses are essential tools in GIS, and should be understood by all users. Raster analyses are widespread and well developed due to many reasons, including the simplicity of the data structure, the ease with which continuous variable may be represented, and the long history of raster analyses.

Map algebra is a concept in which raster data layers are combined via summation and multiplication. Values are combined on a cell-by-cell basis, and may be added, subtracted, multiplied, or divided. Care must be taken to avoid ambiguous combinations in the output that originate from distinct input combinations.

Raster analyses may be local, neighborhood, or global, and general analyses such as buffering and overlay may be applied using raster data sets. Neighborhood operations are particularly common in raster analyses, and may be applied with a moving window approach. A moving window is swept across all points in a data layer, typically multiplying kernal values by data found around a center cell. Window size and shape may be modified at the edges of the data layers. Moving windows may be used to specify a wide range of combinatorial, terrain, and statistical functions.

Cost or frictions surfaces are an important subset of proximity analyses that may be easily applied in raster analyses. A cost surface identifies the travel costs required for movement from a specified set of locations.

Suggested Reading

Band, L. E., Topographic partition of watersheds with digital elevation models, *Water Resources Research*, 1986, 22:15-24.

Berry, J. K., Fundamental operations in computer-assisted mapping, *International Journal of Geographic Information Systems*, 1987, 1:119-136.

Berry, J. K., A mathematical structure for analyzing maps, *Environmental Management*, 1986, 11:317-325.

Bonhame-Carter, G. F., Geographic Information Systems for Geoscientists: Modelling with GIS, Pergamon, Ontario, 1996.

Burrough, P. A. and McDonnell, R. A., Principles of Geographical Information Systems, 2nd Edition. Oxford University Press, New York, 1998.

Cliff, A.D., and Ord, J.K., Spatial Autocorrelation, Methuen, New York, 1987.

Tomlin, C. D., Geographic Information Systems and Cartographic Modeling, Prentice-Hall, New Jersey, 1990.

Study Questions

What is map algebra?

Why must raster layers have compatible cell sizes and orientations for most raster combination operations?

Can you describe the moving window operation in the context of a raster neighborhood operation? What are some examples of moving window operations?

What is meant by high or low spatial covariance in a raster data layer?

Can you describe a cost or friction surface and their components? What are the mechanics of simple and friction cost calculations?

11 Terrain Analysis

Introduction

Elevation and related terrain variables are important in most peoples lives. Terrain determines the natural availability and location of surface water, and hence soil moisture and drainage. Terrain in large part determines water quality through control of sediment entrainment and transport, and elevation and slope define flood zones, watershed boundaries, and hydrologic networks. Terrain also strongly influences the location and nature of transportation networks, and the cost and methods of house and road construction. Terrain data are used in shading hardcopy and digital maps and thereby depict a three-dimensional perspective in two-dimensional media (Figure 11-1). Terrain variables and their importance have

Figure 11-1: An example of a terrain-based image of the western United States. Shading, based on local elevation, emphasizes terrain shape. Topographic features are clearly identified, including the Central Valley of California at the left of the image, and to the center and right the parallel mountains and valleys of the Basin and Range region. (courtesy USGS)

been identified for a number of spatial analyses (Table 11-1).

Given the importance of elevation and terrain in resource management and the difficulties of manual terrain analysis, it is not surprising that considerable effort has been directed at the development of terrain analysis in GIS. It is difficult to perform consistent terrain analyses in a non-digital environment. For example, slope is of vital importance in many resource management problems. However slope calculation over large areas from a hardcopy map are slow, error prone, and inconsistent. Elevation differences over a horizontal distance are difficult to measure. Further, these

Table 11-1: A subset of commonly used terrain variables. (adapted from Moore *et al.*, 1993).

Variable	Description	Importance
Height	Elevation above base	Temperature, vegetation, visibility
Slope	Rise relative to horizontal distance	Water flow, flooding, erosion, travel cost, construction suitability, geology, insolation, soil depth
Aspect	Downhill direction of steepest slope	Insolation, temperature, vegetation, soil characteristics and moisture, visibility
Upslope area	Watershed area above a point	Soil moisture, water runoff volume and timing, pollution or erosion hazards
Flow length	Longest upstream flow path to a point	Sediment and erosion rates
Upslope length	Mean upstream flow path length to a point	Sediment and erosion rates
Profile curvature	Curvature parallel to slope direction	Erosion, water flow acceleration
Plan curvature	Curvature perpendicular to slope direction	Water flow convergence, soil water, erosion
Visibility	Site obstruction from given viewpoints	Utility location, viewshed preservation

measurements are slow and likely to vary among human analysts.

Both the data and methods exist for the extraction of important terrain variables via a GIS. Digital elevation models (DEMs), described in Chapter 7, have been developed for most of the World. These data have been produced at considerable cost, and there is a continuing process of DEM renewal and improvement.

Most terrain analyses are performed using a raster data model. While the TIN (Chapter 2) and other data models have been developed to store and facilitate terrain analyses, and while in some instances there are distinct advantages when TIN or other data models are adopted, raster data structures are most common in terrain analyses. Raster data sets are often quite simple, and so are easy to understand, manipulate, and program. Most DEMs are easily ingested into raster data sets because of the simple structure. Raster data sets also facilitate the easy and uniform calculation of slope, aspect, and other important terrain variables that may be determined by comparing cell values. Many terrain analysis functions may be specified by an appropriate moving window function, and hence are compatible with a raster data model. Mathematical operations may be applied to a set of elevation values specified by a window. The results from these mathematical operations in turn provide important information about terrain characteristics that are helpful in spatial analysis and problem solving.

Slope and Aspect

Slope and aspect are two commonly used terrain variables. They are required in many studies of hydrology, conservation, site planning, and infrastructure development, and are the basis for many other terrain analysis functions. Road construction costs and safety are sensitive to slope, as are most other construction activites. Watershed boundaries, flowpaths and direction, erosion modeling, and viewshed determination all use slope and/or aspect data as input. Soils and vegetation are influenced by slope and aspect in many parts of the world; knowledge of slope or aspect may be useful in mapping both vegetation and soil resources, thereby increasing the accuracy and specificity of spatial data, and at times reducing costs.

Slope is defined as the change in elevation (a rise) with a change in horizontal position (a run). Seen in cross section, the slope is related to the rise in elevation over the run in horizontal position (Figure 11-2). Slope is often reported in degrees, between zero (flat), and 90 (vertical). The slope is equal to 45 degrees when the rise is equal to the run. The slope in degrees is calculated from the rise and run through the tangent trigonometric function. By definition, the tangent of the slope angle (ϕ) is the ratio of the rise over the run, as shown in (Figure 11-2). The inverse tangent of a measured rise over a run gives

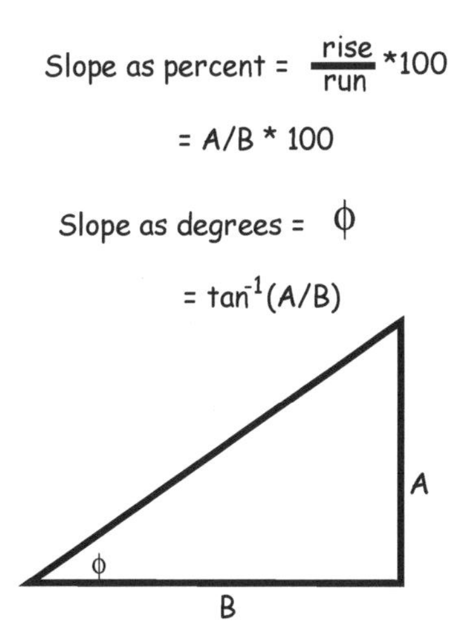

To convert from percent slope to degrees, apply formula,
e.g. 3% = how many degrees?

$A/B * 100 = 3$, then $A/B = 3/100 = 0.03$
$= \tan^{-1}(0.03) = 1.72$ degrees

Figure 11-2: Slope formula.

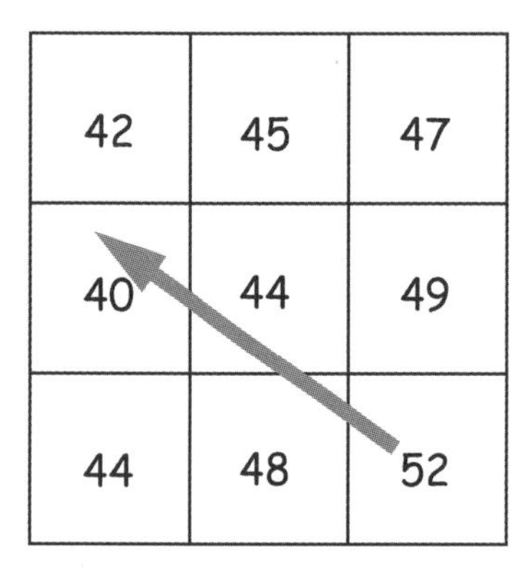

Figure 11-3: Aspect is often defined as the direction of the steepest slope.

the slope angle. A steeper rise or shorter run lead to a higher ϕ and hence steeper slope.

Slope may also be expressed as a percent, calculated by 100 times the rise over the run (Figure 11-2). Slopes expressed as a percent range from zero (flat) to infinite (vertical). A slope of 100 percent occurs when the rise equals the run.

Calculating slope from a raster data layer is a bit more complicated than in the cross-section view shown in Figure 11-2. Slope at a point in the landscape is typically measured in the steepest direction of elevation change. This steepest direction often does not fall in parallel to the raster rows or columns. Consider the cells depicted in Figure 11-3. Higher elevations occur at the lower right corner, and lowest elevations occur towards the upper left. The direction of steepest slope trends from one corner towards the other, but does not pass directly through the center of any cell. How do we obtain values for the rise and run? Which elevations should be used to calculate slope? Intuitively we should use some combination of a number of cells in the vicinity of the center cell, perhaps all of them. A number of researchers have investigated methods for calculating slope, but before we discuss these methods we should describe some general characteristics of slope calculation using raster data sets.

Elevation is often represented by the letter Z in slope, aspect, and other terrain analyses. These terrain functions are usually calculated using a symmetrical moving window. A 3 by 3 cell window is most common, although 5 by 5 and other odd-numbered windows are also used. Each cell in the window is assigned a subscript, and the elevation value found at a window location referenced by a subscripted Z value.

Figure 11-4 shows an example of a 3 by 3 cell window. The central cell has a value of 44, and is referred to as cell Z_0. The upper left cell is referred to as Z_1, the upper center cell as Z_2, and so on through cell Z_8 in the lower right corner.

Slope at each cell center is most commonly calculated from the formula:

$$s = \operatorname{atan}\sqrt{\left(\frac{dZ}{dx}\right)^2 + \left(\frac{dZ}{dy}\right)^2} \tag{11.1}$$

where s is slope, atan is the inverse tangent function, Z is elevation, x and y are the respective coordinate axes, and dZ/dx and dZ/dy are calculated for each cell based on the elevation values at a cell and surrounding cells. The symbol dZ/dx represents the rise (change in Z) over the run in the x direction, and dZ/dy the rise over the run in the y direction. These formula are combined to calculate the slope for each cell based on the combined change in elevation in the x and y directions.

Many different formulas and methods have been proposed for calculating dZ/dx and dZ/dy. The simplest, shown in Figure 11-4 and at the top of Figure 11-5, use cells adjacent to Z_0:

$$dZ/dx = (Z_5 - Z_4)/(2C) \tag{11.2}$$

$$dZ/dy = (Z_2 - Z_7)/(2C) \tag{11.3}$$

where C is the cell dimension and the Zs are defined as in Figure 11-4. This method uses the "four nearest" cells, Z_4, Z_5, Z_2, and Z_7, in calculating dZ/dx and dZ/dy. These four cells share the largest common border with the center. This "four nearest" method is perhaps the most obvious and provides reasonable slope values under many circumstances.

A common alternate method is known as a 3rd-order finite difference approach (Figure 11-5, bottom). This method for calculating dZ/dx and dZ/dy differs mainly in the number and weighting it gives to cells in the vicinity of the center cell. The four nearest cells are given a higher weight than the "corner" cells, but data from all eight nearest cells are used.

Several other methods have been developed that tests have shown may be better for calculating slope under certain conditions. These methods are judged as better because they on average produce more accurate slope estimates when compared to carefully collected field or map measurements of slope. However, no method has proved best under all terrain conditions.

Figure 11-5 shows the application of both the adjacent four nearest cells and the 3rd-order finite difference approaches to calculating dZ/dx, dZ/dy and then slope. The methods are shown with the respective kernals used in the moving windows. Note that the different kernals yield slightly different slope values, even when using identical input data.

Comparative studies using these and other methods have shown the two methods described here to be among the best for calculating slope and aspect when applied over a wide range of conditions. The four nearest cell method was among the best for smooth terrain, and the 3rd-order finite difference approach was among the best when applied to rough terrain. Alternative methods typically performed no better than the two methods described above.

Aspect is also an important terrain variable that is commonly derived from digital elevation data. Aspect is used to define the direction water will flow, the amount of sunlight a site may receive, and to determine what portion of the landscape is visible from any viewing point. The aspect at a point is the steepest downhill direction. The direction is typically reported as an azimuth angle, with zero in the direction of geo-

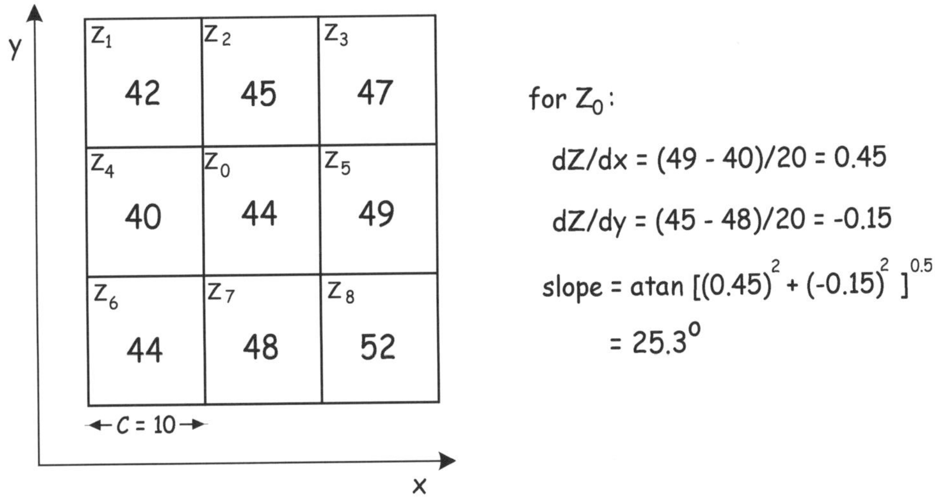

for Z_0:

$$dZ/dx = (49 - 40)/20 = 0.45$$

$$dZ/dy = (45 - 48)/20 = -0.15$$

$$\text{slope} = \text{atan}\,[(0.45)^2 + (-0.15)^2\,]^{0.5} = 25.3^{\circ}$$

Figure 11-4: Slope calculation based on cells adjacent to the center cell.

Four nearest cells

elevation values

42	45	47
40	44	49
44	48	52

← C = 10 →

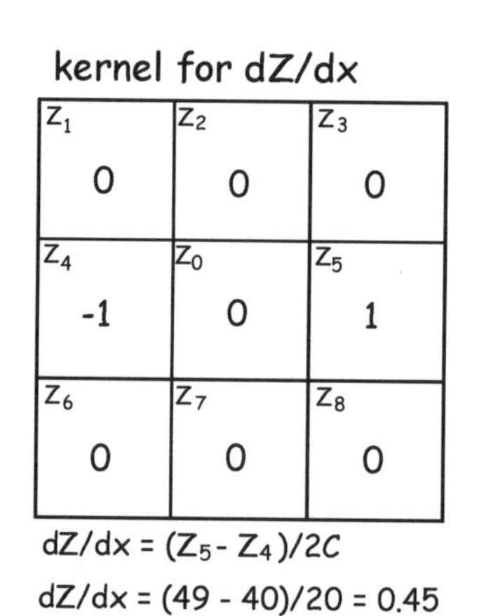

kernel for dZ/dy

Z_1 0	Z_2 1	Z_3 0
Z_4 0	Z_0 0	Z_5 0
Z_6 0	Z_7 -1	Z_8 0

$dZ/dx = (Z_5 - Z_4)/2C$

$dZ/dx = (49 - 40)/20 = 0.45$

$dZ/dy = (Z_2 - Z_1)/2C$

$dZ/dy = (45 - 48)/20 = -0.15$

$$\text{slope} = \text{atan}\left[(0.45)^2 + (-0.15)^2\right]^{0.5} = 25.3^\circ$$

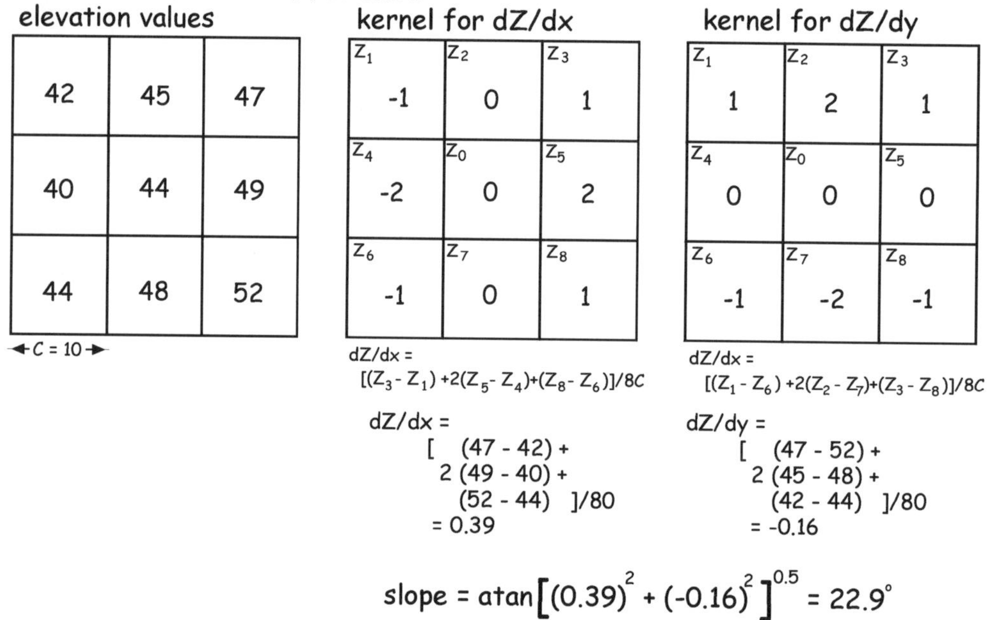

$$\text{slope} = \text{atan}\left[(0.39)^2 + (-0.16)^2\right]^{0.5} = 22.9^\circ$$

Figure 11-5: Four nearest cells (top) and 3rd-order finite difference (bottom) methods used in calculating slope. *C* is cell size and dZ/dx and dZ/dy are the changes in elevation (rise) with changes in horizontal position (run). Note that different slope values are produced by the different methods.

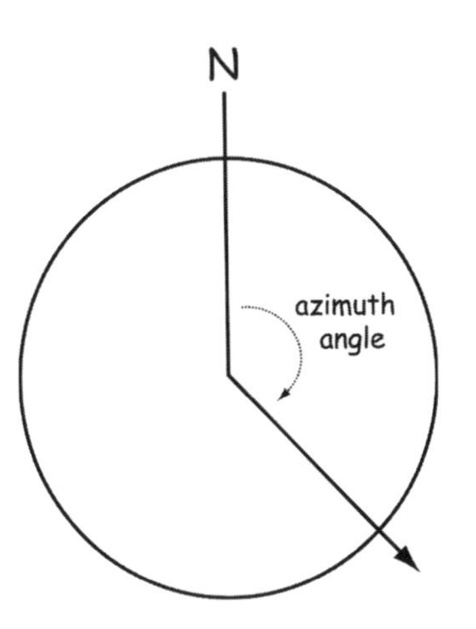

Figure 11-6: Aspect may be reported as an azimuth angle, measured clockwise in degrees from north.

graphic North, and the azimuth angle increasing in a clockwise direction. Aspects defined this way take values between 0 and 360 degrees. Flat areas by definition have no aspect because there is no downhill direction.

Aspect is most often calculated using dZ/dx and dZ/dy:

$$\text{aspect} = \text{atan}\left[\frac{\left(\frac{dZ}{dy}\right)}{\left(\frac{dZ}{dx}\right)}\right] \quad (11.4)$$

where atan is the inverse tangent function, and dZ/dy and dZ/dx are defined as above.

As with slope calculations, estimated aspect varies with the methods used to determine dZ/dx and dZ/dy. Tests have shown the four nearest cell and 3rd-order finite difference methods yield among the most accurate results, with the 3rd-order method among the best under a wide range of terrain conditions.

Hydrologic Functions

Digital elevation models are used extensively in hydrologic analyses. Water is basic to life, commerce, and comfort, and there is a substantial investment in water resource monitoring, gathering, protection, and management. Spatial analyses of water resources include a set of commonly used hydrologic functions.

Flow direction is used in many hydrologic analyses. Excess water at a point on the Earth will flow in a given direction. The flow may be either on or below the surface, but is most often in the direction of steepest descent. This is the same as the local aspect.

Flow directions may be stored as compass angles in a raster data layer. Acceptable values are from zero to 360 if the angle is expressed in degrees azimuth. Alternately, flow direction may be stored as a number indicating the adjacent cell to which water flows, taking a value from 1 to 8 or some other unique identifier for each adjacent cell. Flow direction is often calculated by estimating the aspect, which is the direction of steepest descent from a center cell to immediately adjacent cells (Figure 11-7).

Flow direction may deviate from the local aspect on flat areas or where two directions have the same and steepest descent. The flow direction will then be undefined for the local neighborhood. When the flow direction is not well defined for a cell the neighborhood may be successively increased until an unambiguous flow direction is defined.

Random errors in DEM elevation values often create spurious *pits* or *sinks*. Pits are cells that are lower than all surrounding cells, and they often cause problems when determining flow direction. Pit cells have no direction of steepest descent. Pits may exist in the real terrain surface, e.g., in *karst* regions where sinkholes occur on the surface due to collapsed subterranean caverns. Pits are found in most DEMs due to small random errors over locally flat or valley bottom surfaces. Many hydrologic functions give erroneous results when pits occur, so pits often must be filled prior to further analyses.

A *watershed* is an area that contributes flow to a point on the landscape. The uphill area that drains to any point on a landscape is the watershed for that point. Water falling anywhere in the upstream area of a watershed will pass through that point. Watersheds may be quite small, e.g., on a ridge or high on a slope the watershed may be only a few square meters in area. Local high points have watersheds of zero area because all water drains away. Watersheds may also be quite large, including continental areas that drain large rivers such as the Amazon or Mississippi River. Any point in the main channel of a large river has a large upstream watershed.

Watersheds may be easily identified once a flow direction surface has been determined (Figure 11-8). Flow direction is followed "uphill" from a point, until a downhill flow direction is reached. Each uphill cell may have many contributing cells, and the flow into each of these cells is also followed uphill. The uphill list is accumulated recursively until all cells contributing to a point have been identified, and thus the watershed is defined.

A *drainage network* is the set of cells through which surface water flows. Drainage networks are also based on the flow direction surface. Streams, creeks, and rivers occur where flow directions converge. Thus, the convergence of the flow direction may be used to produce a map of likely stream location, prior to field mapping of stream location (Figure 11-7). A drainage network may be simply defined as any cell which has a contributing watershed larger than some locally-defined threshold area. This is only an approximation, because a drainage network defined this way does not incorporate subsurface properties such as soil texture, depth, porosity, or subsoil water movement. Nonetheless a drainage network derived from terrain data alone is often a useful first approximation. The watershed for each cell may be calculated, and the area compared to

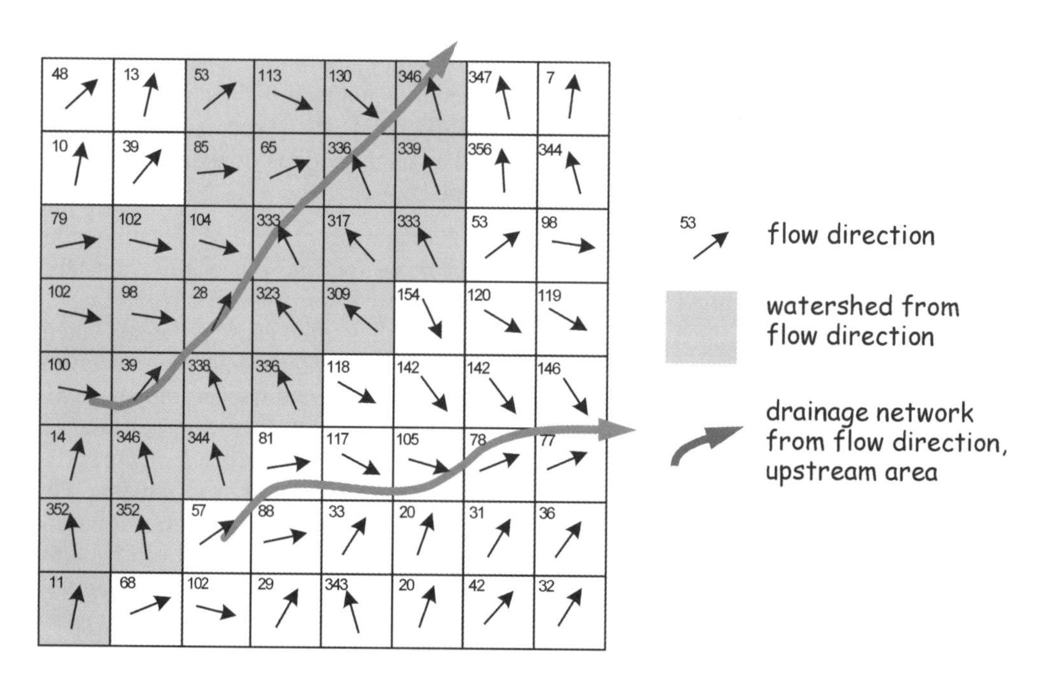

Figure 11-7: Flow direction, watershed, and drainage network shown for a raster grid. Elevation data are used to define the flow direction for each cell. These flow directions are then used to determine a number of important hydrologic functions.

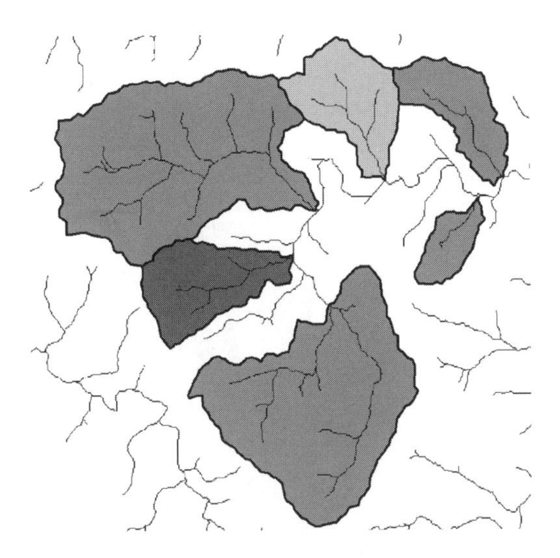

Figure 11-8: Drainage network and watersheds derived from a DEM. Flow direction was determined from local aspect. Upstream watersheds and the probable drainage paths were then determined from flow direction.

the threshold area. The cell is marked as part of the drainage network if the area surpasses the threshold.

There are many other hydrologically-related functions that may be applied. These generally provide data required for specific hydrologic models. Surface and subsurface flow modeling are old and active areas of research. Detailed discussions of these spatial hydrologic functions and analyses are included in the literature listed at the end of this chapter, and in the text of Chapter 12.

Viewsheds

The *viewshed* for a point is the collection of areas visible from that point. Views from any non-flat location are blocked by terrain. Elevations will hide points if the elevations are higher than the line of sight between the viewing point and target point (Figure 11-9).

Viewsheds and visibility analyses are quite important in many instances. High-voltage power lines are often placed after careful consideration of their visibility, because most people are averse to viewing them. Communications antennas, large industrial complexes, and roads are often located at least partly based on their visibility, and viewsheds are specifically managed for many parks and scenic areas.

A viewshed is calculated based on cell-to-cell intervisibility. A line of sight is drawn between the view cell and a potentially visible target cell (Figure 11-9). The elevation of this line of sight is calculated for every intervening cell. If the slope to a target cell is less than the slope to a cell closer to the viewpoint along the line of sight, then the target cell is not visible from the viewpoint. Specialized algorithms have been developed to substantially reduce the time required to calculate viewsheds, but in concept lines of sight are drawn from each viewpoint to each cell in the digital elevation data. If there is no intervening terrain the cell is classified as visible. The classification results in a set of areas that are visible, and a set that are not (Figure 11-10). Viewsheds for line or area features are the accumulated viewsheds from all the cells in those features.

Shaded Relief Maps

A *shaded relief map* is a depiction of the brightness of terrain reflections given a terrain surface and sun location. Although shaded relief maps are rarely used in analy-

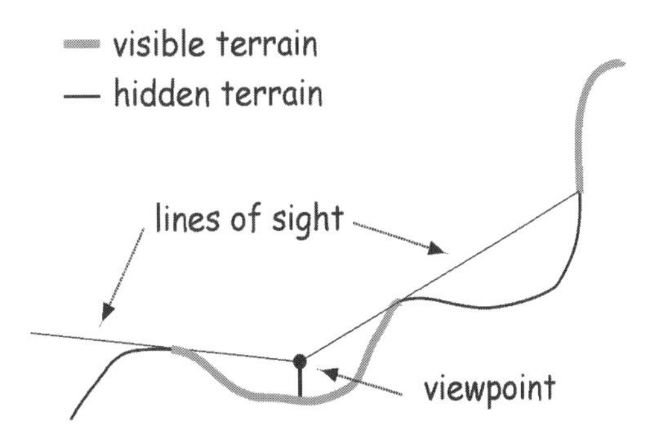

Figure 11-9: Mechanics of defining a viewshed.

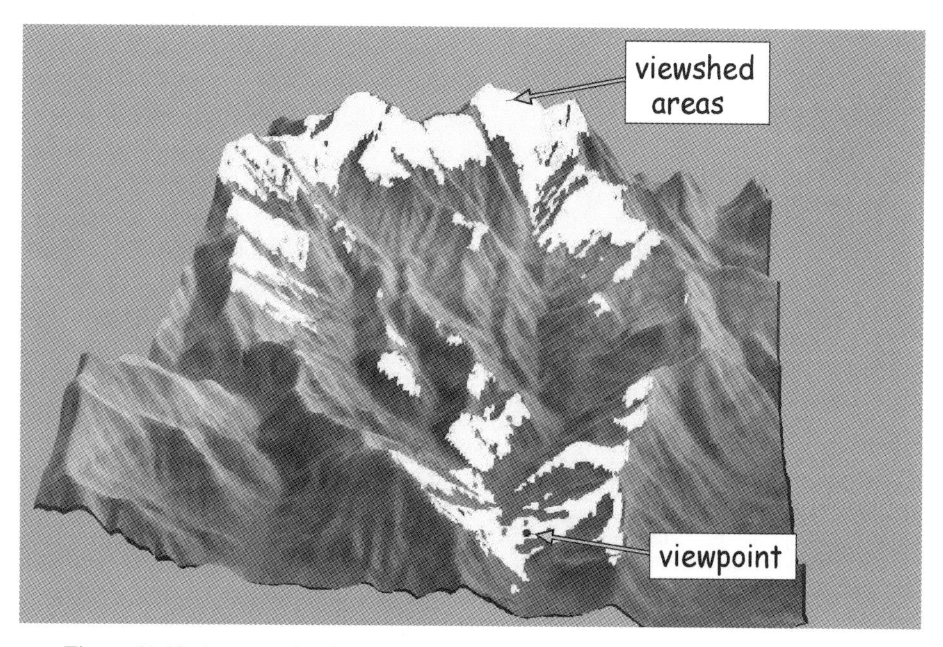

Figure 11-10: An example of a viewpoint, and corresponding viewshed.

ses, they are among the most effective ways to communicate the shape and structure of terrain features, and many maps include relief shading (Figure 11-1).

Shaded relief maps are developed from digital elevation data and a model of light reflectance. An artificial sun is "positioned" at a location in the sky and light rays projected onto the surface depicted by the elevation data. Light is modeled that strikes a surface either as a direct beam, from the sun to the surface, or from background "diffuse" sunlight. Diffuse light is scattered by the atmosphere, and illuminates "shaded" areas, although the illumination is typically much less than that from direct beam. Direct beam sunlight striking at perpendicular angles provides the brightest return, and hence appear light. As the angle between the direct beam and the ground surface deviates from perpendicular the brightness decreases. Diffuse sunlight alone provides a relatively weak return, and hence appears dark. Combinations of direct and diffuse light result in a range of gray shades, and this range depends on the terrain slope and angle relative to the sun's location. Hence, subtle variations in terrain are visible on shaded relief maps.

A shaded relief map is in some ways related to the visibility surface. Visibility to the sun is determined; if a cell is visible from the sun, the slope and aspect values are used to assign the cell brightness. The brightness at a cell depends on the angle at which the sunlight strikes visible cells. Sun visibility is calculated as with a surface visibility map, however the sun is considered at an infinite distance and above, rather than on, the surface of the Earth.

Summary

Terrain analyses are commonly performed with the framework of a GIS. These analyses are important because terrain governs where and how much water will accumulate on the landscape, how much sunlight a site receives, and the visibility of human activities.

Slope and aspect are two of the most used terrain variables. Both are commonly calculated via trignometric functions applied in a moving window to a raster DEM. Several kernels have been developed to calculate changes of elevation in x and y directions, and these component gradients are combined to calculate slope and aspect.

Terrain analyses are also used to develop and apply hydrologic functions and models. Watershed boundaries, flow directions, flow paths, and drainage networks may all be defined from digital elevation data.

Viewsheds are another commonly applied terrain analysis function. Intervisibility may be computed from any location on a DEM. A line of sight may be drawn from any point to any other point, and if there is no intervening terrain then the two points are intervisible. Viewsheds are often used to analyze the visiblity of landscape alterations or additions, for example, when siting new roads, powerlines, or large buildings.

Finally, relief shading is another common use of terrain data. A shaded relief map is among the most effective way to transmit information regarding terrain. Terrain shading is often derived from DEMs and depicted on maps.

Suggested Reading

Ayeni, O. O., Optimum sampling for digital terrain models, *Photogrammetric Engineering and Remote Sensing*, 1982, 48:1687-1694.

Band, L. E., Topographic partition of watersheds with digital elevation models, *Water Resources Research*, 1986, 22:15-24.

Berry, J. K., Fundamental operations in computer-assisted mapping, *International Journal of Geographic Information Systems*, 1987, 1:119-136.

Berry, J. K., A mathematical structure for analyzing maps, *Environmental Management*, 1986, 11:317-325.

Bonham-Carter, G. F., Geographic Information Systems for Geoscientists: Modelling with GIS, Pergamon, Ontario, 1996.

Burrough, P. A. and McDonnell, R. A., Principles of Geographical Information Systems, 2nd Edition. Oxford University Press, New York, 1998.

DeFloriani, L. and Magillo, P., Visibility algorithms on triangulated digital terrain models, *International Journal of Geographical Information Systems*, 1994, 8:13-41.

Dozier, J. and Frew, J., Rapid calculation of terrain parameters for radiation modeling from digital elevation data, *IEEE Transactions on Geoscience and Remote Sensing*, 1990, 28:963-969.

Dubayah, R. and Rich, P. M., Topographic solar radiation models for GIS, *International Journal of Geographical Information Systems*, 1995, 405-419.

Fisher, P. F., Reconsideration of the viewshed function in terrain modelling, *Geographical Systems*, 1996, 3:33-58.

Hodgson, M. E., What cell size does the computed slope/aspect angle represent?, *Photogrammetric Engineering and Remote Sensing*, 1995, 61:513-517.

Horn, B. K., Hill shading and the reflectance map, *IEEE Proceedings on Geosciences*, 1981, 69:14-47.

Jenson, S.K., Applications of hydrologic information automatically extracted from digital elevation models, *Hydrologic Processes*, 1991, 5:31-44.

Jones, N. L., Wright, S. G., and Maidment, D. R., Watershed delineation with triangle-based terrain models, *Journal of Hydraulic Engineering*, 1990, 116:1232-1251.

Moore, I.D., and Grayson, R.B., Terrain-based catchment partitioning and runoff prediction using vector elevation data, *Water Resources Research*, 1991, 27:1177-1191.

Moore, I. D., Turner, A, Jenson, S., and Band, L., GIS and land surface-subsurface process modelling, (in) Environmental Modeling with GIS, Goodchild, M. F., Parks, B. O., and Steyaert, (eds.), Oxford University Press, New York, 1993.

Martz, L.W., and Garbrecht, J., The treatment of flat areas and depressions in automated drainage analysis of raster digital elevation models, *Hydrological Processes*, 12:843-856.

Skidmore, A. K., A comparison of techniques for calculating gradient and aspect from a gridded digital elevation model., *International Journal of Geographical Information Systems*, 1989, 3:323-334.

Tomlin, C. D., Geographic Information Systems and Cartographic Modeling, Prentice-Hall, New Jersey, 1990.

Wilson, J and Gallant, J. eds., Terrain Analysis: Principles and Applications, Wiley, New York, 2000.

Wood, R., Sivapalan, M., and Robinson, J., Modeling the spatial variability of surface runoff using a topographic index, *Water Resources Research*, 33:1061-1073.

Study Questions

What is a digital elevation model, and why are they used so often in spatial analyses?

How are digital elevation data created?

Can you define slope and aspect, and how they are derived from digital elevation data?

What other hydrologic functions are commonly calculated from digital elevation models?

What is a viewshed, when are they used, and how are they calculated?

12 Spatial Models and Modeling

Introduction

A model is a description of reality. It may be a static reproduction that represents the basic shape and form of an object, a conceptual description of the key elements and processes, or a sophisticated replica. Our interests in spatial models are restricted to computer-based models of spatial phenomena. These models describe the basic properties and/or processes for a set of spatial features such that the models may be used to study spatial objects, or to study spatially-related phenomena.

Our computer-based models use spatial data, and are developed and run using some combination of GIS, general and specialized computer programming languages, and spatial and non-spatial analytical tools. Spatially explicit models are one of the main benefits of geographic information systems technologies, and many spatial models are based on data in a GIS. Spatial data may also be prepared in a GIS, and exported to a model developed and run outside a GIS.

While there may be as many classes of models as there are modelers, for our purposes we split spatial models into three broad classes: *cartographic models*, *spatio-temporal models*, and *network models*. *Cartographic models* involve the application of spatial operations such as buffers, interpolation, reclassification, and overlay to solve problems. Suitability analyses, defined here as the classification of land according to their utility for specific uses, are among the most common cartographic models.

Spatio-temporal models are dynamic in both space and time. They differ from cartographic models in that time passes explicitly within the running of the model, and changes in time-driven processes within the model cause changes in spatial variables. The dispersion of oil after a spill might be analyzed via a spatio-temporal model. Currents, winds, wave action, the physics of oil separation and evaporation on exposure to air might be combined in a model to predict the temporally-changing location and quality of an oil slick.

Finally, *network models* may be either temporally dynamic or static, but they are constrained to modeling the flow and accumulation of resources through a network. Detailed automobile traffic flow models might be used to analyze congestion or plan road construction. Traffic flow is modeled through a network, mediated by spatially and temporally varying constraints on speed and direction.

Most cartographic models are temporally static because they represent spatial features at a fixed point or points of time. Data in base layers are mapped for a given time. These data are the basis for spatial operations that may create new data layers. For example, we may be interested in identifying the land that is currently most valuable for agriculture. Costs of production may depend on the slope (steeper is costlier), soil type (ranges of fertility), current land cover (built-up is unsuitable, forests more expensive to clear), or distance to roads or markets, while agricultural production may depend on soil types, topography (neither flooded nor drought-prone), and the ability to irrigate. Spatial data on elevation, soil properties, current land use, roads, market location, and irrigation potential may be combined to rank sites by production value. We may use a mathematical relationship for specific calculations of average costs and revenues, e.g., agronomists may have developed the relationships between soil types and average corn production in the region, and we may use a cost per mile for transport based on local rates. These spatial data are combined in a cartographic model to assign a land value for each parcel in a study region. The model is temporally static in that the values for the spatial variables do not change during the model, e.g., soil fertility or distance to roads is fixed during the analyses.

Cartographic models are generally not temporally dynamic, even though they may be used to analyze change. For example, we may wish to analyze vegetation change over a 10-year period, based largely on vegetation maps produced at the start and end of the period. Each data layer represents the vegetation boundaries at a fixed point in time. The model is static in that the polygon boundaries for a given layer do not change. There may be two vegetation data layers, each corresponding to a different point in time, and the vegetation boundaries are mapped as found at each time interval. Our cartographic model includes a temporal component in that it compares vegetation change through time, but the cartographic model does not generate new boundaries of polygons or any other characteristics of spatial features. Boundaries may be a composite of those lines that exist in the input data layers, but new lines at new coordinate locations are not generated. Most spatial modeling or models conducted in the framework of GIS have been cartographic models that are temporally static in this manner.

Spatio-temporal models differ from cartographic models by the inclusion of time-driven processes within the framework of the model. These processes are typically quite detailed and include substantial computer code to represent important sub-processes. The evaporation of oil when predicting the fate of an oil spill is a good example of a sub-process in a spatial model. Oil evaporation rates depend on many factors, including oil viscosity, component oil fractions, wind speed, temperature, wave height and action, and sunlight intensity. These processes may be modeled by suitable functions applied to spatially defined patches of oil. The sub-model may estimate evaporation of various components of the oil in the patch, and update the characteristics of oil in that patch. Oil chemistry and viscosity may change due to more rapid evaporation of lighter components, in turn affecting future evaporation calculations. Spatial features may change through time due to the coded process, e.g., the boundary defining an oil spill may vary as the model progresses.

Spatio-temporal models are typically more limited than cartographic models in the range and number of spatial themes included, but provide a more mechanistic representation of time varying processes. Substantial effort goes into developing sub-models of important processes. Model components and structures focus on one or a few key output spatial variables, and input data themes are included only as they are needed by these sub-process models. These temporally dynamic models explicitly calculate the changes in the output spatial variables through time. Feature boundaries, point feature locations, and attribute variables that reflect the spatial and aspatial characteristics of key output variables may change within the model run, typically multiple times, and with an explicit temporal frequency.

Network models allow us to analyze the flow of resources within a set of connected features. Resources are broadly defined and may be automobiles when modeling traffic flow, goods when determining optimal delivery routes, or customers when using network analysis to locate a new business. The cost of travel along the network is used to select routes or locations, determine network loads, and calculate travel times. Networks are used in a broad range of fields, particularly in transportation and utility management and in many business applications.

Cartographic Modeling

A *cartographic model* provides information through a combination of spatial data sets, functions, and operations. These functions and operations often include reclassification, overlay, interpolation, terrain analyses, and buffering and other proximity functions. Multiple data layers are combined via these operations, and the information is typically in the form of a spatial data layer. Map algebra, described in Chapter 10, is often used to specify cartographic models for raster data sets.

Suitability analyses are perhaps the most common examples of cartographic models. These analyses rank land according to their utility for various uses. Suitability analyses often involve the overlay, weighting, and rating of multiple data layers to categorize lands into various classes. Relevant data layers are combined and the resultant polygons are classified based on the combination of attributes. Figure 12-1 illustrates a simplistic cartographic model for the identification of potential park sites. Suitable sites are those that are A) near lakes, B) near roads, and C) not wetlands. Three input data layers are used, containing lakes, roads, and hydric status for a common study area. Spatial operations are applied to spatial data layers, including reclassification, buffering, and overlay. These result in a suitability layer. This suitability layer may then be used to

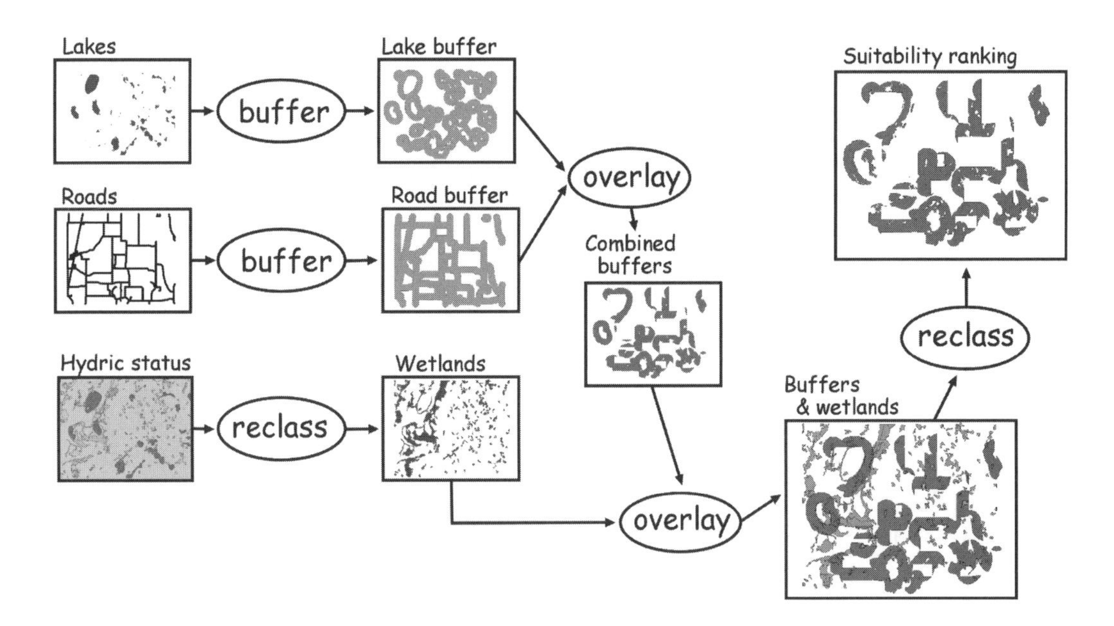

Figure 12-1: An example of a cartographic model. The model identifies suitable sites based on the proximity to roads and lakes and the absence of wetlands.

narrow sites for further evaluation, identify owners, or otherwise aid in park site selection.

Cartographic models have been used for a variety of applications. These include land use planning, transportation route and corridor studies, the design and development of water distribution systems, modeling the spread of human disease or introduced plant and animal species, building and business site selection, pollution response planning, and endangered species preservation. Cartographic models are so extensively used because they provide information useful to managers, the public, and policy makers, and help guide decisions requiring the consideration of spatial location across multiple themes.

Cartographic models are often succinctly represented by *flowcharts*. A flowchart is a graphic representation of the spatial data, operations, and their sequence of use in a cartographic model. Figure 12-2 illustrates a flowchart of the cartographic model illustrated in Figure 12-1. Suitable sites need to be identified that are not wetlands, near roads, and near lakes. Data layers are represented by rectangles, operations by ellipses, and the sequence of operations by arrows. Operations are denoted inside each ellipse. Flowcharts are often required by an agency or organization to document a completed spatial analysis. Because a consistent set of symbols aids in the effective communication of the cartographic model, a standard set of symbols and flowcharting methods may be required.

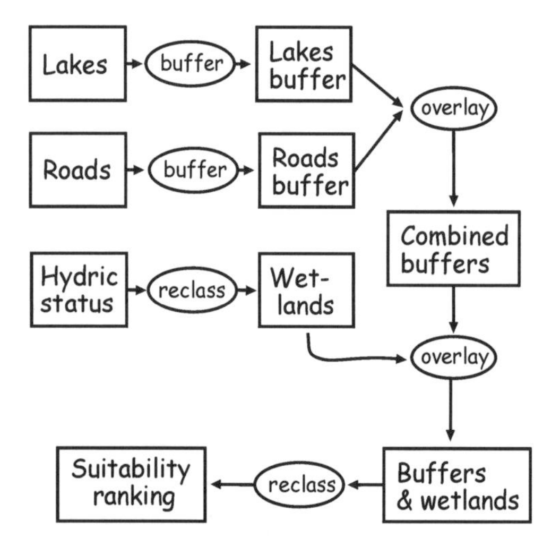

Figure 12-2: A flowchart depicting the cartographic model in Figure 12-1. The flowchart is a rapidly-produced shorthand method of representing a sequence of spatial operations.

Flowcharts are useful during the development and application of a cartographic model. Flowcharts aid in the conceptualization and comparison of various competing approaches and may aid in the selection of the final cartographic model. A sketch of the flowchart is often a useful and efficient framework for documenting the application of the cartographic model. File locations, work dates, and intermediate observations can be noted with reference to the flowchart, or directly on a copy of the flowchart.

Cartographic modeling often produces a large number of "intermediate" or temporary data layers that are not required in the final output or decision-making. Our example in Figure 12-2 illustrates this. The needed information is contained entirely within the suitability ranking data layer. This layer summarizes the ranking of lands based on the provided criteria. Five other data layers were produced during this cartographic modeling. Buffer, recoded, and overlay layers were necessary intermediate steps, but in this analysis their utility was temporary. Once the layers were included in subsequent operations they were no longer needed. This proliferation of spatial data layers is common in cartographic modeling, and this may cause problems as the new layers and other files proliferate in the computer workspace. Frequent cleaning of the workspace is helpful, as this removes un-needed files.

Much of the power of cartographic modeling comes from the flexibility of basic spatial analysis functions. Spatial functions and operations constitute a set of tools that may be mixed and matched in cartographic models. Overlay, proximity, reclassification, and most other spatial analysis tools are quite general. These tools may be combined in an infinite number of ways, by selecting different tools and by changing their sequence of application. For example, differences in distances, thresholds, and reclassification tables may be specified. These variations will result in different output data layers, even when using the same input data layers. With a small set of tools and data layers we may create a huge number of cartographic models. Designing the best cartographic model to solve a problem - the selection of the appropriate spatial tools and the specification of their sequence - is perhaps the most important and often the most difficult process in cartographic modeling.

Designing a Cartographic Model

Most cartographic models are based on a set of criteria. Unfortunately, these criteria are often initially specified in qualitative terms, such as "the parcel must be large". A substantial amount of interpretation is often required in translating the criteria in a suitability analysis to a specific sequence of spatial operations. In our present example we must quantify what is meant by "large". General or qualitative criteria may be provided and these must be converted to specific, quantitative measures. The conversion from a qualitative to quantitative specification is often an iterative process, with repeated interaction between the analyst developing and applying the cartographic model and the manager or decision-maker who will act on the developed information.

We will use a home-site selection exercise to demonstrates this process. The problem consists of ranking sites by suitability for home construction. The area to be analyzed has steep terrain and is in a seasonally cold climate. There are four criteria:

a) Slopes should not be too steep. Steep slopes may substantially increase costs or may preclude construction.

b) A southern aspect is preferred, to enhance solar warming.

c) Soils suitable for on-site septic systems are required. There is a range of soil types in the study area, with a range of suitabilities for septic system installation.

d) Sites should be far enough from a main road to offer some privacy, but not so far as to be isolated.

These criteria must be converted to more specific restrictions prior to the development and application in a cartographic model. The decision-maker must specify what sort of classification is required. Is a simple binary classification needed, with suitable and unsuitable classes, or is a broader range of classes needed? If a range of classes is specified, is an ordinal ranking acceptable, or is an interval/ratio scale preferred? These questions are typically answered via discussions between the analyst and the decision-makers. Each criterion may then be defined once the type and measurement scale of the results are specified. It may be fairly simple to establish the local slope limit that prohibits construction. For example, conversations with local building experts may identify 30 degrees as a threshold beyond which construction is infeasible. Further work is required to quantify how slopes affect construction costs. Similar refinements must be made for each criterion. We must quantify the range and any relative preferences for southern aspects, relative soil suitabilities, what defines a main road, and what constitutes short and long distances.

A second key consideration involves the availability and quality of data. Do the required data layers exist for the study area? Are the spatial accuracies, spatial resolution, and attributes appropriate for the intended analysis? How will map generalizations affect the analysis, e.g., will inclusions of different soil types in a soil polygon lead to inappropriate results? Is the minimum mapping unit appropriate? If not, then the requisite data must be obtained or developed, or the goals and cartographic model modified.

Once the criteria are precisely defined and appropriate data obtained, we often face the more difficult problem of assigning the relative weightings among criteria. How important is slope relative to aspect? Will an optimum aspect offset a moderately steep site? How important is isolation relative to other factors? Because the criteria will be combined in a suitability data layer, the relative weightings given each criterion will influence the results. Different relative weights are likely to result in different suitability rankings. It is often difficult to assign these relative weights in an objective fashion, particularly when suitability depends on the non-quantifiable measures.

The assignment of relative weightings is easiest when the importance of the various criteria may be expressed on a common scale. In our example, we may be able to assign a monetary value to slope effects due to increased construction costs. Soil types may be categorized based on their septic capacity. Different septic systems may be required for different types of soils, either through larger drain fields or the specification of mound vs. field systems. Costs could be estimated based on the variable requirements set by soil type. Nuisance cost for noise and distance cost in lost time or travel might also be quantified monetarily. The reduction of all criteria to a common scale removes differential weighting among criteria. Criteria may be equally weighted in combination by virtue of the common scale.

There are many instances where a common measurement scale does not exist and it is not possible to develop one. Many rankings are based on variables that are difficult to quantify. Personal values may define the distances from a road that constitute "isolated" vs. " private", or what is the relative importance of slope vs. construction cost. In such cases the scales and weights may be defined in conference with the decision-maker. Expert opinion, group interviews, or stakeholder meetings may be required when there are multiple or competing parties. Multiple model runs may be required, each run with a different set of relative weights within and among criteria. These multiple runs may reveal criteria that are important or unimportant under all viable weightings, and those that are influential under a limited range of weightings.

Cartographic Models: a Detailed Example

Here we provide a detailed description of the steps involved in specifying and applying a cartographic model. We use a refinement of the general criteria for home site selection described in the prior section. These general criteria are listed on the left side and refined criteria are shown on the right side of Table 12-1. These refined criteria may have been defined after further discussion with the decision-makers, local area experts, and a review of available data and methods.

These criteria are to be weighted equally, and to produce a binary classification. Sites are considered suitable if they meet all criteria, and are unsuitable if they do not meet one or more of the criteria.

In our example we will apply the cartographic model described by the flowchart in Figure 12-2 to a small watershed in a mountainous study area. Application of the refined criteria require three base data layers - elevation, soils, and roads. For this example we assume the three data layers are available at the required positional and attribute accuracy, clipped to the study area of interest. The need for new data layers often becomes apparent during the process of translating the initial, general criteria to specific, refined criteria, or during the development of the flowchart describing the cartographic model. Once data availability and quality have been assured we may complete the final flowchart.

Figure 12-3 contains a flowchart of a cartographic model that may identify suitable sites. Spatial data layers are shown as rectangles, with a descriptive data layer name included within the rectangle. Spatial operations or functions are contained in ellipses, and arrows define the sequence of data layers and spatial operations. The three base data layers, elevation, soils, and roads, are shown at the top of the flowchart.

There are three main branches in the flowchart in Figure 12-3. The left-most

Table 12-1 Original and refined criteria for cartographic model example.

General Criteria	Refined Criteria
Slopes not too steep	Slopes < 30 degrees
Southern aspect preferred	90 < Aspect < 270
Soils suitable for septic system	Specified list of septic-suitable soil units
Far enough from road to provide privacy, but not isolated	300 meter < distance to road < 2000 meters

branch addresses the terrain-related criteria, the center branch addresses the soils criterion, and the right branch applies the road distance criteria. All three branches join in the cartographic model, producing a final suitability classification.

The left branch of the cartographic model is shown in detail in Figure 12-4. This and subsequent detailed figures show a thumbnail of the spatial data layers at each step in the process. Data layer names are adjacent to the spatial data layer. The first two criteria involve terrain-related constraints. Suitable sites are required to possess a restricted set of slopes and aspects. These criteria require the calculation of slope and aspect data layers, and the classification of these layers into areas that do and do not meet the respective criteria. The elevation data layer is shown at the top of Figure 12-4, low elevations in black through higher elevations in lighter shades. There are two main stream systems in the study area, one running from west to east in the northern portion of the study area, and one running from south to north. Highland areas are found along the north, west, and east margins of the study area.

Slope and aspect are derived from the elevation data layer (Figure 12-4). Lower slope values are shown in light shades, higher slope values are shown in dark shades, and aspects are shown in a range of light to dark shades from 0 to 360 degrees. Slope and aspect layers are reclassified based on the threshold values specified in the criteria listed in table 12-1. A reclassification table is used to assign values to the **slope_suit** variable based on the slope layer. Cells with a **slope_val** less than 30 are assigned a **slope_suit** of 1, while cells with a **slope_val** of 30 or higher are given a value **slope_suit** of 0. Aspect values are also reclassified using a table.

Slope and aspect layers are combined in an overlay, converted from raster to vector, and reclassified to produce a suitable terrain layer (Figure 12-5). Raster to vector conversion is chosen because two of the three base data layers are in a vector format, and because future complex selections might be better supported by the attribute data structure used for vector data sets. This conversion creates polygons that have the attributes of the input raster data layer. Note this conversion takes place after the raster layers have been reclassified into a small number of classes, and after the data have been combined to a single layer in an overlay. Raster to vector conversion proceeds more quickly

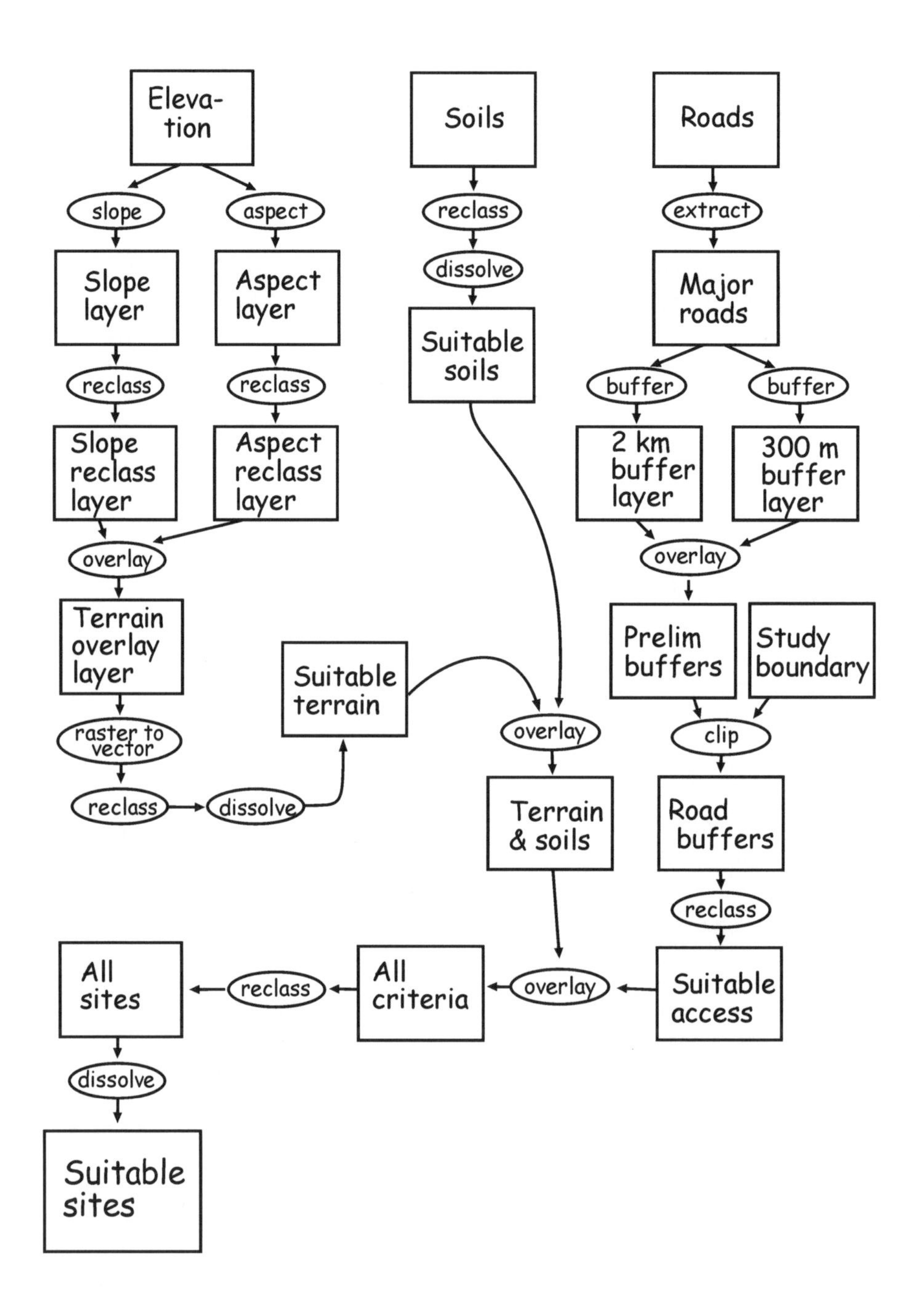

Figure 12-3: Flowchart for the homesite suitability cartographic model. Three basic data layers are entered. A sequence of spatial operations is used to apply criteria and produce a map of suitable sites.

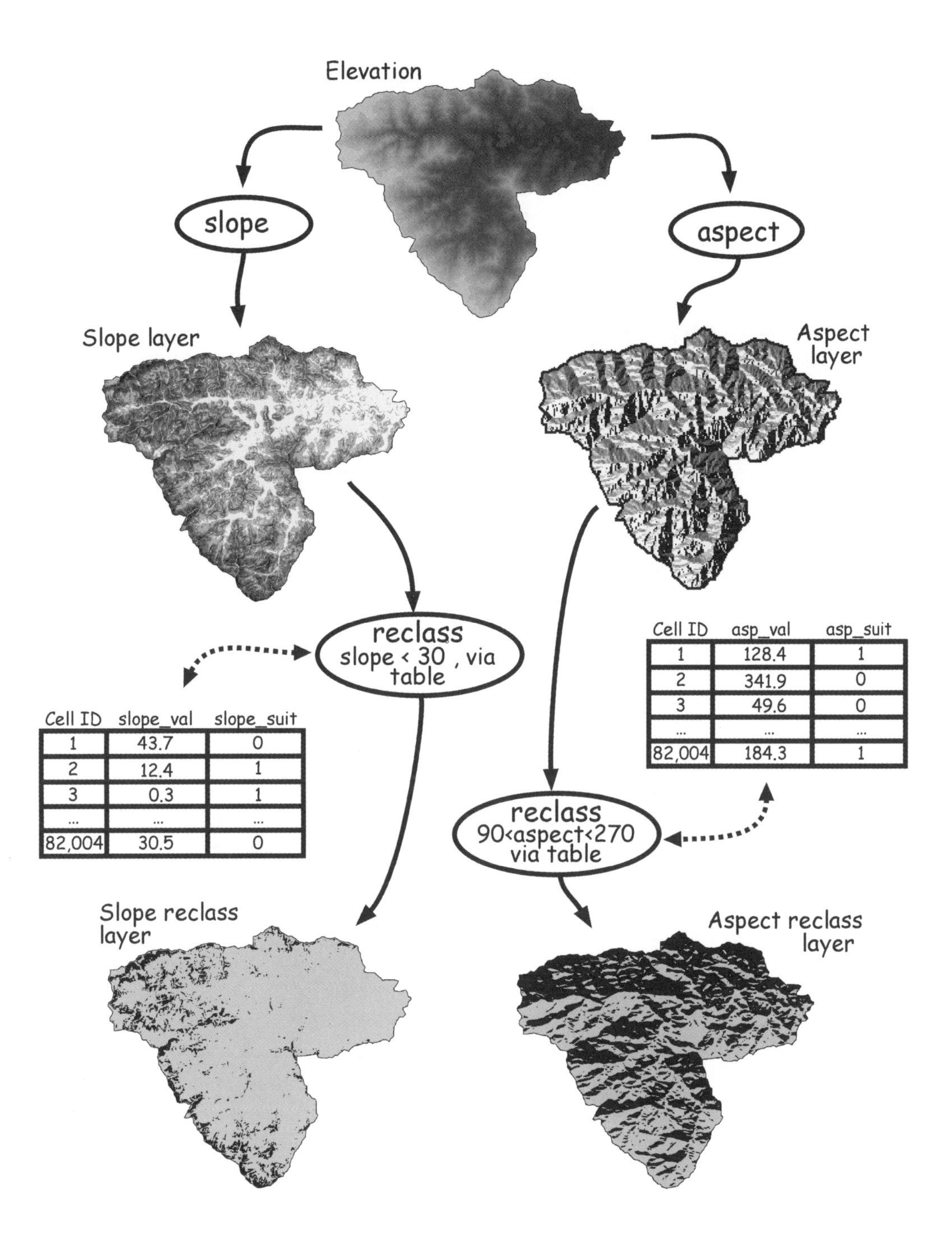

Figure 12-4: A detailed depiction of the left-most branches of the cartographic model shown in Figure 12-2. Slope and aspect are derived from an elevation data layer for the study region. Both layers are then reclassified using a table assignment. Slope values < 30 are reclassified as suitable (gray), all other slopes as unsuitable (black). Aspect values between 90 and 270 are reclassified as suitable (gray), all others as unsuitable (black).

after the number of raster classes has been reduced and the data combined in a single terrain-suitability layer. The terrain overlay must then be reclassified to identify those areas that meet both the slope and aspect criteria (see the terrain suitability coding in Figure 12-5). Those polygons with a 1 for both **slope_suit** and **asp_suit** are assigned a value of 1 for **terrain_suit**. All others are given a value of 0, indicating they are unsuitable home sites based on the slope and/or aspect criteria. Because we wish to reduce the number of redundant polygons where possible, a dissolve is applied after the reclassification. This substantially reduces the size of the output data set, and speeds future processing. Reclassified, dissolved terrain data are saved in a layer labeled **Suitable Terrain** (Figure 12-5).

The central branch of the cartographic model is shown in Figure 12-6. Digital county soil surveys are available that map homogenous soil mapping units. Attribute

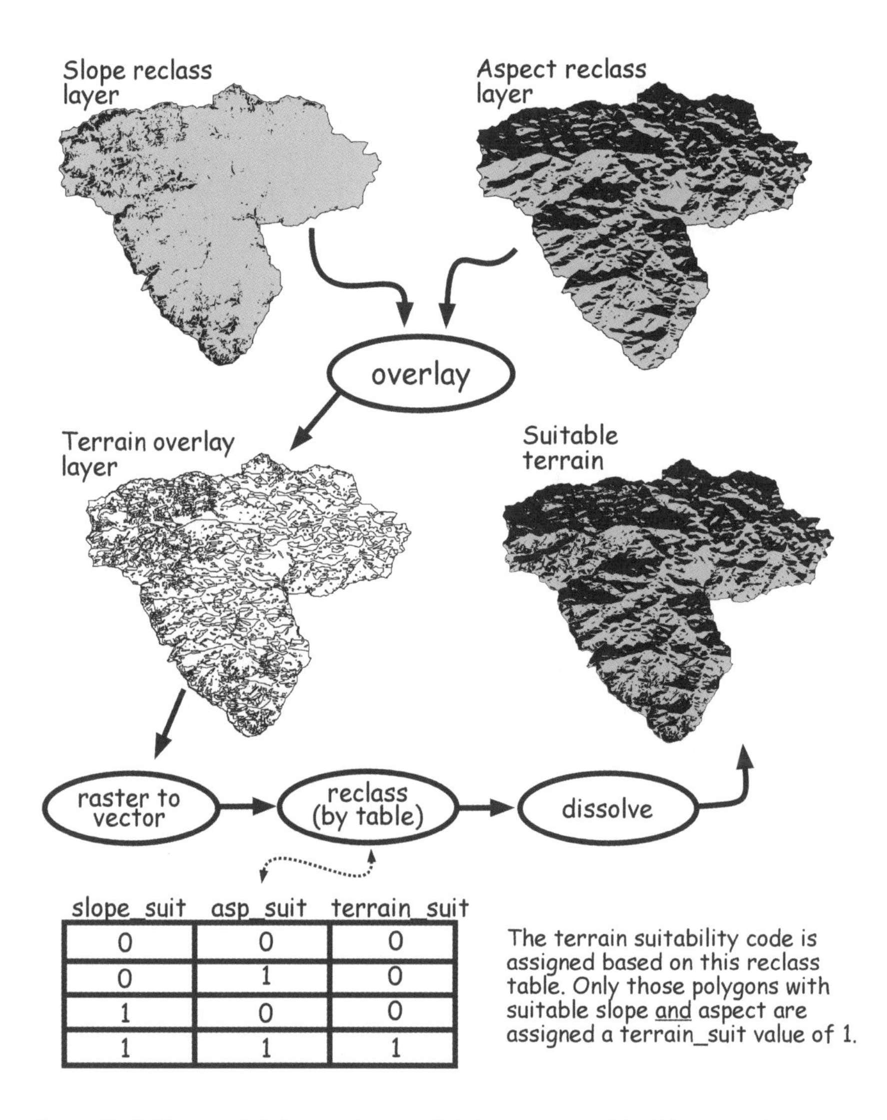

slope_suit	asp_suit	terrain_suit
0	0	0
0	1	0
1	0	0
1	1	1

Figure 12-5: The recoded slope and aspect data layers are combined in an overlay operation, and the result reclassified. Suitable terrain is shown in gray, unsuitable in black.

data are attached to each polygon, including soil type and soil suitability for septic systems. Soils data for the study area may be reclassified based on these septic suitability attributes. A reclassification table assigns a value of 1 to the variable soil_suit if the soil type is suitable for septic systems, 0 if the soil type is not (Figure 12-6).

After reclassification there may be many adjacent soil polygons with the same soil_suit value. These are grouped using a dissolve operation (data between reclass and dissolve not shown in the figure, see Chapter 9 for an example). The dissolve removes boundaries between like polygons, thereby substantially reducing the number of polygons and hence the number of entries in the attribute table. This may be particularly

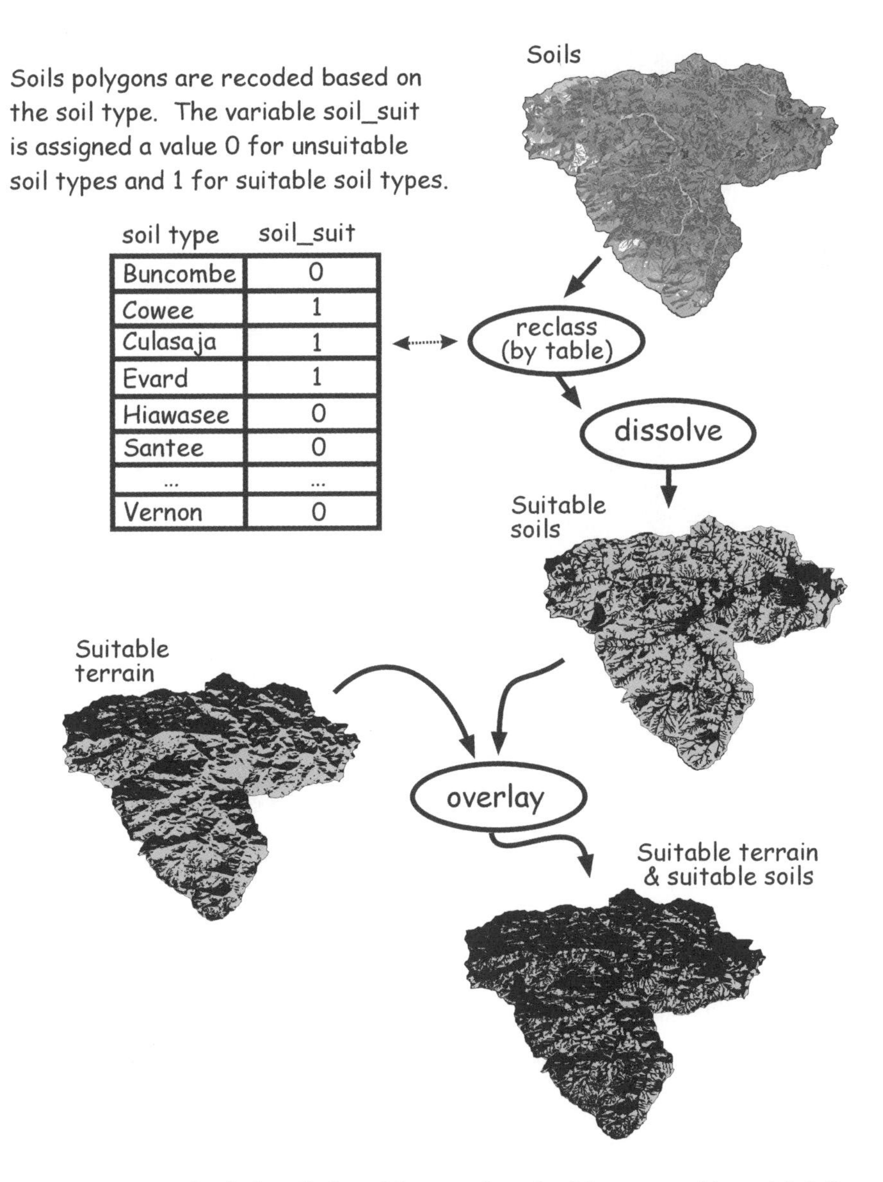

Figure 12-6: A detailed prediction of the center branch of the cartographic model. Soils data are reclassified into those suitable for septic systems and those not, and then combined with the suitable terrain data layer to identify sites acceptable based on both criteria.

important with complex data sets such as county soils data, or with converted raster data, as these often have thousands of entries, many of which will be removed after the dissolve.

The right branch of the cartographic model is presented in Figure 12-7. The Roads data layer is obtained and Major roads extracted. This has the effect of removing all minor roads from consideration in further analyses. What constitutes a major road has been defined prior to this step. In this case all divided and multi-lane roads in the study area were selected. Two buffers are applied, one at a 300 meter distance and one at a two kilometer distance from major

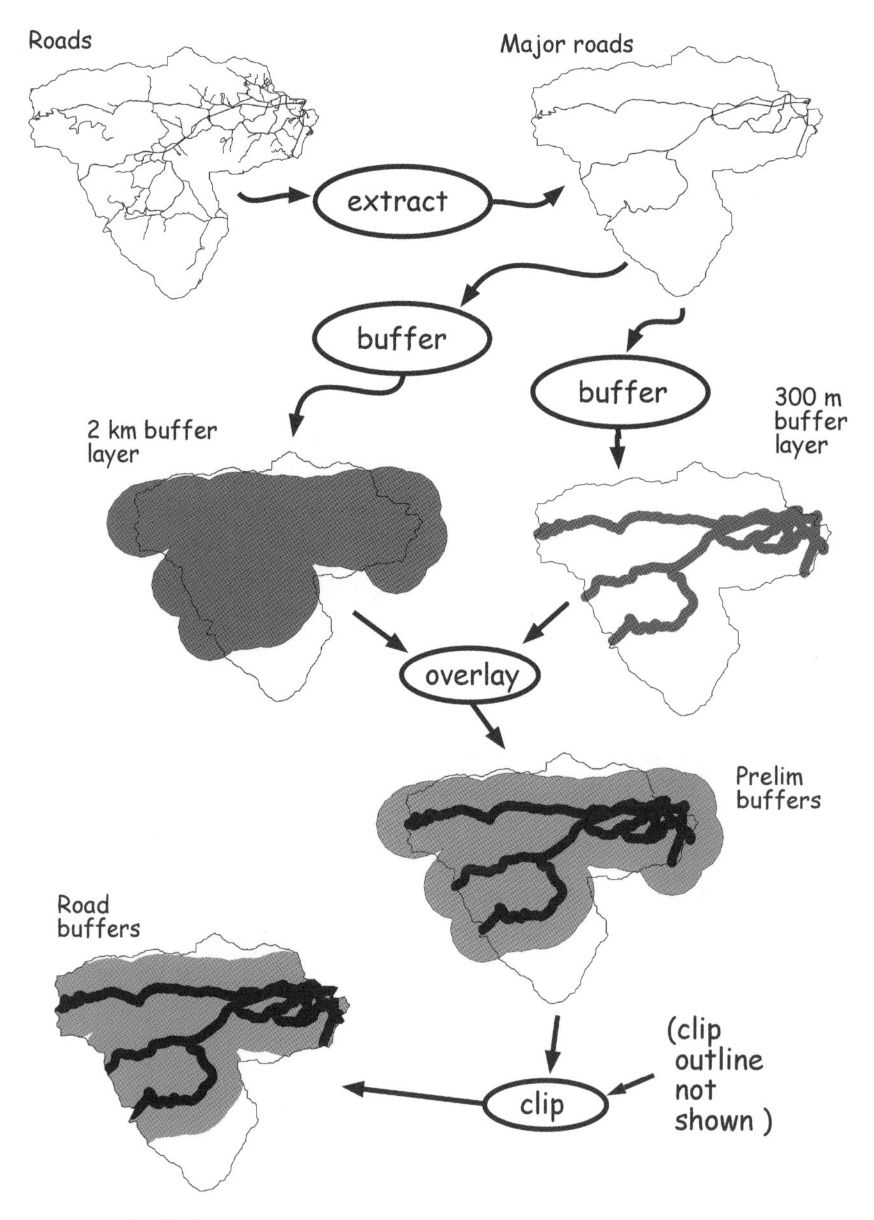

Figure 12-7: A detailed representation of the right branch of the cartographic model shown in Figure 12-3. Roads are buffered at 300 meters and 2 kilometers, and these buffers overlain. The buffers are clipped to the study region, and suitable areas more than 300 meters and less than 2 kilometers from roads identified.

roads. These buffers are then overlain. Because the buffer regions extend outside the study area, the buffers must be clipped to the boundary of the study area. These data are then reclassified into suitable and unsuitable areas, resulting in the **Road buffers** layer (lower left, Figure 12-7).

All data layers are combined in a final set of overlays and reclassifications (Figure 12-8). The **Suitable access** layer, derived from the roads data and criteria, is combined with the **Terrain & soils** layer. The **All criteria** layer contains the required spatial data to identify suitable vs. unsuitable sites. This overlay layer must be reclassified based on the road, soil and terrain suitability variables, classifying all potential sites into a suitable or unsuitable class. A final dissolve yields the final digital data layer, **Suitable sites**.

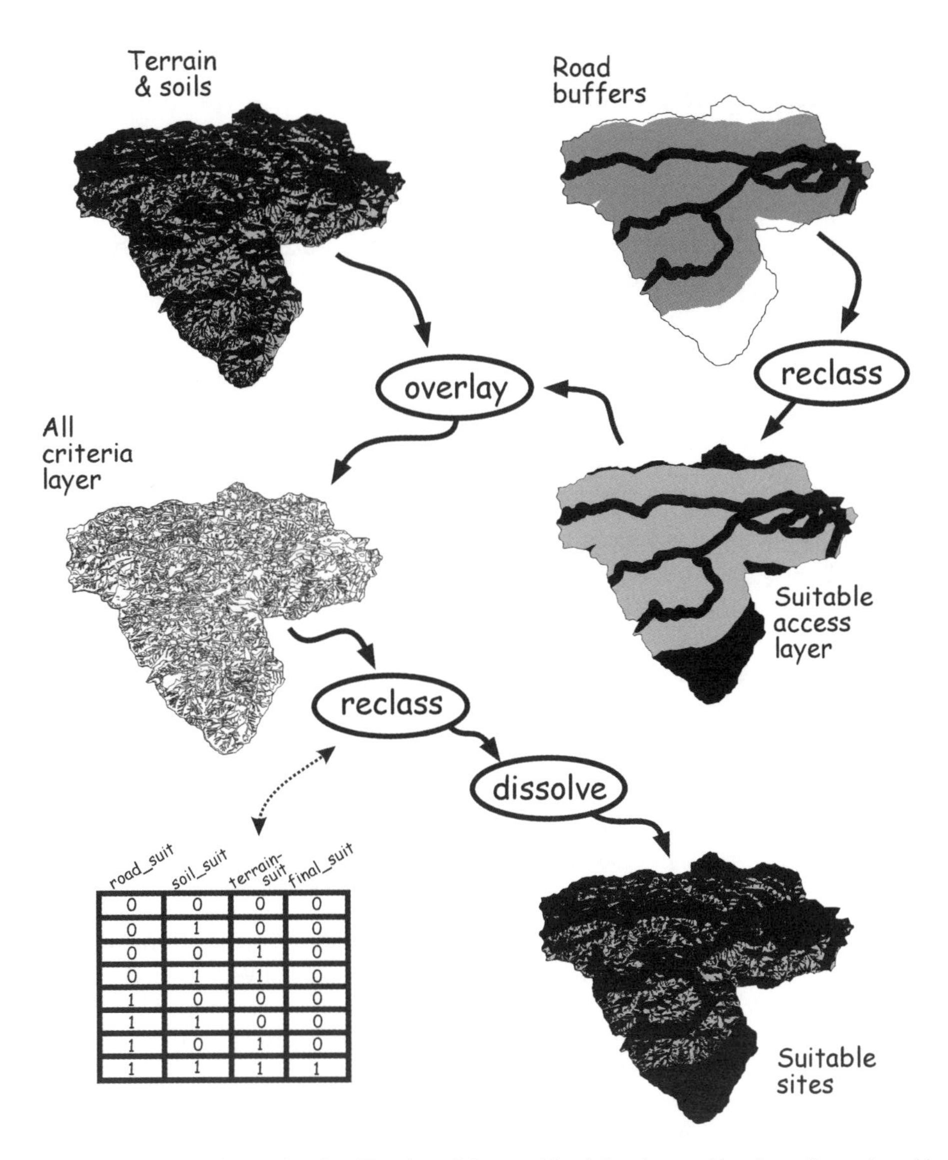

road_suit	soil_suit	terrain-suit	final_suit
0	0	0	0
0	1	0	0
0	0	1	0
0	1	1	0
1	0	0	0
1	1	0	0
1	0	1	0
1	1	1	1

Figure 12-8: The overlay and reclassification of the combined data layers. Terrain, soils, and road buffer data are combined in an overlay. These data are reclassed based on the suitability criteria. A final dissolve is applied to reduce the number of polygons, resulting in a final layer of suitable sites.

This example analysis, while simple and limited in scope, illustrates both the flexibility and complexity of spatial data analysis via cartographic models. In some respects the cartographic model was simple because only three input spatial data layers were required and a small set of spatial data operations were used. Reclass, overlay, and dissolve operations were used repeatedly, with buffer, slope, aspect and raster-to-vector conversion also applied. The modeling is flexible in that the spatial operations, e.g., each recode may be tailored in each application to provide the desired intermediate outputs. Finally, this example illustrates the complexity that can be obtained with cartographic modeling, in that over 20 different instances of a spatial operation were applied, in a defined sequence, resulting in at least fifteen intermediate data layers as well as the final result.

Spatio-temporal Models

Spatio-temporal models have been developed and applied in a number of disciplines. This is an active area of research, as there are many fields of study and management that require analysis and predictions of spatially and time varying phenomena. We will briefly discuss some basic characteristics of spatio-temporal models. We will then describe their differences from other models, some basic analysis approaches, and describe two examples of spatio-temporal models.

Spatio-temporal models use spatially-explicit inputs to calculate or predict spatially-explicit outputs (Figure 12-9). Rules, functions, or some other processes are applied using spatial and often non-spatial data. Input variables such as elevation, vegetation type, human population density, or rainfall may be used as inputs to one or more mathematical equations. These equations are then used to calculate a value for one or more spatial locations. The values are often saved in a spatial data format, e.g., as a layer in a GIS.

Spatio-temporal models involve at least a three dimensional representation of one or more key attributes – variation in planar (x-y) space and through time. A fourth dimension may be added if the vertical (z) direction is also modeled. We arbitrarily treat spatially-variable network analyses separately, because networks are constrained to a subset of two-dimensional space. Spatio-temporal models may also be classified by a number of other criteria – whether they treat continuous fields vs. discontinuous objects, if they are process-based or rely on purely fit models, and if they are stochastic or deterministic. Combinations of these model characteristics lead to a broad array of spatio-temporal model types.

Models of continuous phenomena predict values that vary smoothly across time or space. Air temperature, precipitation, soil moisture, and atmospheric pollutants are examples of continuous variables that are predicted using spatio-temporal models. Soil moisture this month may depend on soil moisture last month and the temperature, precipitation, and sunshine duration in the intervening period. All these factors may be entered in spatial data layers, and the soil moisture predicted for a set of points.

Models of discrete phenomena predict spatial or attribute characteristics for discontinuous features. Boundaries for vegetation types are an example of features that are often considered discrete. An arc identifies the separation between a grassland and a forest. A spatial model may consider the current position of the forest and grassland as well as soil type, fire prevention, and climatic data to predict the encroachment of forest on grassland sites. The boundaries between new forests and grasslands are always discrete, although their positions shift through time.

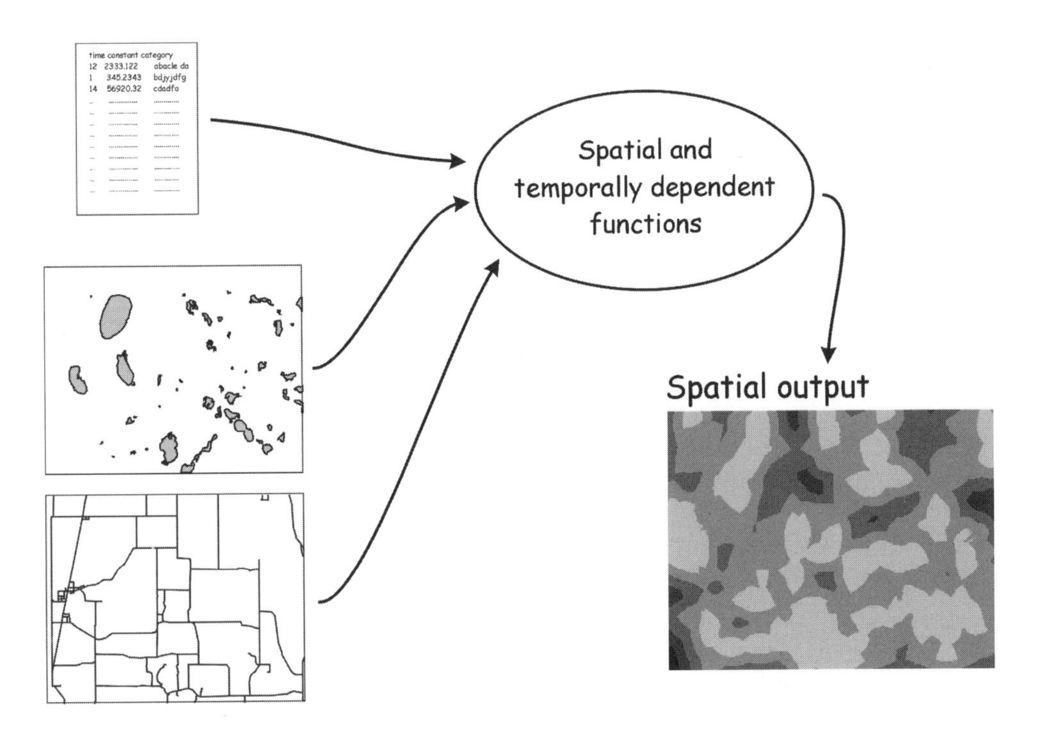

Figure 12-9: Spatio-temporal models combine spatial and aspatial data with time-variant functions to produce spatial output.

Models are considered process-based if their workings in some way represent a theoretical or mechanistic understanding of the processes underlying the observed changes, and models are purely fit models when they do not. We may predict the amount of water flowing in a stream by a detailed spatial representation of the hydrologic cycle. Many processes may be explicitly represented by equations or sub-routines in a spatial model. For example, rainfall location and intensity may be modeled through time for each raster cell in a study area. We may then follow the rainwater as it infiltrates into the soils and joins the stream system through overland flow, subsurface flow, and routing through stream channels. Calculations for these processes may be based on slope, topography, and channel characteristics. These processes are tied together in space and calculations performed at each point on the landscape, and model calculations for a point may in turn affect adjacent, downslope locations.

Rainfall might be modeled differently using a purely-fit, statistical approach. A purely fit model might simply measure precipitation in the previous hour and average the precipitation for the previous week and previous month, and predict stream flow at a point. Processes such as evaporation or subsurface flow are not explicitly represented, and the output may be a statistical function of the inputs. The model may be more accurate than a process-based approach, in that the predicted outflow at any point in the stream may be closer to measured values than those derived from a process-based model, or the output may be poorer, in that the measurements may be farther from predictions. Process modelers argue that by incorporating the structure and function of the system into a process model we may better predict under new conditions, e.g., for extreme drought or rainfall events never seen before. They also argue that process models aid in our understanding a system and in generating new hypotheses about system function.

Besides continuous vs. discrete and process vs. fit, models may be stochastic or deterministic. A deterministic model provides the same outputs every time it is given exactly the same inputs. If we enter a set of variables into a model without modifying the model, it will always produce exactly the same results. A stochastic model will not. Stochastic models often have random generation or some other variability generation procedures that change model results from run-to-run, even when using exactly the same inputs.

A disease spread process is a good example of a phenomenon that might be modeled with a stochastic process. Disease may occur at a set of locations, and may be spread through the atmosphere, water, or carried by animals or humans to initiate new disease centers. The infection and growth at new centers might be determined stochastically. A random number might be generated, and the new center "dies" in the spatial model if the random number is below some threshold value. Thus, the map of disease locations after different model runs may differ, even though the runs were initiated with identical input conditions.

The target location of the spatial model output is usually, but not always, the location of the inputs. For example, a demographics model may use a combination of current population in a census tract, housing availability and cost, job opportunities and location, general migration statistics, and age and marital status of those currently in the census tract to predict future population for the census tract. This model has a target location, the census tract, that is the same as the location for most of the input data.

In contrast, the target location of the model outputs may be different than the location of the inputs. Consider a fire behavior model. This model might predict the location of a wildfire based on current fire location, wind speed, topography (fires burn faster upslope than down), and vegetation type and condition. Fire models often incorporate mechanisms for fire spread beyond the current burn front of a fire. Embers are often lifted above a fire by the upwelling heated air. These may be blown well in advance of a fire front, starting spot fires at some distance. In this case the target location for a calculation in the spatial model is not the same as the location of inputs.

Example 1: Process-based Hydrologic Models

Water flows downhill. This simple knowledge was perhaps sufficient until humans began to build houses and roads, and populations grew to dominate most of the Earth's surface. Land scarcity has led humans to build in low-lying areas, and farming, wetland drainage, and upstream development have all contributed to more frequent and severe flooding.

Humans have been improving their understanding of water movement since the beginning of civilization, but the need for quantitative, spatially-explicit water flow models has increased substantially over the past few decades. Water models are needed because demands for water resources are reaching the naturally available supply in many regions of the world. Water models are also needed because population pressures have driven farms, cities, and other human developments into flood-prone areas, and the same developments have increased the speed and amount of rainfall runoff, thereby increasing flood frequency and severity. These factors are spurring the development of spatio-temporal hydrologic models. The models are often used to estimate stream water levels, such that we may better manage water resources and avoid loss of property or life due to flooding.

Many spatio-temporal hydrologic models predict the temporal fluctuations in soil moisture, lake or stream water levels, and discharge in hydrologic networks. The network typically consists of a set of connected rivers and streams, including impoundments such as lakes, ponds, and reservoirs (Figure 12-10). This network typically has a branching pattern. As you move upstream from the main discharge point for the network, streams are smaller and carry less water. Water level or discharge may be important at fixed points in the hydrologic network, at fixed points on land near the network, or at all points in the landscape. The hydrologic network is often embedded in a watershed, defined as the area that contributes downslope flow to the network.

Spatially-explicit hydrologic models are almost universally dependent on digital elevation data. DEMs are used to define watershed boundaries, water flow paths, the speed of downslope movement, and stream location (Chapter 12). Slope, aspect, and other factors that effect hydrologic systems may be derived from DEMs. For example, evaporation of surface water and transpiration of soil water depend on the amount of solar radiation. Site solar radiation depends on the slope and aspect at each point, and in mountainous terrain it may also depend on surrounding elevations – sunrise is later and sunset earlier in valley locations, and north facing slopes in the northern hemisphere receive significantly less solar radiation than south facing slopes. Site-specific variables representing slope and aspect are used when estimating the evaporation of water from the Earth surface, or the water use by plants.

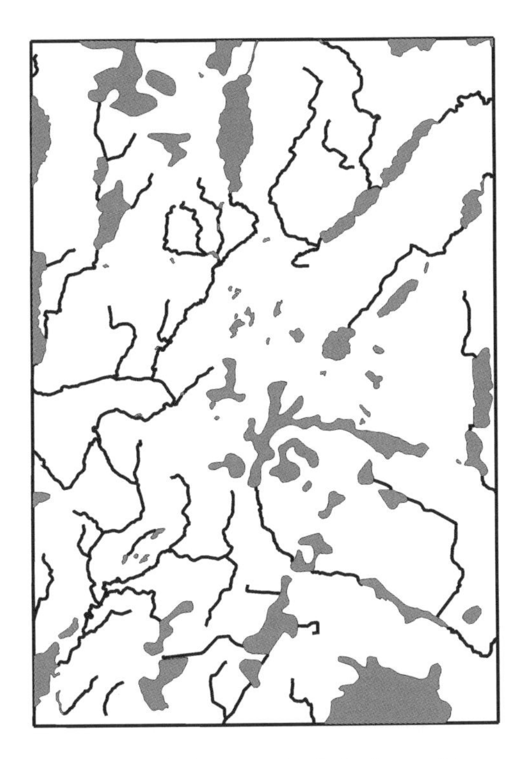

Figure 12-10: An example of a hydrographic network. Lakes and/or rivers form an interconnected network. Water may be routed from upland areas to and through this network.

Slope and aspect are often used to define an important spatial data layer in hydrologic modeling – flow direction. This layer defines the direction of water flow at important points on the surface. If a raster data structure is used, flow direction is calculated for every cell. If a vector data structure is used, flow direction is defined between adjacent or connected vector elements.

Many spatio-temporal hydrologic models adopt a raster data structure. Raster data structures preserve variation in surface elevation that drives water movements. As described in Chapter 2, raster data sets have a relatively simple structure and so are easily integrated into hydrologic models. The connection between adjacent cells is explicitly recorded in raster data sets, so flow between cells is easily represented.

Most raster-based hydrologic models represent water flow through each grid cell. Water falls on each cell via precipitation. Precipitation either infiltrates into the soil or flows across the surface, depending on the surface permeability at the cell. For example, little water infiltrates for most human-made surfaces, such as parking lots or buildings. These sites have low permeability, so most precipitation becomes surface flow. Conversely, nearly all precipitation infiltrates in sandy and many undisturbed forest soils. Downslope water flow is calculated in the model, depending on a number of factors at each cell. Slope and flow direction determine the rate at which water flows downhill. Downslope flow eventually reaches the hydrologic network and is routed via the network to the outlet. Mathematical functions describing cell-specific precipitation, flows, and discharge may be combined to predict the flow quantity and water level at points in the watershed and through the network (Figure 12-11).

Spatio-temporal hydrologic models often require substantial data development. Elevation, surface and subsurface permeability, vegetation, and stream network location are often required prior to the application of a hydrologic model. DEM data may require substantial extra editing because terrain largely drives water movement. For example, local sinks occur much more frequently in DEMs than in real surfaces. Sinks may occur during data collection or during processing. Sinks are particularly troublesome when they occur at the bottom of a larger accumulation area. Water may flow into the sink but may not flow out, depending on how water accumulation is modeled. Local spikes may push water incorrectly to surrounding cells, although they typically cause fewer problems than sinks. Both sinks and spikes must be removed prior to application of some hydrologic models.

Example 2: A Stochastic Model of Forest Change

Many human or natural phenomena are analyzed through spatially-explicit stochas-

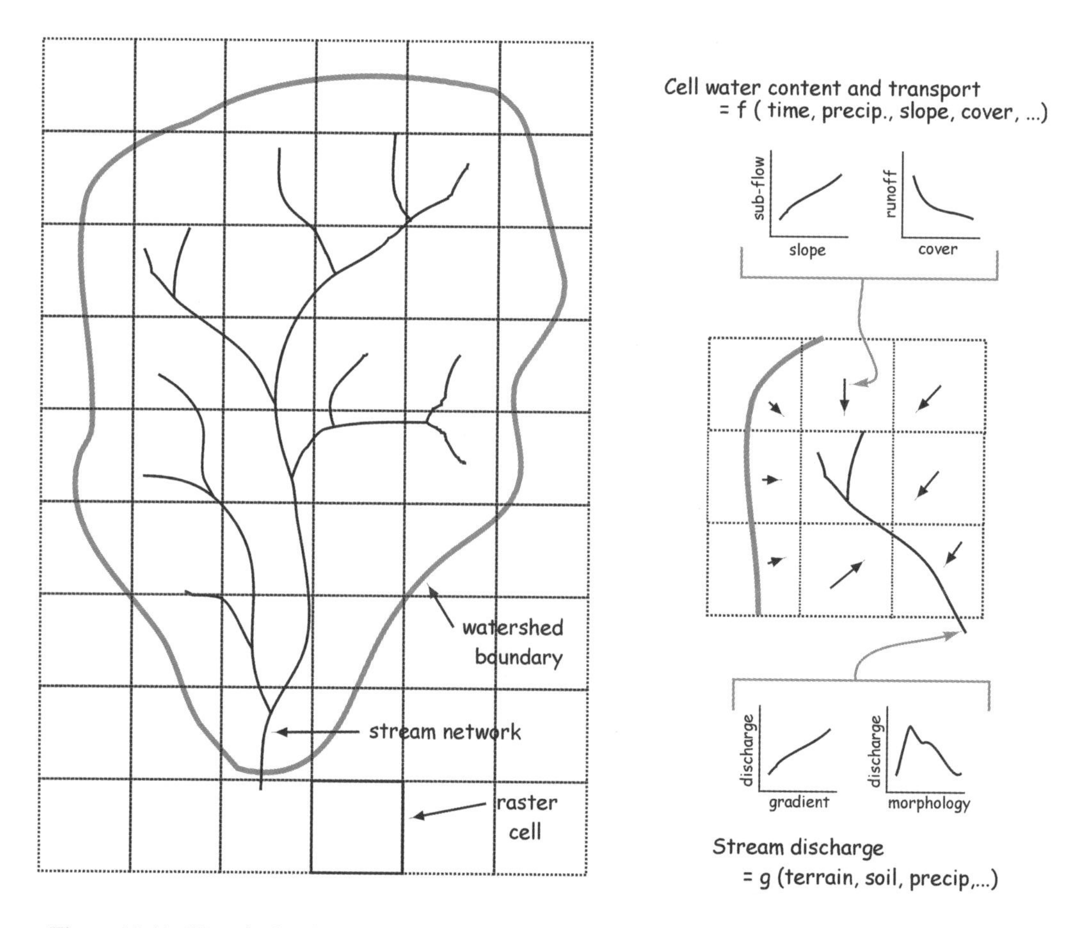

Figure 12-11: Watershed and stream network hydrology may be modeled in a raster environment. Cell characteristics for a watershed are modeled, and water accumulation and flow driven through the system. Soil water, stream levels, and stream discharge depend on spatially- and time-dependent functions.

tic models. Disease spread, the development of past societies, animal movement, fire spread, and a host of other important spatial phenomena have been modeled. All these phenomena have a random element that substantially affects their behavior. Events too obscure or complex to predict may cause large changes in the system action or function. For example, windspeed or dryness on a given day dramatically affects fire spread, yet windspeed is notoriously difficult to predict. Spatially-explicit, stochastic models allow us to analyze the relative importance of component inputs and processes and the nature and variability of system response. We will discuss one spatial-stochastic model that incorporates techniques used in a wide range of models.

Forest vegetation changes through time. Change may be caused by the natural aging and death of a group of trees, replacement by other species, or may be due to periodic disturbances – fire, windstorms, logging, insects, or disease outbreaks. Because trees are long-lived organisms, the composition and structure of forests often change on temporal scales exceeding a normal human life span. Human actions today may substantially alter the trajectory of future change. We often have need to analyze how past actions have led to current forest conditions, and how present actions will alter future conditions.

Forest disturbance and change are important spatial phenomena for many reasons. Humans are interested in producing wood and fiber, preserving rare species, protecting clean water supplies and fish spawning areas, protecting lives and property from wildfires, and enjoying forest-based recreation.

Forest change is inherently a spatial phenomenon. Fires, diseases, and other disturbances travel across space. The distribution of current forests largely affects the location and species composition of future forests. Seeds disperse through space, aided by wind and water or carried by organisms. Physical and biotic characteristics that largely determine seed and seedling survival and subsequent forest growth are variable in space. Some plants are better adapted to grow under existing forests, while others are aided by disturbances that open the canopy. Some species change soil or understory conditions in ways that prevent other species form growing beneath them. Plant succession, the replacement of one group of plants or species by another through time, is substantially affected by the current forest distribution and structure. It is not surprising that many process-based models of forest change incorporate spatial data.

Forests are extremely heterogeneous in space, and this complicates our understanding and predictions of forest change. Tree species, size, age, soils, water availability, and other factors change substantially over very short distances. Each forest stand is different, and we struggle to represent these differences. Given the long time scales, broad spatial scales, and inherent spatial variability of forests, many organizations have developed models based on spatial data, models that are in some way integrated into GIS.

LANDIS – a Forest Succession Model

LANDIS (LANdscape DISturbance) is an example of a spatially-explicit, process-based forest dynamics model. LANDIS has been developed by Dr. Dave Mladenoff and colleagues and has been applied to forest biomes across the globe. LANDIS incorporates natural and human disturbances with models of seed dispersal, plant establishment, and succession through time to predict forest composition over broad spatial scales and long temporal scales. LANDIS is notable for the broad areas it may treat at relatively high resolution, and long temporal scales. LANDIS has been used to model forest dynamics at a 30 meter resolution, over tens of thousands of hectares and five centuries.

LANDIS integrates forest succession, fire, windthrow, and logging to study changes in forest composition through time

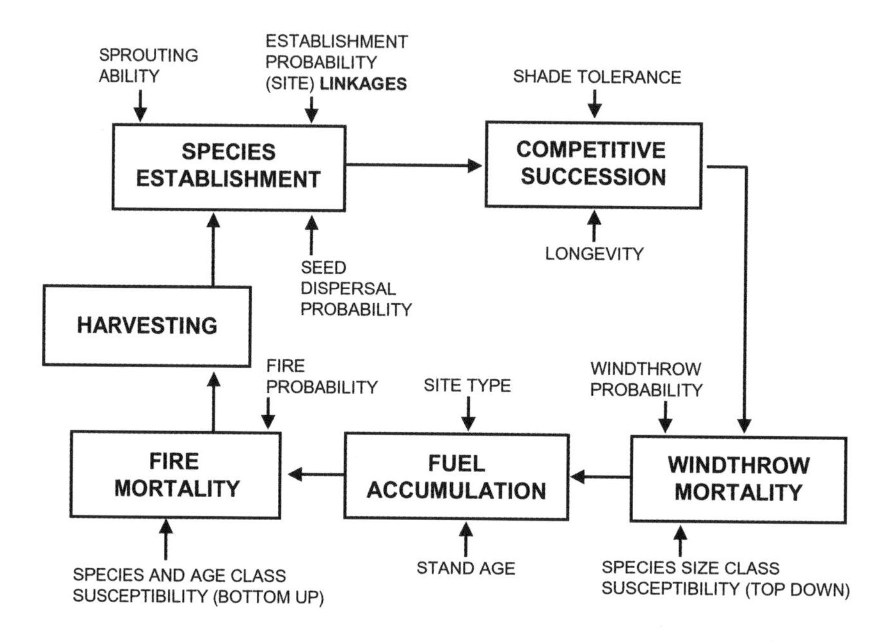

Figure 12-12: The major processes represented in LANDIS, a spatio-temporal forest succession model. (courtesy D. Mladenoff)

(Figure 12-12). Succession is the replacement of species through time. Succession is a common occurrence in forests, for example when fast-growing, light demanding tree species colonize a disturbed site, and are in turn replaced by more shade-tolerant, slower growing species. These shade-tolerant species may be self-replacing in that seeds germinate and seedlings survive and grow, albeit slowly, beneath a closed forest canopy. Small gaps from branch breakage or individual tree deaths allow small patches of light to reach these shade-tolerant seedlings, enabling them to eventually reach the upper canopy. This may result in a stable, self-replacing stand over long time periods. Fire, windthrow, logging or other disturbance events may open up existing stands to a broader range of species. LANDIS simulates large, heterogeneous landscapes, incorporates the interactions of dominant tree species, and includes spatially explicit representations of ecological interactions. The model has been optimized to simulate millions of acres in reasonable run times, less than a day on desktop computers at the time of this writing.

LANDIS Design Elements

The design of LANDIS is driven by the overall objectives for the model, simulating forest disturbance and succession through time. LANDIS also satisfies a number of other requirements. LANDIS readily integrates satellite data sets and other appropriate spatial data, and simulates the basic processes of disturbance, stand development, seed dispersal, and succession in a spatially explicit manner. These requirements led to the adoption of a number of specific design features in the model.

LANDIS is an object-oriented model. Specific features or processes are encapsulated in objects, and object-internal processes are isolated as much as possible from other portions of the model. As an example, there is a SPECIE object that encapsulates most of the important information and processes for each tree species included in the model. Each instance of a SPECIE has a name, e.g., "Aspen", and other characteristics such as longevity, shade tolerance, or age to maturity, as well as methods for birth, death, clear, and other actions or characteristics. Because these characteristics and processes are encapsulated in a SPECIE object, they may be easily changed as new data or a better understanding of forest succession processes become available. Many models are incorporating this object-oriented design, including spatially explicit models, as it simplifies maintenance and modifications.

LANDIS uses a raster data model which eases the entry of classified satellite imagery, elevation, and other data sets reflecting short-range environmental and forest species variation. Interactions such as seed dispersal, competitions, and fire spread are explicitly modeled for each species occupying each grid cell.

LANDIS tracks the presence of age classes (cohorts) for a number of species in each cell and through time. The model begins with an initial condition, the distribution of species by age class across the landscape. Ten-year age classes are currently represented. The longevity, age of initial seed production, seed dispersal distance, shade tolerance, fire tolerance, and ability to sprout from damaged stumps or roots is recorded for each species. On undisturbed sites cohorts pass through time until they reach their longevity. Older cohorts die and disappear from the cell. Younger cohorts appear, depending on the availability of seed.

The spatially explicit representation of seed sources and dispersal is an improvement of LANDIS over many earlier forest succession models. Previous models typically assumed constant or random seed availability. LANDIS is representative of spatially-explicit models, in that the specific location in characteristics of a process are represented. Disturbed sites may be occupied by seedlings from a disturbed cell or nearby cells, or by sprouting from trees in a cell prior to disturbance. Cells cycle through the species establishment, succession, disturbance, and mortality processes (Figure 12-13).

The effects of site characteristics on species establishment and interactions are represented in LANDIS. For example, establishment coefficients are used to represent the interaction between site characteristics and species establishment. Establishment coefficients vary by land type, for example, Jack pine, *Pinus banksiana*, has a higher establishment coefficient on coarse

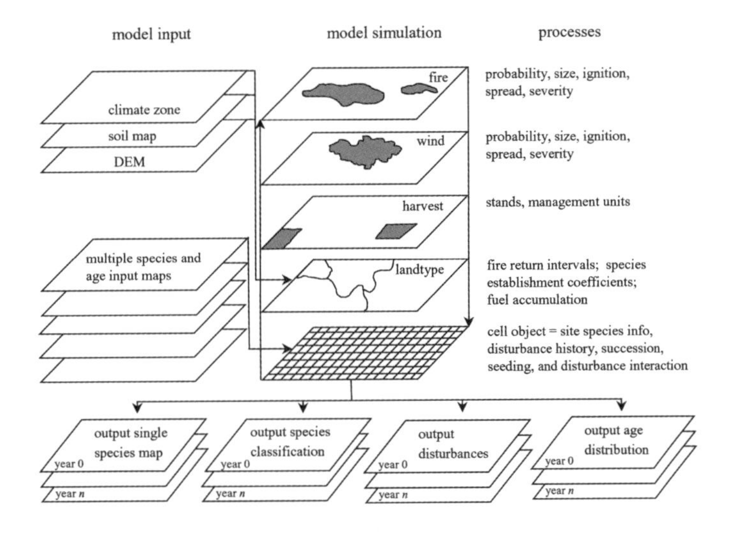

Figure 12-13: Basic spatial data and processes represented in LANDIS

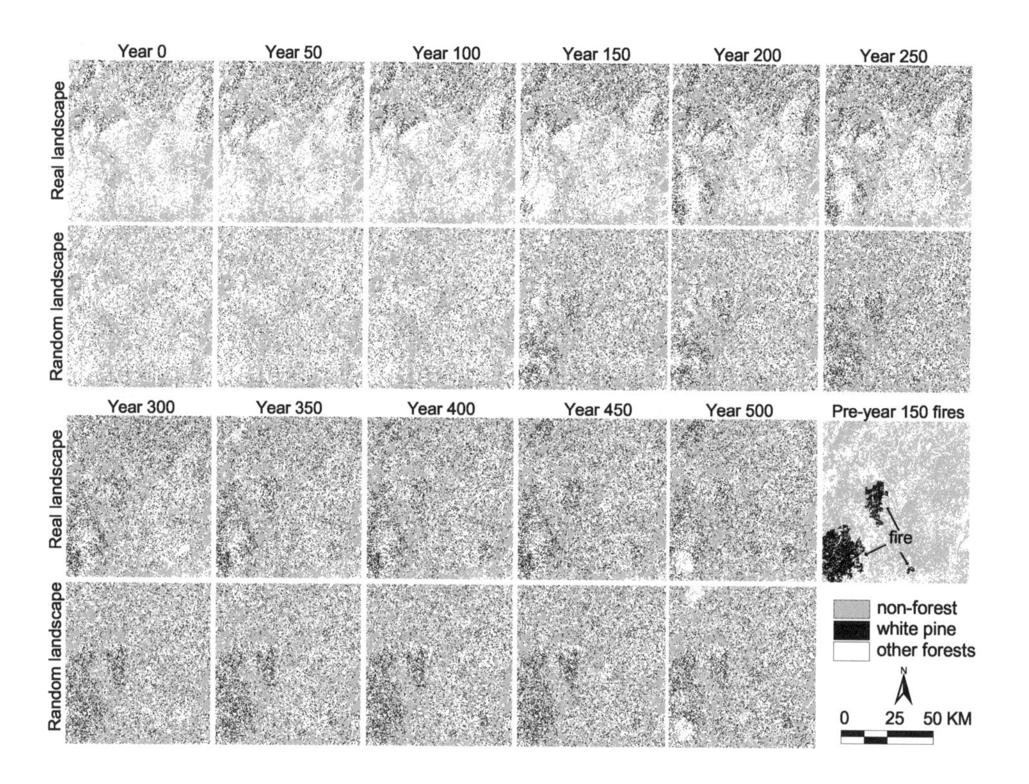

Figure 12-14: Changes in the spatial distribution of white pine, a forest tree species, through time as predicted by LANDIS. This exemplifies the prediction of a feature of interest both spatially and temporally, and is representative of many analytical tools in use or under development.

sandy soils than any other overstory tree species, while black spruce, *Picea mariana* L., has the highest establishment coefficient on wet, boggy soils. Fire severity also varies by land type, as may seedling survival. Elevation, aspect, soils, and other spatial data are used as input to the spatial model.

Fire and wind disturbances are simulated based on historical records of disturbance sizes, frequencies, and severities. Disturbances vary in these properties across the landscape. For example, wind disturbances may be more frequent and severe on exposed ridges, and fires less frequent, less intense, and smaller in wetlands. Disturbances are stochastically generated, but the variability depends on landscape variables, e.g., fires are generated more frequently on dry upland sites.

LANDIS has been applied to a number of forest science and management problems. These include the effects of climate change on forest composition and production, the impacts of changing harvesting regimes on landscape patterns, and regional forest assessments (Figure 12-14).

Hundreds of other spatially-explicit temporally dynamic models have been developed, and many more are currently under development. As spatial data collection technologies improve and GIS systems become more powerful, spatio-temporal models are becoming standard tools in geographic science, planning, and in resource management.

Network Models

Networks are common in our lives. Roads, powerlines, telephone and television cables, and water distribution systems are all examples of networks we utilize many times each day. As networks are crucial to civilization, they need to be effectively managed. These networks also represent substantial investments, and their expansion and management merits considerable attention. Spatial analysis tools have been developed to help us use and maintain networks.

A network may be defined as a set of connected features, often termed *centers*. These features may be centers of demand, centers of supply, or both (Figure 12-15). Centers are connected to at least one and possibly many *network links*. Links interconnect and provide paths between centers. Traveling from one center to another often requires traversing many separate links.

Network models are used to represent and analyze the cost, time, delivery, and accumulation of resources along links and between the connected centers. Resources flow to and from the centers through the networks. In addition, resources may be generated or absorbed by the links themselves.

The links that form the networks may have attributes that affect the flow. For example, there may be links that slow or speed up the flow of resources, or a link may allow resources to flow in only one direction. Link attributes are used to model flow characteristics of the real network, for example, travel on some roads is slower than oth-

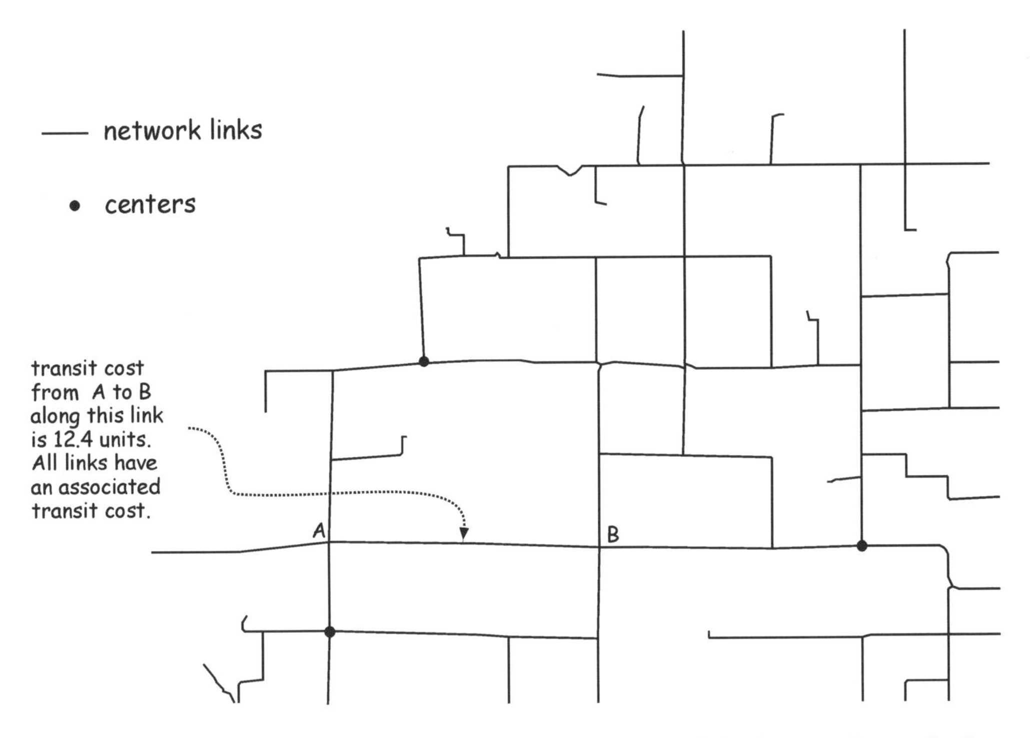

Figure 12-15: Basic network elements. Centers are connected by a set of links. Costs may be associated with traversing the links. Network analysis typically involves moving resources or demands among centers.

ers, or cars may legally move in one direction only on a one-way street.

The concept of a *transit cost* is key to many network analysis problems. A transit cost reflects the price one pays to move a resource through a segment of the network. Transit costs are typically measured in time, distance, or monetary units, e.g., it costs 10 seconds to travel through a link. Costs may be constant such that it always takes 10 seconds to traverse the link regardless of direction or time of day. Alternatively, costs may vary by time of day or direction, e.g., it takes 15 seconds to traverse an arc during morning and evening rush hours, but 10 seconds otherwise, or it may take twice as long to travel north to south than to travel south to north.

We will discuss three types of problems that are commonly analyzed using networks: route selection, resource and territory allocation, and traffic modeling. There are many types of problems, however these three are among the most common and provide an indication of the methods and breadth of network analyses.

Route selection involves identifying a "best" route based on a specified set of criteria. Route selection is often applied to find the least costly route that visits a number of centers. Two or more centers are identified within a network, including starting and ending centers. These centers must all be visited by traversing the network. There are usually a very large number of alternative routes, or pathways, which may be used to visit all centers. The best route is selected based on some criteria, usually the shortest, quickest, or least costly route. Further restrictions may be placed on the route, for example, the order in which centers are visited may be specified.

Route selection may be used to improve the movement of public transportation through a network. School buses are often routed using network analyses. Each bus must start and finish at a school (a center)

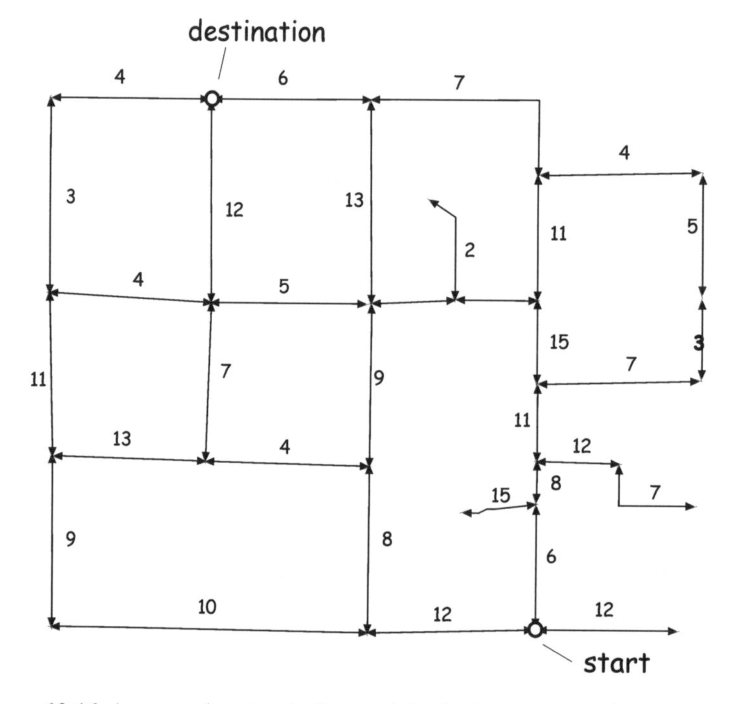

Figure 12-16: An example network. Start and destination centers and costs for link traversal are shown.

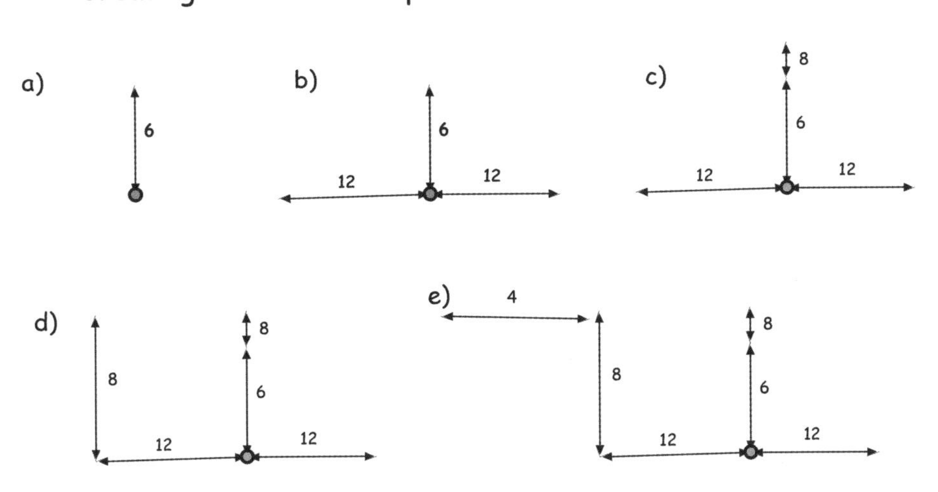

Figure 12-17: Steps in the identification of the least-cost path.

and pick up children at a number of stops (also centers). The shortest path or time route may be specified. Alternate routes are analyzed, and the "best" route selected.

Selection of the best route involves an algorithm that recursively follows a least-cost set of arcs from the current node. A set of interconnected network links are identified, as well as start and destination centers (Figure 12-16). The route from start to destination locations is typically built iteratively. One route-finding algorithm adds the least-cost link at each step. Multiple paths are tested until a path connects the start and destination centers.

This simple method begins at the start center. Paths are extended by adding the link that gives the lowest total cost for all paths currently pursued. The initial set of candidate links consists of any connecting to the starting point. The lowest cost link is added, as shown in Figure 12-17a. The link with a value of six is chosen. Now the set of candidate links consists of any link connected to this selected link (the two links with costs of 15 and 8, respectively), plus any connected to the starting point. All paths are examined, and the link added which gives the lowest total path length. In Figure 12-17b, two links are added. Note that the links added are not connected to the initially selected link. This would have given a total cost of 14 or 21, while the selected links give a lower path cost of 12. Now, the set of candidate links are those connected to any of the selected links or to the start point. Since all links from the start point have been selected, only those connected to candidate links are examined. Of these, the lowest cost path is added. The link with a cost of 8 that is attached to the initially selected link is chosen (Figure 12-17c). The candidate set expands accordingly, and is evaluated again. Verify that the links shown in Figure 12-17d and Figure 12-17e are then selected. This method is used until the destination is reached, and the least-cost path identified (Figure 12-18).

Many different pathfinding algorithms have been developed, most of which are much more sophisticated than the one described here. Note that the described pathfinding algorithm has a rapidly expanding

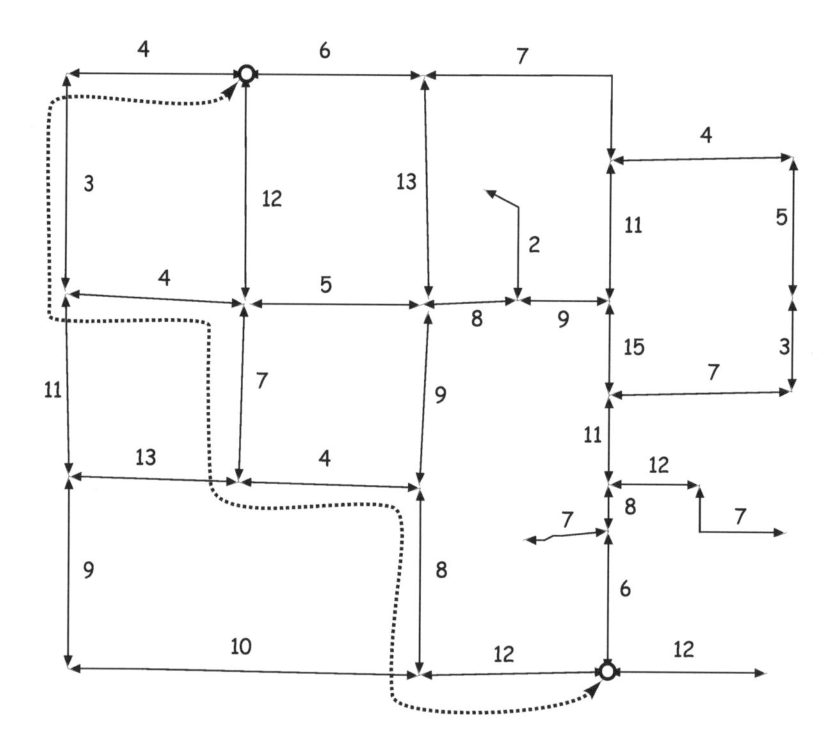

Figure 12-18: Least cost path for the example route finding algorithm described in the text.

number of links to evaluate at each step. Computational burdens increase accordingly. A subset of all possible candidate paths may be examined because it becomes too computationally time-consuming to examine all possible paths. Most pathfinding algorithms periodically review the total cost accumulated thus far for each candidate path and stop following the highest cost or least promising paths.

There are many variations on this route-finding problem. There may be multiple centers that must be visited in a specific order. Centers may add to or subtract from a carrier, e.g., some centers might represent houses with children, and other centers may represent schools. Houses must be visited to pick up children, but a bus has a fixed capacity. These children must be transported to the school, and there may be time constraints, e.g., children cannot be picked up before 7 a.m. and must be at school by 7:55 a.m. Network-based route selection has been successfully used to solve these and related problems.

Resource allocation problems involve the apportionment of a network to centers. One or more *allocation centers* are defined in a network. Territories are defined for each of these centers. Territories encompass links or non-allocation centers in the network. These links or non-allocation centers are assigned to only one allocation center. The features are usually assigned to the nearest center, where distance is measured in time, length, or monetary units.

Resource allocation algorithms may be similar to route finding algorithms in that the distance out from each center is calculated along each path. Each center or arc is

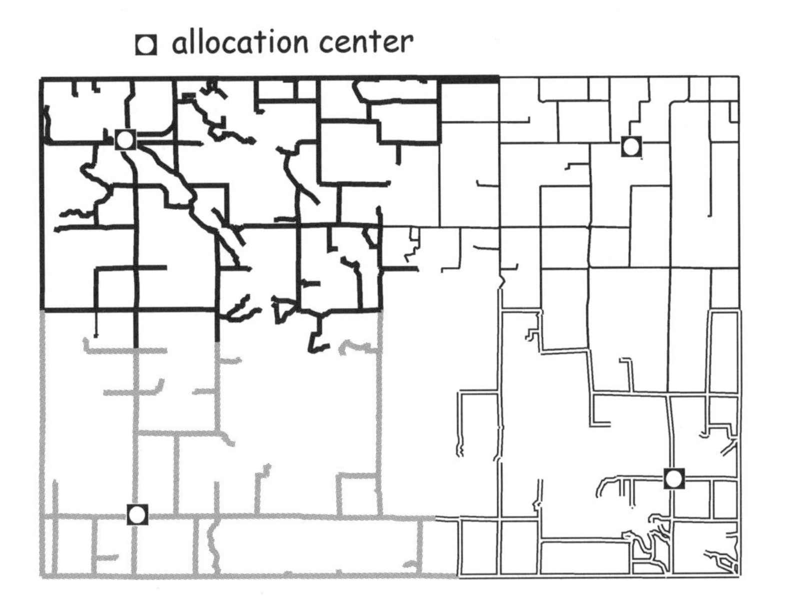

Figure 12-19: Allocation of network links to distinct centers. Network links or resources are assigned to the "nearest" center, where distance may be determined by physical distance, or by cost, travel time, or some other factor.

assigned to the nearest or least-cost center. The route finding method is exhaustive in resource allocation, in that all routes are pursued, not just the least-cost route. The routes are measured outward from each allocation center. Each network link or non-allocation center is assigned to the "nearest" allocation center (Figure 12-19).

Variations on resource allocation include setting a *center capacity*. The center capacity sets an upper limit on resources that may be encompassed by a territory. Links are assigned to the nearest center, but once the capacity is reached, no more are added. Maximum distance also serves to limit the range of the territory from the center. Both of these restrictions may result in some unassigned areas, portions of the network that are not allocated to a center.

Resource allocation analyses are used in many disciplines. School districts may use resource allocation to assign neighborhoods to schools. The type and number of dwellings in a district may be included as nodes on a network. The number of children along each link are added until the school capacity is reached. Resource allocation may also be used to define sales territories, or to determine if a new business should be located between existing businesses. If enough customers fall between the territories of existing business centers, a new business between existing business centers may be justified.

Traffic modeling is another oft-applied network analysis. Streets are represented by a network of interconnected arcs and nodes. Attributes associated with arcs define travel speed and direction. Attributes associated with nodes identify turns and the time or cost required for each turn. Illegal or impossible turns may be modeled by specifying an infinite cost. Traffic is placed in the network, and movement modeled. Bottlenecks, transit times, and under-used routes may be identified, and this information used to improve traffic management or build additional roads.

Traffic modeling through networks is a sub-discipline in its own right. Due to the cost and importance of transportation and traffic management, a great deal of emphasis has been placed on efficient traffic managment. Transportation engineers, computer scientists, and mathematicians have been modeling traffic via networks for many years. An in depth discussion of network analyses for traffic management may be found in literature listed at the end of this chapter.

Summary

Spatial analysis often involves the development of spatial models. These models may help us understand how phenomena or systems change through space and time, and may be used to solve problems. In this chapter we described cartographic models, spatio-temporal models, and network models.

Cartographic modeling typically involves the combination of several data layers to satisfy a set of criteria. Data layers are combined through the application of a sequence of spatial operations, including overlay, reclassification, and buffering. The cartographic model may be specified with a flowchart, a diagram representing the data layers and sequence of spatial operations. Cartographic models are static in time relative to the other two model types.

Spatio-temporal models explicitly represent the changes in important phenomena through time within the model. These models are typically more detailed, and less flexible than cartographic models, in part because spatial-temporal models often include some representation of process. For example, many spatio-temporal models have been developed to model the flow of water through a region, and these models incorporate equations regarding the physics of water transport movement. Models may be stochastic or deterministic, process-based or statistical, or they may have a combination of these characteristics.

Network models may be temporally dynamic or static, but they are constrained to model the flow of resources through a connected set of linear and point features. Traffic flow, oil and gas delivery, or electrical networks are examples of features analyzed and managed with network models. Route finding, allocation, and flow are commonly modeled in networks.

Suggested Reading

Cliff, A. D. and Ord, J. K., Spatial Processes: Models and Applications, Pion, London, 1981.

Goodchild, M.F., Steyaert, L. T., and Parks, B. O., GIS and Environmental Modeling: Progress and Research Issues, GIS World Books, Fort Collins, 1996.

He, H.S., Mladenoff, D. J., and Boeder, J., An object-oriented forest landscape model and its representation of tree species, *Ecological Modeling*, 1999, 119:1-19.

Huevelink, G. B. M. and Burrough, P. A, Error propagation in cartographic modelling using Boolean logic and continuous classification, *International Journal of Geographical Information Systems*, 1993, 7:231-246.

Johnston, C., GIS in Ecology, Blackwell Scientific, Boston, 1998.

Krzanowski, R and Raper, J., Spatial Evolutionary Modelling, Oxford University Press, Oxford, 2001.

Mladenoff, D. J., and He, H. S., Design, behavior and application of LANDIS, an object-oriented model of forest landscape disturbance and succession, in Mladenoff, D. J. and Baker, W. L. (eds.), Advances in spatial modeling of forest landscape change: approaches and applications, Cambridge University Press, Cambridge,1999.

Monmonnier, M., How To Lie With Maps, Chicago Press, Chicago, 1993.

Moore I.D., Gessler, P. E., Nielsen, G. A., and Peterson, G. A., Soil attribute prediction using terrain analysis, *Soil Science*, 1993, 57:443-452.

Rossiter, D. G., A theoretical framework for land evaluation, *Geoderma*, 1996, 72:165-190.

Turner, M.G., and R.H. Gardener, eds., Quantitative Methods in Landscape Ecology, Sprinter-Verlag, New York, 1991.

Wilson, J and Gallant, J. eds., Terrain Analysis: Principles and Applications, Wiley, New York, 2000.

Study Questions

What is a cartographic model?

Can you provide an example of a cartographic model, including the criteria and a flowchart of the steps used to apply the model?

Why must the criteria be refined in many cartographic modeling processes?

What do we mean when we say that most cartographic models are temporally static?

Describe the main characteristics that distinguish spatio-temporal models from cartographic models.

Can you describe/define network models? What distinguishes them from other spatial or temporal models?

What are the common uses for network models? Why are these models so important?

Can you define centers, destinations, links, and other components of network models?

Describe an example of an algorithm by which a network model computes a shortest path.

13 Spatial Interpolation

Introduction

Interpolation is used to estimate values at unsampled locations. An obvious question is why interpolate? Why not just measure the value at all locations? Interpolation is often required because we cannot measure everywhere. Time and/or money are often limiting and we cannot afford complete coverage. It is impossible to sample every location for a continuous variable because there are an infinite number of locations in any study area. Practical constraints usually limit samples to a small subset of the possible lines, polygons, points, or raster cell locations. In addition, interpolation may be required when changing to a smaller cell size in a raster data set. The "sampling" frequency is set by the original raster, and values must be estimated for the new, smaller cells. Finally, interpolation may be used to replace missing or erroneous data.

Interpolation typically translates from lower to the same or higher dimensions, e.g., we typically generate points or lines from point data, or areas from point, line, or area data. Interpolation methods allow us to extend the information we have collected, most often to "fill in" between sampled locations.

Interpolation may be required for other reasons. Besides cost, some areas may be difficult or impossible to visit because of access or safety restrictions. The owner of a parcel may not allow access. It may be impossible to sample vegetation types in part of a park because lions may eat the sampling crew, elephants trample them, or malarial mosquitoes bite them.

Interpolation may be required due to missing or otherwise unsuitable samples. If it is difficult, expensive, or the wrong season for sampling, it may be impossible to recover lost samples. Samples may be discovered to be unreliable or suspect once the measuring crew has returned to the office. These suspect or outlier points may be dropped from the data set, even if they have been collected. The missing point may be crucial to the analysis and the missing values estimated by interpolation. Finally, data may have been collected in the past, and conditions at the sampling location may have changed. Temperature on some past date may have been measured across a network of sites. If we wish to obtain the temperature on that date for a non-measured site, we may interpolate.

An individual sample consists at least of the coordinates of the sample location and a measurement of the variable of interest at the sample location. Coordinates should be measured to the highest accuracy and precision practical, given cost and time constraints and the intended use of the data. Sample variables should be measured using accurate, standardized, repeatable methods.

Sampling

Interpolation is based on a sample of known points. The aim is to estimate the values for a variable at unknown locations based on values measured at sampled locations. Planning will improve the quality of the samples, and usually lead to a more efficient and accurate interpolation.

We control two main aspects of the sampling process. First, we may control the location of the samples. Samples must be spread across our working area. However we may choose among different patterns in dispersing our samples. The pattern we choose will in turn affect the quality of our interpolation. A poor distribution of sample points may increase errors or may be inefficient, resulting in unnecessary costs.

Sample number is the second main aspect of the sampling process we may control. One might believe the correct number is "as many as you can afford", however this is not always the case. A law of diminishing returns may be reached, and further samples may add relatively little information for substantially increased costs. Unfortunately, in most practical applications the available funds are the main limiting factor. Most surfaces are undersampled, and additional funds and samples would almost always increase the quality of the interpolated surface. To date there have been relatively few studies or well-established guidelines for determining the optimum sample number for most interpolation methods.

There are times when we control neither the distribution nor number of sampling points. This often occurs when we are working with "found" variables, for example, the distribution of illness in a population. We may identify the households where a family member has contracted a given illness. We may control neither the number nor the distribution of samples, but we may wish to use these "samples" in an interpolation procedure.

Sampling Patterns

There are a number of commonly applied sampling patterns. A *systematic sampling pattern* is the simplest (Figure 13-1a). Samples are spaced uniformly at fixed x and y intervals. The intervals may not be the same in both directions, and the x and y axes are not required to align with the northing and easting grid directions. The sampling pattern appears as points placed systematically along parallel lines.

Systematic sampling has an advantage over other sampling patterns by way of ease in planning and description. Field crews quickly understand how to lay out the sample pattern, and there is little subjective judgement required.

Systematic sampling may suffer from a number of disadvantages. It is usually not the most statistically efficient sampling pattern because all areas receive the same sampling intensity. If there is more interest or variation in certain portions of the study area this preference is not addressed by systematic sampling. The difficulty and cost of traveling to the sample points is another potential disadvantage. It may be difficult or impossible to stay on line between sampling points. Rough terrain, physical barriers, or lack of legal access may preclude sampling at prescribed locations. In addition, systematic sampling may introduce a bias, particularly if there are patterns in the measured variable that coincides with the sampling interval. For example, there may be a regular succession of ridges and valleys associated with underlying geologic conditions. If the systematic sampling interval coincides with this pattern there may be a bias in sample values. A terrain location might be oversampled, for example, a preponderance of samples might come from valley locations. This bias might in turn result in inaccurate interpolations for values on ridge locations.

Random sampling (Figure 13-1b) may avoid some, but not all, of the problems that

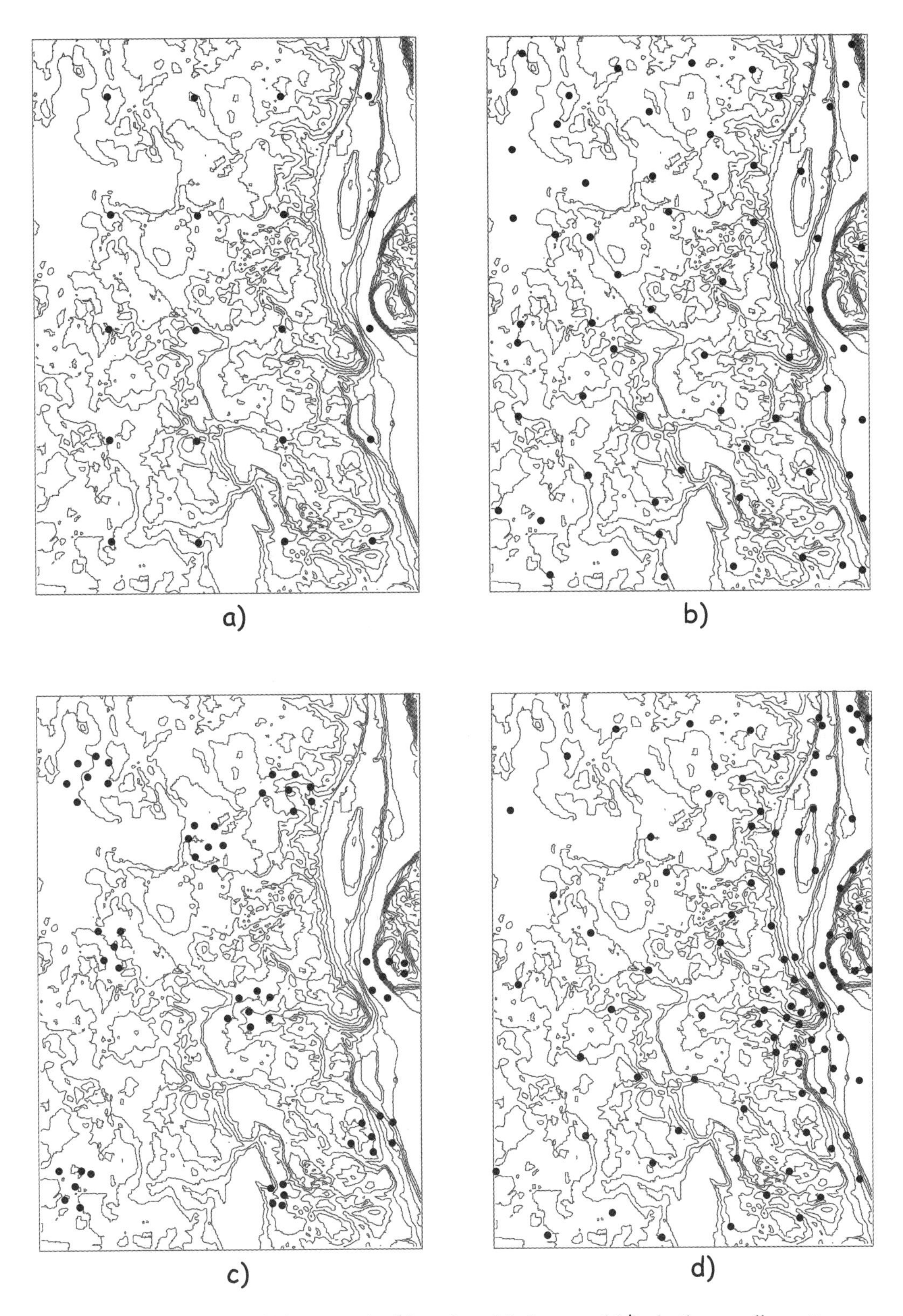

Figure 13-1: Examples of (a) systematic, (b) random, (c) cluster, and (d) adaptive sampling patterns. Sample points are shown as solid circles. Contours for the surface are shown as lines. Sampling methods differ in the distribution of sample points.

affect systematic sampling. Random sampling entails selecting the point location based on a random number generation process. Typically both the easting and northing coordinates are chosen by independent random processes. These may be plotted on a map and/or listed, and then visited with the aid of a GPS or other positioning technology to collect the sample. The points do not have to be visited in the order in which they were selected, so in some instances travel distances between points will be quite small. On average the distances will be no shorter than with a systematic sample, so travel costs are likely to be at least no worse than with systematic sampling.

Random samples have an advantage over systematic samples in that they are unlikely to match any pattern in the landscape. Hence, the chances for biased sampling and inaccurate predictions are less likely.

Random sampling is only a slight improvement over systematic sampling as it does nothing to distribute samples in areas of high variation. More samples than necessary may be collected in uniform areas, and fewer samples than needed in variable areas. In addition, random sampling is more complicated and hence more difficult to explain. More training may be required for sampling crews when implementing random sampling. Random sampling is seldom chosen when sampling over large areas, due to these disadvantages and relatively few advantages over alternative sampling strategies.

Cluster sampling is a technique that groups samples (Figure 13-1c). Cluster centers are established, with a cluster of samples arranged around each center. The distances between samples within a cluster are generally much smaller than the distances between cluster centers.

Reduced travel time is the primary advantage of cluster samples. Because groups of sample points are found in relatively close proximity, the travel times within a cluster are generally quite small. A sampling crew may travel several hours to reach a cluster center, but only a few minutes between each sample within a cluster. Cluster sampling is often used in natural resource surveys that entail significant off-road travel because of the reduction in travel times.

Cluster sampling may be applied in several variants. Cluster centers may be located at random or systematically. Samples within a cluster may also be placed at random or systematically around the cluster center. Both approaches have merit, although it is more common to locate cluster centers at random and distribute samples within a cluster according to some systematic pattern. This is a common approach used by the U.S. Forest Service to conduct national surveys of forest production and forest conditions, and by prospectors during mineral exploration.

Adaptive sampling is a final method we will describe. Adaptive sampling is characterized by higher sampling densities where the feature of interest is more variable (Figure 13-1d). Samples are more frequent in these areas, and less frequent in "flatter", less variable areas. Adaptive sampling greatly increases sampling efficiency because small-scale variation is better sampled. Large, relatively homogenous areas are well represented by a few samples, reserving more samples for areas with higher spatial variation.

Adaptive sampling requires some method for estimating feature variation while in the field, or it requires repeat visits to the sampling areas. Sample density is adaptively increased in areas of high variation. In some instances it is quite obvious where the variation is greatest while in the field. For example, when measuring elevation it is obvious where the terrain is more variable. Sample density may be increased based on field observations of steepness.

If there is no method of identifying where the features are most variable while in the field, then sample density cannot be increased "on the spot". Samples may be returned to the office or lab for analysis and a preliminary map produced. Sample locations are then selected based on local variation. The list or map of coordinate locations may be taken to the field and used as a guide in collecting samples.

Interpolation Methods

There are many different interpolation methods. While methods vary, all methods combine the sampled values and positions to estimate values at unmeasured locations. Often, mathematical functions are used that incorporate both distance from interpolation points to sample points and the values at those sample points. Methods differ in the mathematical functions used to weight each observation, and the number of observations used. Some interpolators use every observation when estimating values at unsampled locations, while other interpolators use a subset of samples when estimating values, for example, the three points nearest an unmeasured location.

Different interpolation methods will often produce different results, even when using the same input data. This is due to the differences in the mathematical functions and number of data points used when estimating values for the unsampled locations. Each method may have unique characteristics, and the overall accuracy of an interpolation will often depend on the method and samples used.

Accuracy is often judged by the difference between the measured and interpolated values at a number of withheld sample points. These withheld points are not used when performing the interpolation, but are checked against the interpolated surface. No single interpolation method has been shown to be more accurate than all others for every application. Each individual or organization should test several sampling regimes and interpolation methods before adopting an interpolation method.

Interpolation methods may produce one or more of a number of different output types. Interpolation is usually used to estimate values for a raster data layer. Other methods produce *contour lines*, more generally known as *isolines*, lines of uniform value. Contour lines are less frequently produced by interpolation methods, but are a common way of depicting a continuous surface. At least one interpolation method defines polygon boundaries.

Interpolation to a raster surface requires estimates of unmeasured values at the center of each grid cell. Raster layer boundaries and cell dimensions are specified, in turn defining the location of each raster cell. The interpolation method uses the sample values to estimate values for each cell in the raster data layer.

Contour line generation is more involved, and may require an iterative process. The location of a set of known levels is determined. For example, a set of points where temperature is exactly 10 degree centigrade may be estimated. These points are connected to form a line. Other sets of points may be estimated for 12, 14, 16, and other temperatures. Points for any given temperature are joined with the restrictions that lines of different temperatures do not cross. These contour lines depict the changes in temperature (or any other plotted variable) across the landscape.

We will describe the most common interpolation methods and apply them all to a single data set to facilitate comparisons. The left side of Figure 13-2 shows contour lines for an elevation surface, and the right side shows a set of sample points from the same area. These sample points will be used to demonstrate the application of various interpolation methods in the following sections of this chapter. Contours for each interpolated surface will be shown.

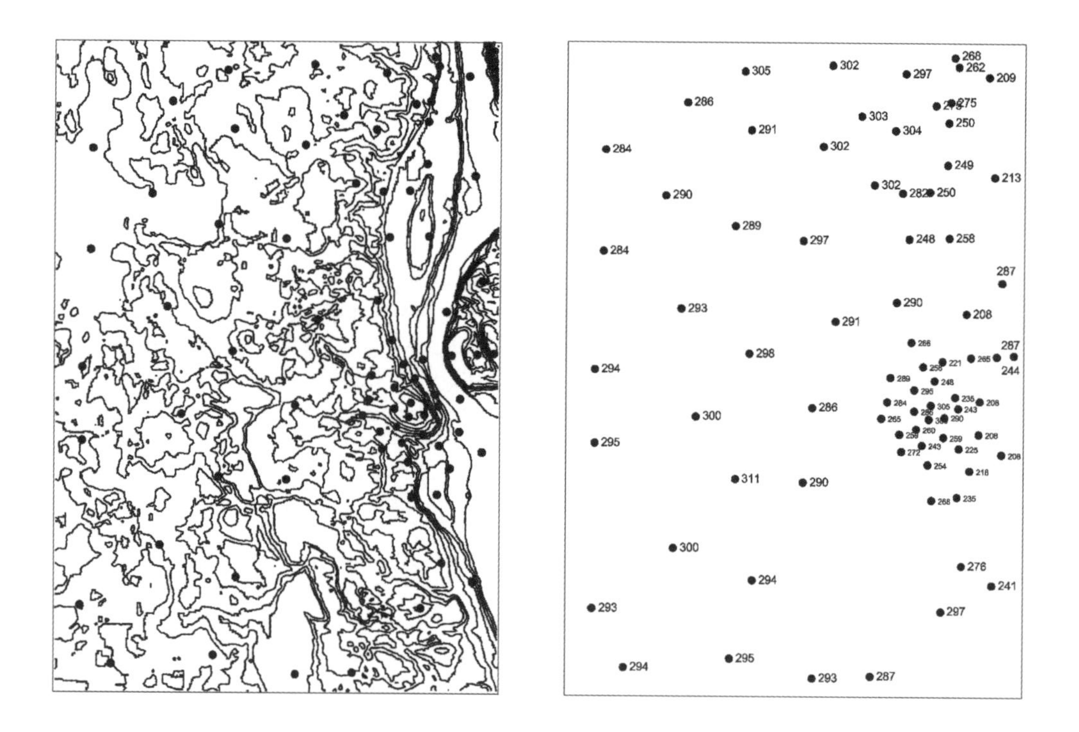

Figure 13-2: Contour lines (left) and sample points (right) for an elevation surface. These points will be used later in this chapter to demonstrate interpolation methods.

Note that the comparisons and figures are only to illustrate different interpolation methods. They are not to establish the relative merit or accuracy of the various methods. The best interpolation method for any given application depends on the characteristics of the variable to be estimated, the cost of sampling, available resources, and the accuracy requirements of the users. The relative performance of interpolators has been determined for some variables in some locations. Enough comparisons have been conducted to establish that no interpolation method is superior for all data sets or conditions. Further experience and studies will establish the likely best method for each application.

Nearest Neighbor Interpolation

Nearest neighbor interpolation, also known as Thiessen polygon interpolation, assigns a value for any unsampled location that is equal to the value found at the nearest sample location. This is conceptually the simplest method, in the sense that the mathematical function used is the simple equality function, and only one point, the nearest point, is used to assign a value to an unknown location. The nearest neighbor interpolator defines a set of polygons, known as Thiessen polygons. All locations within a given Thiessen polygon have an identical value for the Z variable (in this and other chapters, Z will be used to denote the value of a variable of interest at an x and y sample location). Z may be elevation, size, production in pounds per acre, or any other variable we may measure at a point. Thiessen polygons define a region around each sampled point that have a value equal to the value at the sampled point. The transition between polygon edges is abrupt, that is, the variable jumps from one value to the next across the Thiessen polygon boundary.

Figure 13-3 shows contours of an elevation surface, sample points, and Thiessen polygons based on the sample points. The left side of Figure 13-3 shows contours derived from a 7.5 minute DEM in central Minnesota. The contour interval is 10 meters, and the land is gently undulating, with an abrupt decrease in elevation near the right edge. A river runs from top to bottom on the right edge of the DEM. Cliffs are indicated by the convergence of contours along the river's edge. Sample points are indicated by filled circles. Sampling is sparse, in most cases there are more than 500 meters between samples, however sampling is denser in portions of the map near the river. Thiessen polygons are shown on the right of Figure 13-3. Elevation values are assigned to the nearest point sample. Elevation within a polygon is estimated to be equal to the value of the sample point found near the center of the polygon. Polygons are smaller where sampling density is highest.

Thiessen polygons provide an *exact interpolator*. For an exact interpolator, the interpolated surface equals the sampled values at each sample point. The value for each sample location is preserved, so there is no difference between the true and interpolated values at the sample points. This does not mean an exact interpolator is perfect; Thiessen polygon method and other exact interpolators are in error at non-sample locations.

An independent error measure is required if we are to obtain an accurate estimate of the interpolation accuracy (or error). Accuracy estimates may be obtained with a withheld samples technique, where the sur-

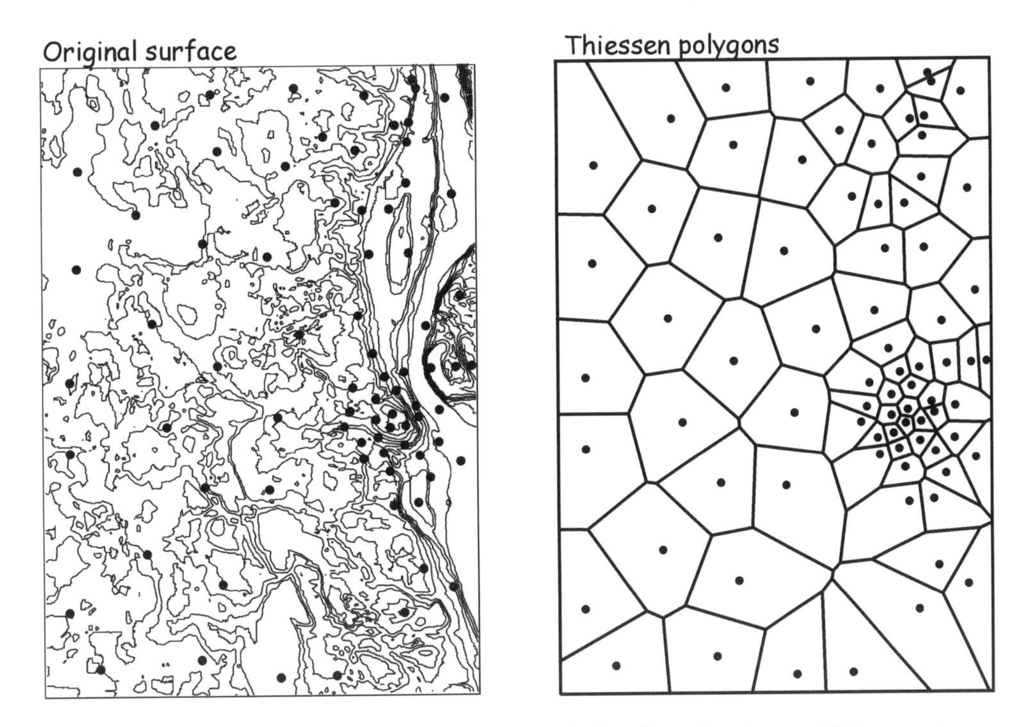

Figure 13-3: Contours and sample points for a surface (left) and sample points and Thiessen polygons (right).

face is fit withholding one data point, and the error determined at that point. The sample is replaced, a new sample selected and withheld, and the surface fit and error again determined. This is repeated for each data point. A less efficient testing method entails collecting an independent set of sample points that are withheld from the interpolation process. Their measured values are then compared to the interpolated values, and the mean error, maximum error, and perhaps other error statistics identified.

Fixed Radius – Local Averaging

Fixed radius interpolation may be viewed as a bit more complex than nearest neighbor interpolation, but less complex than most other interpolation methods. A raster grid is specified in a region of interest. Cell values are estimated based on the average of nearby samples.

The samples used to calculate a cell value depend on a *search radius*. The search radius defines the size of a circle that is centered on each cell. Any sample points found inside the circle are used to interpolate the value for that cell (Figure 13-4). Points that fall within the circle are averaged, those outside the circle ignored.

Figure 13-5 shows a perspective view of fixed radius sampling. Note that there is a sample data layer, shown at the top of Figure 13-5, vertically aligned with the interpolated surface. The surface is a raster data layer with interpolated values in each raster cell. The fixed-radius circle is centered over a raster cell. The average is calculated for all samples contained within the sample circle, and this average is placed in the appropriate output raster cell. The process is repeated for each raster cell in the surface. The fixed-radius circles are shown corresponding to three raster cells, containing three, zero, and one sample points, respectively. Circles may contain no points, in which case a zero or no data value must be placed in the raster cell. The radius for the circle is typically much larger than the raster cells width. This means

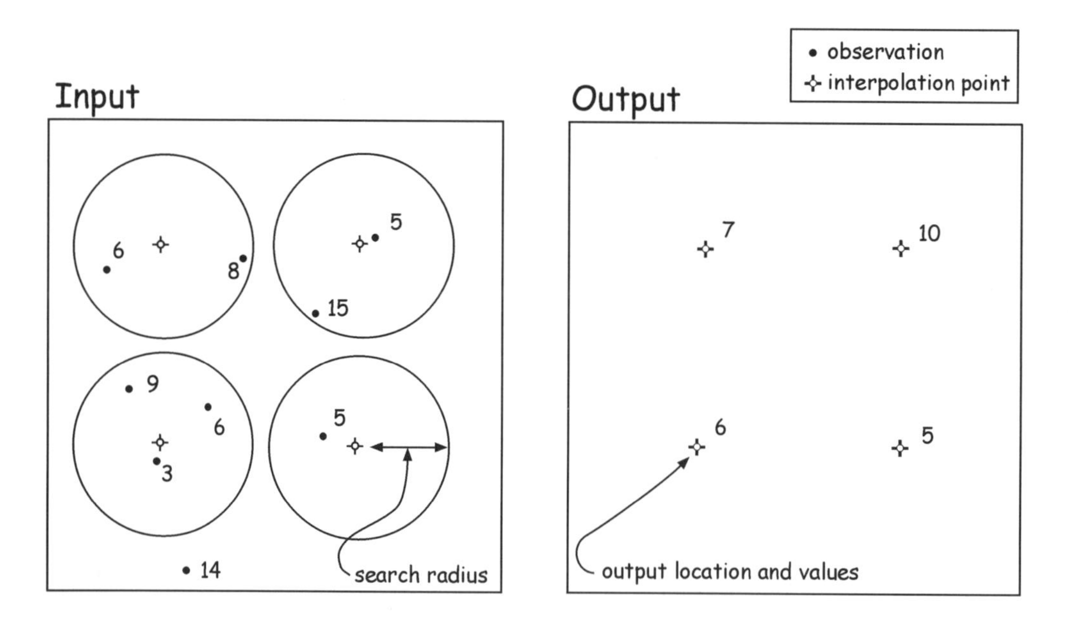

Figure 13-4: A diagram and example of a fixed radius interpolation. Values within each sampling circle are averaged to estimate an output value for the corresponding point.

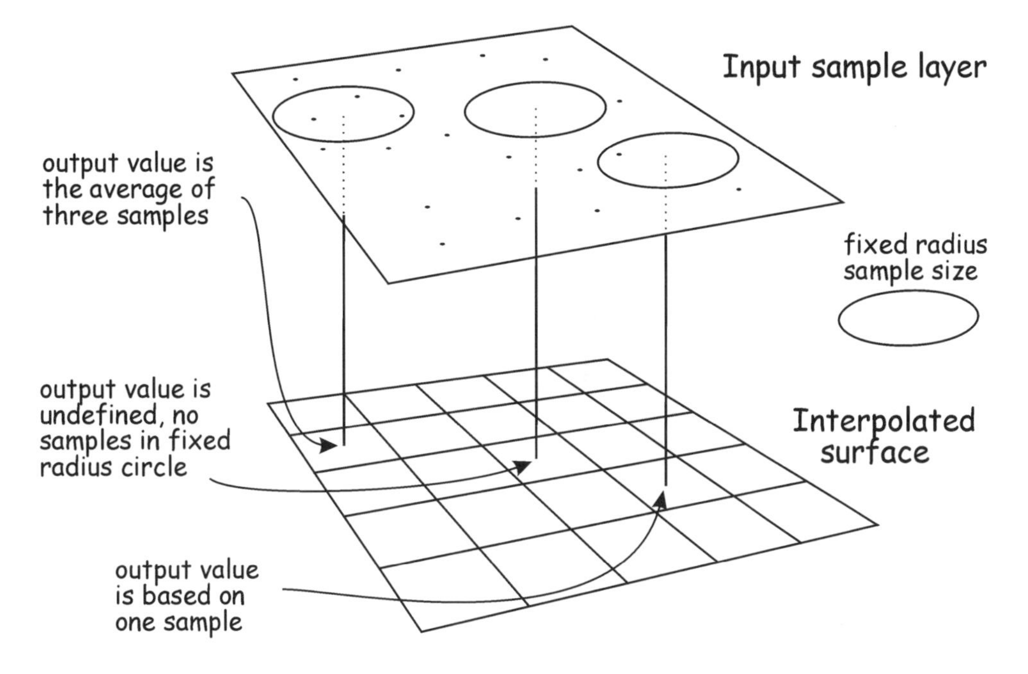

Figure 13-5: A perspective diagram of fixed radius sampling. A circle is centered on each raster cell location. Samples within the circle contribute to the value assigned to each corresponding raster cell. (adapted from Mitchell, 1999)

circles overlap for adjoining cells, which causes neighboring cell values to be similar.

Fixed radius interpolators are often used to create a moving average of the samples. Each sample point may correspond to a sum or density value, and may be averaged spatially to interpolate the values for nearby cells. For example, an agronomist may measure corn production in bushels per acre at several points in a county. These may be converted to a raster surface by averaging the measurements of bushels per acre that fall within the circles centered on each cell.

The fixed radius interpolator tends to smooth the sample data (Figure 13-6). Large or small values sampled at a given point are maintained when only that one sample point falls within a search radius for a cell. Large or small values are brought toward the overall sample mean when they occur within a search radius with other sample points. Smaller search radii are more likely to preserve the extreme values, but this comes at the cost of a higher frequency of empty cells. Fixed radius interpolators are not exact interpolators as they may average several points in the vicinity of a sample, and so they are unlikely to place the measured value at sample points in the interpolated surface.

The search radius affects the values and shape of the interpolated surface. Too small a search radius results in many empty cells. The cell value is typically set to a value that indicates no data are present. Too large a search radius may smooth the data too much. In the extreme a search radius may be defined that includes all sample points for all cells. This would result in a single interpolated value repeated for all cells. Some intermediate search radius should be chosen. If many cell values change with a small change in the search radius, this may be an indication that the samples are too sparse, and more sample points may be required

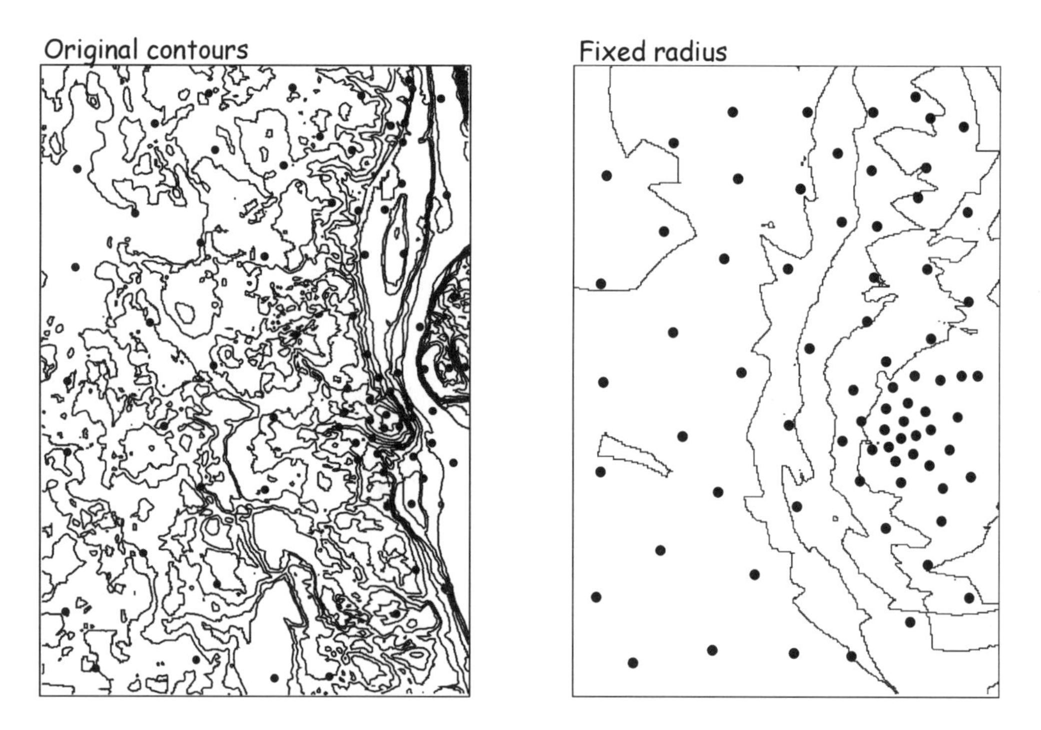

Figure 13-6: Contours and sample points (left) and a fixed radius interpolation (right).

Inverse Distance Weighted Interpolation

The inverse distance weighted (IDW) interpolator estimates the value at unknown points using the distance and values to nearby known points. IDW reduces the contribution of distant points. The weight of each sample point is an inverse proportion to the distance, thus the name. The farther away the point, the less weight the point has in helping define the value at an unsampled location. Values are estimated by:

$$Z_j = \frac{\sum_i \frac{Z_i}{d^n_{ij}}}{\sum_i \frac{1}{d^n_{ij}}} \tag{13.1}$$

where Z_j is the estimated value for the unknown point at location j, d_{ij} is the distance from known point i to unknown point j, Z_i is the value for the known point i, and n is a user-defined exponent. Any number of points greater than two may be used, up to all points in the sample. Typically some fixed number of close points is used, e.g., the three nearest sampled points will be used to estimate values at unknown locations. Note that the farther away the point (larger d_{ij}), the smaller the weight ($1/d_{ij}$), so the less influence a point has on the estimate of the unknown point.

Figure 13-7 illustrates an IDW interpolation calculation. The three nearest samples are used. Each measured value is weighted by the inverse of the distance. These weighted values are added. The result is divided by the sum of the weights to "scale" the weights to the measurement units. This

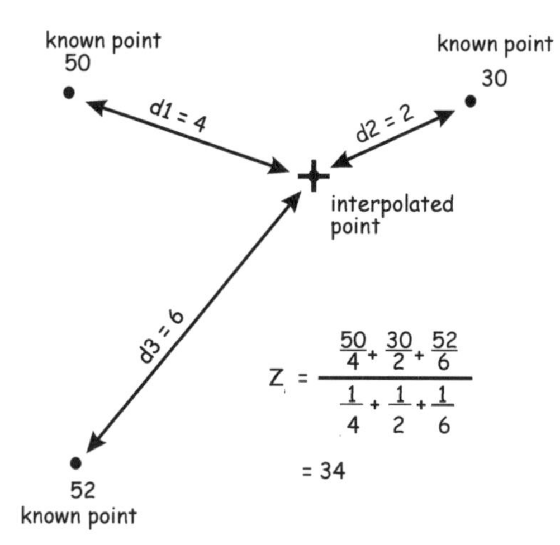

Figure 13-7: An example calculation for a linear inverse distance weighted interpolator.

produces an estimate for the unsampled location.

IDW is an exact interpolator. Interpolated values are equal to the sampled values at each sampled point. As a d_{ij} becomes very small (sample points near the interpolated location), the $1/d_{ij}$ becomes very large. The contribution from the nearby sample point dwarfs the contributions from all other points. The values $1/d_{ij}$ are very near zero for all i values except the one very near the sampled point, so the values at all other points are effectively multiplied by zero in the numerator of the IDW equation. The sum in the denominator reduces to the weight $1/d_{ij}$. The weights on the top and the bottom of the IDW equation become more similar, and the fraction approaches one. This is equivalent to substituting the measured sample value for the interpolated value at the sample point. Thus, at a sampled point the IDW interpolation formula reduces to:

$$\frac{\frac{Z_i}{d_{ij}^n}}{\frac{1}{d_{ij}^n}} \tag{13.2}$$

By simple division this is reduced mathematically to Z_i, the value measured at the sampling location.

Inverse distance weighting results in smooth interpolated surfaces. The values do not jump discontinuously at edges, as occurs with Thiessen polygons. While IDW is easily and widely applied, care must be taken in evaluating the particular n and j selected. The effects of changing n and j should be tested in an oversampled case, where adequate withheld points can be compared to interpolated points. The IDW, and all other interpolators, should be applied only after the user is convinced the method provides estimates with sufficient accuracy. In the case of IDW, this may mean testing the interpolator over a range of n and j values, and selecting the combination that most often gives acceptable results.

The size of the user-defined exponent, n, affects the shape of the interpolated surface (Figure 13-8). When a larger n is specified the closer points become more influential. Higher exponents result in surfaces with higher peaks, lower valleys, and steeper gradients near the sampled points. Contours become much more concentrated near sample points when $n = 2$ (Figure 13-8c) when compared to the $n = 1$ (Figure 13-8b). These closer contours reflect steeper gradients near the known data points.

The number of points, i, used to estimate an interpolated point, j, also affects the estimated surface (Figure 13-8). Both bottom panels (c and d) use an exponent of $n = 2$ but differ in the number of nearby sample points used in each interpolation. Panel c, on the bottom left, interpolates using the nearest six points, while panel d, on the bottom right of Figure 13-8, uses the twelve nearest points. These panels show complex patterns and no distinct trends. In some regions the gradients are steeper, in others shallower. A larger number of sample points tends to result in a smoother interpolated surface. However the effects depend also on the distribution and values of known data points, and the impacts of changing sample number are difficult to generalize.

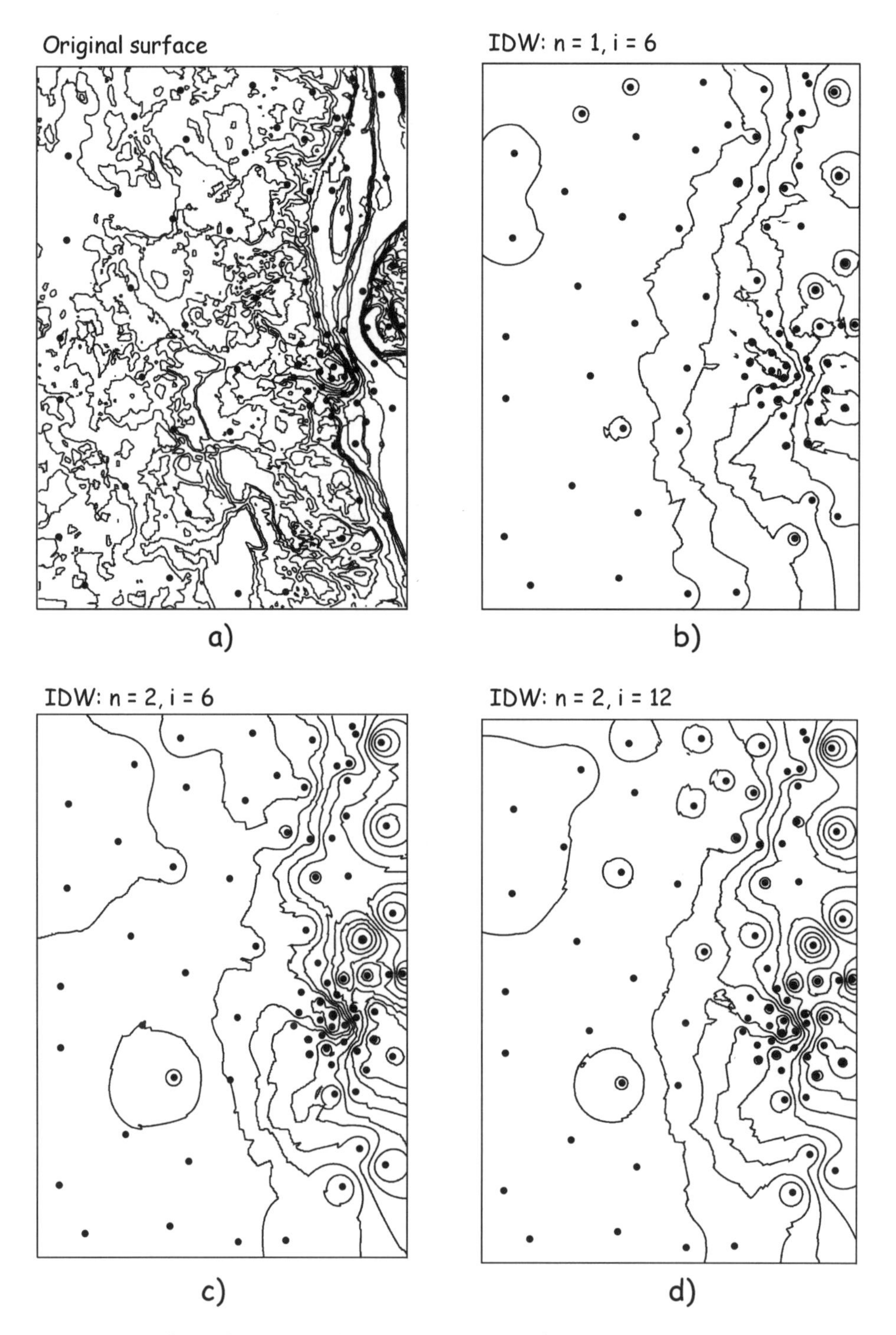

Figure 13-8: The effects of exponent order, n, and sample size, i, on the interpolated values for an inverse distance weighted interpolator. Local influences are stronger as the exponent increases and the number of sample points decreases.

Trend Surface Interpolation

Trend surface interpolation involves fitting a statistical model, or trend surface, through the measured points. The surface is typically a polynomial in the X and Y coordinate system. For example, a second order polynomial model would be:

$$Z = a_0 + a_1x + a_2y + a_3x^2 + a_4y^2 + a_5\,xy \quad (13.3)$$

Where Z is the value at any point x and y, and each a_i is a coefficient estimated in a regression model. Least-squares methods, described in most introductory statistical textbooks, are used to estimate the best set of a_i values. The a_i values are chosen to minimize the average difference between the measured Z values and the surface.

There must be at least one more sample point than the number of estimated a_i coefficients due to statistical constraints. This does not pose a practical problem for most applications, because the best polynomial models are often second or third order and have fewer than ten parameters. More than ten sample points are typically collected to ensure adequate coverage of a study region.

Trend surfaces are not exact interpolators in that the surface typically does not pass through the measured points. This means that there is an error at each sample location, measured as the difference between the interpolated surface and the measurement. Trend surfaces may be preferred to other interpolation methods, however, because the surface may more accurately represent the shape of real surfaces. Trend surfaces typically represent smooth surfaces well (Figure 13-9), and do not have the "bulls-eye" artifact that may appear with excessive local influence in IDW interpolation.

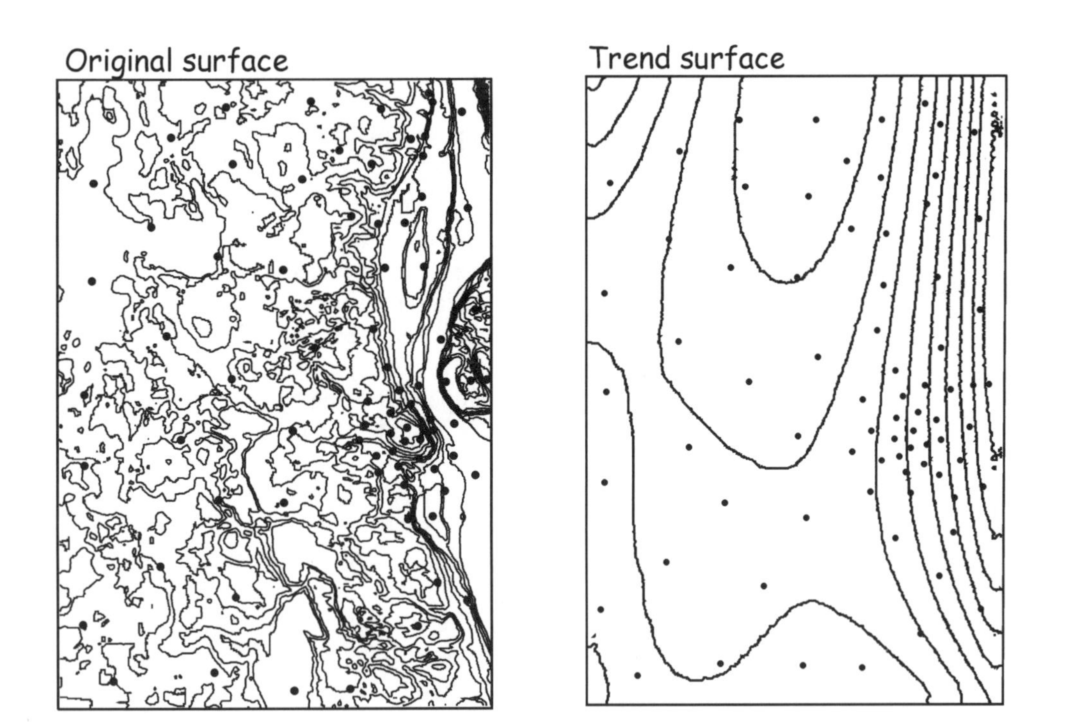

Figure 13-9: Sample points and contours from the original surface (left), and sample points and derived contours from a third-order trend surface fit to the sample points (right).

Trend surface interpolation typically does not perform well when there is a highly convoluted surface, e.g., elevation in a dissected mountain region. Even high-order polynomials may not be sufficiently flexible to fit these complex, convoluted surfaces. However trend surfaces are often among the most accurate methods when fitting smoothly varying surfaces, such as mean daily temperature over large areas.

Splines

A *spline* is a flexible ruler that was commonly used by draftsmen to create smooth curves through a set of points. A road location may have been surveyed at a set of points. To produce a smoothly bending line to represent the road, the draftsman carefully plotted the points, and the spline ruler was bent along a path defined by the set of points. A smoothly curving line was then drawn along the edge of the spline.

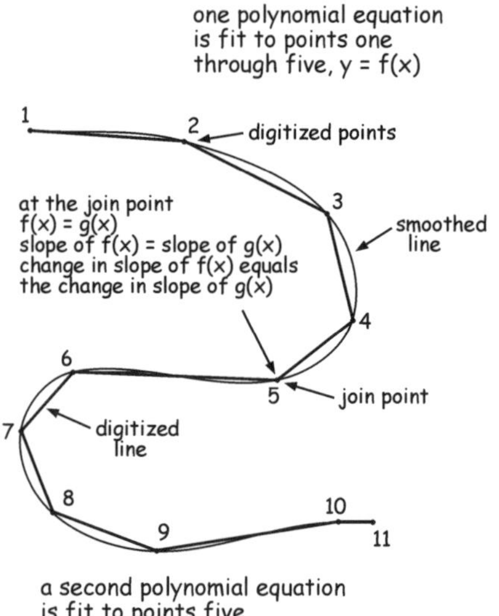

Figure 13-10: Diagram of a two-dimensional (line) spline. Segments are fit to portions of a line. Segments are constrained to join smoothly at knots, where they meet.

Spline functions, also referred to as splines, are used to interpolate along a smooth curve. These functions serve the same purpose as the flexible ruler in that they force a smooth line to pass through a desired set of points. Spline functions are more flexible because they may be used for lines or surfaces and they may be estimated and changed rapidly. The sample points are analogous to the drafted points in that these points serve as the "guides" through which the spline passes.

Spline functions are constructed from a set of joined polynomial functions. Line functions will be described here, but the principles also apply to surface splines. Polynomial functions are fit to short segments. An exact or a least-squares method may be used to fit the lines through the points found in the segment. For example, a third-order polynomial may be fit to a line segment (Figure 13-10). A different third-order polynomial will be fit to the next line segment. These polynomials are by their nature smooth curves within a given segment.

Splines are typically first, second, or third order, corresponding to the order of the polynomial used to fit each segment. Segments meet at *knots*, or *join points*. These join points may fall on a sampled point, or they may fall between sampled points.

Constraints are set on spline functions to ensure the entire line remains smooth at the join points. These constraints are incorporated into the mathematical form of the function for each segment. These constraints require the slope of the line and the change in slope of the line be equal across segments on either side of the join point. Typically, spline functions give exact interpolation (the splines pass through the sample points) and show a smooth transition (Figure 13-11). Strictly enforcing exact interpolation can sometimes lead to artifacts at the knots or between points. Large loops or deviations may occur. The spline functions are often modified to allow some error in the fit, particularly when fitting surfaces rather than lines. This usually removes the artifacts of spline fits, while maintaining the smooth and continuous interpolated lines or surfaces.

Kriging

Kriging is a statistically-based estimator of spatial variables. It differs from the trend-surface approach in that interpolation models are based on regionalized variable theory, and includes three main components. The first component in the kriging model is the spatial trend, an increase or decrease in a variable that depends on direction, e.g., temperature may decrease toward the northwest. The second component describes the local spatial autocorrelation, that is, the tendency for points near each other to have similar values. The third component is random, stochastic variation. These three components are combined in a mathematical model to develop an estimation function. The function is then applied to the measured data to estimate values across the study area.

Much like IDW interpolators, weights in kriging are used with measured sample variables to estimate values at unknown locations. With kriging the weights are chosen in a statistically optimal fashion, given a specific kriging model and assumptions about the trend, autocorrelation, and stochastic variation in the predicted variable.

Kriging methods are the centerpiece of geostatistics. Geostatistics was initially developed in the early 1900s by D.G. Krige and Georges Matheron for use in mining. Prospecting samples may be expensive to obtain or process, and accurate predictions may be quite difficult, but valuable. Krige and Matheron sought to develop estimators that would incorporate trends, autocorrelation, and stochastic variation and also provide some estimate of the local variance in the predicted variable.

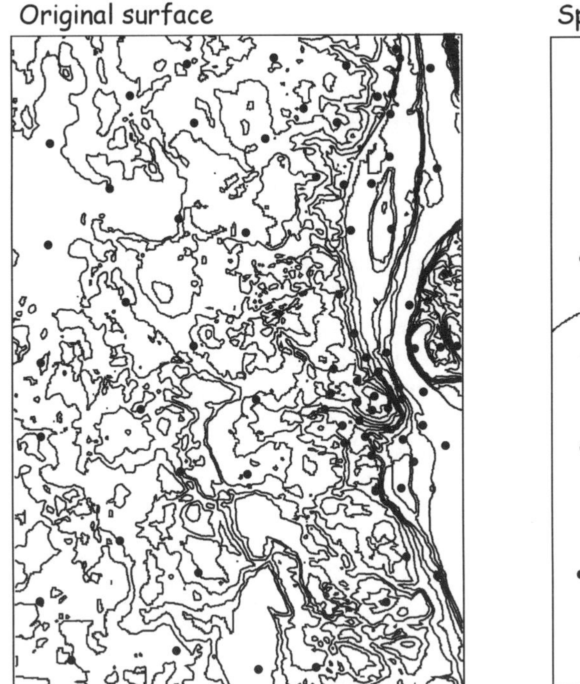

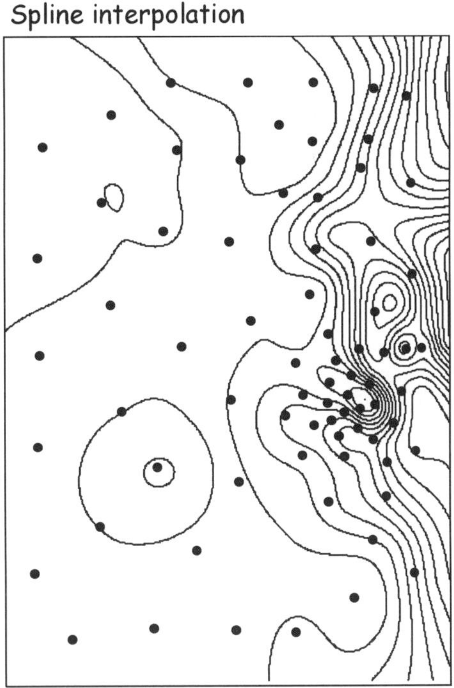

Figure 13-11: Contours and samples from an original elevation surface (left) and contours derived from a spline-fit surface (right).

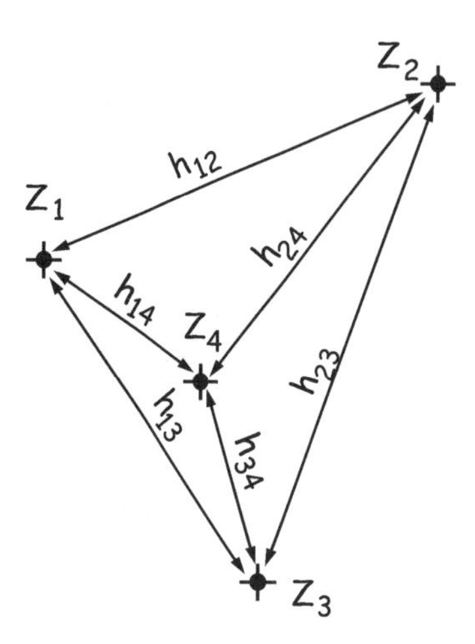

Figure 13-12: Lag distances, used in calculating semivariances for kriging.

Kriging uses the concept of a *lag distance*. Consider the sample set shown in Figure 13-12. Each value for the variable Z is shown plotted over a region. Individual points may be listed as Z_1, Z_2, Z_3, etc., to Z_k, when there are k sample points. The lag distance for a pair of points is the distance between them, and is by convention denoted by h. The lag distance is calculated from the x and y coordinate values for the sample points, based on the Pythagorean formula. In our example in Figure 13-12 the lag (horizontal) distance between the locations of sample points Z_1 and Z_2 is approximately 6 units. The difference in values measured at those points, $Z_1 - Z_2$, is equal to 6. Each pair of sample points is separated by a distance, and also has a difference in the values measured at the points. For example, Z_1 is three units from Z_4, and Z_1 is five units from Z_3. Each pair has a given difference in the Z values, e.g., Z_1 minus Z_4 is 11. Every possible set of pairs Z_a, Z_b, defines a distance h_{ab}, and is different by the amount $Z_a - Z_b$. The distance h_{ab} is known as the lag distance between points a and b, and in general there is a subset of points in a sample set that are a given lag distance apart.

Kriging methods hinge on spatial autocorrelation. Surfaces with low and high autocorrelation are shown in (Figure 13-13). The figure shows two surfaces, Layer 1, with a high autocorrelation, and Layer 2, with a low autocorrelation. Scatter diagrams of sample pairs separated by a uniform, short lag distance are shown to the right of each corresponding layer. Higher autocorrelation, as shown in Layer 1, indicates points near each other are alike. A sample from a surface with high autocorrelation provides substantial information about the values at nearby locations (Figure 13-13, top). Samples from a surface with low autocorrelation do not provide much information at values in the vicinity of the sample point (Figure 13-13, bottom).

Geostatistical prediction uses the key concept of a *semivariance*. A semivariance is the variance based on nearby samples, and is defined mathematically as:

$$\gamma(h) = 1/2n * \sum (Z_a - Z_b)^2 \qquad (13.4)$$

Z_a is the variable measured at one point, Z_b is the variable measured at another point h distance away, and n is the number of pairs that are approximately the distance h apart.

Note that when nearby points are similar, the difference $(Z_a - Z_b)$ is small, and so the semivariance is small. High spatial autocorrelation means points near to each other have similar Z values. When spatial autocorrelation is high, the semivariance for small h is low. The semivariance may be calculated for any h. For example, when h=1, the semivariance, $\gamma(h)$ may be equal to 0.3; when h=2 then $\gamma(h)$ may be 0.5; when h=3 then $\gamma(h)$ may be 0.8. We may calculate a semivariance for any distance h, provided there are sufficient point pairs that are h distance apart to give a good estimate of the semivariance at that distance.

Note that the lag distance h is an approximate distance. By approximate, we mean the lag distance is within some small tolerance of h. Most samples are distributed across the landscape with some imprecision, so it is unlikely we will have many pairs of points that are exactly a given distance h_i apart. For example, we may wish to calculate the semivariance for points that are 112 meters apart. If we are inflexible and only use point pairs that are exactly 112 meters apart (within the precision of our measurement system), we may have only a few, or perhaps even no points that meet this strict criterion. We may specify a tolerance, Δh, which expands the number of points available for a semivariance calculation. By allowing a tolerance, distances that are plus or minus that tolerance from the given lag distance are used to calculate a spatial variability. For example, we might set a tolerance for h of 10 units. Any pair of points between 102 and 122 units apart are used to calculate the semivariance for the lag distance $h = 112$.

We may plot the semivariance over a range of lag distances (Figure 13-14), and this plot is known as a *variogram* or *semivariogram*. A variogram summarizes the spatial autocorrelation of a variable. Note that the semivariance is usually small at small lag distances, and increases to a plateau as the lag distance h increases. This is the typical form of a variogram. The *nugget* is the initial semivariance when the autocorrelation typically is highest. The nugget is shown at the left of the diagram in Figure 13-14, the semivariance at a lag distance of zero. This is the intercept of the variogram. The *sill* is the point at which the variogram levels off. This is the "background" variance, and may be thought of as the inherent variation when there is little autocorrelation.

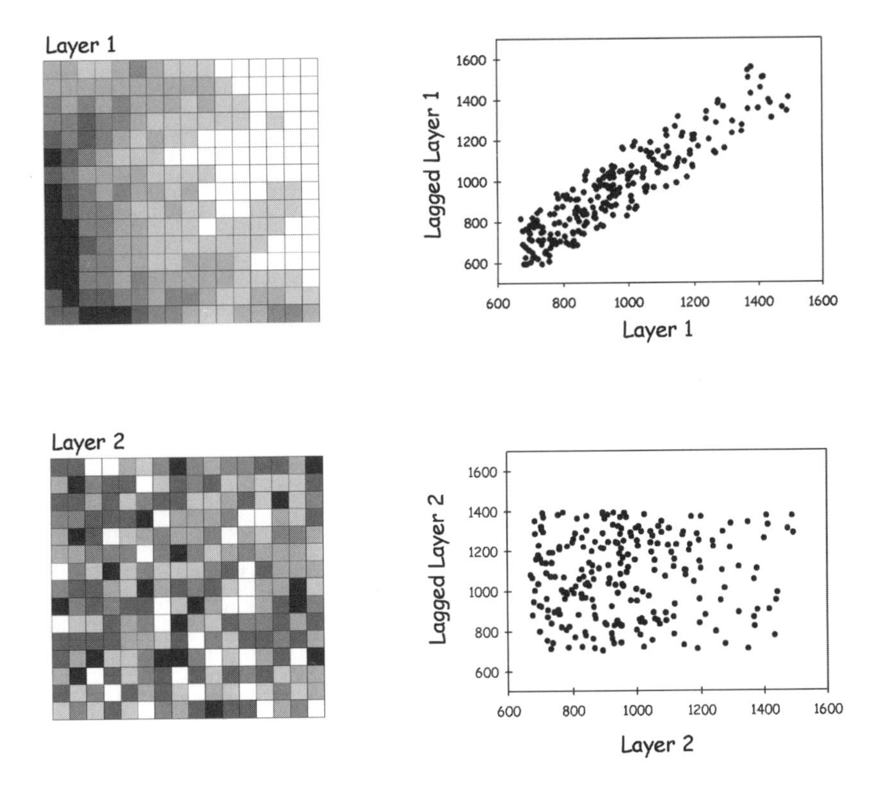

Figure 13-13: Spatially autocorrelated (top) and spatially uncorrelated (bottom) data layers. Plots of sample pairs with a lag distance $h = 1$ (right panels) show similar values for the autocorrelated Layer 1, and unrelated values for uncorrelated Layer 2.

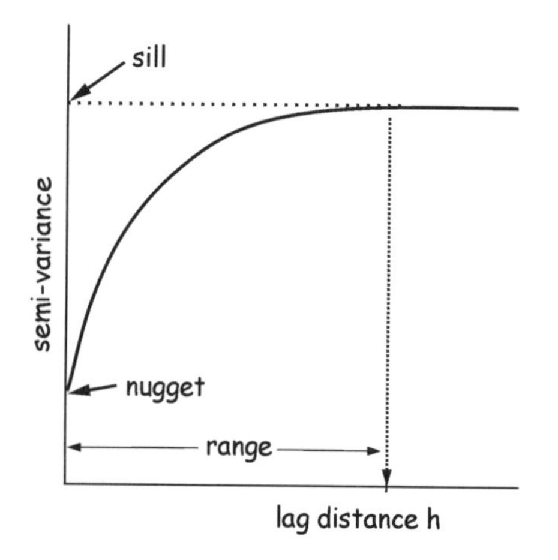

Figure 13-14: An idealized variogram, with the nugget, sill, and range identified.

The *range* is the lag distance at which the sill is reached. The nugget, sill, and range will differ among spatial variables.

A set of sample points is used to estimate the shape of the variogram. First, a set of lag distances h_1, h_2, h_3, etc., are defined, each with a given tolerance. The semivariance is then calculated for each lag distance. An example is shown in Figure 13-15. Remember, each of these points is calculated from Equation 13.4 for a given lag distance apart. A line may then be fit through the set of semivariance points, and the variogram estimated. This line is sometimes called the variogram model.

Interpolation is among the most important applications of the variogram model (Figure 13-16). There are many variations and types of kriging models, but the simplest and most commonly applied rely on using the variogram to estimate "optimal" weights for interpolation. Weights are optimal in the sense that they minimize the error in a prediction, and are unbiased, given a specific data set and model. The calculation of optimal weights requires some rather involved mathematics, beyond our present scope. As stated earlier, kriging is similar to IDW interpolation in that a weighted average is calculated. However kriging uses the minimum variance method to calculate the weights, rather than applying some arbitrary and perhaps more imprecise weighting scheme as with IDW.

Interpolation with kriging and other geostatistical methods can be a complex and nuanced process. There is a wide range of models that may be fit, and these in part depend on the characteristics of the data. Different data characteristics indicate particular modeling methods or model forms, e.g., if there are trends in the data, or directional differences in the variance. These considerations are beyond the scope of our present discussion, and the interested reader is referred to more complete treatments, such as Isaaks and Srivastava (1989), and Burrough and McDonnell (1998) listed under suggested reading at the end of this chapter.

Summary

Interpolation allows us to estimate values at locations where they have not been measured. Interpolation is common because our budgets are limited, samples may be lost

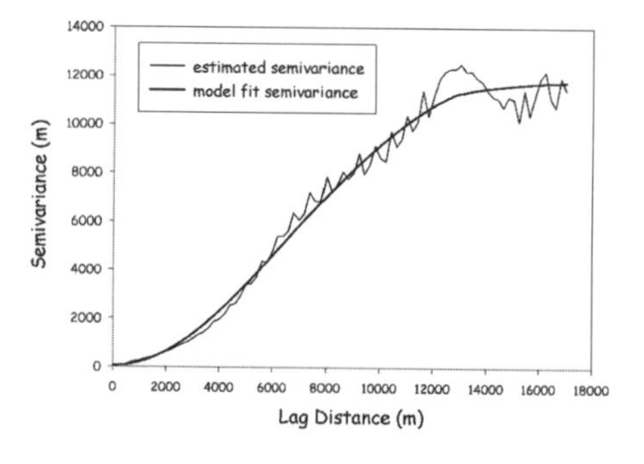

Figure 13-15: A variogram, a plot of calculated and fit semivariance vs. lag distance.

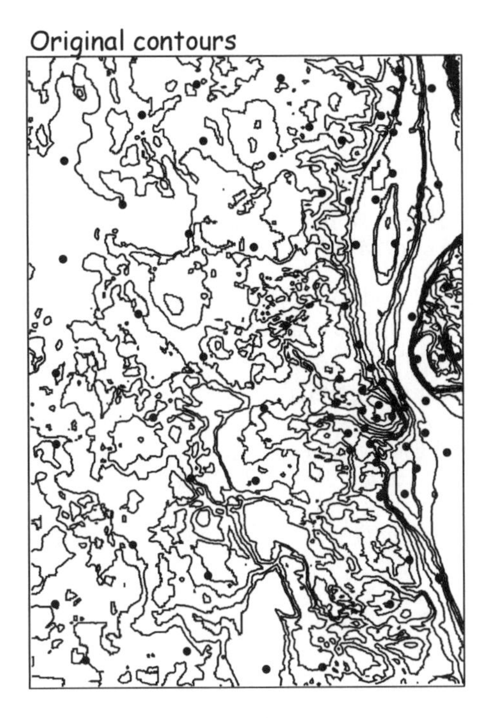

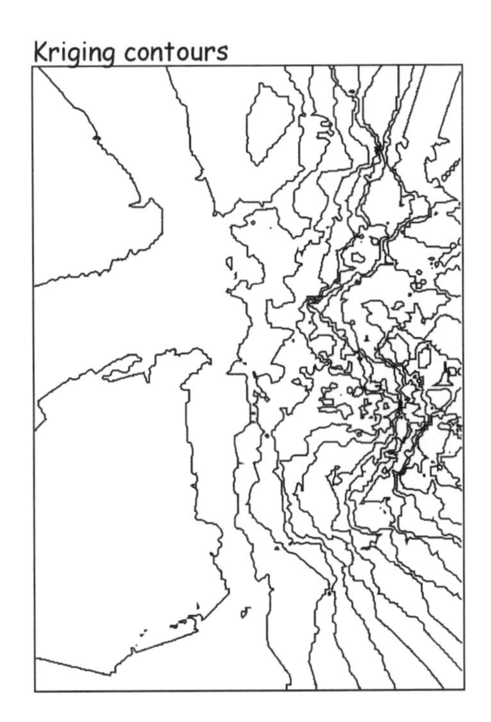

Figure 13-16: Contours and sample points from an elevation surface (left) and contours from a kriged surface, interpolated using the sample points as input (right).

or found wanting, or because time has passed since data collection. We may also interpolate when converting between data models, for example, when calculating a raster grid from a set of contour lines, or when resampling a raster grid to a finer resolution.

Interpolation involves collecting samples at known locations and using rules and equations to assign values at unsampled locations. There are many ways to distribute a sample, including a random selection of sample locations, a systematic pattern, clustering samples, adaptive sampling, or a combination of these. The sampling regime should consider the cost of travel and collecting samples, as well as the nature of the spatial variability of the target feature and the intended use of the interpolated surface.

Sample values are combined with sample locations to interpolate values at unsampled locations. There are many interpolation methods, but the most common are Thiessen (nearest neighbor) polygon, local averaging, inverse distance weighted, trend surface, and kriging interpolation. Each of these methods has advantages and disadvantages relative to each other, and there is no method that is uniformly best. Each method should be tested for the variables of interest, and under conditions in the study area of interest. The best tests involve comparisons of interpolator estimates against withheld sample points.

Suggested Reading

Ayeni, O. O., Optimum sampling for digital terrain models, *Photogrammetric Engineering and Remote Sensing*, 1982, 48:1687-1694.

Burgess, T. M. and Webster, R., Optimal sampling strategies for mapping soil types. I. Distribution of boundary spacing, *Journal of Soil Science*, 1984, 35:641-654.

Burrough, P. A. and McDonnell, R. A., Principles of Geographical Information Systems, Oxford University Press, New York, 1998.

Cressie, N. A. C., Statistics for Spatial Data, Wiley, New York, 1991.

DeGruijter, J. J. and Ter Braak, C. J. F., Model-free estimation from spatial samples: a reappraisal of classical sampling theory, *Mathematical Geology*, 1990, 22:407-415.

Dubrule, O., Comparing splines and kriging, *Computers and Geosciences*, 1984, 10:327-338.

Hutchinson, M. F., Interpolating mean rainfall with thin plate smoothing splines, *International Journal of Geographical Information Systems*, 1995, 9:385-404.

Isaaks, E. H. and Srivastava, R. M., An Introduction to Applied Geostatistics, Oxford University Press, New York, 1989.

Lam, N. S., Spatial interpolation methods: a review, *American Cartographer*, 1983, 10:129-149.

Laurini, R. and Thompson, D., Fundamentals of Spatial Information Systems, Academic Press, London, 1992.

Mark, D. M., Recursive algorithm for determination of proximal (Thiessen) polygons in any metric space, *Geographical Analysis*, 1987, 19:264-272.

Mitasova, H. and Hofierka, J., Interpolation by regularized spline with tension: application to terrain modeling and surface geometry analysis, *Mathematical Geology*, 1993, 25:657-669.

Mitchell, A. The ESRI Guide to GIS Analysis, ESRI Press, Redlands, 1999.

Varekamp, C., Skidmore, A. K., and Burrough, P. A., Using public domain geostatistical and GIS software for spatial interpolation, *Photogrammetric Engineering and Remote Sensing*, 1996, 62:845-854.

Study Questions

Why perform a spatial interpolation?

Can you describe four different sampling patterns, and provide the relative advantages or disadvantages of each? Which do you think is used most in practice, and why?

Can you define contours or isolines? How are they useful?

Can you describe the Thiessen polygon method of spatial interpolation? Why is it considered one of the simplest methods?

Describe and compare fixed radius and inverse distance interpolation.

How is trend surface interpolation different from the previous three methods mentioned above? We say it is an inexact interpolator. What do we mean by that?

Can you describe kriging interpolation, including the basic process, and the components, including the semivariance and variogram and their components?

14 Data Standards, Quality, and Documentation

Introduction

A standard is an established or sanctioned measure, form, or method. It is an agreed-upon way of doing, describing, or delivering something. The adoption of spatial data and analysis standards is an important endeavour, given the range of organizations producing and using spatial data and the ease with which the spatial data may be manipulated using GIS software packages. Data standards allow a common understanding of the components of a spatial data set, how it was developed, and the utility and limitations of these data.

Several types of standardization are desirable when using GIS. *Data standards* are used to format, assess, document, and deliver spatial data. *Analysis standards* may ensure the most appropriate methods are used, and that the spatial analyses provide the best information possible. *Professional* or *certification standards* may establish the education, knowledge, or experience of the GIS analyst, thereby improving the likelihood that the technology will be used appropriately.

We have progressed further in defining spatial data standards than in defining analysis and professional standards. This is in part because of the newness of the technology, and in part because GIS are used in such a wide range of disciplines. Urban planners, conservationists, civil and utility engineers, business people, and a number of other professions use GIS.

The development of professional and analysis standards that are meaningful across all disciplines may not be appropriate. Standard methods for one discipline may be inappropriate for another. For example, digital data collection or analysis methods for cadastral surveyors may be different than those for foresters. Field measurement techniques, data reduction, and positional reporting for cadastral surveys often require accuracies measured in centimeters (0.5 inches) or less, while relatively sparse attribute information is recorded. Conversely, forest inventory methods may allow relatively coarse-scale positional measurements, to the nearest few meters, but require standard methods for measuring a large set of attributes. Because professional and analysis standards are often discipline-specific, and due to the importance of data standards and progress made in establishing them, the remainder of this chapter will focus on spatial data standards.

Spatial Data Standards

Spatial data standards may be defined as methods for structuring, describing, and delivering spatially-referenced data. Spatial data standards may be categorized into four areas: media standards, format standards, accuracy standards, and documentation standards. All are important, although the last two are substantially more complex.

Media standards refer to the physical form in which data are transferred. They may specify a CD-ROM, tape type and format, or some proprietary drive or other media type. Physical or "device-level" formatting may be specified for the media. CD-ROM has become a common data delivery medium, and standardized formats are specified by the International Standards Organization (ISO). The ISO-9600 standard defines the physical and device-level formatting. Other international standards may be applied to other devices, and media written to one of the standards may be read by any device that adheres to the standards.

Format standards specify data file components and structures. The number of files used to store a spatial data set is established, as well as the basic components contained in each file. The order, size, and range of values for the data element contained in each file are defined. Information such as spacing, variable type, and file encoding may be included.

Format standards aid in the practical task of transferring data between computer systems, either within or between organizations. Government agencies such as the USGS spend substantial time and resources to develop spatial data, with the primary purpose of delivering these data to end users. Many businesses and other private organizations exist primarily to develop and deliver spatial data, either in addition to analyses, or as a specific product. Producers and users may not use the same hardware or GIS software. The interchange between different software systems is aided by general, standard forms in which data may be delivered.

Spatial data accuracy standards document the quality of the positional and attribute values stored in a spatial data set. Knowledge of data quality is crucial to the effective use of GIS, but we are often remiss in our assessment and reporting of spatial data quality. Many of us are conditioned by looking at paper maps. Intrinsically, we know the positions are not plotted perfectly, and there are certain allowances granted. Map generalization, described in Chapter 4, is one such allowance. We know the roads are not plotted at map scale. In most instances we would not be able to see the roads because their widths would be too narrow. Small towns are often plotted as circles of a fixed size, and we generally don't believe these to be the true shape of the town. There is a certain fuzziness we allow in the relative and absolute geographic position of objects in maps.

While we typically do not transfer this philosophy to our digital spatial data, we often pay less attention to documenting spatial data accuracy than is warranted. This is due in part to the cost of adequately estimating the errors in our spatial data sets. Field sampling is quite expensive, and we are tempted to spend the additional funds collecting additional data. Data production or analysis are often pushed to the limits of available time or monies, and the documentation of data accuracy may be given scant attention. Adherence to spatial data accuracy standards ensures we assess and communicate spatial data quality in a well-defined, established manner.

Documentation standards define how we describe spatial data. Data are derived from a set of original measurements taken by specific individuals or organizations at a specified time. Data may have been manipulated or somehow transformed, and data are stored in some format. Data documentation describes the source, development, and form of spatial data, and documentation standards are an agreed-upon way of providing the documentation. When documentation standards are used they ensure a complete description of the data origin, methods of development, accuracy, and delivery formats. Standard documentation allows the data steward to maintain the data, and any potential user to assess the appropriateness of these data for an intended task.

Data quality is a good example of an area that benefits from the adoption of documentation standards. There are many ways to describe data quality. We may describe the positional error in many ways, e.g., the average distance error, the biggest distance error, the percentage of points that are above an error threshold, or the total area that is misclassified. Attribute error may also be described in many different ways. The producer may describe the spatial data quality in one manner, but this may not provide sufficient information for the user to judge if the data are acceptable for an intended application.

Spatial data quality is often not well-documented, perhaps due to the recency of GIS. GIS have become widespread in the past two decades. Although academic research on spatial data quality has a much longer history, the study of spatial data accuracy has intensified substantially during this same period. The comparison of various sampling and accuracy measurement methods, the development of standardized statistics, and the establishment of standard methods for calculating and reporting spatial data accuracy all take more time to develop. However, standards for documenting spatial data accuracy exist and should be used.

Data Accuracy

An accurate observation is close to the truth. When the concept of accuracy is applied to spatial variables, it is a measure of how often or by how much our data values are in error. It may be expressed as how often a value is wrong, e.g., for categorical data, four percent of the fields listed as row crops are perennial grasses. Alternatively, accuracy may be expressed as a number or probability distribution of the size of an error, e.g., the average positional error is 12.4 meters for power pole locations, or more than 90 percent of the digitized geodetic monuments are within 3.2 meters of their surveyed locations. We use spatial data accuracy to describe how faithfully our digital representation reflects the true shape, location, and characteristics of the phenomena represented in a GIS.

Spatial data always contain some error, as there are many steps in the process of data collection, reduction, organization, and manipulation where inaccuracies may be introduced. Errors may be caused by how we conceptualize the features, our methods of data collection and analysis, human error, mis-understanding information regarding equipment or data collection methods, or data may simply be out of date. Each of these causes a difference between what currently exists "on the ground" and the representation in our spatial data.

Inadequacies in our spatial data model are a common cause of spatial data error. We may use a raster data set with a fixed cell size, and this limits our positional accuracy. We can locate objects no better than one-half the cell resolution, on average. The raster model assumes a homogeneous pixel. If more than one category or value for a variable is found in the pixel, then the attribute value may be in error. Consider a raster data layer containing land use data and a 100 meter cell dimension. A small woodlot less than one-half hectare in size may not be represented if it is embedded in an urban or agricultural setting – it is too small and so may be "included" as part of the land use around it. This "generalization" or "inclusion" error may also occur in vector data sets. Any feature smaller than the minimum mapping unit may not be represented.

Vector data sets may poorly represent gradual changes, so there may be increased attribute error near vector boundaries. Digital soils data are often provided in a vector data model, yet the boundaries between soil types are often not discrete, but change over a zone of a few to several meters.

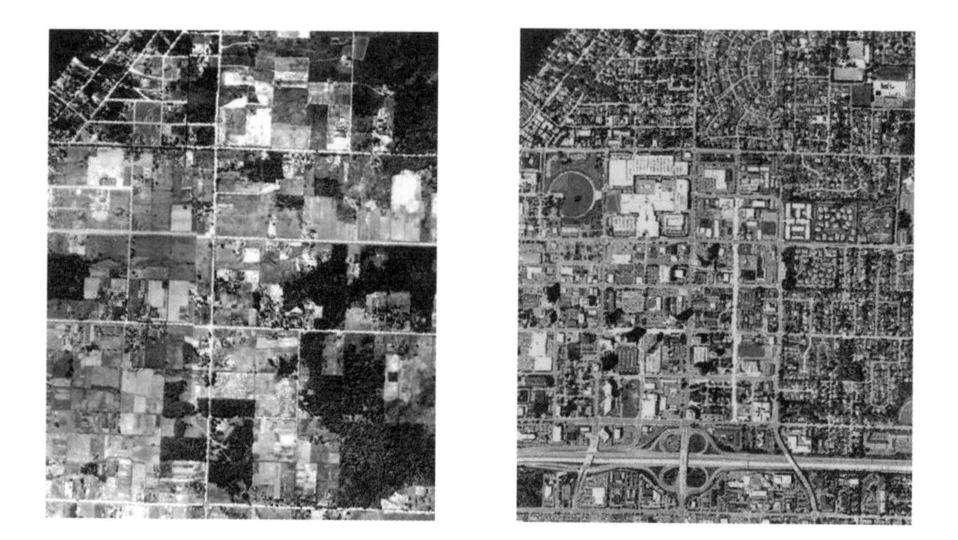

Figure 14-1: Spatial data may be in error because of the passage of time. Road maps based on 1936 photographs (left) from the city of Bellvue, Washington, are likely to be in error in 1997 (right). (courtesy Washington Department of Natural Resources)

Errors are often introduced during spatial data collection. Many positional data are currently collected using GPS technologies. The spatial uncertainty in GPS positions described in Chapter 5 is incorporated into the positional data. Feature locations derived from digitized maps or aerial photographs also contain positional errors due to optical, mechanical, and human deficiencies. Lenses, cameras, or scanners may distort images, positional errors may be introduced during registration or errors may be part of the digitization process. Blunders, fatigue, or differences among operators in abilities or attitudes may result in positional uncertainty.

Spatial data accuracy may be degraded during laboratory processing or data reduction. Mis-copies during the transcription of field notes, errors during keyboard entry, or mistakes during data manipulation may alter coordinate values used to represent a spatial data feature. Improper representation in the computer may cause problems, for example, round-off errors when multiplying or storing large numbers. These errors plus mistakes or improper laboratory techniques may also alter attribute values and introduce errors.

Data may also be in error due to changes through time (Figure 14-1). Features in our spatial dataset represent real objects and in many instances these objects change through time. Vegetation boundaries may be altered by fire, logging, construction, conversion to agriculture, or a host of other human or natural disturbances. A vegetation data layer more than a few years old is likely to contain substantial error due to vegetation change. Even in instances where positions are static, attributes may change through time. A two-lane gravel road may be paved or widened. If the spatial data representing this road are not changed, then the attributes will be erroneous. Few objects are truly static, as even the hardest rocks erode through time. However, the relative rate of change varies among objects so each object type or theme should have a recommended update interval. Elevation, geology, and soils usually change at very slow rates relative to the lifetimes of organizations or the humans working in them. These data layers may be considered static for most uses. Vegetation, population, land use, or other factors change at faster rates. Differences between values in a spatial data layer and true values may be due to these changes.

Documenting Spatial Data Accuracy

Spatial data accuracy must be documented because it helps determine the utility of a data set. We can not properly evaluate if a data set is appropriate for an analysis without knowledge of data accuracy. Therefore, each data set must come with some description of how close the recorded data values are to true conditions.

We must unambiguously identify true conditions if we are to document spatial data accuracy. For example, a road segment may be completely paved, or not. The data record for that road segment is accurate if it describes the surface correctly, and inaccurate if it does not. However, in many cases the truth is not completely known. The locations for the above roads may be precisely surveyed using the latest carrier-phase GPS methods. Road centerlines and intersections may be known to the nearest 0.5 centimeters. While this is a very small error, this represents some ambiguity in what we deem to be the truth. Establishing the accuracy of a data set requires we know the accuracy of our measure of truth. In most cases, the truth is defined based on some independent, higher order measurements. In our roads example, we may desire that our data layer be accurate to 15 meters or better. Gauged on this scale, the 0.5 centimeter accuracy from our carrier-phase GPS measurement may be considered the truth.

There are at least four ways we describe spatial data accuracy: *positional accuracy, attribute accuracy, logical consistency*, and *completeness*. These four components may be complemented with information on the *lineage* of a data set to define the accuracy and quality of a data set. These components are described in turn below.

Positional accuracy describes how close the locations of objects represented in a digital data set correspond to the true locations for the real-world entities. In practice, truth is determined from some higher-order positioning technology.

Attribute accuracy summarizes how different the attributes are from the true values. Attribute accuracies are usually reported as a mean error or quantile above a threshold error for attributes measured on interval/ratio scales, and as percentages or proportions accurate for ordinal or categorical attributes.

Logical consistency reflects the presence, absence or frequency of inconsistent data. Tests for logical consistency often require comparisons among themes, for example, all roads and buildings occur on dry land, and not in wetlands or water bodies. This is different than positional accuracy in that both the wetland and building locations may contain positional error. However, these errors do not cause impossible or illogical juxtapositions. Logical consistency may also be applied to attributes, e.g., wetland soils erroneously listed as suitable for construction without special construction methods.

Completeness describes how well the data set captures all the features it is intended to represent. A buildings data layer may omit certain structures, and the frequency of these omissions reflects an incomplete data set. Data sets may be incomplete because of generalizations during map production or digitizing. For example, a minimum mapping unit may be set at two hectares when compiling a vegetation map. Isolated small pastures scattered through the forest may not be represented because they are only slightly larger than this minimum mapping unit, and erroneously they are not represented in the data layer. Thus, the data set is inherently incomplete because it does not represent these pastures. Completeness often refers to how well or often a data set contains those features it purports to represent. In our example above, completeness would define how often features greater than two hectares in size are included in the data set.

Lineage describes the sources, methods, timing, and persons responsible in the development of a data set. Lineage helps establish bounds on the other measures of accuracy described above, because knowledge about certain primary data sources helps define the accuracy of a data set. We know data digitized from a 1:24,000 scale USGS topographic map are unlikely to have positional accuracies much better than 15 meters, or that maps produced for Los Angeles in 1960 are likely to omit many new roads.

Accuracy is most reliably determined by a test. A test involves a comparison of true values to the values represented in a spatial data set. This requires the collection of a set of sample locations, with due consideration of the number of samples and their distribution. "True" values are collected at these sample locations. Corresponding values are collected for the digital spatial data. The true and data values are compared, errors calculated, and summary statistics generated.

The truth source, sampling method, method for calculating error, and summary statistics chosen will depend on the type and component of spatial data that are to be evaluated. Positional data will be assessed using different methods than attribute data. Nominal attribute data, for example, the type of land cover, will be assessed differently than a measurement recorded on a continuous range, e.g., the soil nitrogen content.

Positional Accuracy

Accuracy measures how close an observation is to the true value. Positional accuracy indicates how far a spatial feature is from the true location. Database representations of a position are accurate if the distance between the true and recorded values is small. Small is defined subjectively, but may at least be quantified.

Positional accuracy for spatial features may be reported many ways. We may report accuracy for any single observation, e.g., the position for the center of the lightpole at 14th Avenue and Walker Street has a positional accuracy of 2 centimeters. By this we mean the difference between the true and database values is 2 centimeters. We may also report accuracy as an average for a group of observations, e.g., the mean positional accuracy for all lightpoles is 5 centimeters. We consider an individual position accurate when the distance between the spatial database and true locations is small. We describe a group of locations as accurate if the average distance between spatial database and true positions is small.

Precision refers to how repeatable a process or measurement may be. Precision is usually defined in terms of how far the set of repeat measurements are from the average measurement. A precise measurement system provides predictable, consistent results. Precise digitizing means we may repeatedly place a point in the same location. Because precision is related to the proximity of a set of measurements, we cannot describe the precision of a single measurement.

Accuracy and precision are two related but distinct concepts that are often confused. The confusion may be due to the desire to obtain both accurate and precise data; with a repeated process, poor precision often leads to poor accuracy. However, precision and accuracy are two different characteristics, and may change independently.

A set of measurements may be precise, but inaccurate. Repeat measurements may be well clustered, but they may not be near the true value. A *bias* may exist, defined as a systematic offset in coordinate values. Thus, a precise set of measurements may be relatively inaccurate. A less precise process may not introduce this bias and may be more accurate.

Figure 14-2 illustrates the difference between accuracy and precision in point location. Four digitizing sessions are shown. The goal is to place several points at the center of the cloverleaf intersection in Figure 14-2. The upper left panel shows a digitizing process that is both accurate and precise. Points, shown as white circles, are clustered tightly and accurately over the intended location.

The upper right panel of Figure 14-2 shows a set of points that are precise, but not accurate. The digitized points are tightly clustered, but are shifted up and to the right from the true location. This might be due to an equipment failure or some problem in registration, e.g., the operator made some blunder in photo registration and introduced a bias.

The lower left panel of Figure 14-2 shows a set of points that are accurately but imprecisely digitized. The average for these points is quite near the desired position, the center of the cloverleaf intersection, even though individual points are widely scattered. These points are not very close to the mean value and so precision is low, even though accuracy is high. Sensitive equipment, an operator with unsteady hands, or equipment malfunction could all result in this situation.

The panel at the lower right of Figure 14-2 shows a set of points that are both imprecise and inaccurate. The mean value is

high average accuracy,
high precision

low average accuracy,
high precision

high average accuracy,
low precision

low average accuracy,
low precision

Figure 14-2: Accuracy and precision. Points (white circles) are digitized to represent the center of the cloverleaf intersection. Average accuracy is high when the average of the points falls near the true location, as in the panels on the left side of the figure. Precision is high when the points are all clustered near each other (top panels). A group of points may be accurate, but not precise (lower left), or precise, but not accurate (upper right). We typically strive for a process that provides both accuracy and precision (upper left), and avoid low accuracy and low precision (lower right).

not near the true location, nor are the values tightly clustered.

The threshold that constitutes high accuracy or precision is often subjectively defined. We may deem a value accurate because we have knowledge of previous attempts. A duffer may consider as accurate any golf shot that lands on the green. This definition of accuracy may be based on thousands of previous attempts. However, for a professional golfer anything farther than 2 meters from the hole may be an inaccurate shot. In a similar fashion, high spatial accuracy for a land surveyor is different than that of a federal land manager. Cadastral surveys require the utmost in accuracy because people tend to get upset when there is material permanent trespass, as when a neighbor builds a garage on their land. Lower accuracy is acceptable in other applications, e.g., the boundaries defining vegetation type may be off by tens of meters in a statewide assessment of land cover. Unknown accuracy leads to uncertain analysis, eroded confidence in recommendations, and frequent challenges.

Where possible, positional accuracy should be quantified. We should measure and report a number or set of numbers that gives the user a physical measurement of the expected accuracy. Some indication of the mean and the spread of positional errors should be reported. In addition, some indication of the attribute accuracy should be provided. If the attributes are continuous variables then the accuracy may be reported as a mean error, the spread or distribution of the errors, or some statistics such as a threshold frequency that combines information on both the mean and distribution of errors. If the attribute is categorical then the error may be best reported as an error frequency, the number of times a wrong category is assigned.

The mean error and an error frequency threshold are the statistics most often used when documenting positional data accuracy. Consider a set of wells that are represented as point features in a spatial data layer. We wish to document the accuracy of the coordinates used to define well location. Suppose that after we have digitized our well locations we gain access to a GPS system that effectively gives us the true coordinate locations for each well. We may then compare these well locations to the coordinate locations in our database. We begin by calculating the distance from our true and database coordinates for each well. This leaves us with a list of errors, one associated with each well location.

We typically quantify positional accuracy with a measurement of distance error (Figure 14-3). Distance error may be defined as the difference between a true and recorded point location. Distance is measured using the Pythagorean formula with the true and database coordinates. Distances are always positive because of the form of the formula.

We may compute the mean error by summing the errors and dividing the sum by the number of observations. This gives us our average error, a useful statistic somewhere near the midpoint of our errors. We are often interested in the distribution of our errors, and so we also commonly use a frequency histogram to summarize our spatial error. The histogram is a graph of the number of error observations by a range of error

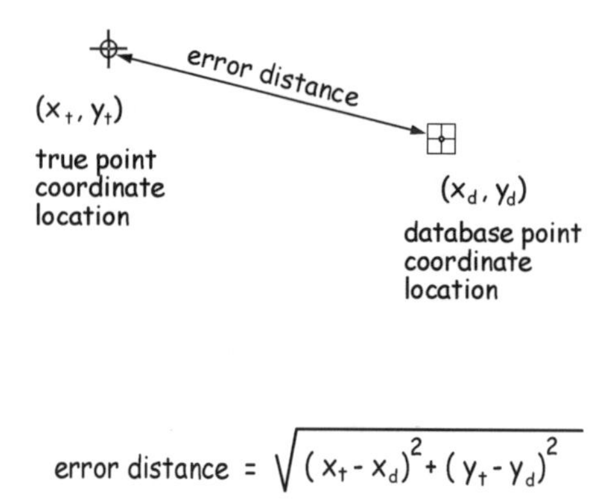

Figure 14-3: Positional errors are measured by the Pythagorean distance between a true and database coordinate for a location.

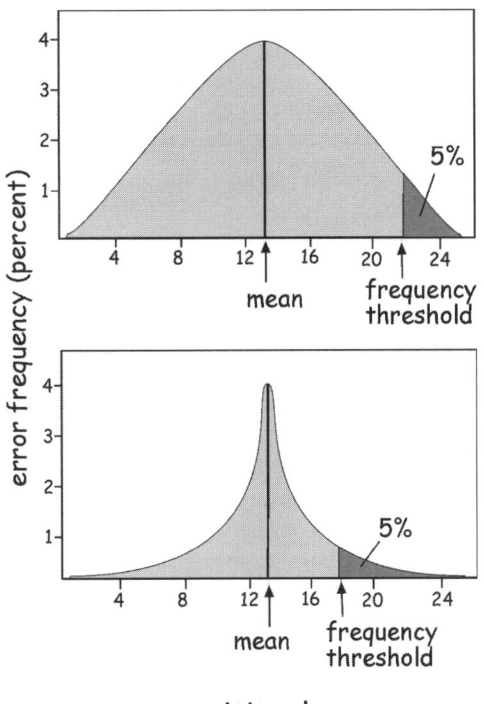

Figure 14-4: Mean error and frequency thresholds are often used to report positional error.

values, e.g., the number of error values between 0 and 1, between 1 and 2, between 2 and 3, and so on for all our observations. The graph will indicate the largest and smallest errors, and also give some indication of the mean and most common errors.

Examples of mean error values and error frequency histograms are shown in Figure 14-4. Each plot shows the frequency of errors across a range of error distances. For example the top graph shows that approximately 1 error in 100 has a value of near 4.5 meters, and the mean error is near 13 meters.

The mean error value does not indicate the distribution, or spread of the errors. Two data sets may have the same mean error but one may be inferior; the data set may have more large errors. The bottom graph in Figure 14-4 has the same mean error, 13 meters, as the top graph. Note that the errors have a narrower distribution, meaning the errors are clumped closer to the mean than in the top graph, and there are fewer large errors. Although the mean error is the same, many would consider the data represented in the bottom graph of Figure 14-4 as more accurate.

Because the mean statistic alone does not provide information on the distribution of positional errors, an error frequency threshold may be reported. An error frequency threshold is a value above or below which a proportion of the error observations occur. Figure 14-4 shows the 95% frequency threshold for two error distributions. The threshold is placed such that 95% of the errors are smaller than the threshold and approximately 5% are larger than the threshold. The top graph shows a 95% frequency threshold of approximately 21.8 meters. This indicates that approximately 95% of the positions tested from a sample of a spatial database are less than or equal to 21.8 meters from the true locations. The bottom panel in Figure 14-4 has a 95% frequency threshold at 17.6 meters. This means 5% of the errors in the second tested database are larger than 17.6 meters from their true location. If we are concerned with the frequency of large errors, this may be a better summary statistic than the mean error.

Note that the frequency threshold is sensitive to both the mean and spread of positional errors. A higher mean will lead to a higher frequency threshold, all other things being equal. A broader distribution of errors will lead to a larger frequency threshold, all other things being equal. The threshold is often chosen when reporting a single summary number because it contains information on the mean and distribution of positional errors. Either a larger mean error or a broader distribution of errors will result in a larger threshold. Figure 14-4 illustrates a case where the mean errors are equal, but the data set in the upper graph is more scattered, with a broader distribution and hence a larger frequency threshold.

A Standard Method for Measuring Positional Accuracy

The Federal Geographic Data Committee of the United States (FGDC) has described a standard for measuring and reporting positional error. They have done so because it is such an important characteristic of every digital data set and there is a need for a standardized vocabulary and method for documenting spatial accuracy. This standard is known as the National Standard for Spatial Data Accuracy (NSSDA). The NSSDA specifies the number and distribution of sample points, and prescribes the statistical methods used to summarize and report positional error. Separate methods are described for horizontal (x and y) accuracy assessment and vertical (z) accuracy assessment, although the methods differ primarily in the calculation of summary accuracy statistics. There are five steps in applying the NSSDA:

•Identify a set of test points from the digital data set under scrutiny.

•Identify a data set or method from which "true" values will be determined.

•Collect positional measurements from the test points as they are recorded in the test and "true" data sets.

•Calculate the positional error for each test point and summarize the positional accuracy for the test data set in a standard accuracy statistic.

•Record the accuracy statistic in a standardized form that is included in the metadata description of a data set. Also include a description of the sample number, true data set, the accuracy of the true data set, and the methods used to develop and assess the accuracy of the true data set.

Test Points

Test points must be identified as a first step in conducting a positional accuracy assessment. These test points must be clearly identifiable in both the test data set and in the field or in the "truth" data set. Points that are precisely, unambiguously defined are best. For example, we may wish to document the accuracy of roads data compiled from medium-scale sources and represented by a single line in a digital layer. Right-angle road intersections are preferred over other features because the position represented in the data base may be precisely determined. The coordinates for the precise center of the road intersections may also be determined from a higher accuracy data set, e.g., from digital orthophotographs or field surveys. Other road features are less appropriate for test points, including road intersections at obtuse angles or acute curves because there may be substantial uncertainty when matching the data layer to true coordinates. Obtuse road intersection points may be easily identified in the digital data layer, however the corresponding point on an orthophotograph may be difficult to define. Matching points on a curve is even more difficult.

The source of the true coordinate position depends on the data set to be tested. Data from a large-scale map of known accuracy may be appropriate when assessing the accuracy of data derived from a very small scale map. We may use data derived from 1:20,000-scale maps to assess the accuracy of corresponding features derived from 1:2,000,000-scale maps, provided the accuracies of the 1:20,000-scale maps have been tested and found acceptable. Higher accuracy data sets such as those derived from larger scale maps must be tested against true coordinates that have been derived from more accurate sources, e.g., precise GPS surveys. Small features are required for test points, e.g., utility poles, benchmarks, sidewalk corners, or other precisely surveyed locations.

Figure 14-5 shows an example set of test points, a data layer, and an image backdrop.

We wish to assess the accuracy of the roads data layer. In this example the roads data were developed from manual digitization of a 1:62,500-scale map. Previous experience indicates these data are likely to have average positional errors in excess of 20 meters, and we wish to test accuracy to at least the nearest 30 meters, that is, errors larger than 30 meters are cause for concern. We have selected digital orthophotoquads (DOQs, described in Chapter 7) as our source of "true" coordinates. We recognize that the DOQ-derived coordinates will contain some error. Information provided with the DOQs indicates an average positional error below 2 meters. DOQs were selected because 1) we expect errors in the roads layer to be more than an order of magnitude larger than true positions derived from the DOQs, 2) acceptable levels of error are set at 30 meters, and 3) the DOQs are available for the entire area of interest.

The display of road locations on top of the DOQ shows there are substantial differences in true positions of features and their representations in the roads data layer. Any right-angle intersection is a prospective test point. Errors appear to be particularly large near the top of the image (Figure 14-5), but this information should not be used when selecting the test points. Points should be

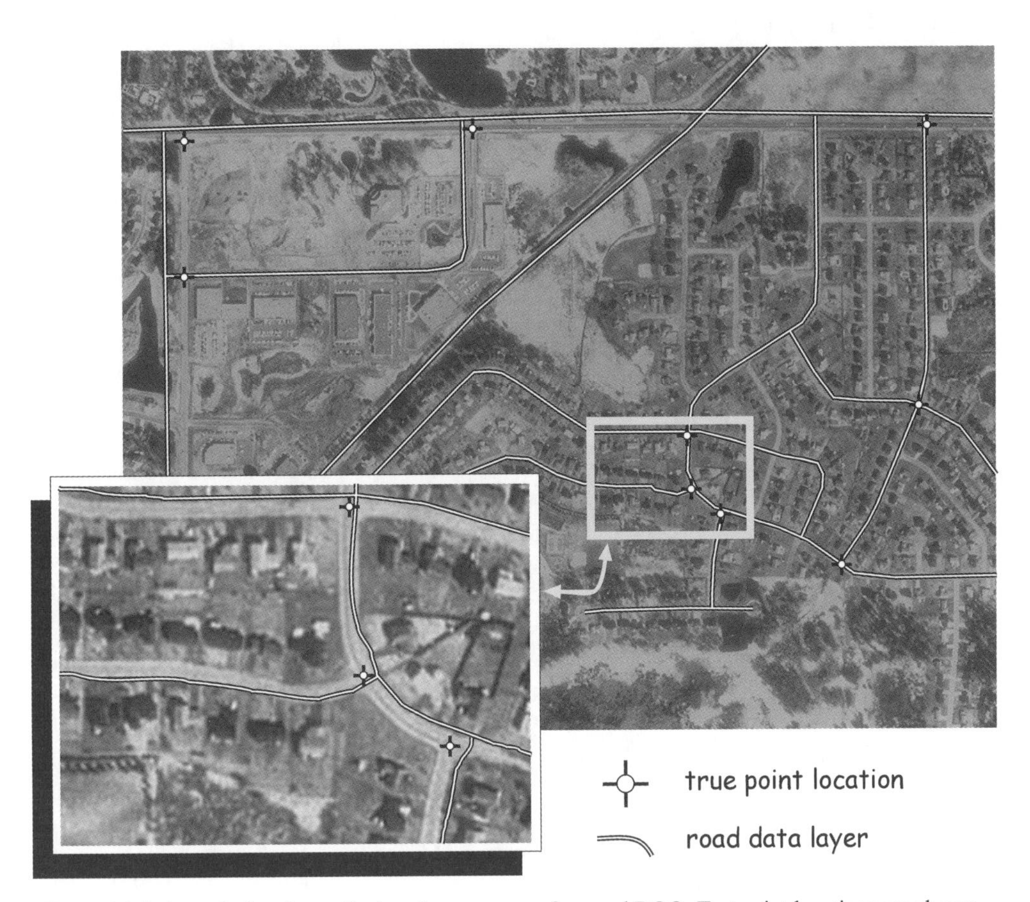

Figure 14-5: A roads data layer displayed over a georeferenced DOQ. Test point locations are shown. The true coordinate values may be derived from the DOQ and the data coordinate values from the roads layer. Differences in these locations would be used to estimate the positional accuracy of the roads data layer.

selected by some random process, and not where error appears largest or smallest.

The inset in the lower left of Figure 14-5 shows the true point locations relative to the road intersections. We know the digitizing technicians were instructed to record the road centerline locations. The intersection of two road centerlines falls mid-way between the road edges. These true locations would be identified on the DOQs, perhaps by pointing a cursor at a georeferenced image displayed on a computer monitor. The data coordinates would then be extracted for the corresponding road intersection in the roads data layer, and these two coordinate pairs, the true x, y and data x, y would be one test point used in accuracy calculations.

According to the NSSDA we should select between twenty and thirty test points such that they are well distributed throughout the test data layer (Figure 14-6). Statistical science has established that accuracy estimates improve substantially up through 15 to 20 sample points. The added value of each additional point declines up to 30, and the marginal gain beyond 30 test points does not often justify the added cost of collecting additional test points. Test points should be distributed as evenly as possible throughout the data layer to be tested. Each quadrant of the tested data layer should contain at least 20 percent of the test points, and test points should be spaced no closer than one-tenth the longest spanning distance for the tested data layer (d, in Figure 14-6).

There is often a trade-off between the quality of the "true" point coordinates and the distribution of test points. Well-defined points may be absent from portions of the test area, sparse in others, or concentrated in a small area. There may be a choice between a poorly-distributed set of accurately defined points, or a well-distributed set of points,

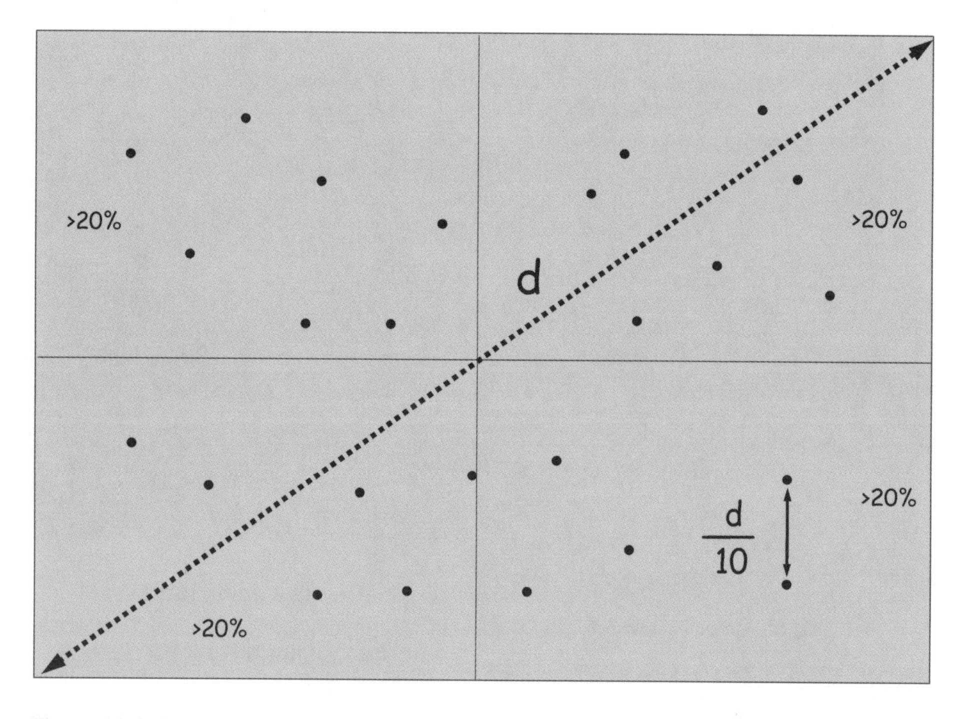

Figure 14-6: Recommended sampling for spatial data accuracy. Samples should be well distributed with at least 20% of the samples in each quadrant of the study region. Samples should also be well-spaced, with samples no closer than one-tenth the diagonal dimension of the study area. (Adapted from LMIC, 1999)

some with less accurate true locations. The best solution is to identify well-defined points in areas lacking high quality test points, and collecting the true coordinates for these new test points. If well-defined points do not exist, e.g., if a quadrant of the study area is covered by closed-canopy forest with few uniquely identifiable features, then there may be no alternative to accepting a sub-optimal distribution. For most data sets it is possible to set a minimum acceptable quality for true coordinates, and strive to distribute them as evenly as possible.

Accuracy Calculations

The calculation of point accuracies and summary statistics are the next step in accuracy assessment. First, the coordinates of both the true and data layer positions for a feature are recorded. These coordinates are used to calculate a positional difference, known as a positional error, based on the distance between the true coordinates and the data layer coordinates (Figure 14-7). This is a portion of the image inset shown in Figure 14-5. The true coordinates fall in a different location than the coordinates derived from the data layer. Each test point yields an error distance **e**, shown in Figure 14-7 and defined by the equation:

$$e = \sqrt{(x_t - x_d)^2 + (y_t - y_d)^2} \qquad (14.1)$$

where x_t, y_t are true coordinates and x_d, y_d are the data layer coordinates for a point.

The squared error differences are then calculated, and the sum, average, and root mean square error (**RMSE**) statistics determined for the data set. As previously defined in this book, the **RMSE** is:

$$RMSE = \sqrt{\frac{e_1^2 + e_2^2 + ..e_n^2}{n}} \qquad (14.2)$$

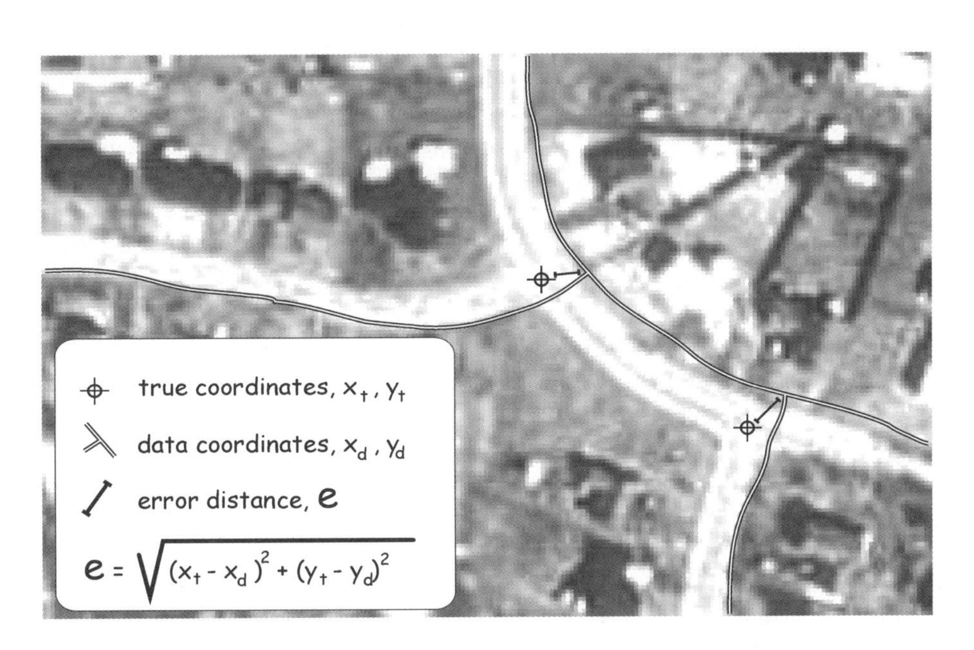

Figure 14-7: True and test point locations in an accuracy assessment. True and test point locations do not coincide. The differences in coordinates between true and test points are used to calculate an error distance.

where e is defined as in equation (14.1) above, and n is the number of test points used.

The RMSE is not the same as the average distance error, nor a "typical" distance error. The RMSE is a statistic that is useful in determining probability thresholds for error. The RMSE is related to the statistical variance of the positional error. If we assume the x and y errors are normally distributed, that is, they follow a bell-shaped Gaussian curve commonly observed when sampling, then the RMSE tells us something about the distribution of distance errors. The RMSE is a mathematical combination of the x and y variables. We can use knowledge about the RMSE that we get from our sample to determine what is the likelihood of a large or small error. A large RMSE means the errors are widely spread, and a small RMSE means the errors are packed tightly around the mean value.

Statistical theory allows us to establish fixed numbers that identify error thresholds. Because we have two variables, x and y, if we make appropriate assumptions, we can fix an error threshold at a given value. An error threshold is commonly set for 95%. When we fix a 95% error threshold, this means we identify the specific number such that 95% of our errors are expected to be less than or equal to the threshold. Statistical theory tells us that when we multiply the RMSE by the number 1.7308 and assume a Gaussian normal distribution, we obtain the 95% threshold. We will not cover the statistical assumptions and calculations used in deriving these constants, nor the theory behind them. A thorough treatment may be found in the references listed at the end of this chapter.

Accuracy calculations may be summarized in a standard table, shown in Table 14-1. The example shows a positional accuracy assessment based on a set of 22 points. Data for each point are organized in rows. The true and data layer coordinates are listed, as well as the difference and difference squared for both the x and y coordinate directions. The squared differences are summed, averaged and the RMSE calculated, as shown in the summary boxes in the lower right portion of Table 14-1. The RMSE is multiplied by 1.7308 to estimate the 95% accuracy level, listed as the NSSDA. Ninety-five percent of the time the true horizontal errors are expected to be less than the estimated accuracy level of 12.9 meters listed in Table 14-1.

Errors in Linear or Area Features

The NSSDA as described above treats only the accuracies of point locations. It is based on a probabilistic view of point locations. We are not sure where each point is, however we can specify an error distance r for a set of features. A circle of radius r centered on a point feature in our spatial data layer will include the true point location 95% of the time. Unfortunately there are no established standards for describing the accuracy or error of linear or area features.

Previous work in cartography and GIS used the concept of an *epsilon band* to characterize uncertainty in line position. An epsilon band may be defined that has a very high probability of containing the line (Figure 14-8). Within this epsilon band the line location is uncertain. The concept of an epsilon band is congruent with our model of point positional errors if we remember that most vector line data are recorded as a sequence of nodes and vertices. Lines are made up of point locations. Nodes and vertices defining line locations contain some error. The epsilon band may be thought of as having a high probability of encompassing the true line segments. Larger epsilon bands are associated with poorer quality data, either because the node and vertex locations are poorly placed and far from the true line location, or because the nodes or vertices are too widely spaced.

In some instances we may assume the well-defined point features described in our accuracy test above may also represent the accuracy for nodes and vertices of lines in a data layer. Nodes or vertices may be used as test points, provided they are well defined and the true coordinates are known. How-

Table 14-1: An accuracy assessment summary table.

ID	x (true)	x (data)	x differ-ence	(x differ-ence)^2	y (true)	y (data)	y differ-ence	(y differ-ence)^2	sum x diff^2 + y diff^2
1	12	10	2	4	288	292	-4	16	20
2	18	22	-4	16	234	228	6	36	52
3	7	12	-5	25	265	266	-1	1	26
4	34	34	0	0	243	240	3	9	9
5	15	19	-4	16	291	287	4	16	32
6	33	24	9	81	211	215	-4	16	97
7	28	29	-1	1	267	271	-4	16	17
8	7	12	-5	25	273	268	5	25	50
9	45	44	1	1	245	244	1	1	2
10	110	99	11	121	221	225	-4	16	137
11	54	65	-11	121	212	208	4	16	137
12	87	93	-6	36	284	278	6	36	72
13	23	22	1	1	261	259	2	4	5
14	19	24	-5	25	230	235	-5	25	50
15	76	80	-4	16	255	260	-5	25	41
16	97	108	-11	121	201	204	-3	9	130
17	38	43	-5	25	290	288	2	4	29
18	65	72	-7	49	277	282	-5	25	74
19	85	78	7	49	205	201	4	16	65
20	39	44	-5	25	282	278	4	16	41
21	94	90	4	16	246	251	-5	25	41
22	64	56	8	64	233	227	6	36	100
								Sum	1227
								Average	55.8
								RMSE	7.5
								NSSDA	12.9

ever the errors at intervening locations are not known, e.g., midway along a line segment between two vertices. The error along a straight line segment may be at most equal to the largest error observed at the ends of the line segments. (Figure 14-9). If the data line segment is parallel to the true line segment then the errors are uniform along the full length of the segment. Vertices that result in converging or crossing lines will lead to mid-point errors less than the larger of the two errors at the endpoints (Figure 14-9). These observations are not true if a straight line segment is used to approximate a substantially curved line. However, if the line segments are sufficiently short, e.g., the interval along the line is small relative to the radius of a curve in the line, and the positional errors are distributed evenly on both sides of the line segments, then the NSSDA methods described above will provide an approximate upper limit on the linear error.

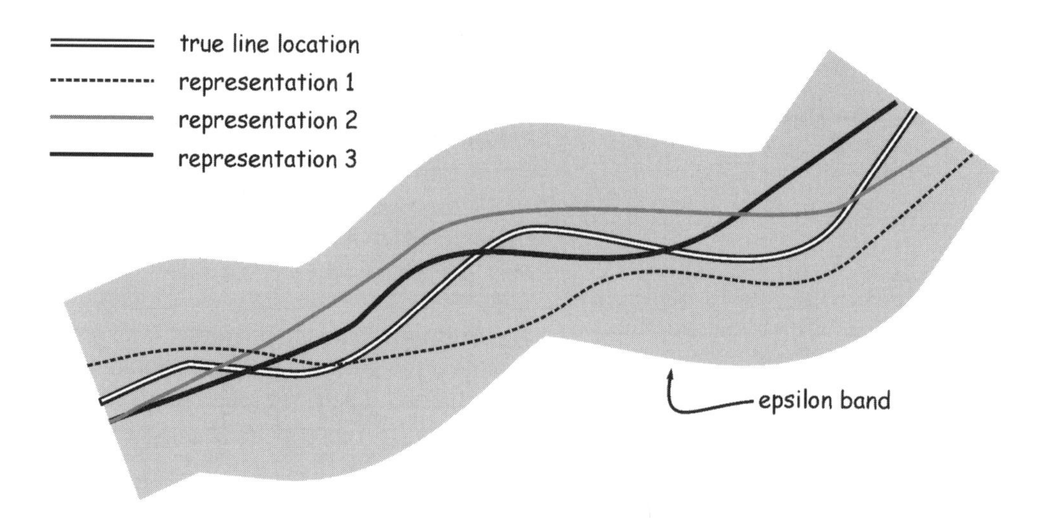

Figure 14-8: The concept of an epsilon band. A region to either side of a line is specified. This region has a high probability of containing the true lines. Multiple digital representations may exist, e.g., multiple tries at digitizing a line. These representations will typically lie within the epsilon band.

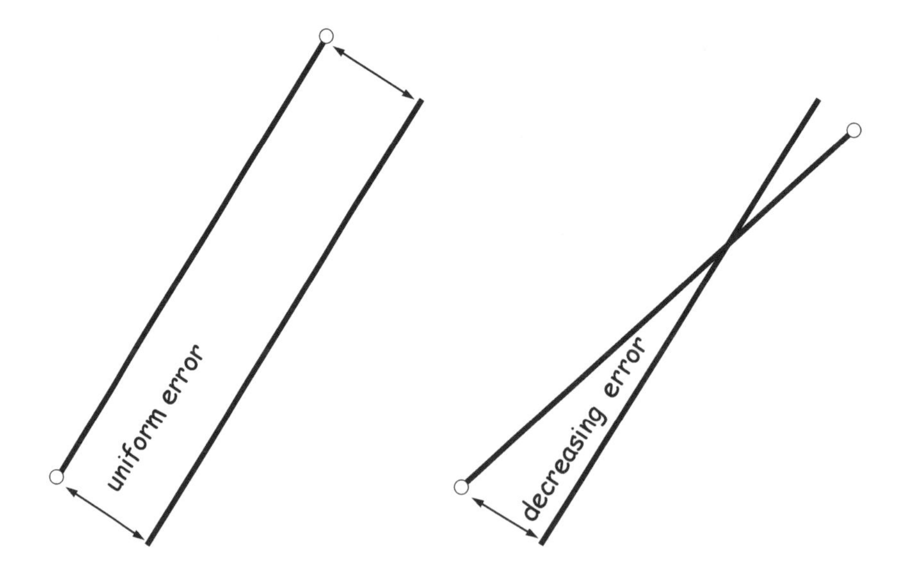

Figure 14-9: Errors for straight line segments are either the same (left) or less than (right) the maximum error observed at the end points. If nodes or vertices are sufficiently close such that the true line segments are approximately linear, then error assessments on point features may provide an approximation of the line positional errors.

Attribute Accuracy

Unlike positional accuracy, there is no national standard for measuring and reporting attribute accuracy. Accuracy for continuous variables may be calculated in an analogous manner to positional accuracy. Accuracy for each observation is defined as the difference between the true and database values. Rather than calculate a positional difference by the method of Phythagoras, the attribute difference may be defined as the number of times the observed attribute value is different from the true value. A set of test data points may be identified, the true attribute value determined for each of those test data points, the difference calculated for each test point, and the accuracy summarized.

Accuracy of categorical attribute data may be summarized using an *error table* and associated accuracy statistics. Points may be classified as correct, that is the categorical variable matches the true category for a feature, or it may be incorrect. This occurs when the true and layer category values are different. Error tables, also known as error matrices, confusion matrices, and accuracy tables, are a standard method of reporting error in classified remotely-sensed imagery. They have more rarely been used for categorical attribute accuracy assessment.

An error table summarizes a two-way classification for a set of test points (Figure 14-10). A categorical variable will have a fixed number of categories. These categories are listed across the columns and along the rows of the error table. Each test feature is tallied in the error table. The true category and the value in the data layer are known for each test feature. The test feature is tallied in the error table based on these values. The true values are entered via the appropriate column and the data layer values are entered via the appropriate row. The table is square, because there are the same number of categories in both the rows and columns. Correctly classified features are tallied on the diagonal – the true value and data layer value are identical, so they are noted at the intersection of the categories. Incorrectly assigned category values fall off the diagonal.

Error tables summarize the main characteristics of confusion among categories. The diagonal elements contain the test features that are correctly categorized. The diagonal sum is the total number correct. The proportion correct is the total correct divided by the total tested. The percent correct can be obtained by multiplying the proportion correct by 100.

Per category accuracy may be extracted from the error table. Two types of accuracy may be calculated, a *user's accuracy* and a *producer's accuracy*. The user relies on the data layer to determine the category for a feature. The user is most often interested in how often a feature is mis-labeled for each category. In effect, the user wants to know how many features classified as a category (the row total) are truly from that category (the diagonal element for that row). Thus, the user's accuracy is defined as the number of correctly assigned features (the diagonal element) divided by the row total for the category. The producer knows the true identity of each feature, and is often interested in how often these features are assigned the correct category. The producer's accuracy is defined as the diagonal element divided by the column total.

Error Propagation in Spatial Analysis

While we have discussed methods for assessing positional and attribute accuracy, we as yet have not established how we determine how accuracy in the input layers affects the accuracy of output when we apply a spatial operation. Clearly, input error affects output values in most calculations. A large elevation error in DEM cells will likely affect slope values. Output for the slope function will be in error. If slope is then combined with other features from other data layers, these errors may in turn propagate through the analysis. How do we assess the propagation of errors and their impacts on spatial analysis?

There are currently no widely applied, general methods for assessing the effect of positional errors on spatial models. Research is currently directed at several promising avenues, however the range of variables and conditions has confounded the development of general methods for assessing the impacts of purely positional errors on spatial models.

Several approaches have been developed to estimate the impacts of attribute errors on spatial models. One approach involves assessing errors in the final result irrespective of errors in the original data. For example, we may develop a cartographic model to estimate deer density in a suburban environment. The model may depend on the density of housing, forest location, type, and extent, the location of wetlands, and road location and traffic volumes. Each of these data sources may contain positional and attribute errors.

Questions may arise regarding how these errors in our input data affect the model predictions for deer density. Rather than trying to identify how errors in the input propagate through to affect the final model results, we may opt to perform an error assessment of our final output. We would perform a field survey of deer density and compare the values predicted by the model with the values observed in the field. For example, we might subdivide the study area into mutually exclusive census areas. Deer

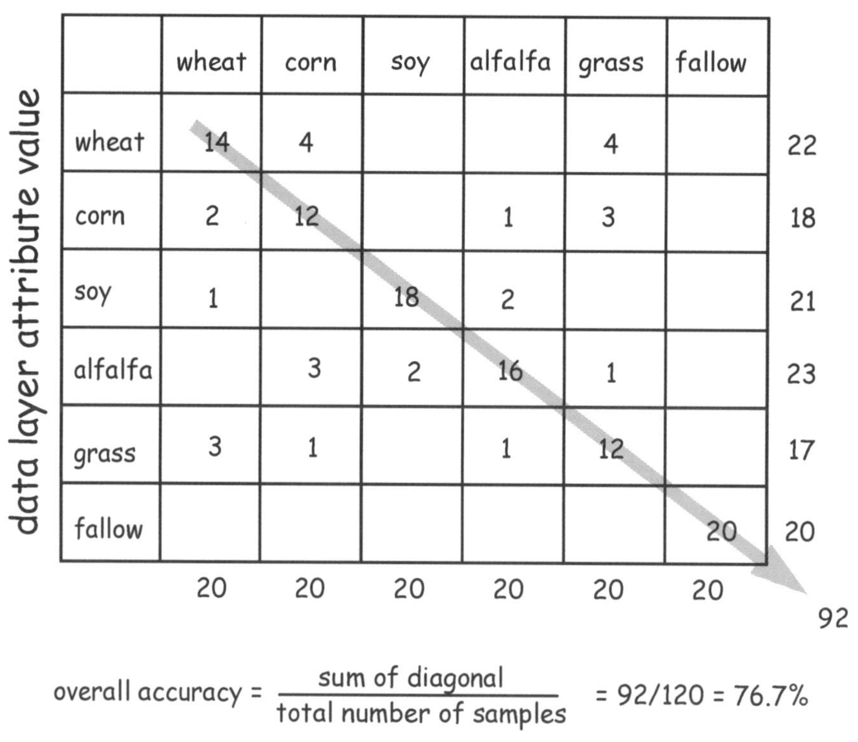
true value

data layer attribute value

	wheat	corn	soy	alfalfa	grass	fallow	
wheat	14	4			4		22
corn	2	12		1	3		18
soy	1		18	2			21
alfalfa		3	2	16	1		23
grass	3	1		1	12		17
fallow						20	20
	20	20	20	20	20	20	92

$$\text{overall accuracy} = \frac{\text{sum of diagonal}}{\text{total number of samples}} = 92/120 = 76.7\%$$

Figure 14-10: An error table succinctly summarizes the attribute accuracy for categorical variables.

might be counted in each census area and the density calculated. We have replicated values from each census area, so we may calculate a mean and a variance, and the difference between modeled and observed values might be compared relative to the natural variation we observe among different census areas. We could also census an area through time, e.g., on successive days, months, or years, and compare the difference between the model and observed values for each sample time.

It may not be possible or desirable to wait to assess accuracy until after completing a spatial analysis. Input data for a specific spatial analysis may be expensive to collect. We may not wish to develop the data and a spatial model if errors in the input preclude a useful output. After model application, we may wish to identify the source of errors in our final predictions. Improvements in one or two data layers may substantially improve the quality of our predictions, for example, better data on forest cover may increase the accuracy for our predictions of deer density.

Error propagation in spatial models is often investigated with repeated model runs. We may employ some sort of repeat simulation model that adds error to data layers and records the impacts on model accuracy. These simulation models often employ a standard form known as a *Monte Carlo* simulation. The Monte Carlo method assumes each input spatial value is derived from a population of values. For example, land cover may range over a set of values for each cell. Further, model coefficients may also be altered over a range. In a cartographic model, the weights when combining layers may be allowed to range over a specified interval.

A Monte Carlo simulation controls how these input data or model parameters are allowed to vary. Typically, a random normal distribution is assumed for continuous input values. If all variables save one are held constant, and several model runs performed at different, random selections of the variable, we may get an indication of how a variable affects the model output. We may find that the spatial model is insensitive to large changes in most of our input data values. For example, predicted deer density may not change much even when land cover varies over a wide range of values. However, we may also find a set of input data, or a range of input data or coefficients, that substantially control model output.

A Monte Carlo or similar simulation is a computationally intensive technique. Thousands of model runs are often required over each of the component units of the spatial domain. The computational burden increases as the models become more involved, and as the number of spatial units increases.

Metadata: Data Documentation

Metadata are information about spatial data. Metadata describes the content, source, lineage, methods, developer, coordinate system, extent, structure, spatial accuracy, attributes, and responsible organization for spatial data. The Federal Geographic Data Committee (FGDC) of the United States has defined a Content Standard for Digital Geospatial Metadata (CSDGM) to describe the content and format for metadata. The CSDGM ensures that spatial data are clearly described so that they may be used effectively within an organization. The use of the CSDGM also ensures that data may be described to other organizations in a standard manner, and that spatial data may be more easily evaluated by and transferred to other organizations.

The CSDGM consists of a standard set of elements that are presented in a specified order. The standard is exhaustive in the information it provides, and is flexible in that it may be extended to include new elements for new categories of information in

4. Spatial_Reference_Information:
  4.1 Horizontal_Coordinate_System_Definition:
    4.1.2 Planar:
      4.1.2.2 Grid_Coordinate_System:
        4.1.2.2.1 Grid_Coordinate_System_Name:
                                  Universal Transverse Mercator
        4.1.2.2.2 Universal_Transverse_Mercator:
          4.1.2.2.2.1 UTM_Zone_Number: 10-19
      4.1.2.4 Planar_Coordinate_Information:
        4.1.2.4.1 Planar_Coordinate_Encoding_Method:
                                  coordinate pair
        4.1.2.4.2 Coordinate_Representation:
          4.1.2.4.2.1 Abscissa_Resolution: 2.54
          4.1.2.4.2.2 Ordinate_Resolution: 2.54
        4.1.2.4.4 Planar_Distance_Units: meters
    4.1.4 Geodetic_Model:
      4.1.4.1 Horizontal_Datum_Name: North American Datum 1927
      4.1.4.2 Ellipsoid_Name: Clark 1866
      4.1.4.3 Semi-major_Axis: 6378206.4
      4.1.4.4 Denominator_of_Flattening_Ratio: 294.98
  4.2 Vertical_Coordinate_System_Definition:
    4.2.1 Altitude_System_Definition:
      4.2.1.1 Altitude_Datum_Name:
                      National Geodetic Vertical Datum of 1929
      4.2.1.2 Altitude_Resolution: 1
      4.2.1.3 Altitude_Distance_Units: feet or meters
      4.2.1.4 Altitude_Encoding_Method: attribute values
    4.2.2 Depth_System_Definition:
      4.2.2.1 Depth_Datum_Name: Mean lower low water
      4.2.2.2 Depth_Resolution: 1
      4.2.2.3 Depth_Distance_Units: meters or feet
      4.2.2.4 Depth_Encoding_Method: attribute values

Figure 14-11: Example of a small portion of the FGDC recommended metadata for a 1:100,000 scale derived DLG data set.

the future. There are over 330 different elements in the CSDGM. Some of these elements contain information about the spatial data, and some elements describe or provide linkages to other elements. Elements have standardized long and short names and are provided in a standard order with a hierarchical numbering system. For example, the western-most bounding coordinate of a data set is element 1.5.1.1, defined as follows:

1.5.1.1 West Bounding Coordinate – western-most coordinate of the limit of coverage expressed in longitude.

Type: real

Domain: -180.0 < = West Bounding Coordinate < 180.0

Short Name: westbc

The numbering system is hierarchical. Here, 1 indicates it is basic identification information, 1.5 indicates identification information about the spatial domain, 1.5.1 is for bounding coordinates, and 1.5.1.1 is the western-most bounding coordinate. There are numbers for the other bounding coordinates, e.g. 1.5.1.2 is for the eastern-most bounding coordinate.

There are 10 basic types of information in the CSDGM:

1) identification, describing the data set,
2) data quality,
3) spatial data organization,
4) spatial reference coordinate system,
5) entity and attribute,
6) distribution and options for obtaining the data set,
7) currency of metadata and responsible party,
8) citation,
9) time period information, used with other sections to provide temporal information, and
10) contact organization or person.

The CSDGM is a content standard and does not specify the format of the metadata. As long as the elements are included, properly numbered and identified with correct values describing the data set, the metadata are considered in conformance with the CSDGM. Indentation and spacing are not specified. However, because metadata may be quite complex, there are a number of conventions that are emerging in the presentation of metadata. These conventions seek to ensure that metadata are presented in a clear, logical way to humans, and are also easily ingested by computer software. There is a Standard Generalized Markup Language (SGML) for the exchange of metadata. An example of a portion of the metadata for a 1:100,000 scale digital line graph data set are shown in Figure 14-11.

Metadata are most often created using specialized software tools. Although metadata may be produced using a text editor, the numbering system, names, and other conventions are laborious to type. Furthermore there are often complex linkages between metadata elements, and some redundant information. Software tools may ease the task of metadata entry by reducing redundant entries, ensuring correct linkages, and checking elements for contradictory information or errors. For example the metadata entry tool may check to make sure the western-most boundary is west of the eastern-most boundary. Metadata are most easily and effectively produced when their development is integrated into the workflow of data production.

Although not all organizations adhere to the CSDGM metadata standard, most organizations record and organize a description and other important information about their data, and many organizations consider a data set incomplete if it lacks metadata. All U.S. government units are required to adhere to the CSDGM when documenting and distributing spatial data.

Summary

Data standards, data accuracy assessment, and data documentation are among the most important activities in GIS. We cannot effectively use spatial data if we do not know its quality, and the efficient distribution of spatial data depends on a common understanding of data content.

Data may be inaccurate due to several causes. Data may be out of date, collected using improper methods or equipment, or by unskilled or inattentive persons. An accuracy assessment or measurement applies only to a specific data set and time.

Inaccuracies may be reported using many methods, including a mean value, a frequency distribution, or a threshold value. Accuracy should be recognized as distinct from precision. Accuracy is a measure of error, a difference between a true and represented value. Precision is a measure of the repeatability of a process. Imprecise data collection often leads to poor accuracy.

Standards have been developed for assessing positional accuracy. Accuracy assessment and reporting depend on sampling. A set of features are visited in the field, and the true values collected. These true values are then compared to corresponding values stored in a data layer, and the differences between true and database values quantified. An adequate number of well-distributed samples should be collected. Standard worksheets and statistics have been developed.

Data documentation standards have been developed in the United States. These standards, developed by the Federal Geographic Data Committee, are known as the Content Standard for Digital Geospatial Metadata. This standard identifies specific information that is required to fully describe a spatial data set.

Suggested Reading

Blakemore, M., Generalization and error in spatial data bases, *Cartographica*, 1984, 21:131-139.

Bolstad, P., Gessler, P., and Lillesand, T., Positional uncertainty in manually digitized map data, *International Journal of Geographical Information Systems*, 1990, 4:399-412.

Dunn, R., Harrison, A. R., and White, J. C., Positional accuracy and measurement error in digital databases on land use: an empirical study, *International Journal of Geographical Information Systems*, 1990, 4:385-398.

Fisher, P., Modelling soil map unit inclusions by Monte Carlo simulation, *International Journal of Geographical Information Systems*, 1991, 5:193-208.

Goodchild, M. F. and Gopal, S., The Accuracy of Spatial Databases, Taylor and Francis, London, 1989.

Guptill, S.C. and Morrison, J.L. eds., Elements of Spatial Data Quality, Elsevier, New York, 1995.

Heuvelink, G., Error Propogation in Environmental Modeling with GIS, Taylor and Francis, London, 1999.

Lodwick, W. A., Monson, W., and Svoboda, L., Attribute error and sensitivity analysis of map operations in geographical information systems, *International Journal of Geographical Information Systems*, 1990, 4:413-427.

Thapa, K. and Bossler, J., Accuracy of spatial data used in geographic information systems, *Photogrammetric Engineering and Remote Sensing*, 1992, 58:841-858.

Walsh, S. J., Lightfoot, D. R., and Butler, D. R., Recognition and assessment of error in geographic information systems, *Photogrammetric Engineering and Remote Sensing*, 1987, 53:1423-1430.

Study Questions

Why are standards so important in spatial data? Can you describe processes or activities which are greatly helped by the existence of standards?

What are the several ways in which spatial data standards can be defined? What are the differences between media, format, accuracy, and documentation standards?

What are the differences between accuracy and precision?

How do mean and frequency thresholds differ in the way they report positional error?

What are some of the primary causes of positional error in spatial data?

Can you describe each of these with reference to documenting spatial data accuracy: positional accuracy, attribute accuracy, logical consistency, and completeness?

What is the NSSDA, and how does it help us measure positional accuracy?

What are the basic steps in applying the NSSDA?

What are the constraints on the distribution of sample points under the NSSDA, and why are these constraints specified?

What are good candidate sources for test points in assessing the accuracy of a spatial data layer?

Can you complete an NSSDA summary assessment table?

How are errors in nominal attribute data often reported?

What are metadata, and why are they important?

15 New Developments in GIS

Introduction

As every meteorologist knows, predicting the future is fraught with peril. Near-term predictions may be safe; if it is raining now, it will probably be raining in 10 minutes. However, the farther one reaches into the future, the more likely he'll be wrong. This chapter is devoted to a description of methods and technologies that are on the verge of widespread adoption, and some speculations on future trends.

Many changes are based on advances in computers and other electronic hardware that support spatial data collection and analyses. Computers are becoming smaller and less expensive. This is true for both general purpose machines and for specialized computers, such as ruggedized, portable computers suitable for field data collection. The wizards of semiconductors continue to dream up and then produce impossibly clever devices. Given current trends, we should not be surprised if at some future time a pea-sized device holds all the published works of humankind. Computers may gain personalities, recognize us as individuals, respond entirely to voice commands, routinely conjure three-dimensional images that float in space before our eyes. These and other developments will alter how we manipulate spatial data.

Changes in GIS will also be due to increased sophistication in GIS software and users, and increased familiarity and standardization. Change will be driven by new algorithms or methods, e.g., improved data compression techniques that speed the retrieval and improve the quality of digital images. Specialized software packages may be crafted that turn a multi-day, technically complicated operation into a few mouse clicks. These new tools will be introduced as GIS technologies continue to evolve and will change the way we gather and analyze spatial data.

GPS

GPS receivers will continue to shrink, cost less, and measure with increasing accuracy for some time to come, and these improvements will spur even more widespread adoption of this technology (Figure 15-1). Microelectronic miniaturization is helping shape the GPS market. As GPS use grows and manufacturing methods improve, single chip GPS systems are not far off. GPS receivers smaller than a deck of cards are available, including some that may be plugged into personal digital assistants

Figure 15-1: A GPS unit as an accessory to a PDA. (courtesy Nexian, Inc.)

(PDAs). Many vendors are well on a path to system integration, and it will become more common to embed the antenna, receiver, supporting electronics, power supply, and differential correction radio receivers in a single piece of equipment. Some of these integrated systems are smaller than most GPS antennas of a decade ago, and systems will continue to shrink. A wristwatch sized GPS is not far off.

Miniaturization substantially reduces the limitations imposed by the physical size of the GPS electronics. As receivers shrink in size and cost, it becomes practical to collect positional information on smaller individual objects. While GPS is unlikely to solve the universal problem of the lost TV remote control, small GPS receivers will be deployed to collect spatial data for a substantially increasing number of objects. For example, in the shipping industry a few years ago it was uneconomical to track objects smaller than a cargo ship. Now trucks or containers are routinely followed. In the near future it may be common to positionally track individual packages.

GPS miniaturization means we will directly collect much more data in the field than in times past. A city engineer may study traffic patterns by placing special-purpose GPS receivers into autos. How long does the average commute take? How much of the time is spent sitting at stop signs or lights, and where is the congestion most prevalent? How is traffic affected by weather conditions? Analyses of traffic networks will become substantially easier with small-unit GPS. Disposable GPS receivers may be pasted, decal-like, on windshields by the thousands, to transmit their data back to a traffic management center.

Ubiquitous, inexpensive, or free differential correction signals will substantially improve the accuracy commonly achieved with GPS. The free GPS correction signal broadcast by the U.S. Coast Guard is restricted to areas near navigable waters, leaving much of the U.S. uncovered, and the same is true in most other countries. National or regional governments will establish more complete coverage. If not, commercial solutions will be further developed and made less expensive. Meter-level positional accuracy will be available in the field in real-time, in turn supporting many new GPS applications.

Mobile Real-time Mapping

Wireless communication is a technology that will substantially lengthen the reach of geoprocessing. While radios and telephones have provided communications across space for some time, early telephones needed a physical connection, and until recently mobile communications suffered from a narrow bandwidth, the inability to transmit large data volumes. Wireless digital communications allows the transmittal of large quantities of data in real time, irrespective of device location.

The impacts of wireless technologies will be most substantial where accurate positioning information is required rapidly or with high frequency. When combined with GPS, portable computers, and spatial data software, wireless technologies in particular may aid in a number of emergency service, military, or public safety operations, and they may also be used in transportation or other applications. One can imagine self-driving automobiles that navigate via a combined GPS/GIS, using systems to avoid collisions via wireless communications with "nearby" automobiles.

Wildfire control offers perhaps the best examples of real-time mapping via the integration of GPS, GIS, and wireless communications. Wildfires in North America cause millions of dollars in damage each year and cost millions of dollars to control (Figure 15-2). Humans have increasingly built their homes among the trees, and a long history of fire suppression has led to large quantities of dead wood and other fuels for many fire-prone areas. This results in more damaging wildfires because fires are burning hotter and faster and because there are more homes in harm's way. Substantial efforts are expended towards detecting, planning, and fighting fires, and in a bad fire season resources may be stretched beyond their limits. Spatial data collection and analysis tools may aid substantially in fighting fires.

Consider the problem facing a chief of firefighting operations for a large fire. There will be fire crews spread out near the perimeter of the fire as well as planes, helicopters, trucks, and other equipment to fight the blaze. Perhaps the fire is occurring in steep terrain, with buildings distributed across the potential burn area, a road system for access, and water sources for fighting the fire. The chief must integrate this relatively "static" information with that on fuels, wind, fire location and intensity. Crucial information such as the location of the fire front may be updated irregularly, and may be of unknown quality if more than a few hours old, yet human lives and millions of dollars in property often hinge on the decisions of the fire boss. Technological innovations in the form of handheld GIS/GPS systems with wireless communications promise to dramatically increase the effectiveness and safety of wildfire control (Figure 15-3).

Figure 15-2: Wildfires are a significant problem in much of North America, causing millions of dollars in damage most years. Spatial data collection and analysis methods have been added to the arsenal of tools used to fight fires. (courtesy Tom Patterson, NPS)

Figure 15-3: A GPS receiver combined with GIS software and spatial data in a field computer may be used to map important features in real time. In turn, these may be transmitted via a wireless communications link. Here a fire perimeter is mapped via a helicopter. Captured on a field computer (inset), spatial data may be transmitted in near real time to a command center where decisions may be made based on the most current positional data. (courtesy Tom Patterson, NPS)

Recent work by the U.S. National Park Service (NPS) provides a good example of GIS/GPS innovation. The NPS, in cooperation with the U.S. Interagency Fire Center, is developing a lightweight, portable, integrated geographic data collection systems to aid in real-time fire mapping and communication. Prior to 1990, geographic tools used in tactical fire planning were primarily limited to hardcopy maps. Portable systems based on notebook computers and hand-held GPS receivers have proven the utility of rapid field mapping, but the systems were bulky, heavy, and at times fragile.

The current systems are based on handheld computers and personal GPS receivers. The combined units are small and light, weighing less than one pound and fitting into a pocket on a standard-issue flight suit. Screen color, resolution, size, and brightness allow the handheld computers to be read in bright sunlight and to display areas up to several hundred hectares at the required detail for fire mapping. System memory of 8 to 32 megabytes is currently ample for most needed mapping and communications software, and to store sufficient spatial data. These data include background images such as DRGs and DOQs, and recent field data such as fire or damage assessment. The equipment is fairly rugged and inexpensive, and standard batteries provide power for 10-12 hours or more, longer than the typical working day (Figure 15-4).

Tom Patterson of the U.S. National Park Service has led one effort to develop and test these real-time mapping systems. The capabilities of these systems were demonstrated in their application to the Viejes Fire, which in early January, 2001, burned 10,353 acres of the Cleveland National Forest in California. The fire caused substantial property damage, required the evacuation of hundreds of residents, and exposed firefighting crews from the State of California, the U.S. Forest Service, and local fire departments to significant personal danger.

GPS technical specialists used mobile mapping systems to map the fire perimeter and control activities in real time (Figure 15-4). The specialists were deployed with fire crews for fireline mapping, and also carried the systems on daily helicopter surveys of fire mapping and assessment. Field and helicopter data were transmitted from the collection units to tactical decision makers via wireless communications, and the system performed well. Based on these and other experiences, the U.S. National Interagency Fire Center has established training courses to disseminate this information and speed the adoption of this technology in national firefighting efforts.

Mobile wireless mapping has also been used in rapid assessment of earthquake damage and in military and law enforcement. In most cases these methods reduce the interval between data collection and analysis, a critical improvement for many applications.

Figure 15-4: Fire crew locations or their activities may be field-entered using handheld GPS/GIS. Data may be transmitted immediately to a command center for improved decision making, and also used later to map the characteristics or impacts of a fire. (courtesy Tom Patterson, NPS)

Improved Remote Sensing

Spatial data collection will be substantially improved with the continuing advances in image acquisition and processing. More satellites, higher spatial and temporal resolution, improved analog and digital aerial cameras, and new sensor platforms will all increase the array of available data. We will be able to sense new phenomena, and also locate already measured features with increased precision and accuracy.

The shuttle radar topography mission (SRTM) is a good example of improved remote sensing (Figure 15-5). A radar scanner was deployed on board the Space Shuttle in February 2000. The scanner acquired terrain height data for approximately 85% of the Earth surface. Data were collected at a range of spatial resolutions and horizontal accuracies, but much of the data will be available at a 30-meter horizontal resolution in the U.S. and at a 3 arc-second horizontal resolution (approximately 90 meters) for the rest of the world. These elevation data are to be processed over the next several years and will yield uniform, global terrain data that may be combined with other spatial data.

Improved data quality over much of the world is one of the main advantages of SRTM data. Spatial resolution and accuracy expected from the SRTM are approximately

Figure 15-5: New data will enable improved spatial analyses. Here improved elevation data from the Shuttle Radar Topography Mission are combined with land surface imagery from the Landsat Enhanced ThematicMmapper. (courtesy NASA)

Figure 15-6: The Helios, an experimental, solar-powered aircraft that may serve as a platform for collecting aerial imagery and other environmental data. (courtesy NASA)

equal to those of terrain data currently available for much of North America and Europe. Therefore, SRTM terrain data will not likely provide a substantial improvement, except for a few areas where data are missing, particularly poor, or suffer from edge ambiguities. In contrast, SRTM data will be markedly better for much of Asia, South America, and Africa.

Satellites with improved spatial, spectral, and temporal resolution will be developed and launched. Currently available imagery provides approximately one-meter resolution in black-and-white and three meter color imagery. The Ikonos satellite was the first to offer high resolution, however there are recent deployments or planned launches for at least three more systems, including the Quickbird satellite by Earthwatch Inc., the EROS A1 by Imagesat International, Orbview 4 by Orbital Imaging Corporation, in addition to the SPOT 5 satellite by SPOT Image Corporation. A growing number of options for high resolution imagery will increase availability and reduce costs, spurring the wider use of satellite image data for spatial data entry.

These higher spatial resolutions will aid many new endeavors. Detailed land surveys often use imagery with effective resolutions of a few centimeters. Property and asset management, involving the inventory and assessment of roadways, powerlines, culverts, or other infrastructure are often based on high-resolution photographs and site visits. News organizations, business intelligence and strategic planning, and resource management applications all provide incentives for improvement in spatial and spectral resolution.

Advances in full-sized and miniature aircraft, known as remotely piloted vehicles (RPVs) may lead to increased availability of a broader range of aerial imagery. NASA, other government laboratories, and aerospace firms have been developing pilotless vehicles to collect imagery and data about the Earth's surface and atmosphere. The NASA Helios is a solar-powered aircraft designed to stay in the upper atmosphere for weeks to months (Figure 15-6). As such, this ultralight flying wing may be positioned to collect aerial imagery as well as atmospheric samples, without the need to return to Earth due to human limitations or to refuel. NASA is also developing the ALTOS, a platform best characterized as an unmanned jet. This may be able to collect aerial imagery from higher altitudes, longer time periods, or larger areas than with manned flight, and may compete with satellites as a platform for image data collection.

RPVs may also be small helicopters or airplanes outfitted with cameras and positioning, control and telemetry electronics (Figure 15-7). RPVs will increasingly be used to provide imagery for coordinate data collection, and as additional data associated with a spatial feature stored in a database. These small RPVs are often trucked to a site, where they are flown via remote-control by an operator on the ground. RPVs are used most typically to obtain imagery of a specific set of targets. Small RPVs are currently used primarily for qualitative data collection, e.g., tower or bridge inspection, promotional photographs, or visual surveys for real-estate appraisals, but some RPVs have been used in mapping applications, e.g., in forest inventories or for construction site mapping. GPS receivers have been fit on RPVs, providing precise position and photo orientation information. These data may be combined with standard photogrammetric methods to produce detailed two and three-dimensional coordinates. Small RPVs, while limited in the size of an area that may be covered, are able to provide extremely high resolution imagery tailored to the specific needs of an individual project. They offer the utmost flexibility in scheduling, and may be flown under a low cloud ceiling, increasing the likelihood of image acquisition.

Figure 15-7: A scale RPV fitted with a camera, to collect low-altitude aerial imagery. (courtesy Skycam, Ltd.)

Internet Mapping

The internet will have an increasingly important impact on GIS, particularly in expanding the number and breadth of spatial data uses and in improving the update and distribution of spatial information. Maps are popping up in web-pages all over the planet. Internet maps provide the common map functions in that they serve as directional aids, to demonstrate proximity, and to help provide a mental image of spatial locations.

Many internet applications allow users to compose maps on a web page. The individual user has some control over the data layers shown, the extent of the mapped area, and the symbols used to render the map. The internet is different from other technologies because it allows a broad array of people to custom-produce maps. Each user may choose her own data and cartographic elements to display. The user is largely free from any data development chores and thus needs to know very little about data entry, editing, or the particulars of map projections, coordinates, or other details required for the production of accurate spatial data. Typically the map itself is the end product, and may be used for illustration, or to support analyses that will be performed entirely within the user's head.

These internet mapping applications are particularly appropriate when a large number of users need to access a limited number of data layers to compose maps. The internet users may select the themes, variables, and symbolization, in contrast to a static map graphic, where a website cartographer defines the properties of a map.

Many internet mapping applications are built for users with little knowledge of spatial data, maps, and analysis. This severely restricts the suite of spatial operations allowed. Most internet mapping is currently limited to creating simple displays. In some instances the user may perform a limited number of analyses, but in general the low user knowledge requirements come at the expense of limited flexibility.

The internet is also useful for the timely distribution of important public interest data. These data have often been unavailable, outdated, or in a form that was not easily accessible. Crime data are an important example. Many citizens have a notion of where the high crime areas are located in a community. New or prospective residents, or interested news organizations or businesses may not

Use the interface below to find the location you are interested in. Alternatively you can use the advanced interactive mapping interface, linked at the bottom of this page.

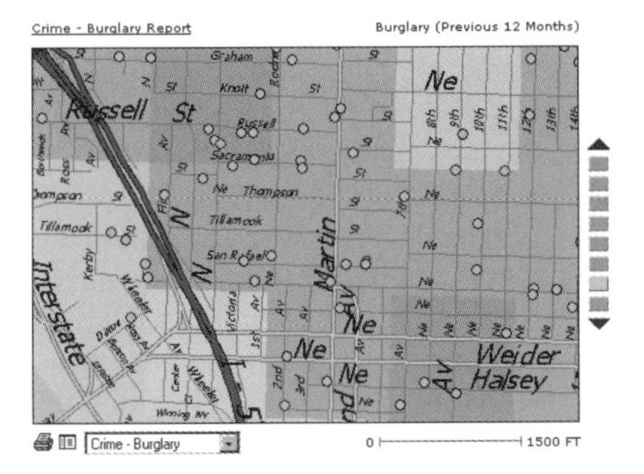

Note that **you must select an address or intersection** in order **to get detailed report information**. You can tell you have selected an address or intersection when you see a red dot appear in the center of the map.

Figure 15-8: Important spatial data, such as crime locations and type, are easily and widely distributed, as in this innovative application by the City of Portland. (copyright, City of Portland, 2001, used with permission)

possess this knowledge and therefore may have interest in crime statistics routinely collected by police. Most police departments summarize and report crime via paper documents, and while available to the public, access is often cumbersome and updates infrequent. In practice, these crime data may not reach much of the interested public. Internet mapping changes this.

As exemplified by the City of Portland in their CrimeMapper program, current crime data are published on web pages by position, crime type, and time (Figure 15-8). CrimeMapper is a partnership between the Portland Police Bureau and the City of Portland, Bureau of Information Technology, Corporate Geographic Information Systems. Choropleth maps, point position maps, and other depictions of crime by type may be displayed on a backdrop of city streets, rivers, and other identifying features. Tabular information by neighborhood or area may be accessed via a graphical query. CrimeMapper substantially increases the ease of access and usefulness of these data. These data may be particularly helpful in community-based or neighborhood crime watch organizations, and internet delivery of these and other crime data will become widespread as they are perceived to substantially improve public safety.

In addition to on-screen map composition, data download over the internet has become the standard distribution method for most government and many commercial organizations. Users may browse data sets by location, name, or other features, and download them to their local computer. Bulk downloads of "base" data are quite common.

A third major use of the internet is as a tool to aid in the cooperative entry or update of detailed data layers that cover broad areas and are under the jurisdiction of a large number of organizations. Property line (cadastral) data are a good example of spatial data that are gathered in aggregate over large areas, but by independent units each working in a small area. Most counties, cities, parishes, or equivalent jurisdictional units in the U.S. are developing digital cadastral data.

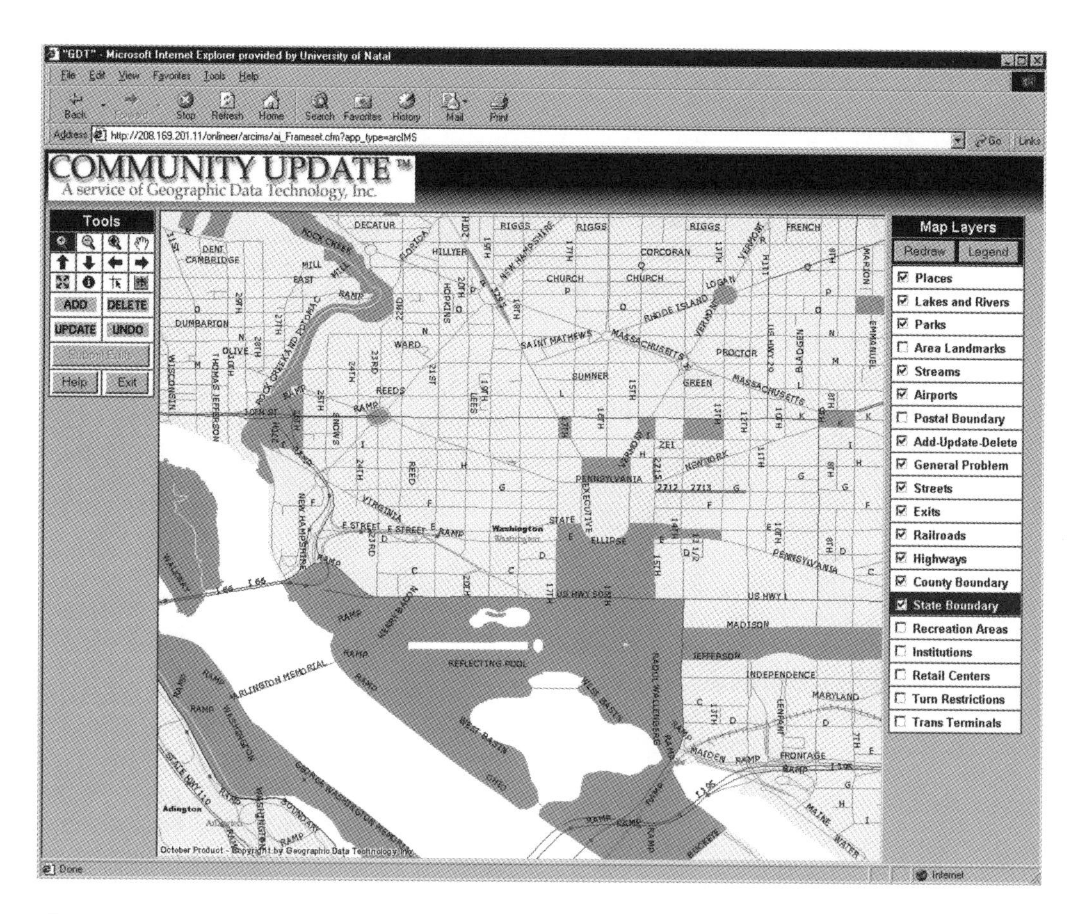

Figure 15-9: Internet-based applications have been developed for data entry. This allows those closest to the primary data to enter it, while centralizing management and access. (courtesy GDT, Inc.)

Property transfer and tax records are more easily modified and managed when using digital spatial data. There are thousands of local governments undertaking this process and most are developing data with little interaction between adjacent local governments. The lack of interaction is not surprising, given the goals of each government. Interaction may require significant additional cost while not providing an obvious tangible benefit to each individual unit.

There may be much to gain from a combined data set, but often these combined data sets are not produced until after all counties have completed their data. Seamless cadastral data may be valuable for statewide emergency planning, national insurance companies, logistics or shipping industries, large landholders such as forestry companies, national public health services, or GIS in support of sales forces with non-overlapping territories. These organizations must combine and in many respects re-compile cadastral data to create a uniform, seamless format. Much time and effort may be spent converting each data set to a uniform format, or altering or modifying attribute data.

Local governments may be enticed to cooperate when they receive immediate tangible benefits. Many local governments are not large enough to support a GIS and the spatial analysis professionals needed to manage the system. Data entry, editing, and management tools delivered over the internet at little or no cost may spur the use of GIS technologies, particularly if these data entry tools are easy to learn and use. The data integrator providing the internet-based mapping tools may generate revenues by selling the data to businesses, thereby adding value to the collective data set.

The Community Update (CU) initiative, coordinated by Geographic Data Technologies, Incorporated, is a good example of a web-based, distributed-data input application (Figure 15-9). Governments enter digital cadastral data on a CU website under a cooperative agreement. Data are entered by cooperators in a password-protected data space. Data are developed using a standard set of tools, and an integrated cadastral data layer developed. These data are maintained and provided back to qualified users.

The internet may be used to provide a standard set of tools through which each organization enters data. This homogenizes data entry and avoids many of the problems described above when data are entered without coordination among adjacent municipalities. The web-based application may provide the limited number of editing tools needed for basic spatial data entry. These may be provided in a template or in a custom-built form. The editing tools are provided on a web-based graphics page, perhaps with appropriate background data such as DOQs or digital line data. Each individual organization or unit may then enter their data through the web page display, e.g., digitizing on-screen to record line or point locations. Data are recorded and saved in a uniform manner, greatly easing the development of an integrated, broad-area data layer.

Visualization

Spatial data are increasingly being used in *visualization*. Visualization involves the realistic computer rendering of entities. These entities may be buildings, people, automobiles, or other real entities. Any real or imagined feature may be depicted in a visualization.

Visualization typically renders two-dimensional views that give a three-dimensional impression. Differential shading, shadows, and perspective distortion are all used to give the impression of depth (Figure 15-10).

Spatial data visualization are increasingly being used in activities where it is too expensive, time consuming, or impractical to transport people to the real terrain locations. Pilot training is a good example. Visualizations are used widely in training jet pilots because a flight simulator is less expensive to build and operate than a real jet. While the flight simulator does not completely replace training in a real cockpit, it substantially reduces the cost, saving tens of thousands of dollars in training each pilot. Renderings of real terrain from airports all over the world are possible, so a simulated twenty-minute flight may take off from Quito, Ecuador and land in Oslo, Norway.

Visualizations based on real terrains were initially introduced into high cost or high value activities, but are now routinely used in an expanding number of fields. Common home computer software renders realistic terrains, and real estate and site tours are increasingly being offered that do not require participants to leave an office.

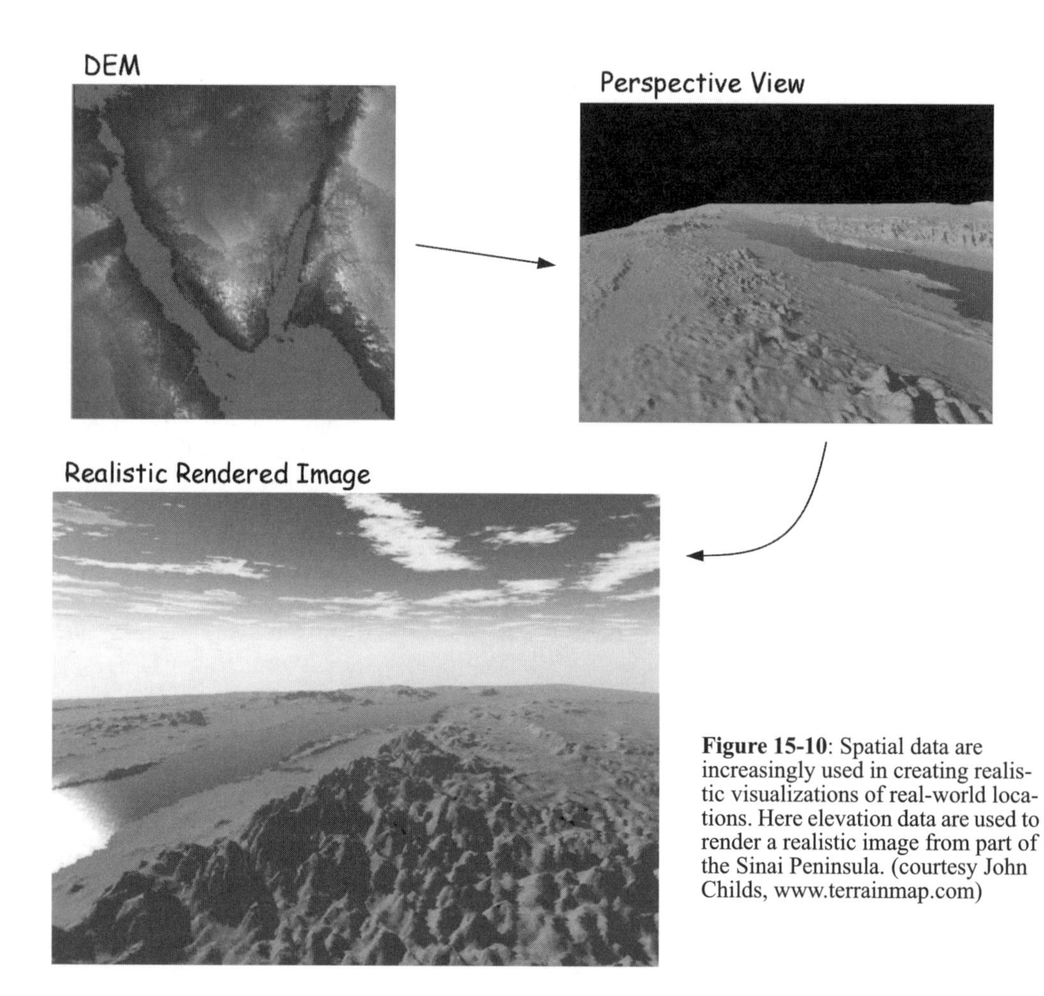

Figure 15-10: Spatial data are increasingly used in creating realistic visualizations of real-world locations. Here elevation data are used to render a realistic image from part of the Sinai Peninsula. (courtesy John Childs, www.terrainmap.com)

Open Standards for GIS

Open standards in computing seek to reduce barriers to sharing data and information. Spatial data structures may be very complex, perhaps more than many other kinds of data. Data may be raster or vector, real or binary, or represent point, line or area features. In addition, different software vendors may elect to store their raster imagery using different formats, and data may be delivered on different physical media, or formatted different ways. If a person orders an image in one format, but her computing system does not support the physical media on which the data are written, or does not understand the file structures used to store the image, then she may not be able to use these data. Incompatible systems are generally described as non-interoperable, and open standards seek to remove this non-interoperability.

The development of open standards in computing is driven by the notion that the larger user community benefits when there are no technical barriers that inhibit the free exchange of data and methods. Open standards seek to establish a common framework for representing, manipulating, and sharing data. Open standards also seek to provide methods for vendors and users to certify compliance with the standard. Standards have been developed in a number of endeavors, for example, the ISO 9600 format for CD ROMs is a defined standard allowing any manufacturer, data developer, or user to build, read, write, or share data on CD devices.

Businesses and many other organizations by their nature have a proprietary interest in the spatial data entry, storage, and methods they produce. Many vendors survive by the revenue their GIS products generate, and so have a strong interest in protecting their investments and intellectual property. However, the developers also may spur adoption of their GIS packages and speed up the development of complementary software by making the internal workings of some portions of their GIS packages public knowledge, for example, by publishing the data structures and formats used to store their spatial data. Thus, these vendors also have a strong interest in making parts of their system open to the public.

Open standards for spatial data are the responsibility of the OpenGIS Consortium. The OpenGIS consortium has developed a framework to ensure interoperability. They do this by defining a general, common set of base data models, types, domains, and structures, a set of services needed to share spatial data, and specifications to ease translation among different representations that are compatible with the OpenGIS standards. Data developed by a civil engineer and stored in a raster format on a Unix version of Arc/Info should be readily accessible to a soil scientist using MAPINFO on an OS-X Apple system.

Open standards in GIS are relatively new. While most of the large software vendors, data developers, and government and educational organizations are members of the OpenGIS consortium, some components of the standard are still under development. Increased emphasis on compliance to the OpenGIS standards will be apparent in the future.

Summary

GIS are a dynamic collection of tools, methods, and perceptions in how to use spatial data. As such they will continue to evolve. What becomes standard practice in the future may be quite different from the methods we apply today. However the fundamental set of knowledge will remain unchanged. We will still gather spatial and attribute data, adopt a spatial data model to conceptualize real world entities, and use map coordinates to define positions in space. The coordinates are likely to remain based on a standard set of map projections, and we will combine the spatial data of various classes of entities to solve spatial problems. This book is an attempt at providing you with the intellectual foundation to effectively use spatial analysis tools. I hope it has provided enough information to get you started, and has sparked your interest in learning more.

Suggested Reading

The Internet is the best source for information about new developments and trends in spatial data acquisition, analysis, and output. In contrast to previous chapters, nearly all the suggested readings are websites. We apologize that many links may be short-lived, but the reader is directed to search for similar and additional sites for the most current information.

www. spatialnews.com

www.nasa.gov, general NASA entry point

www. gis.com, an ESRI-sponsored website, general information

www.usgs, public domain data from the USGS

edcwww.cr.usgs.gov, another common USGS entry point

www.jpl.nasa.gov/srtm, shuttle radar topography mission information

www.OpenGIS.org, open GIS consortium description

www.westindianspace.com, for Imagesat International high resolution satellite data

www.orbimage.com, OrbView high resolution satellite data

www.digitalglobe.com, QuickBird high resolution satellite data

Study Questions

Which of the described new technologies is likely to have the largest impact in GIS over the next five years? Why?

What are areas of spatial data entry, analysis, output, or storage that are in dire need of innovation or new and better methods? What is a major bottleneck to further advancement of spatial information science and technology?

Appendix A: Glossary

Terms used in GIS and Spatial Data Development and Analysis

Accuracy: The nearness of an observation or estimate to the true value.

Adjacency: Two area objects that share a bounding line are topologically adjacent.

Affine coordinate transformation: A set of linear equations used to transform from one Cartesian coordinate system to another. The transformation applies a scaling, translation, and rotation.

Arc: A line, usually defined by a sequence of coordinate points.

Area feature: A polygon, collection of contiguous raster cells, or other representation of a bounded area. The feature is characterized by a set of attributes and has an inside and an outside.

ASCII: American Standard Code for Information Interchange. A set of numbers associated with a symbol used in information storage and processing. Numbers are between 0 and 255 and may be represented by a single byte of data.

Attribute: Non-spatial data associated with a spatial feature. Crop type, value, address, or other information describing the characteristics of a spatial feature are recorded by the attributes.

AVHRR: Advanced Very High Resoluton Radiometer. A satellite system run by the National Oceanographic and Atmospheric Administration to collect visible, thermal, and infrared satellite images of the globe each day. The system has up to a 1.1 km resolution.

Bearing: A direction, usual specified as a geographic angle measured from some base line, e.g., true north.

Benchmark: A monumented, precisely surveyed location for which coordinates are known to a high degree of accuracy.

Bit: A binary digit. A bit has one of two values, on or off, zero or one. This is the smallest unit of digital information storage and the basic building block from which all other computer data and programs are represented.

Buffer: A buffer area is a polygon or collection of cells that are within specified proximities of a set of features. A buffer operation is one that creates buffer areas.

Byte: A unit of computer storage consisting of 8 binary digits. Each binary digit may hold a zero or a one. A byte may store up to 256 different values.

C/A code GPS: Coarse acquisition code, a GPS signal used to make for rapid, relatively low-accuracy positional estimates. Accuracies without further corrections are typically from a few to tens of meters.

CAD/CAM: Computer Aided Design/Computer Aided Mapping. Software used primarily by design engineers and utilities managers to produce two and three dimensional drawings. Related to GIS in that coordinate information are input, manipulated, and output. These systems typically do not store map-projected coordinates, and do not have sophisticated attribute entry and manipulation capabilities.

Carrier-phase GPS: Relatively slow but accurate signal used to estimate position. Position may be determined to within a few centimeters or better.

Cartesian coordinate system: A right-angle two or three-dimensional coordinate system. Axes intersect at 90 degrees, and the interval along each axis is linear.

Cartographic modeling: The combination of spatial data layers through the application of spatial operations.

Cartographic object: A digital representation of a real-world entity.

Cartometric map: A map produced such that the relative positions of objects depicted are spatially accurate, within the limits of the technology and the map projection used.

Centroid: A centeral point location for an area feature, often defined as the point with the lowest average distance to all points that define the area boundary.

Code-phase GPS: see C/A code.

COGO: Coordinate Geometry, the entry of spatial data via coordinate pairs, usually obtained from field surveying instruments.

Conformal projection: A map projection is conformal when it preserves shape for some portions of the map.

Conic projection: A map projection that uses a cone as the developable surface.

Connectivity: A record or representation of the connectedness of linear features. Two linear features or networks are connected if they may be traversed without leaving the network.

Continuous surface: A variable or phenomena that changes gradually through two-dimensional space, e.g., elevation or temperature.

Contour line: A line of constant value for a mapped variable.

Control points: Point locations for which map projection and database coordinate pairs are known to a high degree of accuracy. Control points are most often used to convert digitized coordinates to standard map projection coordinates.

Coordinates: A pair or triplet of numbers used to define a position in space.

Cylindrical projection: A map projection that uses a cylinder as the developable surface.

Data model: A method of representing spatial and aspatial components of real-world entities on a computer.

Datum: A set of coordinate locations specifying horizontal positions (for a horizontal datum) or vertical positions (for a vertical datum) on the Earth surface.

DBMS: Database Managment System, a collection of software tools for the entry, organization, storage, and output of data.

Declination: The angle between the bearing towards True North and the bearing towards Magnetic North.

DEM: Digital Elevation Model, a raster set of elevations, usually spaced in a uniform horizontal grid.

Developable surface: A geometric shape onto which the Earth sphere is cast during a map projection. The developable surface is typically a cone, cylinder, plane, or other surface that may be mathematically flattened.

Differential GPS: GPS positioning based on two receivers, one at a know location and one at a roving, unknown location. Data from roving receivers are corrected by the difference error computed at the known location.

Digitize: To convert paper or other hardcopy maps to computer-compatible and stored data.

Digitizing table: A device with a flat surface and input pointer used to digitize hardcopy maps.

DLG: Digital Line Graph, vector data developed and distributed by the United States Geological Survey.

DOQ: Digital Orthophoto Quadrangle, an orthographic photograph provided in digital formats by the USGS. Most tilt and terrain error have been removed from DOQs.

DRG: Digital Raster Graphics, a digital version of USGS fine- to medium-scale maps.

Easting: The axis approximately parallel to lines of equal latitude in UTM and a number of other standard map projections.

Electromagnetic spectrum: A range of energy wavelengths, from X-rays through radar wavelengths. The electromagnetic spectrum is typically observed at wavelengths emitted by the Sun or by objects on Earth, covering wavelengths from the visible to the thermal infrared region.

Ellipsoid: A mathematical model of the shape of the Earth that is approximately the shape of a flattened sphere, formed by rotating an ellipse.

Endlap: The end-to-end overlap in aerial photographs taken in the same flight line.

Epsilon band: A band surrounding a linear feature that describes the positional error relative to the feature location.

ETM+: Enhanced Thematic Mapper, a scanner carried on board Landsat 7, providing image data with resolutions of 30 meters for visible through mid infrared, 15 meter pancromatic, and 60 meter for thermal wavelengths.

Facet: A triangular face in a TIN.

Feature: An object or phenomenon in the landscape. A digital representation of the feature is often called a cartographic feature.

FIPS: Federal Information Processing Standards code - a set of numbers for defined political or physical entities in the United States. There are FIPS codes for each state, county, and other features.

Friction surface: A raster surface used in calculating variable travel costs through an area. The friction surface represents the cost per unit distance to travel through a cell.

Flatbed scanner: An electronic device used to record a digital image of a hardcopy map or image.

FTP: File Transfer Protocol, a standard method to transfer files across a computer network.

Generalization: The simplification of shape or position that inevitably occurs when features are mapped.

Geoid: A measurement-based model of the shape of the Earth. The geoid is a gravitational equipotential surface, meaning a standard surface of equal gravitational pull. The geoid is used primarily as a basis for specifying terrain or other heights.

Geographic North: The northern axis of rotation of the Earth. By definition true north lies at 90° latitude.

GIS: A geographic information system. A GIS is a computer-based system to aid in the collection, maintenance, storage, analysis, output, and distribution of spatial data and information.

Graticule: Lines of latitude and longitude drawn on a hardcopy map or represented in a digital database.

GRS80: Geodetic Reference Surface of 1980, an ellipsoid used for map projections in much of North America.

Hierarchical data model: A method of organizing attribute data that structures values in a tree, typically from general to more specific.

Ikonos: A high resolution imaging satellite system. Ikonos provides 1-meter panchromatic and 3-meter multispectral image data.

Interval/ratio scale: A measurement scale that records both order and absolute difference in value for a set of variables.

IDW: Inverse Distance Weighted interpolation, a method of estimating values at unsampled locations based on the value and distance to sampled locations.

JPEG: An image compression format.

Kriging: An interpolation method based on geostatistics, the measurement of spatial autocorrelation.

Landsat: A NASA project spanning more than three decades and seven satellites that proved the capabilities of space-based remote sensing of land resources.

Latitude: Spherical coordinates of Earth location that vary in a north-south direction.

LIS: A Land Information System, a name originally applied for GIS systems specifically developed for property ownership and boundary records managment.

Longitude: Spherical coordinates of Earth location that vary in an east-west direction.

Map algebra: The combination of spatial data layers using simple to complex spatial operations.

Meridian: A line of constant longitude.

Magnetic North: The point where the northern lines of magnetic attraction enter the Earth. Magnetic North does not occur at the same point as "True" or Geographic North. In the absence of local interference a compass needle points towards magnetic north. The magnetic north pole is currently located in northern Canada.

Moving window: A usually rectangular arrangement of cells that shifts in position across a raster data set. At each position an operation is applied using the cell values currently encountered by the moving window.

MSS: Multi-spectral Scanner, an early satellite imaging scanner carried by Landsat satellites.

MODIS: Moderate Resolution Imaging Sensor. A later generation imaging scanner that is part of NASA's Mission to Planet Earth. Provides high spectral resolution, frequent global coverage, and moderate spatial resolution of from 250 to 1000 meters.

NAD27: North American Datum of 1927, the adjustment of long-baseline surveys to establish a network of standardized horizontal positions in the early 20th century.

NAD83: North American Datum of 1983. The successor to NAD27, using approximately an order of magnitude more measurements and improvements in analytical models and computer power. The current network of standard horizontal positions for North America.

NAVD29: North American Vertical Datum of 1929, an adjustment of vertical measurements to establish a network of heights in the early 20th century.

NAVD88: North American Vertical Datum of 1988, the successor vertical datum to NAVD29.

Network: A connected set of line features, often used to model resource flow or demand through real-world networks such as road or river systems.

Network center: A location on a network the provides or requires resources.

NLCD: National Land Cover Data set, a Landsat Thematic Mapper (TM) based classification of landcover for the United States.

NOAA: National Oceanic and Atmospheric Administration, the U.S. goverment agency that oversees the development of national datums.

Node: An important point along a line feature, where two lines meet or intersect.

Nominal scale: A measurement scale that indicates the difference between values, but does not reflect rank or absolute differences.

Northing: The axis in the approximately north-south direction in UTM and other standard coordinate systems.

Normal forms: A standard method of structuring relational databases to aid in updates and remove reduncancy.

NWI: National Wetlands Inventory data compiled by the U.S. Fish and Wildlife Service over most of the United States. These data provide first-pass indications of wetland type and extent.

Object: See cartographic object.

Object-oriented data model: A data model that incorporates encapsulation, inheritance, and other object-oriented programming principles.

Ordinal scale: A scale the represents the relative order of values but does not record the magnitude of differences between values.

Orthogonal: Intersecting at a 90 degree angle.

Orthographic: Horizontal placement as would be seen from a vertical viewpoint at infinity. There is no terrain or tilt-perspective distortion in an orthographic view

Orthophotograph: A vertical photograph with an orthographic view. Orthophotographs are created by using projection geometry and measurements to remove tilt, terrain, and perspective distortion from aerial photographs.

Overlay: The “vertical” combination of two or more spatial data layers.

Overshoot: A digitized line that extends past a connecting line.

PDOP: Positional Dilution of Precsion, a figure of merit used to represent the quality of the satellite geometry when taking GPS readings. PDOPs between 1 and 6 are preferred for most applications, and lower is better.

Planar topology: The enforcement of intersection for line and area features in a digital data layer. Each line crossing requires an explicit node and intersection.

Pointer: An address stored in a data structure pointing to the next or related data elements. Pointers are used to organize data and speed access.

Polygon: A closed, connected set of lines that define an area.

Precision: The repeatability of a measure or process.

Quad-trees: A raster data structure based on successive, adaptive reductions in cell size within a data layer to reduce storage requirements for thematic area data.

Queries: Requests or searches for spatial data, typically applied via a database management system.

Raster data model: A regular “grid cell” approach to defining space. Usually square cells are arranged in rows and columns.

Resampling: The recalculation and assignment of cell values when changing cell size and/or orientation of a raster grid.

RMSE: Root Mean Square Error, a statistic that measures the difference between true and predicted data values for coordinate locations.

Rubbersheeting: The use of polynomial or other nonlinear transformations to match feature geometry.

Run-length coding: A compression method used to reduce storage requirements for raster data sets. The value and number of sequential occurences are stored.

Sidelap: Edge overlap of photographs taken in flightlines.

Semi-major axis: The larger of the two radial axes that define an ellipsoid.

Semi-minor axis: The smaller of the two radial axes that define an ellipsoid.

Semivariance: The variance between values sampled at a given lag distance apart.

Skeletonizing: Reducing the width of linear features represented in raster data layers to a single cell.

Sliver: Small, spurious polygons at the margins or boundaries of feature polygons that are an artefact of imprecise digitizing or overlay.

Snapping: Automatic line joins during vector digitizing or layer overlay. Nodes or vertices are joined if they are within a specified snap distance.

Spaghetti data model: Vector data model in which lines may cross without intersecting.

Spectrum: see electromagnetic spectrum.

Spherical coordinates: A coordinate system based on a sphere. Location on the sphere surface is defined by two angles of rotation in orthogonal planes. The geographic coordinate system of latitude and longitude is the most common example of a spherical coordinate system.

Spheroid: A mathematical model of the shape of the Earth, based on the equation of a sphere.

Spline: A smoothed line or surface created by joining multiple constrained polynomial functions.

SPOT: Systeme Pour l'Observation de la Terre, a satellite imaging system providing 10 to 20 meter resolution images.

SQL: Structured Query Language, a widely adopted set of commands used to manipulate relational data.

SSURGO: Fine resolution digital soil data corresponding to county level soil surveys in the United States. Produced by the Natural Resource Conservation Service.

STATSGO: Coarse resolution digital soil data distributed on a statewide basis for the United States Most often derived from aggregation and generalization of SSURGO data.

State Plane Coordinates: A standardized coordinate system for the United States of America that is based on the Lambert conformal conic and transverse Mercator projections. State plane zones are defined such that projection distortions are maintained to be less than 1 part in 10,000.

Stereophotographs: A pair or more of overlapping photographs that allow the perception of three dimensions due to a perspective shift.

Thematic layer: Thematically distinct spatial data organized in a single layer, e.g., all roads in a study area placed in one thematic layer, all rivers in a different thematic layer.

TM: Thematic Mapper, a high-resolution scanner carried on board later Landsat satellites. Provides information in the visible, near infrared, mid infrared, and thermal portions of the electromagnetic spectrum.

TIFF: Tagged Image File Format, a widely-supported image distribution format. The Geo-TIFF variant comes with image registration information embedded.

TIGER: Topologically Integrated Geographic Endcoding and Referencing files, a set of structures used to deliver digital vector data and attributes associated with the U.S. Census.

TIN: Triangulated Irregular Network, a data model most commonly used to represent terrain. Elevation points are connected to form triangles in a network.

Topology: Shape-invariant spatial properties of line or area features such as adjacency, contiguity, and connectedness, often recorded in a set of related tables.

Triangulation Survey: Horizontal surveys conducted in a set of interlocking triangles, thereby providing multiple pathways to each survey point. This method provides inherent internal checks on survey measurements.

True North: See Geographic North.

Undershoot: A digitizing error in which a line end falls short of an intended connection at another line end or segment.

USGS: United States Geological Survey - the U.S. government agency responsible for most civilian nationwide mapping and spatial data development.

UTM: Universal Transverse Mercator coordinate system, a standard set of map projections developed by the U.S. Military and widely adopted for coordinate specification over regional study areas. A cylindrical projection is specified with a central meridian for each six-degree wide UTM zone.

Vector data model: A representation of spatial data based on coordinate location storage for shape-defining points and associated attribute information.

Vertex: Points used to specify the position and shape of lines.

WAAS: Wide Area Augmentation System, a satellite-based transmission of correction signals to improve GPS positional estimates, largely through the removal of ionsopheric and atmospheric effects.

WGS84: World Geodetic System, a Earth-centered reference ellipsoid used for defining spatial locations in three dimensions. Very similar to GRS80 ellipsoid. Commonly used as a basis for map projections.

Appendix B: Sources of Geographic Information

General Information

www.gis.com - software, data, education overview from ESRI Inc.

www.gisdatadepot.com - news, data, information on GIS

spatialnews.geocomm.com - general GIS information, news, and data

Spatial Data

U.S.

www.blm.gov/gis - U.S. Bureau of Land Management Spatial Data

www.census.gov/geo/www/tiger/index.html - TIGER data from the U.S. Census Bureau

www.mapping.usgs.gov - U.S. National Mapping Information

www.nsdi.usgs.gov - USGS node of the National Geospatial Data Clearinghouse

www.nwi.fws.gov - U.S. National Wetlands Inventory Data

www.statlab.iastate.edu/soils/nsdaf - SSURGO, STATSGO, MUIR soils data

www.water.usgs.gov - Spatial data for water resources

edcwww.cr.usgs.gov - EROS Data Center portal, general U.S. government site for image and map data

landcover.usgs.gov - NLCD and other landcover data for the U.S.

terraserver.homeadvisor.msn.com/default.asp - Microsoft Inc. site serving USGS topographic map and DOQ data.

164.214.2.59 - National Imagery and Mapping Agency, formerly the U.S. Defense Mapping Agency

U.S. States

agdc.usgs.gov/data - Alaska Geospatial Data Clearinghouse

www.asgdc.state.ak.us - Spatial data for Alaska

www.land.state.az.us/agic/agichome.html - Arizona Geographic Information Council

www.ca.gis.gov - California state government GIS pages and data

www.dnr.state.co.us - Colorado environmental and other spatial data

www.cdphe.state.co.us/hs/gishom.asp - Colorado public health GIS data

magic.lib.uconn.edu - Connecticut Map and Geographic Information Center

gis.smith.udel.edu/fgdc2/clearinghouse - Deleware Spatial Data Clearinghouse www.dep.state.fl.us/gis - Florida Department of Environmental Protection, spatial data

www.gis.state.ga.us - Spatial data from the Georgia state government

www.hawaii.gov/dbedt/gis - Hawaii statewide data and programs

www.idwr.state.id.us/gisdata - Idaho GIS and remote sensing data

www.isgs.uiuc.edu/nsdihome - Illinois state spatial data clearinghouse

www.igsb.uiowa.edu/nrgis/gishome.htm - Iowa Department of Natural Resources GIS library

gisdasc.kgs.ukans.edu/dasc.htm - State of Kansas spatial data

ogis.state.ky.us - Kentucky state government GIS data site

lagic.lsu.edu - Louisiana Geographic Information Center

musashi.ogis.state.me.us/catalog/catalog.asp - State of Maine data catalogue

www.state.ma.us/mgis - Massachusetts state GIS data

www.state.mi.us/dmb/mic/gis - Michigan geographic information clearinghouse

www.deli.dnr.state.mn.us - Minnesota state government spatial data

msdis.missouri.edu - Missouri spatial data

nris.state.mt.us/gis/gis.html - Montana natural resources data

geodata.state.ne.us - Nebraska spatial data clearinghouse

www.state.nj.us/dep/gis - New Jersey Department of Environmental Protection data

www.nysgis.state.ny.us - New York State spatial data clearinghouse

cgia.cgia.state.nc.us/ncgdc - North Carolina spatial data clearinghouse

www.state.nd.us/ndgs/GISInfo/gis.html - North Dakota spatial data clearinghouse

www.sscgis.state.or.us/data/index.html - Oregon spatial data library

www.odf.state.or.us/gis/data.html - Oregon Department of Forestry GIS data

www.pasda.psu.edu - Pennsylvania spatial data

www.edc.uri.edu/rigis - Rhode Island spatial data

water.dnr.state.sc.us/gisdata/index.html - South Carolina Department of Natural Resources

www.sdgs.usd.edu - South Dakota Geological Survey

www.tnrcc.state.tx.us/gis - Texas Natural Resources Conservation Commission, spatial data

www.agrc.utah.gov/sgid.html - Utah statewide geographic information database

geo-vt.uvm.edu/cfdev2/warehouse_new/warehouse.cfm - Vermont GIS data waterhouse

fisher.lib.virginia.edu - Virginia Geospatial and Statistical Data Center

www.wa.gov/gic/washdat.htm - Washington State spatial data clearinghouse

wisclinc.state.wi.us - Wisconsin land information clearinghouse

wgiac.state.wy.us/wgrp.html - Wyoming spatial data, organizations, and information services

Global Data and Data for non-U.S. Countries

www.asian.gu.edu.au - Spatial data for Asia

www.auslig.gov.au - spatial data and information for Australia

www.ciesin.org - Center for International Earth Science Information Network

www.eurogi.org - Umbrella organization for spatial data in Europe

www.eurogeographics.org - European mapping agencies cooperative venture

www.ign.fr - Institut Geographique National, spatial data for France

www.ordsvy.gov.uk - British National Mapping Agency data

edcsnw4.cr.usgs.gov/adds - Spatial data for Africa

geogratis.cgdi.gc.ca - Spatial data for Canada

Images

www.edc.usgs.gov/programs/NSLRSDA.html - U.S. National satellite land remote sensing data archive

www.eurimage.com - Satellite image sales for most of Europe

www.rsi.ca - RADARSAT radar imaging data

www.spaceimaging.com - Ikonos high resolution satellite imagery

www.digitalglobe.com - Quickbird high resolution satellite imagery

www.spot.com - SPOT satellite imagery

landsat.gsfc.nasa.gov - Landsat 7 gateway

modis-land.gsfc.nasa.gov - MODIS land surface data

photojournal.jpl.nasa.gov - NASA images of the Earth, other planets, and astronomy

Software

www3. autodesk.com - Autodesk Map

www.baylor.edu/grass - GRASS GIS

www2.bentley.com - Microstation GIS

www.erdas.com - Imagine image processing and raster GIS software

www.esri.com - ArcGIS and related products

www.idrisi.com - IDRISI and Cartalinx spatial data analysis and entry software

www.intergraph.com - MGE and Geomedia GIS software

www.mapinfo.com - MapInfo GIS

www.microimages.com - TNTmips map and image processing system

News, Journals, Industry Information

www.asprs.org - American Society for Photogrammetry and Remote Sensing, publishers of Photogrammetric Engineering and Remote Sensing

www.elsevier.nl- Elsevier, publishers of Computers and Geosciences

www.geoinfosystems.com - Geospatial solutions journal

www.geoplace.com - GIS news

www.gpsworld.com - developments in GPS

www.journals.tandf.co.uk - Taylor and Francis, publishers of the International Journal of Geographical Information Science, and Geographical and Environmental Modeling

www.wkap.nl - Kluwer, publishers of GeoInformatica

www.urisa.org - Urban and Regional Information Systems Organization, publishers of the URISA Journal

Organizations, Standards

www.aag.org - American Association of Geography

www.fgdc.gov - U.S. Federal Geographic Data Committee

www.igu-net.org - International Geographical Union

www.geog.ubc.ca/cca - Canadian Cartographic Association

www.ngs.noaa.gov - U.S. National Geodetic Survey

www.soc.org.uk - Society of Cartographers

www.urisa.org - Urban and Regional Information Systems Organization

Appendix C: Useful Conversions and Information

Length

1 meter = 100 centimeters
1 meter = 1000 millimeters
1 meter = 3.28083989501 International feet
1 meter = 3.28083333333 U.S. survey feet
1 kilometer = 1000 meters
1 kilometer = 0.62137 miles
1 mile = 5280 feet

distance between two points, x_1, y_1 and x_2, y_2.

x_2, y_2

d

x_1, y_1

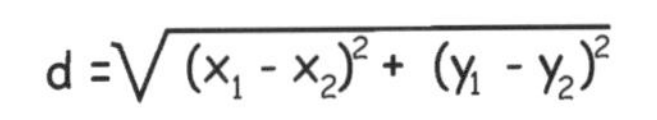

$$d = \sqrt{(x_1 - x_2)^2 + (y_1 - y_2)^2}$$

Area

1 hectare = 10,000 square meters
1 square kilometer = 100 hectares
1 acre = 43,560 square feet
1 square mile = 640 acres
1 hectare = 2.47 acres
1 square kilometers = 0.3861 square miles

Angles

1 degree = 60 minutes of arc
1 minute = 60 seconds of arc
decimal degrees = degrees + minutes/60+seconds/3600
180 degrees = 3.14159 radians
1 radian = 57.2956 degrees

Trigonometric Relationships

H

A

a

90°

B

sine (a) = A/H
cosine (a) = B/H
tangent (a) = A/B
cotangent (a) = B/A
secant (a) = H/A
cosecant (a) = H/B

Scale

Scale value	1 centimeter distance on map equals a distance on the ground of:
1:5,000	50 meters
1:10,000	100 meters
1:25,000	250 meters
1:50,000	500 meters
1:100,000	1000 meters

Scale value	1 inch distance on a map equals a distance on the ground of:
1:6,000	500 feet
1:15,840	1,320 feet
1:24,000	2,000 feet
1:62,500	5,208 feet
1:100,000	8,333 feet

State Plane Zones

UTM Zones - USA

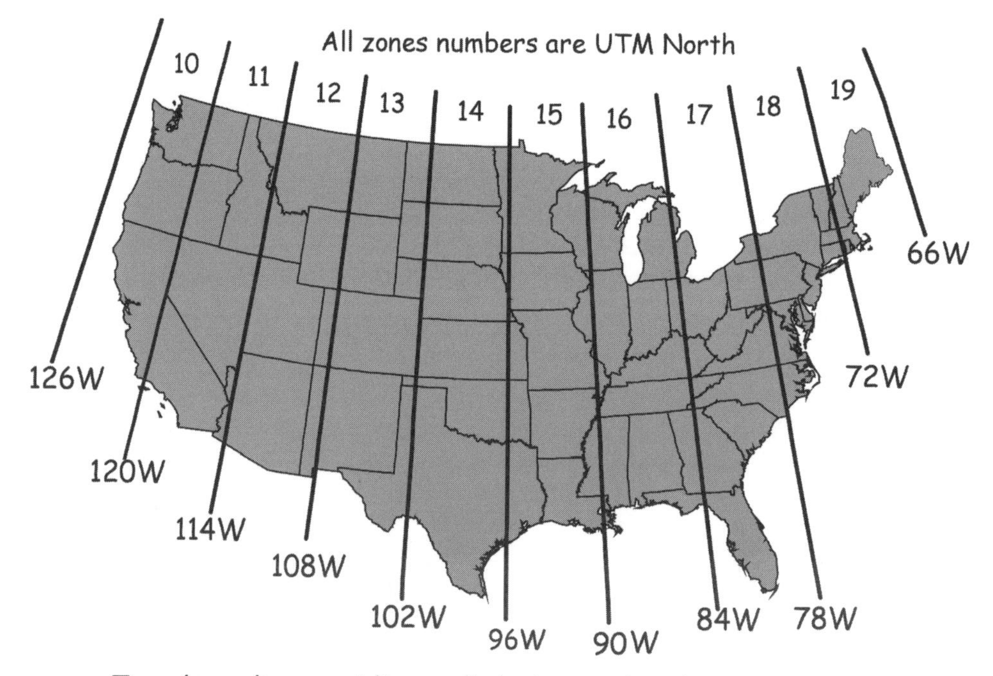

INDEX

A

B

C

D

E

F

G

H

I

K

L

M

N

O

P

Q

R

S

T

U

V

W